# Animal Behavior and Parasitism

ECOLOGY AND EVOLUTION OF
INFECTIOUS DISEASES SERIES

Series Editor: John M. Drake

Infectious diseases are embedded in a complex of social, environmental, and biological forces. This series provides an up-to-date synthesis of contemporary issues in the ecology and evolution of infectious diseases. Topics covered include aspects of population ecology, vector biology, animal behavior and movement, methods for inference and disease forecasting, eco-immunology, evolution, biogeography, and others. Each book makes topical connections to pathogens of current interest. The series is intended for students and researchers working in ecology, evolution, and epidemiology with applications to the disciplines of One Health (animal health, public health, and environmental health), environmental management, health policy, therapeutics, and wildlife conservation.

**Pertussis: Epidemiology, Immunology, & Evolution**
*Edited by Pejman Rohani and Samuel V. Scarpino*

**Population Biology of Vector-Borne Diseases**
*Edited by John M. Drake, Michael B. Bonsall, and Michael R. Strand*

# Animal Behavior and Parasitism

EDITED BY

## Vanessa O. Ezenwa
*Department of Ecology and Evolutionary Biology, Yale University, USA*

## Sonia Altizer
*Odum School of Ecology, University of Georgia, USA*

## Richard J. Hall
*Odum School of Ecology & Department of Infectious Diseases, College of Veterinary Medicine, University of Georgia, USA*

OXFORD
UNIVERSITY PRESS

# OXFORD
## UNIVERSITY PRESS

Great Clarendon Street, Oxford, OX2 6DP,
United Kingdom

Oxford University Press is a department of the University of Oxford.
It furthers the University's objective of excellence in research, scholarship,
and education by publishing worldwide. Oxford is a registered trade mark of
Oxford University Press in the UK and in certain other countries

Published in the United States of America by Oxford University Press
198 Madison Avenue, New York, NY 10016, United States of America

British Library Cataloguing in Publication Data
Data available

Library of Congress Control Number: 2022938819

ISBN 978–0–19–289556–1

DOI: 10.1093/oso/9780192895561.001.0001

Printed and bound by
CPI Group (UK) Ltd, Croydon, CR0 4YY

Links to third party websites are provided by Oxford in good faith and
for information only. Oxford disclaims any responsibility for the materials
contained in any third party website referenced in this work.

# Contents

# List of Contributors

**Jessica L. Abbate** Geomatys Montpellier, France
email: jessie.abbate@gmail.com

**James S. Adelman** Department of
Biological Sciences University of
Memphis, Memphis, TN, USA
email: Jim.Adelman@memphis.edu

**Sonia Altizer** Odum School of Ecology and Center for the Ecology of Infectious Diseases,
University of Georgia, Athens, GA, USA
email: saltizer@uga.edu

**Nili Anglister** School of Zoology Tel
Aviv University, Tel Aviv, Israel
email: nilsanglister@yahoo.com

**Simone Anzà** Department of Behavioral Ecology, Johann-Friedrich-Blumenbach Institute
for Zoology & Anthropology, Georg-August-University Göttingen, Göttingen, Germany
Research Group Primate Social Evolution,
German Primate Center, Leibniz Institute
for Primate Research, Göttingen, Germany
email: simoneanza@gmail.com

**Ben Ashby** Department of Mathematics, Simon
Fraser University, Burnaby, BC, Canada
email: bashby@sfu.ca

**Lewis J. Bartlett** Center for the Ecology of Infectious Diseases, University of Georgia, Athens, GA, USA
email: lewis.bartlett@uga.edu

**Sandra A. Binning** Département de
Sciences Biologiques, Université de
Montréal, Montréal, Québec, Canada
email: sandra.ann.binning@umontreal.ca

**Lauren J. Cator** Department of Life Sciences, Imperial College London, Ascot, UK
email: l.cator@imperial.ac.uk

**Hettie Chapman** Institute for Quantitative and
Theoretical Biology, Heinrich Heine University Düsseldorf, Düsseldorf, Germany
email: hec41@bath.ac.uk.

**Marie Charpentier** Institut des Sciences de l'Evolution de Montpellier (ISEM), Montpellier, France
email: marie.charpentier@umontpellier.fr

**Meggan E. Craft** Department of Ecology,
Evolution, and Behavior, University
of Minnesota, Minneapolis, MN, USA
email: craft004@umn.edu

**Miranda M. Crafton** School of Zoology,
Tel Aviv University, Tel Aviv, Israel
email: mirandacrafton@gmail.com

**Shaun Davis** Department of Entomology,
University of Arizona, Tucson, AZ, USA
email: shaundavis@email.arizona.edu

**Charlotte Defolie** Department for Sociobiology/Anthropology, Johann-Friedrich-Blumenbach Institute for Zoology &
Anthropology, Georg-August-University
Göttingen, Göttingen, Germany; Behavioral Ecology and Sociobiology Unit, German Primate Center, Leibniz Institute for
Primate Research, Göttingen, Germany
email: charlotte.defolie@gmail.com

**Vanessa O. Ezenwa** Department of Ecology and Evolutionary Biology, Yale
University, New Haven, CT, USA
email: vanessa.ezenwa@yale.edu

**Susannah S. French** Department of Biology and the Ecology Center, Utah
State University, Logan, UT, USA
email: susannah.french@usu.edu

**Marie L.J. Gilbertson** Department of Veterinary Population Medicine, University of Minnesota, Minneapolis, MN, USA
email: mgilbertson5@wisc.edu

**Stephanie S. Godfrey** Department of Zoology, University of Otago, Dunedin, New Zealand
email: Stephanie.Godfrey@otago.ac.nz

**Sarah Guindre-Parker** Department of Ecology, Evolution and Organismal Biology, Kennesaw State University, Kennesaw, GA, USA
email: sguindre@kennesaw.edu

**Richard J. Hall** Odum School of Ecology, Department of Infectious Diseases, and Center for the Ecology of Infectious Diseases, University of Georgia, Athens, GA, USA
email: dr.richard.hall@gmail.com

**Dana M. Hawley** Department of Biological Sciences, Virginia Tech, Blacksburg, VA, USA
email: hawleyd@vt.edu

**Geoffrey E. Hill** Department of Biological Sciences, Auburn University, Auburn, AL, USA
email: ghill@auburn.edu

**Carl N. Keiser** Department of Biology, University of Florida, Gainesville, FL, USA
email: ckeiser@ufl.edu

**Rebecca E. Koch** School of Biological Sciences, Monash University, Clayton, VIC, Australia
email: rebecca.adrian@monash.edu

**Patricia C. Lopes** Schmid College of Science and Technology, Chapman University, Orange, CA, USA
email: lopes@chapman.edu

**Virgile Manin** Department of Human Behavior, Ecology and Culture, Max Planck Institute for Evolutionary Anthropology, Leipzig, Germany and Tai Chimpanzee Project Centre Suisse de Recherche Scientifique, Abidjan, Cote d'Ivoire
email: virgile_manin@eva.mpg.de

**Hannah R. Meredith** Department of Epidemiology, Johns Hopkins Bloomberg School of Public Health, Baltimore, MD, USA
email: hrmeredith12@gmail.com

**Janine Mistrick** Department of Ecology, Evolution, and Behavior, University of Minnesota, St. Paul, MN, USA
email: mistr033@umn.edu

**Nadine Müller-Klein** Institute for Evolutionary Ecology and Conservation Genomics, Conservation Genomics and EcoHealth, Ulm University, Ulm, Germany
email: Nadine.Mueller-Klein@uni-ulm.de

**Tatiana Murillo** Behavioral Ecology and Sociobiology Unit, German Primate Center, Leibniz Institute for Primate Research, Göttingen, Germany and Genomic and Applied Microbiology and Göttingen Genomics Laboratory, Institute of Microbiology and Genetics, Georg-August-University Göttingen, Göttingen, Germany
email: tatimurillo@gmail.com

**Stephanie J. Peacock** Department of Ecosystem and Public Health, University of Calgary, Calgary, AB, Canada
email: stephanie.j.peacock@gmail.com

**Alistair Pirrie** Department of Mathematical Sciences, University of Bath, Bath, UK
email: arp59@bath.ac.uk

**Clémence Poirotte** Behavioral Ecology and Sociobiology Unit, German Primate Center, Kellnerweg, Göttingen, Germany.
email: c.poirotte@gmail.com

**Robert Poulin** Department of Zoology University of Otago, Dunedin, New Zealand
email: Robert.Poulin@otago.ac.nz

**Emlyn J. Resetarits** Center for the Ecology of Infectious Diseases, University of Georgia, Athens, GA, USA
email: emlyn.resetarits@uga.edu

**Baptiste Sadoughi** Department of Behavioral Ecology, Johann-Friedrich-Blumenbach Institute for Zoology & Anthropology, Georg-August-University Göttingen, Göttingen, Germany, Research Group Primate Social Evolution, German Primate Center, Leibniz Institute for Primate Research, Göttingen, Germany, and Leibniz Science Campus Primate Cognition, German Primate Center, Göttingen, Germany
email: baptistesadoughi@gmail.com

**Todd Schlenke** Department of Entomology, University of Arizona, Tucson, AZ, USA
email: schlenke@email.arizona.edu

**Allison K. Shaw** Department of Ecology, Evolution, and Behavior, University

of Minnesota, St. Paul, MN, USA
email: ashaw@umn.edu

**Orr Spiegel** School of Zoology, Tel
Aviv University, Tel Aviv, Israel
email: orrspiegel@tauex.tau.ac.il

**Jessica F. Stephenson** Department
of Biological Sciences, University
of Pittsburgh, Pittsburgh, PA, USA
email: Jess.Stephenson@pitt.edu

**Alexander T. Strauss** Odum School of Ecology, University of Georgia, Athens, GA, USA
email: atstrauss@uga.edu

**Jenny Tung** Department of Evolutionary
Anthropology, Department of Biology
and Duke Population Research Institute, Duke University, Durham, NC, USA
Department of Primate Behavior and Evolution, Max Planck Institute for Evolutionary Anthropology, Leipzig, Germany
email: jenny.tung@duke.edu

**Markus Ulrich** Epidemiology of Highly
Pathogenic Microorganisms, Robert
Koch Institute, Berlin, Germany
email: ulrichm@rki.de

**Amy Wesolowski** Department of Epidemiology, Johns Hopkins Bloomberg School
of Public Health, Baltimore, MD, USA
email: awesolowski@jhu.edu

**Lauren A. White** National Socio-
Environmental Synthesis Center, University of Maryland, Annapolis, MD, USA
email: whit1951@umn.edu

**Anna R. Willoughby** Odum School of Ecology
and Center for the Ecology of Infectious Diseases, University of Georgia, Athens, GA,
USA
email: anna.willoughby@uga.edu

**Cali A. Wilson** Odum School of Ecology and
Center for the Ecology of Infectious Diseases,
University of Georgia, Athens, GA, USA
email: cali.wilson@uga.edu

**Jamie C. Winternitz** Department of Animal
Behavior, Bielefeld University, Bielefeld,
Germany
email: jamie.winternitz@uni-bielefeld.de

**Douglas C. Woodhams** Department of
Biology, University of Massachusetts,
Boston, Boston, MA, USA and Smith-
sonian Tropical Research Institute,
Panama
email: douglas.woodhams@umb.edu

**Doris Wu** Epidemiology of Highly Pathogenic
Microorganisms, Robert Koch Institute, Berlin,
Germany
email: dorrywu@gmail.com

# Abbreviations

| | | | | |
|---|---|---|---|---|
| AMP | Antimicrobial peptide | | MBON | Mushroom body output neuron |
| ATLAS | Advanced tracking and localization of animals in real-life systems | | MG | *Mycoplasma gallisepticum* |
| | | | MHC | Major histocompatibility complex |
| ATP | Adenosine triphosphate | | mt | Mitochondrial |
| BBB | Behavior begets behavior | | N-mt | Nuclear mitochondrial |
| Bd | *Batrachochytrium dendrobatidis* | | NPF | Neuropeptide F |
| BLUP | Best linear unbiased predictor | | NSERC | Natural Sciences and Engineering |
| bTB | Bovine tuberculosis | | | Research Council of Canada |
| BCYV | Barley and cereal yellow virus | | OE | *Ophryocystis elektroscirrha* |
| BYDV | Barley yellow dwarf virus | | OHH | Oxidative handicap hypothesis |
| CDR | Call data record | | OID | Ordinary infectious disease |
| CMR | Capture–mark–recapture | | OR | Olfactory receptor |
| CWD | Chronic wasting disease | | OSN | Olfactory sensory neuron |
| DFTD | Devil facial tumor disease | | OXPHOS | Oxidative phosphorylation |
| DH | Definitive host | | PAMP | Pathogen-associated molecular pattern |
| ECSIT | Evolutionarily conserved signaling intermediate in toll pathways | | PaV1 | *Panulirus argus* virus 1 |
| | | | PCH | Parental care hypothesis |
| EIP | Extrinsic incubation period | | PCR | Polymerase chain reaction |
| EPN | Entomopathogenic nematode | | PEERS | Perspectives in ecological research |
| eQTL | Expression quantitative trait loci | | PEMV | Pea enation mosaic virus |
| ERGM | Exponential random graph model | | PHA | Phytohemagglutinin |
| FB | Fan-shaped body | | PIT | Passive integrated transponder |
| GGH | Good genes hypothesis | | PLRV | Potato leaf roll virus |
| GoG | Gambit of the group | | PMS | Parasite-mediated selection |
| GPS | Global positioning system | | PMSS | Parasite-mediated sexual selection |
| HIV | Human immunodeficiency virus | | reQTL | Response expression quantitative trait loci |
| HLA | Human leukocyte antigen | | ROS | Reactive oxygen species |
| HPA | Hypothalamic–pituitary–adrenal | | RRV | Residual reproductive value |
| IH | Intermediate host | | SEIR | Susceptible, exposed, infected, removed/recovered |
| ITS | Internal transcribed spacer | | | |
| LPAIV | Low pathogenic avian influenza virus | | SEIRA | Susceptible, exposed, infected, recovered, asymptomatic |
| LPS | Lipopolysaccharide | | | |
| IBM | Individual-based model | | SNA | Social network analysis |
| ICHH | Immunocompetence handicap hypothesis | | SRB1 | Scavenger receptor B1 |
| IL | Interleukin | | ST | Serovar typhimurium |
| IMA | Inverse matching alleles | | StAR | Steroidogenic acute regulatory protein |
| LH | Lateral horn | | STI | Sexually transmitted infection |
| MA | Matching alleles | | TAH | Transmission avoidance hypothesis |
| MAVS | Mitochondrial antiviral signaling | | TLR | Toll-like receptor |
| MB | Mushroom body | | TSWV | Tomato spotted wilt virus |

| | | | |
|---|---|---|---|
| TYLCV | Tomato yellow leaf curl virus | VIIRS | Visible infrared imaging radiometer suite |
| UAS | Upstream activation sequence | VOC | Volatile organic compound |
| VHF | Very high frequency | | |

# Foreword

No-one who experienced the COVID-19 pandemic could deny the importance of disease and parasites in virtually all aspects of our lives. At the same time, we tend to see pandemics, whether recent or in the distant past, as aberrations, temporary horrors that must be dealt with so that we can return to our usual way of life. We see a photograph of a leaf riddled with holes from caterpillars, or a bird nest crawling with ectoparasites, and have the urge to fix the image, so that it remains unsullied, a representation of its "real" self. Yet the truth is that pandemics, plagues, and infestations are a part of life, not an interruption of it. The caterpillars and fleas belong in those images. To use a phrase from computer science, parasites are not a bug, they are a feature. Parasites have shaped life from its beginning, and they are part of every ecosystem and every organism.

Historically, however, it was difficult for early ecologists and evolutionary biologists to incorporate parasites into their thinking. One noteworthy exception was Charles Elton, the British biologist who helped make ecology a quantitative science. His 1927 work on food webs outlined the "pyramid of numbers," in which consumers are larger than their prey and at higher trophic levels species become less abundant. Both of these rules are violated by parasites, which Elton urged ecologists to take into account, but relatively little attention was paid to the role of parasites in regulating population growth or community composition for virtually the next century.

Another exception to this early neglect came from the famed evolutionary theorist J.B.S. Haldane, who was an early proponent of the importance of studying the evolution of parasites and their interaction with hosts. He was particularly interested in the genetic arms race between disease-causing organisms and their hosts, contrasting predator-prey evolutionary dynamics with those of pathogens. In a 1949 paper, he pointed out that "[i]t is much easier for a mouse to get a set of genes which enables it to resist [bacteria] than a set which enables it to resist cats."

Parasites finally began to receive their due in the late twentieth century, perhaps heralded by a pair of papers by Robert May and Roy M. Anderson in 1979. The scientists presented simple models showing how parasites could regulate populations. Although Anderson and May distinguished between microparasites (viruses, bacteria, and other microbes) and macroparasites (worms, flukes), they demonstrated the importance of both to ecology. The 1980s saw a rush of books and papers examining how parasites influence host ecology, evolution, and behavior, including Peter Price's *Evolutionary Biology of Parasites* and Robert Desowitz's popular volume *New Guinea Tapeworms and Jewish Grandmothers: Tales of Parasites and People*.

Along with the renewed focus on parasites' ability to regulate populations came an emphasis on their effect on host behavior. Bethel and Holmes' 1976 paper and later work on parasite manipulation of host behavior marked a move toward appreciating how profoundly parasites can affect foraging, mate choice, and other behaviors. Janice Moore's landmark *Parasites and the Behavior of Animals*, published in 2002, pointed out that parasite-altered host behavior sometimes enhances the transmission of the parasite and sometimes helps the host defend itself. Distinguishing between the two is not easy, and the difference matters in our consideration of how species interactions evolve. Her work influenced a generation of biologists, some of whose research appears in this volume. Perhaps because of its inherent plasticity, behavior is particularly subject to change when an animal's internal condition is altered by parasites or disease.

Parasites have also been implicated in the evolution of sexual reproduction, where they provide a source of continually varying selection pressure that makes the production of genetically diverse offspring advantageous. The recognition that parasites are an ever-present selective pressure on their hosts is similarly behind the idea that animals may rely on signs of resistance to disease as an indication of "good genes" in potential mates, making parasites crucial in sexual selection. Under such a scenario, females evaluate health using the degree of elaboration of secondary sexual characteristics. These ornaments are condition-dependent, associated with resistance to prevalent parasites. The genes responsible for that resistance fluctuate over time as the parasite–host arms race continues, but females always use the same traits on which to base their mating decisions.

Social behavior, too, has evolved with the constant interplay between host and parasite. Depending on population and group structure, transmission can be enhanced or suppressed. Differences in the number of contacts among individuals within a social group can in turn influence infection dynamics, potentially making the difference between parasite establishment and elimination. Individual differences in behavior can also mediate the spread of a pathogen, with, for instance, more aggressive or bolder individuals being more likely to transmit disease.

This volume explores a range of behaviors that are significantly affected by parasites. At an individual level, infected animals often behave differently from healthy ones, which in turn can influence host ecology and evolution at a variety of scales. Changes in feeding behavior can then affect parasite transmission by altering the contact rate among individuals, which is an important determinant of the evolution of pathogen virulence. Parasite-induced changes in movement behavior over short and long distances have the potential to shift global patterns of animal distribution and abundance. And as host immunity evolves in response to existing parasites, trade-offs among life history characteristics arise, since defense against disease is often costly.

Whether scientists are studying anti-sickness behavior such as social grooming or the details of parasitic manipulation of hosts, new methods in genomics, computational biology, and remote sensing have pushed the frontiers of the field. For example, an animal's location, movements, and contact with other individuals can now be traced using sophisticated modeling and tracking techniques to shed light on the ways that hosts transmit parasites, and make much more targeted predictions about an outbreak's future than had previously been possible.

**Marlene Zuk**
*Professor of Ecology, Evolution and Behavior*
*University of Minnesota*

# Preface

This book was planned in conjunction with a symposium that was originally to be held in May 2020 at the Center for the Ecology of Infectious Diseases, University of Georgia, USA. The goal was for invited participants and the broader community to come together to discuss current research at the interface of animal behavior and parasitism, generate new ideas, and then disperse to write chapters. As we all know now, the COVID-19 pandemic had other plans. So, instead of bringing participants together in person to launch the book in May 2020, we held a virtual symposium one year later, in May 2021, after chapter authors had submitted their contributions to the book. This unexpected detour turned out to be a blessing in disguise. What had been planned as a relatively small in-person conference, turned into a vibrant online event with over 170 participants from 15 different countries, roughly half of whom were graduate students and postdocs. The two-day virtual gathering helped showcase the diverse swath of scientists interested in how behavior and infectious diseases intersect—a diversity that spans disciplinary perspective, geography, gender, race, ethnicity, career stage, and more. The conference exemplified why diverse perspectives are essential in this field and science, more generally. The thought-provoking presentations, insightful discussions, and overall camaraderie at the conference were highlights of our journey with this book and for that we thank all the conference participants.

This book and the associated conference came to fruition despite the myriad challenges of the past two years. For that we have many people to thank.

First, we thank all the chapter contributors for their steadfast work during a very difficult period. To the graduate students from our Fall 2020 Problems in Ecology seminar, we thank you for discussion and critical feedback on initial chapter drafts that helped set the tone for the book. We also thank the many researchers who generously contributed their time as anonymous chapter reviewers. The virtual conference would not have been possible without support provided by the Center for the Ecology of Infectious Diseases at the University of Georgia. In particular, we thank Trippe Ross for his tireless work on conference logistics and planning and Andrea Silletti and Madison Wilson for assistance with proofreading. We are also grateful for financial support from the National Science Foundation (DEB-2022897) that helped facilitate both the conference and book. Finally, the team at Oxford University Press, particularly Ian Sherman and Charles Bath, provided essential guidance throughout the entire process.

Current events have brought the themes of parasitism and host behavior to the foreground of our collective minds. We hope this book provides an accessible background to these tightly coupled phenomena for beginning and seasoned scientists alike. We also hope that the ideas advanced in this volume will stimulate new research that helps push the boundaries of our understanding of how animal behavior and parasitism are interlinked.

Vanessa O. Ezenwa, Sonia Altizer, and
Richard J. Hall
*December 2021*

PART I

# Introduction

# Animal behavior and parasitism: Where have we been, where are we going?

Vanessa O. Ezenwa, Sonia Altizer, and Richard J. Hall

## 1.1 A world with parasites

We live in a world teeming with parasites, defined in this volume as any organism that lives in or on another organism, exploiting it as a resource and causing it harm. Parasites are diverse, spanning bacteria, fungi, viruses, protozoa, helminths, arthropods, and other taxa, and parasitism is one of the most successful lifestyles on earth. A multitude of animal behaviors are driven by interactions with parasites, from mating displays and social interactions, to grooming and self-medication, and decisions about where to move, nest, and forage (1)–(3). In humans, the same is true—our attraction to mates, hygienic behaviors, food choices, and decisions about when and where to gather in groups depend on current and perceived infection risk (4). Infection by parasites can also drive changes in human psychomotor performance, cognition, memory, and personality (5), (6). Although most of us rarely reflect on the impacts of parasites on our daily lives, much of how we interact with the world around us is shaped by the parasites within us and the long-term evolutionary signature of parasites etched onto our genomes.

At the same time, host behavior drives changes in parasite transmission and evolution. Nowhere is this more apparent than in social insects living in large colonies that are prone to rapidly spreading pathogens owing to high contact rates, where individual behavioral responses to infection, and collective behaviors, can protect the most valuable and vulnerable members of the colony. For example, in honeybees, workers remove dead and diseased larvae from the colony to reduce transmission risk (7), line hives with antimicrobial resin to kill bacteria and fungi (8), and raise the temperature of broods when pathogens are detected to create colony-level behavioral fever (9). Likewise, sick ants protect their nestmates from infection by spending more time outside the colony, including leaving the group to die in isolation (10), (11). Such social distancing in response to infection is mirrored in humans, as vividly demonstrated by SARS-CoV-2, the virus responsible for the ongoing COVID-19 pandemic. Control measures adopted to isolate infected patients, limit gatherings, and reduce local and long-distance travel effectively reduced the rate of pathogen transmission and spread in early stages of the pandemic (12).

The myriad connections between animal behavior and parasitism have been the subject of growing research interest since the 1970s and 1980s when fundamental theories linking the two fields of study emerged (e.g., (13)–(15)). These ideas were expanded upon in the 1990s and early 2000s (e.g., (1), (2), (16)–(18)), and since then there has been a steady increase in research effort in the field (Figure 1.1). In 2000, articles including some aspect of parasitism represented only 5% of research

Vanessa O. Ezenwa, Sonia Altizer, and Richard J. Hall, *Animal behavior and parasitism: Where have we been, where are we going?*. In: *Animal Behavior and Parasitism*. Edited by Vanessa O. Ezenwa, Sonia Altizer, and Richard J. Hall, Oxford University Press.

published on animal behavior, behavioral ecology, or ethology, but by 2021, the representation of parasitism in the animal behavior literature nearly doubled to 9% (Figure 1.1). The same pattern holds when viewed from the perspective of the parasite literature, where the representation of animal behavior among articles published on parasitism grew from 0.26% to 0.44% (Figure 1.1). As a result of this sustained research effort, fascinating new connections between animal behavior and disease ecology and evolution continue to emerge. Now, aided by a combination of conceptual and technological advances and the increased integration of ideas across disciplines, the field is on the precipice of a revolution. Transformative methodological advances are enabling the simultaneous collection of physiological, genomic, movement, and other behavioral data in real time, from tracking individual and colony-level behavioral responses to

infection in ants (11), to documenting the consequences of social behavior for immune regulation in primates (19). This volume takes advantage of this confluence of events to bring together emerging research at the interface of animal behavior and parasitism. The chapters in this book build upon foundational research at multiple levels of biological organization, from genes to proteins and individual organisms to communities, to showcase how far our understanding of animal behavior and parasitism has come and explore what the future holds.

It is impossible for any single text to cover all the ways in which animal behavior and parasitism have been studied together so this volume focuses on five key thematic areas that have seen significant recent development: (i) social behavior, (ii) movement behavior, (iii) sexual selection and mating behavior, (iv) behavioral modification, and (v) behavioral defenses. Chapters provide a diverse

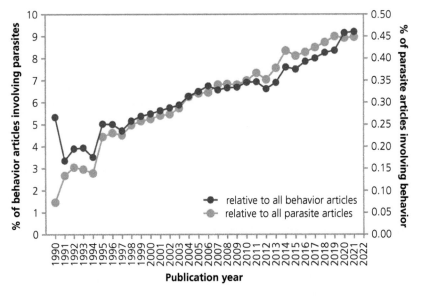

**Figure 1.1** Research focused on the intersection between animal behavior and parasitism increased steadily over the past 30 years. Accounting for the total number of studies referencing either animal behavior (blue line, left axis) or parasitism (green line, right axis) in the title or abstract, the frequency of publications including both topics increased between two- and sixfold from 1990 to 2021. Data are based on a Web of Science search for animal behavior-related publications [search string: (AB=(animal and behavio* or behavio* and ecology or ethology)) OR TI=(animal and behavio* or behavio* and ecology or ethology)], parasite-related publications [search string: (AB=(pathog* or parasit* or bacteri* or virus* or fung* or helminth* or nematod* or trematod* or cestod* or acanthoceph* or protozoa* or ectoparasit*)) OR TI=(pathog* or parasit* or bacteri* or virus* or fung* or helminth* or nematod* or trematod* or cestod* or acanthoceph* or protozoa* or ectoparasit*)], and a combination of the two. The search yielded 127,928 articles on animal behavior, 2.5 million articles on parasitism, and 9,039 articles when the search strings were combined. We divided the number of articles linking behavior and parasitism published each year by either the total number of behavior-related articles or the total number of parasite-related articles. Irrespective of the denominator used, the proportion of behavior–parasite articles increased over time.

and international perspective for a target audience spanning upper-division undergraduates and graduate students learning about parasites and behavior for the first time, to researchers interested in the many intersections of these two fields of study. In this introductory chapter, we develop a simple framework for conceptualizing key links between animal behavior and parasitism, using case studies to help illustrate these fundamental connections. Next, we introduce the chapters of the book, highlighting insights and connections that emerge across chapters. A second introductory chapter explores the concept of feedback between host behavior and parasitism, the idea that behavior of hosts and traits of parasites exert reciprocal influence on one another with consequences for both hosts and parasites (20). We use these behavior–parasitism feedbacks as a unifying sub-theme running throughout the book, serving to link ideas and insights across topics. Finally, perspectives from a diverse and growing animal behavior-parasitism research community that emerged from a virtual symposium on *Research Frontiers in Animal Behavior and Parasitism*, held in May 2021, are captured in two chapters at the end of the book, in an "emerging perspectives" section.

## 1.2 Animal behavior and parasitism: Essential links

Animal behavior and parasitism are linked in at least three fundamental ways. First, behavior affects individual infection risk by altering contact rates with parasites, and susceptibility and/or resistance to infection upon parasite contact (Figure 1.2A). Second, behavior-linked differences between hosts in contact rate and susceptibility often lead to transmission heterogeneities that influence population-level epidemic dynamics (Figure 1.2B). Third, parasite infection has reciprocal effects on behavior. In the short term (i.e., over ecological timescales), infection can modify behavior via energetic and physiological mechanisms (e.g., altered activity due to energy constraints or sickness behavior), and in the long term (i.e., over evolutionary timescales), parasitism can act as a selective agent on host behavior (Figure 1.2C). Parasites also modify animal behavior in the short and

long term by directly manipulating the activity of infected hosts. These three points of intersection between animal behavior and parasitism apply to a range of host behaviors (e.g., foraging behavior, social behavior, mating behavior, movement behavior, etc.) and parasite taxa (e.g., viruses, bacteria, helminths, etc.), and have broad relevance for understanding the ecology and evolution of host-parasite interactions. We use a series of case studies to illustrate each of these three phenomena below.

### 1.2.1 Behavioral risk factors of infection

Inter-individual differences in infection risk are common and many factors, both intrinsic and extrinsic to the host, contribute to this variation. For example, intrinsic factors such as host genotype, sex, and age contribute to differential risk of infection via effects on parasite exposure, susceptibility or both (21). Interestingly, the way in which these factors influence infection risk often depends on behavior (22). Sex-biased parasite infection in vertebrates provides an illustrative example. Across vertebrate taxa, males tend to suffer from higher rates of parasite infection than females (23), (24). This pattern arises because male reproductive and mating behaviors, and the physiological changes associated with them, often lead to differential exposure to parasites and/or susceptibility to infection (25). For example, testosterone is a key modulator of reproductive behavior in male vertebrates, but this hormone can also suppress immune function making males more vulnerable to infection. Indeed, a recent meta-analysis of experimental studies that manipulated testosterone levels in males found evidence for a generalized immunosuppressive effect of testosterone across vertebrate taxa (26). Testosterone also induces changes in behavior that may increase exposure to parasites, and these effects can act in concert with immunosuppression to fuel male biases in infection. For example, in male Grant's gazelle (*Nanger granti*) both levels of endogenous testosterone and testosterone-associated behaviors are independently associated with helminth infection (27).

More generally, some facets of animal behavior are simply inseparable from parasitism. A prime example is social behavior, encompassing the act

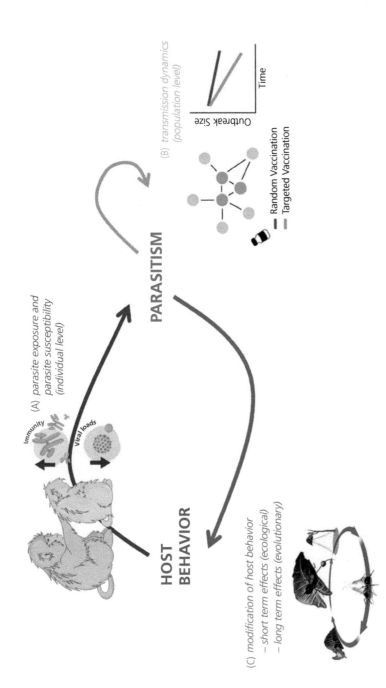

**Figure 1.2** Fundamental links between animal behavior and parasitism. (A) At an individual host level, behavior affects the frequency of exposure to parasites as well as susceptibility to infection given exposure. In rhesus macaques, for example, social behavior influences patterns of immune gene expression and vulnerability to virus infection. (B) Heterogeneities in individual infection patterns created by host behavior have knock-on effects for parasite transmission dynamics, but this phenomenon can be exploited to more effectively manage disease outbreaks, emphasizing the applicability of behavioral knowledge for disease control. For example, outbreak size can be reduced more rapidly by targeted vaccination of individual hosts based on their social network position than by randomly choosing which individuals are vaccinated. (C) Parasites also influence animal behavior on ecological and evolutionary timescales. In some cases, like the fascinating zombie ant fungus, parasite-associated changes in host behavior are the consequence of manipulation by the parasite. The images in (A) and (B) are reprinted from *Social Behavior, Infection, and Immunity*, 2021 (iBook) by Kate Sabey and Vanessa O. Ezenwa. The image in (C) is reproduced with permission from Science Photo Library.

of living in groups and all the social interactions it entails. Many parasites rely on social contact between hosts for successful transmission. Consequently, being more sociable is associated with consistently higher rates of infection across multiple host taxa (16), (28)–(30). Although less well-studied, social behavior also affects susceptibility and resistance to infection (Figure 1.2A). In rhesus macaques (*Macaca mulatta*), for example, males in a socially unstable environment exposed to high levels of aggression were less resistant to a simian immunodeficiency virus challenge and more likely to die of infection compared to individuals in stable social environments (31). Among female macaques, social rank affected how individuals responded immunologically to a challenge with both bacterial and viral immune stimulants, implying the existence of rank-related differences in pathogen susceptibility (19), (32).

## 1.2.2 Transmission heterogeneities and epidemic dynamics

When individuals differ in their propensity to become infected or infect others, there are downstream consequences for epidemiological dynamics. The degree of transmission heterogeneity among hosts can be conceptualized as the individual reproductive number ($v$), a quantity that represents the number of secondary cases a particular infected individual generates in a susceptible host population, with the population average equivalent to the more commonly described basic reproductive number ($R_0$) (33). Superspreaders, or individuals who generate a disproportionately large number of secondary cases, embody the concept of transmission heterogeneity and behavior plays a prominent role in an individual's potential to be a superspreader. Mary Mallon, a cook from New York at the turn of the twentieth century, was probably the first recorded superspreader. Mary was a chronic carrier of the bacteria that causes typhoid fever, and as a direct result of her occupation as a cook, she seeded typhoid outbreaks in multiple private households and commercial establishments in New York, New Jersey, and Maine, earning the alias "Typhoid Mary" (34). We know now that superspreading is characteristic of many diseases of

wildlife and humans, including emerging zoonotic diseases such as SARS-CoV-2.

From an epidemiological perspective, transmission heterogeneity affects key attributes of an outbreak, including the probability of pathogen invasion and the speed and severity of outbreaks when invasion occurs (33). For this reason, knowledge of heterogeneities in animal behavior can be exploited to help design more tractable disease management strategies (Figure 1.2B). As an example, Rushmore and colleagues used detailed observations of wild chimpanzee social behavior to explore the practicality of different vaccination scenarios for controlling pathogen outbreaks in a wildlife population (35). By simulating pathogen transmission on contact networks constructed from their behavioral observations, the authors showed that contact heterogeneity strongly influenced outbreak size. When the first infected individual in the population was highly connected, pathogen outbreaks were larger; this was the case even when the pathogen in question had a low level of infectiousness. However, vaccine strategies that took advantage of this contact heterogeneity effect, targeting the most highly connected individuals in the population, had the dual benefit of effectively controlling large outbreaks and simultaneously requiring fewer individuals to be vaccinated (35).

## 1.2.3 Parasite modification of host behavior

The relationship between behavior and parasitism is rarely unidirectional. Behavior affects parasites, and parasites affect behavior. For example, during infection, the behavioral repertoire of animals frequently changes. This classical sickness behavior, characterized by general inactivity, anorexia, changes in sleep patterns, and reduced social interaction, results from the host's physiological response to infection and is thought to help promote recovery (36), (37). The behavior of uninfected individuals can also change dramatically in response to either the presence of pathogens or infected conspecifics. In colonies of the black garden ant, *Lasius niger*, for example, experimental exposure of a subset of individuals to a pathogenic fungus resulted in changes in the social behavior of both fungus-exposed and unexposed ants (11). These pathogen-induced changes in behavior altered colony social networks in a way that reduced overall fungal spread and protected high value individuals (11), suggesting that animals explicitly use behavior as a strategy to respond to widespread parasite exposure. Indeed, over the long term, and in response to potent selection pressure imposed by parasites, hosts have evolved a considerable suite of behavioral mechanisms that aid in parasite defense (38), (39). These behavioral defenses, collectively termed the behavioral immune system, can be reactive, as is sickness behavior, or proactive, including a wide range of avoidance behaviors that inhibit contact with parasites before it happens (40).

While some behavioral responses of hosts to parasites are in the interest of the host, others are in the interest of the parasite and result from direct manipulation of the host by the parasite. Parasite manipulators are taxonomically diverse, ranging from viruses and fungi to worms, protozoa, and arthropods, and their effects on host behavior range from the bizarre to the more subtle (2). An example of a bizarre manipulator is the zombie ant fungus, *Ophiocordyceps unilaterali*. Carpenter ants (*Camponotus* spp.) infected with this fungus leave their colonies, climb up plant stems overhanging well-used colony trails and lock their jaws onto the vegetation (Figure 1.2C). In this position, the ant dies, the fungus grows, and its spores disperse onto naïve ants walking the trails below. Interestingly, *O. unilaterali* fungal cells appear to behave collectively to achieve this host manipulation. As the fungus triggers the biting behavior of the manipulated ant, its hyphae connect to form networks around host muscle tissue, a strategy that likely promotes nutrient transport among fungal cells and the emergence of the fungal structures that grow out of the ant's body (41).

## 1.3 Introduction to book chapters

The examples above highlight three fundamental connections between parasites and the behavior of their hosts. Considered in combination, the inherent bi-directional nature of host behavior–parasite interactions is inescapable (Figure 1.2). When host behavior affects parasite biology, this effect has repercussions for behavior on both ecological and evolutionary timescales (20). The same processes operate in the reciprocal direction, with parasites affecting behavior (20), As such, the concept of bi-directional feedback between animal behavior and parasitism represents an emerging research frontier that cuts across the many complex ways in which host behavior and parasitism interact. In Chapter 2 of this volume, Hawley and Ezenwa (42) define behavior–parasite feedbacks and identify core conditions that produce these feedbacks. They examine how the study of feedback processes in other fields of biology provides clues as to potential functional outcomes of behavior–parasite feedbacks, and the relevance of the feedback paradigm for understanding host and parasite eco-evolutionary dynamics. Following on from this exploration of feedbacks is a series of chapters (Chapters 3–17) that focus on core themes in the study of animal behavior and parasitism (social behavior, movement behavior, sexual selection and mating behavior, parasite modification of host behavior, and behavioral defenses). Each theme encompasses two to three chapters that approach the subject matter from different perspectives. A final pair of chapters (Chapters 18–19) form a section on emerging perspectives. These two chapters broaden the discourse, focusing on convergent ideas for the future of the field. Importantly, Chapters 3 to 19 each touch upon the concept of behavior–parasite feedback in their own way, and this common thread helps weave a fascinating, multi-dimensional picture of the role of feedbacks in our understanding of animal behavior and parasitism.

### 1.3.1 Chapters 3–5: Social behavior

Few behaviors are as well studied in relation to parasitism as is social behavior. Behavioral ecologists, evolutionary biologists, disease ecologists, and epidemiologists all have a vested interest in understanding the links between social behavior and parasite transmission. In Chapter 3, Sadoughi *et al.* (43) synthesize our current understanding of the costs and benefits of social behavior in a world riddled with parasites. Focusing on primates as a model host taxon, and exploring social behavior from the individual to community-level perspectives, the authors highlight our growing understanding of how social life affects parasitism. In contrast, they note that the reciprocal—how parasites affect social behavior—is less well understood but equally important given that understanding bidirectional relationships between social behavior and parasitism may be central to uncovering both the role of social behavior in epidemiological dynamics and the role of parasites in social evolution. Over the past several years, social networks have emerged as a powerful tool for studying the links between social behavior and parasitism. In Chapter 4, Mistrick and colleagues (44) discuss important methodological considerations for using networks to make inferences about how individual social position or social network topology affects parasite transmission. They also consider how parasite infection can influence individual network position or overall network topology, highlighting ways in which both empirical and modeling studies can be designed to better understand behavior-parasite feedbacks in the context of networks. In Chapter 5, Keiser (45) expands the focus to collective behavior, an emergent property of inter-individual interactions. In particular, he explores how traits of individuals within a group influence the frequency and type of social interactions and the joint repercussions for group-level execution of collective behaviors and the occurrence of disease outbreaks. This exploration helps shape a novel "social fulcrum" hypothesis, the idea that selection may operate on group traits and their joint outcomes on collective behavior and parasitism.

### 1.3.2 Chapters 6–8: Movement behavior

From localized foraging to long-distance tracking of favorable environments, movement is a fundamental component of animal behavior. These movements shape where and when individuals encounter parasites and are often crucial for parasite dispersal. Heterogeneities in movement patterns among hosts can alter transmission dynamics in host populations, allowing some individuals to avoid parasite exposure or recover from infection, while aggregating other individuals at sites with high transmission risk. In turn, physiological costs of parasite infection, and parasite manipulation can alter host movement patterns in ways that are

detrimental or beneficial to parasite fitness, potentially reinforcing heterogeneities in host behaviors. In Chapter 6, Spiegel and colleagues (46) provide a roadmap for embedding parasites within the movement ecology paradigm in which animal movement decisions are influenced by their internal state and external environment. The authors also emphasize how advances in tracking technologies are transforming our ability to study linkages and covariation between individual host behavior and infection. Chapter 7 by Hall *et al.* (47) focuses on the role of long-distance animal migration in shaping host exposure to, and immune defense against, parasites, and how parasites in turn can influence the timing, extent, and success of migratory movement. The authors propose that studying feedbacks between migration and parasite infection can help clarify the role of parasites in maintaining variation in animal migratory strategies and yield fresh insights into how global change is altering movement propensity and pathogen transmission in migratory species. Humans also undertake daily or seasonal movements for work and recreation, and quantifying these movements has been critical for understanding the global spread of pathogens from seasonal influenza to COVID-19. In Chapter 8, Meredith and Wesolowski (48) focus on the study of seasonal mobility patterns in humans, identifying the utility and limitations of different approaches for inferring human movement, and outlining how mobility data can be incorporated into statistical and mathematical models of pathogen transmission and spatial spread.

### 1.3.3 Chapters 9–11: Sexual selection and mating behavior

The idea that parasites play a role in sexual selection by influencing mate choice is nearly as fascinating as the concept of sexual selection itself. Since the introduction of the concept of parasite-mediated sexual selection (PMSS) by Hamilton and Zuk in the 1980s (15), the topic has been subject to intense theoretical and empirical examination. In Chapter 9, Pirrie, Chapman, and Ashby (49) provide a timely review of the current state of the literature on PMSS. They compare the ultimate hypotheses for PMSS, discuss empirical support for these hypotheses, and identify key predictions emerging from mathematical models. Their review also reveals gaps in the existing literature, including a lack of theory on coevolutionary dynamics in the context of PMSS. In Chapter 10, Koch and Hill (50) tackle the question of mechanism. The authors examine major proximate explanations for why mating displays serve as honest signals of parasite resistance, distinguishing between hypotheses that rely on shared physiological costs versus those that rely on shared biochemical pathways. They argue that mechanisms involving shared pathways can offer unique insight, because pathways involving mitochondrial function provide an inescapable link between display production and immune defense. In Chapter 11, Winternitz and Abbate (51) shift the focus to genes. Concentrating on immune genes, which are among the best candidates for genes underpinning PMSS, they examine theoretical models of immune gene-based mate choice and empirical evidence for involvement of immune genes in sexual selection. By systematically surveying the literature, the authors find support for the involvement of diverse immune genes in mate choice in both vertebrate and invertebrate animals.

### 1.3.4 Chapters 12–13: Behavioral modification by parasites

Changes in behavior arising from parasite infection, such as trematode-infected hosts moving into habitats that increase their risk of predation by the parasite's definitive host, or infected mosquitoes preferentially biting uninfected hosts, can substantially alter a parasite's chances of onward transmission, and simultaneously generate heterogeneities in host behaviors. In Chapter 12, Godfrey and Poulin (52) review how parasites manipulate a variety of host behaviors, and by drawing on emerging research on animal personality and social networks, demonstrate how parasite manipulation can influence both individual and group-level behaviors, with consequences for host social organization and collective behavior as well as parasite transmission. Chapter 13 by Cator (53) digs into the mechanisms by which parasites alter feeding behaviors of arthropod vectors, ranging from aphids to mosquitoes. She highlights how research into

coevolutionary interactions between parasites and vectors is necessary to understand whether these behavioral changes can be interpreted as parasite manipulation.

### 1.3.5 Chapters 14–17: Behavioral defenses against parasites

Animals can modify where they spend time, what they eat, and who they associate with in order to reduce exposure to parasites, potentially leading to profound variation in behavior and infection outcomes. In Chapter 14, Lopes and colleagues (54) survey the cues by which animals detect parasite-contaminated food and habitat, or infected conspecifics, and the variety of avoidance behaviors exhibited by diverse non-mammalian taxa. Since individuals adopting parasite avoidance behaviors often forgo foraging and mating opportunities, quantifying the costs and benefits of these behaviors is crucial for understanding how hosts navigate a "landscape of disgust." Thus, Chapter 15 by Poirotte and Charpentier (55) uses a cost-benefit framework to understand the drivers of variation in avoidance behaviors among individuals, and the consequences of this variation for parasite transmission, the ecology and evolution of hosts, and the ecosystem functions they perform.

In addition to avoidance, animals also engage in behavioral defenses that reduce levels of infection or minimize the negative impacts of infection. Being similar in size and mobility to their hosts, parasitoids represent a unique challenge for host populations. In Chapter 16, Davis and Schlenke (56) review how invertebrates both reduce their risk of parasitoid attack (e.g., by initiating offspring protection and escape behaviors) and inhibit internal parasitoid development (e.g., by changing diet). The authors use insights from innovative work on *Drosophila* to explore the neuronal and genetic bases for these anti-parasitoid defenses. In Chapter 17, Stephenson and Adelman (57) investigate the relationships between host behaviors that increase pre-infection fitness (behavioral vigor), and the extent to which hosts can maintain these fitness-enhancing behaviors during active infection (behavioral tolerance) and following recovery from infection (behavioral resilience). Understanding the relationships between behavioral vigor, tolerance, and resilience represents an exciting new avenue for understanding how behavior modulates the fitness consequences of infection.

### 1.3.6 Chapters 18–19: Emerging perspectives

The final two chapters of the book highlight promising future directions for the integrated study of behavior and parasitism. In Chapter 18, Guindre-Parker, Tung, and Strauss (58) describe how integrating tools and concepts from organismal biology, molecular ecology, community and ecosystem ecology can help transform our understanding of behavior–parasitism linkages across scales of biological organization. The authors advocate for a holistic approach that accounts for both interactions between co-infecting parasites and interactions between host (and non-host) species and the environment to explain variation in behavior and infection outcomes. They argue that this approach is especially important in the context of understanding how behavior–parasite associations will respond to anthropogenic change. Finally, Chapter 19 by Resetarits and colleagues (59) surveys the often under-explored topic of parasite behavior. The authors argue that, like host behavior, parasite social, mating, and movement behaviors critically shape the ecology and evolution of host–parasite interactions. They also posit that parasite behaviors are a key parasite trait that helps fuel host behavior–parasite feedbacks.

### 1.4 Concluding remarks

In sum, this book highlights promising new research at the nexus of animal behavior and parasitism that will help advance scientific understanding of these two deeply intertwined processes. A common theme emerging across all chapters is that theoretical and empirical work has frequently focused on one-way linkages between behavior and parasitism (i.e., effects of behavior on parasite risk and transmission, or effects of parasite exposure and infection on behavior). The holistic study of two-way interactions and feedbacks between animal behavior and parasitism has the potential to transform scientific understanding of the ecology

and evolution of parasitism in wildlife and people. We hope that this exciting collection of chapters, highlighting diverse themes, researchers, and approaches will stimulate new ideas and help shape research in these fields in the decades to come.

### References

1. Hart BL. Behavioral adaptations to pathogens and parasites: Five strategies. *Neurosci Biobehav Rev.* 1990;14(3):273–94.
2. Moore J. *Parasites and the Behavior of Animals.* New York: Oxford University Press, 2002, ix, 315 pp.
3. Zuk M. The role of parasites in sexual selection—current evidence and future-directions. *Adv Stud Behav.* 1992;21:39–68.
4. Kramer P and Bressan P. Infection threat shapes our social instincts. *Behav Ecol Sociobiol.* 2021;75(3):47.
5. Nokes C, Grantham-McGregor SM, Sawyer AW, Cooper ES, and Bundy DA. Parasitic helminth infection and cognitive function in school children. *Proc R Soc London B: Biol Sci.* 1992;247(1319):77–81.
6. Martinez VO, de Mendonca Lima FW, de Carvalho CF, and Menezes-Filho JA. *Toxoplasma gondii* infection and behavioral outcomes in humans: A systematic review. *Parasitol Res.* 2018;117(10):3059–65.
7. Al Toufailia H, Evison SEF, Hughes WOH, and Ratnieks FLW. Both hygienic and non-hygienic honeybee, *Apis mellifera,* colonies remove dead and diseased larvae from open brood cells. *Philos Trans R Soc Lond B Biol Sci.* 2018;373:1751.
8. Simone-Finstrom M and Spivak M. Propolis and bee health: The natural history and significance of resin use by honey bees. *Apidologie.* 2010;41(3):295–311.
9. Bonoan RE, Iglesias Feliciano PM, Chang J, and Starks PT. Social benefits require a community: The influence of colony size on behavioral immunity in honey bees. *Apidologie.* 2020;51(5):701–9.
10. Heinze J and Walter B. Moribund ants leave their nests to die in social isolation. *Curr Biol.* 2010;20(3):249–52.
11. Stroeymeyt N, Grasse AV, Crespi A, Mersch DP, Cremer S, and Keller L. Social network plasticity decreases disease transmission in a eusocial insect. *Science.* 2018;362(6417):941–5.
12. Nande A, Adlam B, Sheen J, Levy MZ, and Hill AL. Dynamics of COVID-19 under social distancing measures are driven by transmission network structure. *PLoS Comput Biol.* 2021;17(2):e1008684.
13. Alexander RD. The evolution of social behavior. *Annu Rev Ecol Evol Syst.* 1974;5(1):325–83.

14. Anderson RM and May RM. Population biology of infectious-diseases. 1. *Nature*. 1979;280(5721): 361–7.

15. Hamilton WD and Zuk M. Heritable true fitness and bright birds—a role for parasites. *Science*. 1982;218(4570):384–7.

16. Cote IM and Poulin R. Parasitism and group-size in social animals—a metaanalysis. *Behav Ecol*. 1995;6(2):159–65.

17. Folstad I and Karter AJ. Parasites, bright males, and the immunocompetence handicap. *Am Nat*. 1992;139(3):603–22.

18. Loehle C. Social barriers to pathogen transmission in wild animal populations. *Ecology*. 1995;76(2): 326–35.

19. Snyder-Mackler N, Sanz J, Kohn JN, Brinkworth JF, Morrow S, Shaver AO, *et al*. Social status alters immune regulation and response to infection in macaques. *Science*. 2016;354(6315):1041–5.

20. Ezenwa VO, Archie EA, Craft ME, Hawley DM, Martin LB, Moore J, *et al*. Host behaviour–parasite feedback: An essential link between animal behaviour and disease ecology. *Proc R Soc London B: Biol Sci*. 2016;283(1828).

21. Wilson K, Bjørnstad O, Dobson A, Merler S, Poglayen G, Randolph S, *et al*. Heterogeneities in macroparasite infections: Patterns and processes in Hudson PJ, Rizzoli A, Grenfell BT, Heesterbeck H, and Dobson AP (eds), *The Ecology of Wildlife Diseases*Oxford: Oxford University Press, 2002, 6–44.

22. VanderWaal KL and Ezenwa VO. Heterogeneity in pathogen transmission: Mechanisms and methodology. *Funct Ecol*. 2016;30(10):1606–22.

23. Poulin R. Sexual inequalities in helminth infections: A cost of being a male? *Am Nat*. 1996;147(2):287–95.

24. Schalk G and Forbes MR. Male biases in parasitism of mammals: Effects of study type, host age, and parasite taxon. *Oikos*. 1997;78(1):67–74.

25. Zuk M and McKean KA. Sex differences in parasite infections: patterns and processes. *Int J Parasitol*. 1996;26(10):1009–23.

26. Foo YZ, Nakagawa S, Rhodes G, and Simmons LW. The effects of sex hormones on immune function: A meta-analysis. *Biol Rev*. 2017;92(1):551–71.

27. Ezenwa VO, Ekernas LS, and Creel S. Unravelling complex associations between testosterone and parasite infection in the wild. *Funct Ecol*. 2012;26(1):123–33.

28. Briard L and Ezenwa VO. Parasitism and host social behaviour: A meta-analysis of insights derived from social network analysis. *Anim Behav*. 2021;172: 171–82.

29. Patterson JEH and Ruckstuhl KE. Parasite infection and host group size: A meta-analytical review. *Parasitology*. 2013;140(7):803–13.

30. Rifkin JL, Nunn CL, and Garamszegi LZ. Do animals living in larger groups experience greater parasitism? A meta-analysis. *Am Nat*. 2012;180(1):70–82.

31. Capitanio JP, Mendoza SP, Lerche NW, and Mason WA. Social stress results in altered glucocorticoid regulation and shorter survival in simian acquired immune deficiency syndrome. *PNAS*. 1998;95(8):4714–19.

32. Sanz J, Maurizio PL, Snyder-Mackler N, Simons ND, Voyles T, Kohn J, *et al*. Social history and exposure to pathogen signals modulate social status effects on gene regulation in rhesus macaques. *PNAS*. 2020;117(38):23317–22.

33. Lloyd-Smith JO, Schreiber SJ, Kopp PE, and Getz WM. Superspreading and the effect of individual variation on disease emergence. *Nature*. 2005;438(7066):355–9.

34. Soper GA. The curious career of Typhoid Mary. *Bull N Y Acad Med*. 1939;15(10):698.

35. Rushmore J, Caillaud D, Hall RJ, Stumpf RM, Meyers LA, and Altizer S. Network-based vaccination improves prospects for disease control in wild chimpanzees. *J R Soc Interface*. 2014;11(97):20140349.

36. Hart BL. Biological basis of the behavior of sick animals. *Neurosci Biobehav R*. 1988;12(2):123–37.

37. Kelley KW, Bluthe RM, Dantzer R, Zhou JH, Shen WH, Johnson RW, *et al*. Cytokine-induced sickness behavior. *Brain Behav Immun*. 2003;17:S112–S8.

38. Curtis VA. Infection-avoidance behaviour in humans and other animals. *Trends Immunol*. 2014;35(10): 457–64.

39. Sarabian C, Curtis V, and McMullan R. Evolution of pathogen and parasite avoidance behaviours . *Philos Trans R Soc Lond B Biol Sci* 2018;373:1751.

40. Schaller M. The behavioural immune system and the psychology of human sociality. *Philos Trans R Soc Lond B Biol Sci*. 2011;366(1583):3418–26.

41. Fredericksen MA, Zhang Y, Hazen ML, Loreto RG, Mangold CA, Chen DZ, *et al*. Three-dimensional visualization and a deep-learning model reveal complex fungal parasite networks in behaviorally manipulated ants. *PNAS*. 2017;114(47):12590–95.

42. Hawley DM and Ezenwa VO. Parasites, host behavior, and their feedbacks. In Ezenwa VO, Altizer S, and Hall RJ (eds), *Animal Behavior and Parasitism*. Oxford: Oxford University Press, 2022. DOI: 10.1093/oso/9780192895561.003.0002.

43. Sadoughi B, Anzà S, Defolie C, Manin V, Müller-Klein N, Murillo T, *et al*. Parasites in a social world: Lessons from primates. In Ezenwa VO, Altizer S, and Hall RJ (eds), *Animal Behavior and Parasitism*. Oxford: Oxford University Press, 2022. DOI: 10.1093/oso/9780192895561.003.0003.

44. Mistrick J, Gilbertson MLJ, White LA, and Craft ME. Constructing animal networks for parasite

transmission inference. In Ezenwa VO, Altizer S, and Hall RJ (eds), *Animal Behavior and Parasitism*. Oxford: Oxford University Press, 2022. DOI: 10.1093/oso/9780192895561.003.0004.

45. Keiser CN. Collective behavior and parasite transmission. In Ezenwa VO, Altizer S, and Hall RJ (eds), *Animal Behavior and Parasitism*. Oxford: Oxford University Press, 2022. DOI: 10.1093/oso/9780192895561.003.0005.

46. Spiegel O, Anglister N, and Crafton MM. Movement data provides insights into feedbacks and heterogeneities in host–parasite interactions. In Ezenwa VO, Altizer S, and Hall RJ (eds), *Animal Behavior and Parasitism*. Oxford: Oxford University Press, 2022. DOI: 10.1093/oso/9780192895561.003.0006.

47. Hall RJ, Altizer S, Peacock SJ, and Shaw AK. Animal migration and infection dynamics: Recent advances and future frontiers. In Ezenwa VO, Altizer S, and Hall RJ (eds), *Animal Behavior and Parasitism*. Oxford: Oxford University Press, 2022. DOI: 10.1093/oso/9780192895561.003.0007.

48. Meredith HR and Wesolowski A. Seasonal human movement and the consequences for infectious disease transmission. In Ezenwa VO, Altizer S, and Hall RJ (eds), *Animal Behavior and Parasitism*. Oxford: Oxford University Press, 2022. DOI: 10.1093/oso/9780192895561.003.0008.

49. Pirrie A, Chapman H, and Ashby B. Parasite-mediated sexual selection: To mate or not to mate? in Ezenwa VO, Altizer S, and Hall RJ (eds), *Animal Behavior and Parasitism*. Oxford: Oxford University Press, 2022. DOI: 10.1093/oso/9780192895561.003.0009.

50. Koch RE and Hill GE. Shared biochemical pathways for ornamentation and immune function: Rethinking the mechanisms underlying honest signaling of parasite resistance. In Ezenwa VO, Altizer S, and Hall RJ (eds), *Animal Behavior and Parasitism*. Oxford: Oxford University Press, 2022. DOI: 10.1093/oso/9780192895561.003.0010.

51. Winternitz JC and Abbate JL. The genes of attraction: Mating behavior, immunogenetic variation, and parasite resistance. In Ezenwa VO, Altizer S, and Hall RJ (eds), *Animal Behavior and Parasitism*. Oxford: Oxford University Press, 2022. DOI: 10.1093/oso/9780192895561.003.0011.

52. Godfrey SS and Poulin R. Host manipulation by parasites: From individual to collective behavior.

In Ezenwa VO, Altizer S, and Hall RJ (eds), *Animal Behavior and Parasitism*. Oxford: Oxford University Press, 2022. DOI: 10.1093/oso/9780192895561.003.0012.

53. Cator LJ. Altered feeding behaviors in disease vectors. In Ezenwa VO, Altizer S, and Hall RJ (eds) *Animal Behavior and Parasitism*. Oxford: Oxford University Press, 2022. DOI: 10.1093/oso/9780192895561.003.0013.

54. Lopes PC, French SS, Woodhams DC, and Binning SA. Infection avoidance behaviors across vertebrate taxa: Patterns, processes and future directions. In Ezenwa VO, Altizer S, and Hall RJ (eds), *Animal Behavior and Parasitism*. Oxford: Oxford University Press, 2022. DOI: 10.1093/oso/9780192895561.003.0014.

55. Poirotte C and Charpentier MJE. Inter-individual variation in parasite avoidance behaviors and its epidemiological, ecological, and evolutionary consequences. In Ezenwa VO, Altizer SA, and Hall RJ (eds), *Animal Behavior and Parasitism*. Oxford: Oxford University Press, 2022. DOI: 10.1093/oso/9780192895561.003.0015.

56. Davis S and Schlenke T. Behavioral defenses against parasitoids: Genetic and neuronal mechanisms. In Ezenwa VO, Altizer S, and Hall RJ (eds), *Animal Behavior and Parasitism*. Oxford: Oxford University Press, 2022. DOI: 10.1093/oso/9780192895561.003.0016.

57. Stephenson JF and Adelman JS. The behavior of infected hosts: Behavioral tolerance, behavioral resilience, and their implications for behavioral competence. In Ezenwa VO, Altizer S, and Hall RJ (eds), *Animal Behavior and Parasitism*. Oxford: Oxford University Press, 2022. DOI: 10.1093/oso/9780192895561.003.0017.

58. Guindre-Parker S, Tung J, and Strauss AT. Emerging frontiers in animal behavior and parasitism: Integration across scales. In Ezenwa VO, Altizer S, and Hall RJ (eds), *Animal Behavior and Parasitism*. Oxford: Oxford University Press, 2022. DOI: 10.1093/oso/9780192895561.003.0018.

59. Resetarits EJ, Bartlett LJ, Wilson CA, and Willoughby AR. Parallels in parasite behavior: The other side of the host–parasite relationship. In Ezenwa VO, Altizer S, and Hall RJ (eds), *Animal Behavior and Parasitism*. Oxford: Oxford University Press, 2022. DOI: 10.1093/oso/9780192895561.003.0019.

# CHAPTER 2

# Parasites, host behavior, and their feedbacks

Dana M. Hawley and Vanessa O. Ezenwa

## 2.1 Introduction: What is a behavior–parasite feedback?

Animal behaviors such as foraging, aggression, and group living are key to host survival and reproduction, but they also determine fitness for many parasites (1). Likewise, parasites affect many types of animal behaviors, both on ecological and evolutionary timescales (2), (3). Such unidirectional links between animal behavior and parasitism, and their mechanistic underpinnings, are well documented in the literature, with a myriad of recent conceptual and technical advances fueling recent discoveries (e.g., animal personality/behavioral syndromes (4); social network analysis and animal tracking technology (e.g., (5)); next-generation sequencing (e.g., 6)). However, because relationships between animal behavior and parasite infection are inherently dynamic and often bidirectional, links between the two may be frequently characterized by emergent feedbacks (7). Here we define feedbacks as bidirectional interactions between parasitism and behavior that are self-reinforcing (Figure 2.1), where a parasite trait affects host behaviors, with direct or indirect repercussions for the same type of parasite trait (e.g., prevalence, load, virulence), or vice versa, where a behavioral trait affects a parasite trait with repercussions for the same type of behavior (e.g., mating, foraging, social). For example, behaviors such as gregariousness often augment the risk of hosts acquiring parasites. In turn, parasite infections acquired via social interactions often cause changes in the same social behaviors necessary

for transmission, resulting in positive or negative feedbacks between a given parasite and host behavior (3). Efforts to describe these behavior–parasite feedbacks, including when they operate, over what timescales, and their relative importance as drivers of variation in both host behavior and parasite traits, hold significant promise for transforming our understanding of host behavior–parasite interactions.

Animal behavior and parasite dynamics are often tightly intertwined, each exerting influence on the other (3). Thus, it is often challenging, if not impossible, to determine to what extent existing behavior–parasite feedbacks were initially driven by host versus parasite traits. Nonetheless, one useful exercise to dissect these complex interactions in practice is to consider the general ways in which feedbacks can be initiated. First, behavior–parasite feedbacks can be initiated by core features of animal behavior (e.g., social behavior, mating behavior) that evolved and are maintained by selective pressures other than parasitism (8), (9). In this case, the focal behavior can have strong effects on some aspect of parasitism (e.g., exposure risk, intensity, prevalence), and this effect on parasites can in turn influence behavior, as a result of effects of the parasite on host performance and/or fitness (Figure 2.1A). Second, behavior–parasite feedbacks can be initiated by parasites, arising from interactions between hosts and parasites that occur either prior to potential infection, for example as a result of "fear" of infection, or post-exposure from host or parasite-mediated behavioral responses to infection. In these

Dana M. Hawley and Vanessa O. Ezenwa, *Parasites, host behavior, and their feedbacks.* In: *Animal Behavior and Parasitism.* Edited by Vanessa O. Ezenwa, Sonia Altizer and Richard J. Hall, Oxford University Press. © Oxford University Press (2022). DOI: 10.1093/oso/9780192895561.003.0002

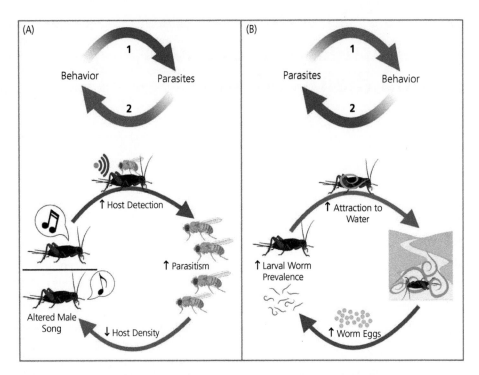

**Figure 2.1** Behavior–parasite feedback loops can be initiated by either host behavior (A) or parasites (B). In (A), the mating vocalizations of male Hawaiian field crickets attract lethal parasitoid flies, which select for male crickets with altered songs that evade detection by parasites (behavior → parasites → behavior). In (B), nematomorph worms manipulate the behavior of their arthropod hosts (here, a cricket), causing their hosts to seek out and jump into water bodies, where the worms complete their life cycle (parasites → behavior → parasites). Created with Biorender.com.

cases, parasites modify host behavior in some way that has downstream consequences for the rate at which hosts encounter parasites or rates of parasite establishment or transmission, generating positive or negative feedback on the parasites that initially modified the host behaviors (Figure 2.1B). In practice, both host behavior and parasites often act together to initiate and maintain feedbacks, and thus the distinction between these two broad categories is often blurred. However, we consider them separately to categorize and thus clarify some of the complexity inherent to behavior–parasite feedbacks in natural systems.

Feedbacks initiated by host behavior (Figure 2.1A) often arise from behaviors related to mating or foraging. Mating behaviors typically result in close physical contact between conspecifics and can thus augment a host's risk of acquiring parasites with diverse transmission modes. In some cases, parasites even use host mating signals to locate and colonize hosts. For example, the lethal parasitoid fly (*Ormia ochracea*) uses the mating calls of male crickets to locate its field cricket (*Teleogryllus oceanicus*) hosts in Hawaii (Figure 2.1A; (10)). This interaction resulted in rapid declines in cricket density, which in turn drove selection for changes in male calling behavior (10)–(13). Similarly, foraging behaviors expose many animal hosts to parasites, but key components of individual foraging behavior, such as foraging frequency and diet selection, are in turn often affected by parasites. In grazing sheep (*Ovis aries*), for example, individuals accumulate gastrointestinal worms as a function of grazing rate. However, worm infection suppresses individual forage intake rates and increases forage selectivity (14). Thus, for behaviors such as foraging, which like mating behaviors, presumably evolved and are maintained independently of parasitism, feedbacks with parasitism likely generate significant inter-individual variation in behavior and may even help explain innovations in foraging (e.g., medicinal plant use (15)).

Many behavior–parasite feedbacks are instead initiated by the parasite (Figure 2.1B), whether pre- or post-exposure. For example, the "ghosts of parasites past" account for the evolution of a suite of behaviors (e.g., avoidance of habitats, objects, or conspecifics that pose potential risks of infection) that serve as general mechanisms of disease prevention in a wide range of taxa (16)–(18). These behaviors, hypothesized to be mediated by generalized "disgust" of cues associated with parasitism (e.g., feces or carcasses) (18), occur prior to infection but have downstream implications for parasites present in a given host's environment, leading to feedbacks. Similarly, the "ghosts of parasites past" often select for mate choice behaviors that reduce infection risk for adults and offspring (19). As an example, major histocompatibility complex (MHC)-based mating preferences are likely to have evolved in many vertebrates to reduce the probability of selecting infection-prone mates and to increase pathogen resistance in offspring (20). These mating preferences also have clear consequences for contemporary parasite transmission dynamics. In Atlantic salmon (*Salmo salar*), individuals allowed to exercise MHC-based mate choice freely produced progeny that were less parasitized by worms when compared to the progeny of artificially mated fish (21). Some of the strongest parasite-initiated feedbacks occur post-exposure, in many cases from parasite manipulation of the behavior of infected hosts (2). Manipulation of host behavior by parasites is typically defined as a change in host behavior associated with infection that has fitness benefits for the parasite but not the host (22), meaning that feedback is inherent to the process of host manipulation. For example, the final larval stage of nematomorph parasites (i.e., Gordian worms) causes its terrestrial arthropod hosts (e.g., crickets, mantids, beetles) to jump into water bodies, where the adult parasite can successfully emerge, killing the host, and facilitating the aquatic sexual reproduction needed for the worm to complete its lifecycle (23). This parasite-initiated manipulation of host behavior enhances worm reproductive success, generating strong positive feedbacks on parasites (Figure 2.1B).

Our approach of classifying feedbacks, which is based on whether they are initiated by host behavior or by parasites, provides a convenient framework for exploring the importance of the feedback process in shaping the relationship between behavior and parasitism. In this chapter, we explore the role that feedbacks may play in driving both stability and change in animal behavior and parasite traits. We begin by considering the basic requirements for feedbacks between animal behavior and parasites to arise, and the timescales over which these feedbacks can operate. We draw from feedback processes documented in other contexts such as cell signaling to consider the types of functional outcomes likely to be generated by interactions between host behavior and parasitism. Using this framework, we discuss current evidence for behavior–parasite feedbacks in nature, and the ways in which biotic or abiotic context may influence the strength and nature of feedbacks in natural systems. We end by exploring outstanding questions and opportunities in the field of behavior–parasite feedbacks.

## 2.2 Basic requirements for behavior–parasite feedbacks

Behavior–parasite feedbacks can arise in diverse ways; however, whether initiated by host behavior or parasites, these feedbacks share two core requirements. First, (A) animal behaviors have to affect parasitism, whether the likelihood of encountering parasites or rates of parasite establishment, reproduction, or transmission. Second, (B) parasitism has to affect host behavior, including the types, quantity, and timing of behaviors expressed by hosts. Importantly, changes in host behavior that arise from parasitism can occur pre-infection, post-exposure, or both. When steps A and B co-occur and result in self-reinforcement (i.e., behavior affects parasitism that then changes behavior, or vice versa), feedbacks will emerge. Conversely, when both steps occur in a given system but are wholly independent (i.e., there are no links between the traits involved in steps A versus B and thus no mutual reinforcement), these bidirectional interactions would not result in feedbacks. As one example of bidirectional interactions that do not produce feedbacks, the same parasitoid fly species

(*O. ochracea*) that responds to mating calls of Hawaiian field crickets, triggering changes in calling behavior (Figure 2.1A), has also been reported to affect fighting intensity in other field cricket species such as *Gryllus integer* and *G. rubens* (24), However, because mating call effects on parasite abundance (behavior → parasites) are independent of parasite effects on fighting intensity (parasites → behavior), this particular set of bidirectional behavior–parasite interactions should not produce feedbacks akin to those depicted in Figure 2.1A .

These broad requirements capture the core conditions that produce behavior–parasite feedbacks, but these feedbacks can operate within or across generations of host and parasite, or some combination of both (7). For instance, feedbacks initiated by host behavior can affect parasite population size (i.e., ecological change) or the frequency of parasite genotypes (i.e., evolutionary change), and lead to either plastic (i.e., ecological) or evolutionary changes in host behavior. As an example of ecological feedback, territorial behavior in male Grant's gazelles (*Nanger granti*) increases gastrointestinal worm parasitism, most likely via effects on both host susceptibility and exposure to parasites (25); in turn, infection depresses the activities required for territory maintenance, a process that likely contributes to high levels of plasticity in male mating behavior (26). In practice, evolutionary feedbacks may be more challenging to detect than ecological ones because the long generation times of most hosts slow down the behavioral responses to selection. Nevertheless, the emergence of novel modes of sexual signaling in field crickets in response to interactions with a parasitoid fly in Hawaii (Figure 2.1A), a region where both the host and parasite are fairly recent invaders, provides a notable example of rapid evolutionary feedback on host behavior (10)–(12). One intriguing possibility is that evolutionary feedbacks on host behavior may be more readily detectable for novel host–parasite interactions such as this one. Further, evolutionary feedbacks on behavior may be particularly likely to occur when parasites affect host reproductive behaviors, which are considered to be among the most rapidly evolving behavioral traits (27).

Given the short generation times of most parasites relative to their hosts, feedbacks initiated by host behavior often result in simultaneous ecological (i.e., population size) and evolutionary (i.e., genotype frequency) changes in parasites. The phenomenon of host manipulation represents a key context in which such dual ecological and evolutionary effects might manifest, because manipulative traits of parasites often affect host behavior in ways that enhance parasite reproduction or transmission. Crucially, these benefits to the parasite may manifest both as increases in parasite population size, including higher intensity within a host or prevalence across hosts, as well as changes in the underlying frequency of parasite genotypes and phenotypes. The trophically transmitted tapeworm *Schistocephalus solidus* manipulates the behavior of both of its intermediate hosts (copepods and sticklebacks), with the effects of *S. solidus* on the first intermediate host (copepods) providing a particularly intriguing example of simultaneous eco-evolutionary feedbacks (Figure 2.2). Activity levels of copepods (*Macrocyclops albidus*) affect the likelihood of predation by the second intermediate host, the three-spined stickleback (*Gasterosteus aculeatus*), with higher fish predation rates on more active copepods (28), (29). To facilitate transmission from the first to second intermediate host, *S. solidus* manipulates copepod activity in ways that decrease the susceptibility of copepods to fish predation when *S. solidus* is not yet infective to fish, but increase copepod activity, and thus susceptibility to fish predation, during periods of parasite infectivity (28), (30). Together, these behavioral changes should increase the prevalence of infective parasites in sticklebacks. However, in addition to this ecological effect on parasite population size, recent work showed that *S. solidus* responds to experimental selection on copepod host manipulation (31). Selection for high host manipulation ability (defined as the magnitude of change in activity between pre-infective and infective copepods for a given parasite strain) increased the size of the difference between the suppressive effects of pre-infective parasite stages and the enhancing effects of infective parasite stages on copepod activity, while selection for low manipulation ability decreased this difference (31). Thus, there was sufficient genetic variation in host manipulation ability in the parasite population to allow for selection-induced changes

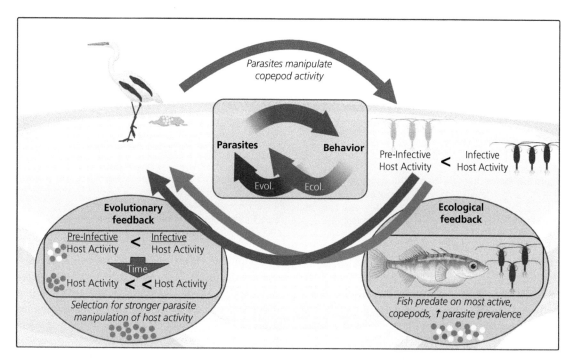

**Figure 2.2** The trophically transmitted tapeworm, *Schistocephalus solidus*, initiates simultaneous ecological (light-blue arrows) and evolutionary (darker blue arrows) feedbacks (parasite → behavior → parasite; inset box) by manipulating activity levels of its copepod intermediate hosts in divergent ways based on parasite infectivity stage. This manipulation leads to ecological feedbacks on parasite prevalence (lower right circle) by ensuring that copepods with infective parasite stages are most active (represented by darker copepod colors) and most likely to be eaten by second intermediate hosts (three-spined sticklebacks), allowing *S. solidus* to ultimately end up in a fish-eating bird and complete its lifecycle. Furthermore, the presence of phenotypic variation in the ability of the parasite to manipulate copepod activity (represented as parasite colors, with darker circles corresponding to stronger manipulation ability) can facilitate the co-occurrence of evolutionary feedbacks (lower left circle). Experimental selection for parasites that most strongly manipulate copepod activity acts on this variation (dark-blue arrows, evolutionary feedback) and results in more divergent activity phenotypes for pre-infective stage copepods versus those with infective parasite stages (31). Created with Biorender.com.

in the frequencies of different phenotypes (31). In a natural context, strong positive reinforcement of host manipulation via successful completion of the parasite life cycle should likewise select for changes in parasite genotype frequency.

## 2.3 Functional outcomes of behavior–parasite feedbacks

Across levels of biological organization, feedbacks are known to produce both stability and change (32), (33). In animal behavior, positive feedback processes feature prominently in theories of sexual selection and cooperation (e.g., (34)) and can lead to the amplification or exaggeration of traits (e.g., Fisherian runaway selection) or to alternative endpoints (e.g., evolution of cooperation). In cell signaling

systems, amplification and bistability are also hallmarks of positive feedback, while negative feedback tends to act as a stabilizing force, for instance, by serving a basic homeostatic function in response to fluctuating signal amplitudes (33). Where positive and negative feedbacks between animal behavior and parasitism occur, these different functional outcomes (e.g., amplification, bistability, homeostasis; Figure 2.3) may be common byproducts of these interactions. Thus, the observed outcomes of host behavior–parasite interactions may themselves provide clues that help identify the presence of feedbacks.

As an example, behavioral homeostasis is a negative feedback process in which an animal uses a behavioral response to correct a physiological imbalance (Figure 2.4A). Herbivores with certain

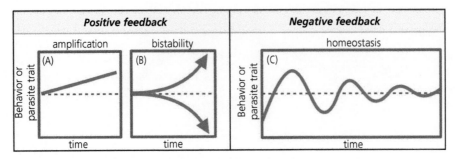

**Figure 2.3** Positive and negative feedbacks between host behavior and parasites can lead to various functional outcomes akin to those proposed for cell signaling systems (33). These outcomes include, but are not limited to: (A) the amplification of host or parasite traits, (B) the emergence of alternative states (bistability), or (C) stabilization of dynamic processes occurring within or between host and parasite individuals (akin to homeostasis). These functional outcomes can be generated via ecological or evolutionary feedbacks. Created with Biorender.com.

nutritional deficiencies or excesses, for instance, often adjust their food selection to overcome the deviation from physiological homeostasis (reviewed in (35)). These dietary adjustments help balance nutrient intake on short timescales (e.g., days) in response to small fluctuations in the availability of specific nutrients. On longer timescales (e.g., months, years), large shifts in nutritional signals, for example due to seasonal changes in resource availability, can also result in negative feedback, with more pronounced behavioral changes (e.g., incorporating new food items into the diet), helping to reestablish homeostasis (35). In the context of host behavior–parasite interactions, reduced food intake (i.e., anorexia) in response to ingested parasites, in sheep for example (14), could be viewed as homeostatic negative feedback, defined here as a change in behavior that helps the host adjust physiologically to the change in parasitism, facilitating a return to an internal state that minimizes the costs of infection (Figure 2.4B). Importantly, if this is a homeostatic feedback process, then understanding links between host foraging behavior, parasite ingestion rates, and the level of physiological imbalance generated by this interaction could be central to predicting when and to what degree behavioral adjustments may be used to reestablish homeostasis, possibly helping to explain some of the observed variation in levels of anorexia induced by ingested parasites (36).

Interestingly, negative feedback triggered by host ingestion of parasites might also lead to the emergence of novel behaviors. As a non-parasite example of novel behaviors in response to ingestion and resulting physiological imbalance, when sheep are fed diets high in tannins, a compound found in many plants that causes illness in grazing herbivores, they learn to selectively ingest polyethylene glycol, which attenuates negative effects of tannins (37). In this case, negative feedback produces a new learned behavioral response—self-medication (Figure 2.4B, orange arrow), and the type of medication behavior described for tannins may also apply to parasites (35). This phenomenon, where feedback produces innovation, could be a key way in which behavior–parasite interactions generate novelty in host and parasite traits. Indeed, self-medication in response to parasite infection has been reported in a number of taxa, ranging from insects and birds to reptiles and mammals (38). Whether negative feedback between foraging behavior and parasites can help explain the origin of some self-medication behaviors and the conditions under which emergence of novel behaviors is the most likely outcome of feedback are intriguing questions. More generally, the exercise of viewing behavior-parasite interactions as feedbacks that can produce a certain range of outcomes (e.g., Figure 2.3) serves as an interesting framework for exploring the origin and maintenance of behavioral trait variation and illuminating the mechanisms (e.g., processes that become unbalanced due to parasite infection, receptors that detect these imbalances to reestablish homeostasis) that link parasites to host behavior.

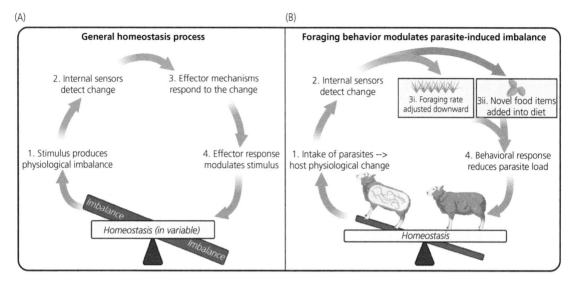

**Figure 2.4** In a wide range of biological systems, negative feedback helps to maintain homeostasis, equilibrating the system after a perturbation such as parasite infection. (A) Typically, a stimulus produces a change or imbalance that is detected by the system, then effector mechanisms are activated to respond to the stimulus, allowing balance to be restored. (B) In host–parasite systems, the process of fecal–oral transmitted parasites inducing anorexia is analogous to homeostatic negative feedback. Foraging behavior results in the intake of parasites which can generate physiological imbalances in hosts. These imbalances trigger immunological, physiological, and neurological mechanisms that induce anorexia (i.e., reduced food intake; (3i)). The reduction in food intake helps resolve the imbalance caused by parasites, allowing hosts to restore homeostasis. In some cases, homeostasis may be restored via novel effector mechanisms (illustrated here with orange arrow). For example, hosts might take advantage of novel food items (i.e., self-medication behavior; (3ii)) to reduce the parasite stimulus and maintain internal homeostasis. Created with Biorender.com.

Ultimately, feedbacks between host behavior and parasites operate along multiple axes that can occur in tandem: both behavior and parasites simultaneously driving feedbacks; feedbacks co-occurring both within and among generations of hosts and parasites (Figure 2.2); and coupled positive and negative feedbacks, all of which can lead to outcomes distinct from those proposed for uncoupled feedbacks (33). For natural systems, some key questions are which types of feedbacks or outcomes are most likely in host behavior–parasite interactions, what biotic and abiotic forces influence the presence of feedbacks, and what role do feedbacks, when they occur, play in shaping host and parasite eco-evolutionary dynamics.

## 2.4 How do behavior–parasite feedbacks operate in nature?

The study of behavior–parasite feedbacks is still largely in its infancy. Nonetheless, there is growing evidence that several broad classes of behaviors can produce negative and positive feedbacks between behavior and parasites (Table 2.1). Here, we consider three sets of behaviors (social/sickness behaviors, mating behaviors, and behavioral resistance) that, in addition to host manipulation and foraging behaviors discussed earlier, represent thematic areas where empirical and theoretical evidence suggest a high likelihood of behavior—parasite feedbacks emerging in natural systems.

### 2.4.1 Social and sickness behavior

Social behaviors such as group living can have diverse and far-reaching impacts on parasitism, and in turn, often respond directly to parasitism (3), (39). Thus, social behaviors serve key roles in feedbacks initiated by host behavior or parasites, both on ecological and evolutionary timescales. For example, social behavior-initiated increases in parasite transmission are widely thought to feed back negatively on host social behavior in evolutionary time, influencing patterns of social evolution in hosts (40).

**Table 2.1** Promising case studies for exploring the way that behavior–parasite feedbacks operate in nature, and the external factors that influence their strength or persistence. Note that these examples are not intended to be exhaustive (see main text for additional exemplar case studies).

| Class of Behavior | Behavior–Parasite Feedback | Exemplar Host–Parasite System |
|---|---|---|
| Foraging | Grazing leads to increased risk of helminth ingestion by sheep; helminth ingestion, in turn, leads to parasite-induced anorexia (i.e., reductions in grazing) | Sheep (*Ovis aries*) and helminths (14) |
| Foraging | Mice orally infected by *Salmonella enterica* serovar Typhimurium (ST) show parasite-induced anorexia; in turn, ST has evolved effector proteins that inhibit anorexia and promote pathogen transmission | Lab mice (*Mus musculus*) and *Salmonella enterica* (77) |
| Host manipulation | Virus-infected aphids prefer to feed on non-infected host plants, while non-infected aphids prefer virus-infected host plants; both types of vector preferences, in turn, promote virus spread to susceptible aphids | Aphids (*Rhopalosiphum padi*) and Barley Yellow Dwarf Virus (79) |
| Host manipulation | Tapeworm parasites manipulate activity levels of copepods (1st intermediate host); this facilitates predation by fish (2nd intermediate host), augmenting tapeworm prevalence and selecting for tapeworms with the strongest ability to manipulate copepod activity | Copepods (*Macrocyclops albidus*) and tapeworm (*Schistocephalus solidus*) (28), (29), (31) |
| Social/sickness behaviors | Social interactions (particularly biting) augment facial cancer transmission between devils; in turn, individuals with DFTD have fewer direct interactions with conspecifics in the mating season | Tasmanian devils (*Sarcophilus harrisii*) and DFTD (50), (80) |
| Social/Sickness behavior | Influenza-like illness in humans is associated with reduced social contacts; in turn, social isolation of infected individuals is predicted to reduce the spread of influenza to susceptible individuals | Humans (*Homo sapiens*) and influenza virus (44) |
| Behavioral resistance | Ants exposed to fungus-contaminated nestmates reduce interactions with high-risk group members and increase social segregation between groups; in turn, this reduces fungal exposure levels within the colony | Ants (*Lasius niger*) and *Metarhizium brunneum* (46) |
| Behavioral resistance | Lice select for efficient preening behavior in pigeons; in turn, preening behavior selects for smaller body size in lice, potentially facilitating escape from preening | Rock pigeons (*Columbia livia*) and feather lice (74) |
| Mating behaviors | Territorial behavior in male gazelles increases helminth burden, and high helminth loads reduce time spent on key behaviors required for the maintenance of territoriality | Grant's gazelle (*Nanger granti*) and helminths (25), (26) |
| Mating behaviors | Parasites likely selected for MHC-disassortative mating in salmon; this selective mating behavior leads to reductions in offspring parasite prevalence and load when compared to artificially bred salmon deprived of the ability to exercise mate choice | Atlantic salmon (*Salmo salar*) and marine nematodes, *Anisakis* spp. (21) |

Furthermore, parasite-initiated ecological changes in host social behavior occur almost universally, both pre- and post-infection (39), with consequences for parasite transmission dynamics. For example, parasite infection often results in immune-mediated changes to host behaviors including inactivity and social withdrawal, collectively termed "sickness behaviors" (41). Such behaviors can reduce the extent to which infected hosts encounter and contact uninfected conspecifics, resulting in strong negative feedbacks for the epidemic dynamics of directly transmitted parasites (42)–(44). Interestingly, these feedbacks may be akin to homeostasis operating between hosts, rather than within individuals (Box 2.1).

Intriguingly, a growing body of work suggests that sickness behavior expression can be highly context dependent, varying with abiotic factors such as photoperiod (48) and biotic factors such as conspecific density (49), and reproductive seasonality (50), (51). Few studies have examined how this variation might result in context-dependent feedbacks, but recent work in Tasmanian devils (*Sarcophilus harrisii*) suggests that behavior–parasite feedbacks caused by a contagious cancer may only arise in the breeding season: individuals with devil facial tumor disease (DFTD) show significant reductions in social interactions consistent with sickness behavior during the breeding season, but social interactions are equivalent regardless of cancer status in the non-breeding season. Whether such differences disrupt the long-term stability of possible homeostatic negative feedbacks that moderate the spread of cancer is not yet known, but could be explored using theoretical models.

Host social behavior and parasitism may also produce positive feedbacks. One potential example is behavior-initiated feedback and "social tolerance," where hosts use social behavior to ameliorate the fitness costs of infection. For social hosts, larger group sizes are often associated with higher risk of acquiring parasites (52); on the other hand, the benefits of group-living, including enhanced resource acquisition, protection from predators, and other forms of social support, can help maintain fitness of infected hosts, which may favor enhanced sociality (53). In particular, if social behavior promotes infected host fitness by increasing host tolerance rather than resistance, then both host and parasite fitness may be enhanced by group living. In this case, the alignment of host and parasite interests may generate positive evolutionary feedbacks that amplify relevant host and parasite traits. Such a scenario could arise because host tolerance, unlike resistance, is predicted to increase parasite prevalence under some conditions (54), and increased parasite prevalence would in turn select for large group sizes that promote tolerance. However, despite a long-standing interest in how parasites affect the evolution of social behavior (e.g., (40)), the core components of this type of social behavior-initiated parasite feedback, where: (a) social hosts are more likely to acquire parasites, and (b) infection either depresses individual sociality as a means to reduce the costs of higher rates of parasitism (negative feedback akin to the parasite-initiated sickness behaviors discussed earlier), or parasitism augments individual sociality as a means to

---

### Box 2.1 Homeostasis across scales

Behavior–parasite feedbacks can manifest at multiple scales of biological organization. For example, negative feedbacks that maintain homeostasis can occur within single host individuals, within social groups, or within populations of parasites and hosts. While the term "homeostasis" describes the maintenance of long-term stability or health in a given system, the mechanisms generating homeostasis are often far from static; instead, dynamic negative feedbacks are often required to maintain homeostasis, regardless of the level of organization (Figure 1). At the within-individual level, negative feedbacks between parasites and host behavior can help maintain physiological homeostasis within individual hosts, thereby limiting the onset of disease (Figure 1A), as occurs in grazing animals that reduce food intake following helminth ingestion (45). However, behavioral changes in response to the risk of parasite infection can also readily generate negative feedbacks at the level of social groups, when these behavioral changes influence social interactions in ways that suppress outbreaks of directly transmitted pathogens (Figure 1B). For example, the rapid behavioral changes exhibited by ants (Lasius niger) when a contagious fungal pathogen is introduced into their colony reduce close social interactions between infected foragers and high-value group members (e.g., nurses and queens) (46), suppressing colony-wide fungal outbreaks and promoting a return to pre-outbreak conditions. Intriguingly, some parasite-initiated feedbacks may even act simultaneously to maintain both within-individual and within-group homeostasis (i.e., stability): for example, negative feedbacks arising from immune-mediated sickness behaviors during infection may suppress parasite loads within infected hosts, as well as reduce the likelihood of spread from that host, contributing to both within-individual and within-group suppression of parasite proliferation. Finally, at the population level, homeostasis can be maintained by antagonistic coevolution, a form of evolutionary feedback that can arise between parasites and host behaviors such as mate choosiness (47). Here, parasites and host behavior constrain each other phenotypically via negative feedbacks, stabilizing the degree of exaggeration in traits such as parasite virulence over evolutionary time (Figure 1C, red). Overall, the potential for feedbacks between host behavior and parasites to maintain stability or homeostasis at several distinct biological scales suggests that these types of interactions may be broadly important components of host–parasite systems on both ecological and evolutionary timescales.

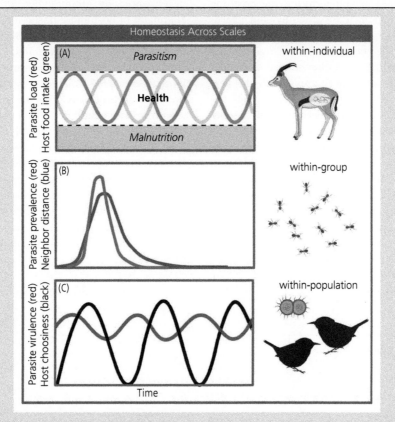

**Box 2.1, Figure 1** Negative feedbacks between host behavior and parasitism can maintain homeostasis across scales. (A) Physiological interactions between food intake and helminth parasite ingestion in grazing mammals helps maintain within-individual homeostasis or health. (B) Behavioral defenses such as social distancing (blue line) in response to parasite introduction into social insect colonies can rapidly suppress parasite transmission (red line), maintaining group-level homeostasis. (C) Antagonistic coevolution between host behaviors such as mate choosiness (black line) and parasites can constrain the evolution of parasite virulence (red line) over evolutionary time, maintaining homeostasis within populations of parasites and hosts (result adapted from (47)). Created with Biorender.com.

tolerate infection (positive feedback), have not yet been robustly demonstrated in the same natural system.

### 2.4.2 Mating behavior

Hosts should be strongly selected to evolve mating behaviors that limit parasite transmission (55), and parasites, in turn, should be strongly selected to counteract such avoidance behaviors, in some cases by becoming less harmful to hosts (56).

Thus, parasite-mediated sexual selection (PMSS), whereby parasites drive the evolution of host mating preferences and sexual signals such as male brightness, is hypothesized to be particularly likely to generate evolutionary feedbacks between host mating behaviors and parasites. For example, precopulatory behaviors in birds often include the direct display and inspection of male genitalia, a potential adaptation to reduce risk of sexually transmitted infections (57), and across taxa including mammals, birds, fish, and insects, the choosy sex often prefers the cues of, or preferentially mates with, uninfected individuals over those infected by parasites (reviewed in (58)). In turn, sexually transmitted infections (STIs) of mammals generally cause lower mortality than parasites transmitted by non-sexual means (59), and some STIs even appear to augment attractiveness of infected hosts to potential mates (e.g., (60)), suggesting that parasites may evolve to counteract such avoidance behaviors in hosts. Thus, empirical patterns are consistent with the idea that parasites shape the evolution of host mating behavior, and vice versa, across a variety of natural systems.

Direct examination of the evolutionary feedbacks generated by host sexual selection and parasitism has almost entirely been limited to theoretical models. While reciprocal evolutionary feedbacks potentially occur for parasites with diverse transmission modes, STIs have been the focus of theoretical work on coevolution between parasite virulence and host mating strategies, largely because STIs are directly tied to host mating behaviors, and often have negative impacts on host reproductive success (e.g., (59)). Ashby and Boots (47) explicitly considered the dynamics of evolutionary feedbacks between mating behavior and STI parasite virulence, and showed that a variety of outcomes are possible from this form of antagonistic coevolution, including stable levels of host mate choice and parasite virulence, sustained cycling (Box 2.1; Figure 1C), and runaway selection (i.e., amplification). Importantly, although prior models predicted the loss of host choosiness (56), accounting for coevolutionary feedbacks between behavior and parasitism in addition to ecological dynamics prevented the loss of host choosiness (47), emphasizing the role that feedback processes operating on evolutionary timescales may play in maintaining variation in both host behavior and parasite traits. For example, when host choosiness causes selection for reduced parasite virulence, this in turn relaxes selection for choosiness; lower choosiness then results in reduced avoidance while mating, potentially allowing parasites to reciprocally evolve to increase virulence (47).

As with social behaviors and parasitism, the strength of feedbacks between parasites and host mating behavior is likely to be highly context dependent. For example, in theoretical models of coevolutionary feedback, the likelihood of a given system outcome varied with factors such as the costs of choosiness for hosts (47), (61), a trait which may be strongly influenced by biotic factors such as conspecific density, operational sex ratio, the spatial distribution of resources, and even predation risk (62). Similarly, for ecological feedbacks between male mating behavior and gastrointestinal parasites in Grant's gazelles, the strength of feedbacks could depend almost entirely on factors such as conspecific density and resource distribution. In this system, adult males acquire and vigorously defend territories in habitat patches preferred by females to enhance their mating success. However, behavioral and physiological changes associated with territoriality increase male parasite burdens, and parasitism depresses territoriality, promoting territory loss (26). This negative feedback loop likely catalyzes cyclical changes in male reproductive status, with individual males switching between territorial and non-territorial states over time. However, the frequency with which parasitized territorial males lose their territories likely also depends in large part on the degree to which there is competition for territories. When adult male density is high, and high-quality habitat patches are rare, switches in male reproductive status may be far more common than when adult density is low and high-quality patches are plentiful, since in the latter case, the energetic costs of maintaining a territory are greatly reduced. Thus, understanding how biotic or abiotic context influences the persistence of host behavior-parasite feedbacks may be crucial to unraveling key aspects of both host and parasite eco-evolutionary dynamics.

### 2.4.3 Behavioral resistance

Choosiness of potential mates is just one of many forms of behavioral resistance to parasites that have the potential to generate behavior–parasite feedbacks. Often termed "behavioral immunity," these forms of resistance include a suite of behavioral strategies that evolved to form the first line of defense against parasites (i.e., avoidance of habitats, objects, or conspecifics that pose potential risk of parasitism (16)). For example, avoidance of infected group-mates outside of a mating context is widespread in nature: in taxa as diverse as ants, crustaceans, fish, and primates, uninfected individuals shun pathogen-exposed or infected conspecifics as a means of reducing their own infection risk (39), (63), (64). Similar to sickness behavior, such avoidance in response to the parasitism status of conspecifics can generate strong negative ecological feedbacks on parasites by suppressing population-level spread. Recent theoretical and empirical studies suggest that avoidance behaviors of uninfected individuals can alone (65), or in combination with active self-isolation of infected individuals (46) suppress pathogen spread, maintaining homeostasis with respect to infection prevalence at the group or colony level (Box 2.1). Thus, active avoidance of infected conspecifics represents an additional avenue, in addition to sickness behaviors of infected hosts (see section 2.4.1), by which parasite-initiated feedbacks can be generated. Whether the two mechanisms differ in terms of the strength of the feedbacks that might be generated, and the way in which the co-occurrence of these two behaviors influences feedback strength, remain open questions.

While parasites can initiate ecological feedbacks by triggering avoidance of infected conspecifics, evolutionary feedbacks can also arise when the same interactions are viewed from the lens of behavior-initiated feedbacks (Figure 2.1B). Such a perspective allows explicit consideration of when and where behavioral resistance strategies such as conspecific avoidance should be favored in hosts. The evolutionary dynamics of host behavioral avoidance have only rarely been considered outside of the context of sexual selection (47), (56), despite the potential for such strategies to have distinct costs and dynamics from those associated with forms of physiological resistance (65), (66). For example, the behavioral costs associated with conspecific avoidance (at the extreme, the loss of the benefits of group living) are likely to vary with social context, potentially driving unique evolutionary dynamics relative to physiological costs of resistance. Recent theoretical work suggests that if the costs of host behavioral avoidance are fixed (e.g., a behaviorally resistant genotype experiences costs associated with reduced activity, whether or not they avoid infected conspecifics), behavior-initiated host–parasite feedbacks can

maintain genetic polymorphism in host behavioral avoidance similar to what has been shown for physiological resistance (65). In contrast, when costs of avoidance arise because of the loss of benefits associated with sociality, as likely occurs for many group-living taxa with the ability to avoid infected conspecifics (3), (64), behavioral polymorphism cannot be maintained and behavioral avoidance will go to fixation when reproductive success of avoiders is higher than that of non-avoiders (65).

The dependence of evolutionary feedbacks on the costs of avoidance and the reproductive success of each behavioral strategy suggests that environmental context, which can strongly influence the relative benefits of sociality and thus the relative costs of avoidance, may play an important role in determining the strength of evolutionary feedbacks of parasites on host behavioral avoidance. These models of behavioral resistance have not yet been extended to consider potential coevolutionary dynamics as is common for models of physiological resistance (e.g., (67)). However, parasites are also likely to coevolve in response to behavioral avoidance (66), as has been considered in models of parasite-mediated sexual selection (see section 2.4.2). Overall, behavioral avoidance of infected conspecifics highlights the idea that multiple behavior–parasite feedbacks, here parasites → behavior → parasites (ecological) and behavior → parasites → behavior (evolutionary), are likely to operate simultaneously in many natural systems.

Host behavioral avoidance of ectoparasites represent another type of behavioral resistance likely to result in simultaneous ecological and evolutionary feedbacks between parasites and behavior. For example, grooming behaviors (termed "preening" in birds), whereby individuals search external structures and remove ectoparasites, have evolved as parasite defense mechanisms in virtually all vertebrate taxa. While often "programmed" (i.e., expressed at an innate level), augmented grooming can be induced in individual hosts via experimental exposure to ectoparasites (68), (69), and in turn, such grooming behaviors reduce parasite load (e.g., (68)). This negative feedback process (parasites → grooming → parasites) typically operates within individuals and could be considered a homeostatic process (Figure 2.4A). However, host grooming behaviors induced by parasitism can simultaneously generate positive evolutionary feedbacks on ectoparasites by amplifying parasite counter-strategies to host defensive behaviors (70). For example, many ectoparasites are dorsoventrally flattened to evade grooming, and many harbor morphological adaptations on their legs or mouthparts that secure them strongly to their hosts and counteract removal attempts (71). Avian lice were recently shown to rapidly (within 60 parasite generations) evolve cryptic coloration in response to behavioral grooming by hosts of distinct plumage coloration (72), suggesting that feedbacks from host behavioral defenses during infection may be key drivers of parasite trait diversification on evolutionary timescales.

These evolutionary counter-adaptations to host defenses by parasites may, in turn, drive behavioral novelty. For example, some ectoparasites respond to host grooming by evolving to selectively migrate to parts of the host body that are difficult to reach via grooming (70), which may in turn generate positive feedback on host grooming by favoring reciprocal allogrooming of areas that individual hosts cannot reach via self-grooming (73). While it is rare to demonstrate reciprocal evolutionary responses by host behavior and parasites in the same natural system, Clayton *et al.* (74) examined both how lice loads affected pigeon fitness when behavioral defenses were compromised, and how altered behavioral defenses, in turn, influenced experimental evolution in lice. By demonstrating both that pigeon survival is compromised when avian preening is prevented, and that preening behavior selects for smaller body sizes in feather lice, the authors demonstrate that ectoparasites and grooming behaviors can generate strong reciprocal evolutionary feedbacks.

## 2.5 Conclusions and future research

The basic requirements for behavior–parasite feedbacks appear to be met for a wide variety of behaviors and systems, yet studies of behavior–parasite feedback loops in natural systems remain relatively rare or incomplete (Table 2.1). In part, this limited progress reflects the logistical challenges associated with studying behavior–parasite feedbacks

in nature. These challenges include: (a) the need to establish causal links between host behavior and parasitism; (b) the ability to conduct longitudinal studies to demonstrate temporal feedbacks on short or longer timescales; and (c) the ability to quantify changes in host behavior and parasite traits, often non-invasively (7). With respect to the third challenge, recent advances in tracking technologies that provide high-resolution data on multiple aspects of host behavior (5) will allow exciting progress to be made in documenting how host behaviors both influence and respond to parasitism (75). For example, high-resolution digital tracking of individual ants following the introduction of a fungal pathogen into colonies allowed Stroeymeyt *et al.* (46) to document rapid and caste-specific behavioral responses to parasite presence that, in turn, altered colony-level epidemic outcomes (Box 2.1, Table 2.1). While such systems provide exciting opportunities for identifying ecological feedbacks between host behavior and parasitism in real time, linking the changes in host behavior uncovered by these technologies with changes in host parasite status remains challenging outside of controlled laboratory settings. Nonetheless, data on behavioral changes in free-living animals in response to parasitism have been obtained from systems for which host infection status is relatively stable over time (e.g., Tasmanian devils and DFTD; (50)), or systems where parasite status can be manipulated, either via medication to reduce parasite loads (e.g., (26)) or via injection of immunogenic substances that stimulate sickness behaviors (e.g., (42), (43)).

While technological advances will improve our ability to study behavior-parasite ecological feedbacks in natural systems, evolutionary feedbacks remain particularly challenging to study outside of laboratory settings. In fact, aside from the intriguing example of Hawaiian cricket mating calls evolving in response to parasitoid wasps (Figure 2.1A), our understanding of evolutionary feedback loops between host behavior and parasites is based either on first principles (e.g., social evolution (76); host manipulation (2)), experimental evolution approaches (e.g., (31)), or theoretical models (e.g., (47)). Applying theoretical models to a broader range of host behavior–parasite interactions holds particular promise in terms of generating predictions about the conditions under which we expect feedbacks to emerge in natural systems, and their likely functional outcomes (Box 2.2). Further, targeting novel host-parasite interactions such as the Hawaiian cricket-parasitoid system may prove especially fruitful, as these novel species interactions may be particularly likely to result in rapid and detectable evolutionary feedbacks between parasites and host behaviors.

---

### Box 2.2 Outstanding questions on host behavior–parasite feedbacks

What types of feedbacks (e.g., behavior vs. parasite initiated) are most common between host behavior and parasites? Are feedbacks stronger in certain host or parasite taxa?

At what biological scales of organization (e.g., individual, social group, population) are host behavior–parasite feedbacks most common?

What range of behavioral and parasite traits are involved in reciprocal feedbacks? For example, beyond load, prevalence, and virulence, what parasite traits (e.g., infectivity, competitive ability) are commonly involved in feedbacks with host behavior? Are some behaviors (e.g., reproductive behaviors) more likely to be involved in evolutionary feedbacks with parasites because they evolve more rapidly?

How do biotic and abiotic factors influence the emergence and persistence of behavior–parasite feedbacks?

Can theoretical models of ecological, evolutionary, or eco-evolutionary feedbacks between host behavior and parasites be used to generate testable predictions for identifying when feedbacks are operating in the wild?

Does the feedback process help generate novelty in host behavior and parasite traits? For example, can the emergence of certain behavioral defenses against parasites (e.g., self-medication) or aspects of host manipulation be accelerated by behavior–parasite feedbacks?

Do feedbacks operate at the community level? For example, can multiple parasite species be involved in the same feedback loop (e.g., parasite 1 → behavior → parasite 2 → parasite 1)? Similarly, can the behaviors of multiple host species be involved in the same feedback loop with a single parasite species (e.g., behavior of host 1 → parasite 1 → behavior of host 2 → behavior of host 1)?

---

Studies of behavior–parasite feedbacks to date have also been limited in the types of parasite traits examined, particularly relative to the suite of diverse behavioral traits considered. Studies of ecological feedbacks almost exclusively examine variation in parasite load or prevalence (e.g., Table 2.1), and studies of evolutionary feedbacks almost exclusively focus on parasite virulence (e.g., (47)). While these traits are clearly important for parasite fitness and are particularly likely to interact with host behavior, studies of behavior–parasite feedbacks should consider a wider variety of parasite traits that may respond to host behavior, including infectivity, competitive ability, morphological adaptations (as seen with ectoparasites to evade host grooming behavior, (71)), and the ability to repress host responses associated with sickness behaviors (e.g., (77). Expanding the types of parasite traits considered will help reveal whether behavior–parasite feedbacks commonly generate or maintain trait innovation in natural systems (Box 2.2). In addition to broadening the traits considered, studies of behavior–parasite feedbacks should continue to broaden the scales of inquiry to understand how these feedbacks operate across parasite and host communities. For example, recent work suggests that host behavioral defenses such as sanitary care can alter competitive interactions between parasite taxa in ant colonies (78). Whether multiple parasite species in a community can be involved in the same feedback loop with host behavior remains largely unconsidered but represents an exciting research frontier (Box 2.2).

Despite our somewhat limited understanding of behavior–parasite feedbacks in natural systems, patterns are already emerging that point to potentially common outcomes from feedback processes across scales. In particular, negative feedbacks between behavior and parasitism that facilitate homeostasis can be generated within and among individuals, and on both ecological and evolutionary timescales (Box 2.1). The prevalence of such homeostatic feedbacks may in part reflect that parasites frequently perturb host systems at both individual (e.g., physiological balance) and population (e.g., outbreak size) scales, and like organ systems, there are inherent mechanisms that facilitate the correction of these imbalances. Furthermore, because behavior is a common effector mechanism that animals use to maintain homeostasis in general (e.g., temperature regulation via shivering or basking), it may not be surprising that homeostatic interactions between behavior and parasites are common and likely to persist over long timescales. Overall, the types of feedbacks that are most likely to emerge from interactions between parasites and behavior in natural systems, and the way such types vary across classes of behavior (foraging, mating, etc), remain exciting open questions (Box 2.2). Further, the degree to which such feedbacks and their outcomes vary with biotic (e.g., host density) and abiotic (e.g., rainfall, temperature) context remains largely unknown, and is an exciting area for future inquiry. Thus, future studies should go beyond simply documenting the types of feedbacks occurring in natural systems by also considering the external factors that affect their strength or persistence.

Perhaps the most important pattern emerging from work to date is that feedbacks between parasites and behavior can occur across a wide variety of host behaviors, parasite transmission modes, and taxonomic groups of both hosts and parasites. Given the challenges in studying them, the number of feedbacks that have already been partly or fully documented in natural systems (Table 2.1) suggests that these interactions are likely to be widespread in nature. Thus, such feedbacks are essential to consider when studying interactions between animal behavior and parasitism, regardless of whether one is explicitly interested in the feedbacks themselves. For example, researchers that aim to study how parasites influence the evolution of host social behavior would be remiss not to also consider how host social behavior influences the ecology and evolution of parasites. Here, our classification of feedbacks by "initiator" (Figure 2.1) can be helpful for researchers approaching the field from a particular perspective, even if the initiator distinction is challenging to operationalize in practice. For example, while a researcher interested in how parasites influence social evolution might want to know whether variation in social group size across populations of the same host species can be explained by parasite pressure, fully understanding this question may require insight into potential feedbacks. This is because any negative selection pressure imposed by parasites

on social behavior (parasites → behavior) will also depend on the extent to which behavior affects parasite transmission (behavior → parasites). Thus, characterizing the reciprocal feedback in this case is essential for disentangling the role parasites might play in generating variation in host behavior.

The coming decades are sure to bring exciting and rapid progress in the study of behavior-parasite feedbacks that helps answer many open questions in the field (Box 2.2). Continued integration of conceptual and methodological advances such as next-generation sequencing into studies of behavior–parasite feedbacks will allow the field to better characterize the importance of feedbacks in maintaining stasis or generating trait change for both parasites and hosts. More broadly, the ecological and evolutionary implications of behavior–parasite feedbacks likely extend well beyond the parasite and host players that generate them. Thus, studying behavior–parasite feedbacks in natural systems is key for uncovering their consequences for the communities and ecosystems in which these host–parasite interactions occur.

## Acknowledgements

We thank Jessica Stephenson and Sebastian Stockmaier for helpful feedback on an earlier version. D. Hawley was supported by NSF grant 1755051.

## References

1. Breed MD and Moore J. *Animal Behavior*, 2nd edn. Cambridge, MA: Academic Press, 2015.
2. Moore J. *Parasites and the Behavior of Animals*. New York, NY: Oxford University Press, 2002.
3. Hawley DM, Gibson AK, Townsend AK, Craft ME, and Stephenson JF. Bidirectional interactions between host social behaviour and parasites arise through ecological and evolutionary processes. *Parasitology*. 2020; 148(3):274–88.
4. Klemme I and Karvonen A. Learned parasite avoidance is driven by host personality and resistance to infection in a fish–trematode interaction. *Proc R Soc B: Biol Sci*. 2016; 283(1838):20161148.
5. Smith JE and Pinter-Wollman N. Observing the unwatchable: Integrating automated sensing, naturalistic observations and animal social network analysis in the age of big data. *J Anim Ecol*. 2021; 90:62–75.

6. Geffre AC, Liu R, Manfredini F, Beani L, Kathirithamby J, Grozinger CM, *et al*. Transcriptomics of an extended phenotype: Parasite manipulation of wasp social behaviour shifts expression of caste-related genes. *Proc R Soc B: Biol Sci*. 2017; 284(1852).

7. Ezenwa VO, Archie EA, Craft ME, Hawley DM, Martin LB, Moore, J, *et al*. Host behaviour–parasite feedback: An essential link between animal behaviour and disease ecology. *Proc R Soc B: Biol Sci*. 2016; 283(1828):20153078.

8. Hofmann HA, Beery AK, Blumstein DT, Couzin ID, Earley RL, Hayes, LD, *et al*. An evolutionary framework for studying mechanisms of social behavior. *Trends Ecol Evol*. 2014; 29(10):581–9.

9. Emlen ST and Oring LW. Ecology, sexual selection, and the evolution of mating systems. *Science*. 1977; 197(4300):215–23.

10. Zuk M, Rotenberry JT, and Tinghitella RM. Silent night: Adaptive disappearance of a sexual signal in a parasitized population of field crickets. *Biol Lett*. 2006; 2(4):521–4.

11. Tinghitella RM, Broder ED, Gallagher JH, Wikle AW, and Zonana DM. Responses of intended and unintended receivers to a novel sexual signal suggest clandestine communication. *Nat Commun*. 2021; 12(1):797.

12. Tinghitella RM. Rapid evolutionary change in a sexual signal: Genetic control of the mutation "flatwing" that renders male field crickets (*Teleogryllus oceanicus*) mute. *Heredity*. 2008; 100:261–7.

13. Tinghitella RM, Broder ED, Gurule-Small GA, Hallagan CJ, and Wilson JD. Purring crickets: The evolution of a novel sexual signal. *Am Nat*. 2018;192(6):773–82.

14. Hutchings MR, Kyriazakis I, Gordon IJ, and Jackson F. Trade-offs between nutrient intake and faecal avoidance in herbivore foraging decisions: The effect of animal parasitic status, level of feeding motivation and sward nitrogen content. *J Anim Ecol*. 1999;68(2):310–23.

15. de Roode JC, Lefèvre T, and Hunter MD. Ecology. Self-medication in animals. *Science*. 2013; 340(6129):150–1.

16. Curtis VA. Infection-avoidance behaviour in humans and other animals. *Trends Immunol*. 2014; 35(10): 457–64.

17. Sarabian C, Curtis V, and McMullan R. Evolution of pathogen and parasite avoidance behaviours. *Philos Trans R Soc Lond B Biol Sci*. 2018;373(1751).

18. Weinstein SB, Buck JC, and Young HS. A landscape of disgust. *Science*. 2018;359(6381):1213–14.

19. Tybur JM and Gangestad SW. Mate preferences and infectious disease: Theoretical considerations and evidence in humans. *Philos Trans R Soc Lond B Biol Sci*. 2011; 366(1583):3375–88.

20. Kamiya T, O'Dwyer K, Westerdahl H, Senior A, and Nakagawa S. A quantitative review of MHC-based mating preference: The role of diversity and dissimilarity. *Mol Ecol*. 2014; 23(21):5151–63.

21. Consuegra S and de Leaniz CG. MHC-mediated mate choice increases parasite resistance in salmon. *Proc R Soc B: Biol Sci*. 2008; 275:1397–403.

22. Godfrey SS and Poulin R. Host manipulation by parasites: From individual to collective behavior. In Ezenwa VO, Altizer S, Hall RJ (eds), *Animal Behavior and Parasitism*. Oxford: Oxford University Press, 2022.

23. Thomas F, Schmidt-Rhaesa A, Martin G, Manu C, Durand P, and Renaud F. Do hairworms (Nematomorpha) manipulate the water seeking behaviour of their terrestrial hosts? *J Evol Biol*. 2002;15(3): 356–61.

24. Adamo SA, Robert D, and Hoy RR. Effects of a tachinid parasitoid, *Ormia ochracea*, on the behaviour and reproduction of its male and female field cricket hosts (*Gryllus spp*). *J Insect Physiol*. 1995; 41(3): 269–77.

25. Ezenwa VO, Stefan Ekernas L, and Creel S. Unravelling complex associations between testosterone and parasite infection in the wild: Testosterone and parasite infection. *Funct Ecol*. 2012; 26(1):123–33.

26. Ezenwa VO and Snider MH. Reciprocal relationships between behaviour and parasites suggest that negative feedback may drive flexibility in male reproductive behaviour. *Proc R Soc B: Biol Sci*. 2016; 283(1831).

27. Anholt RRH, O'Grady P, Wolfner MF, and Harbison ST. Evolution of reproductive behavior. *Genetics*. 2020; 214(1):49–73.

28. Wedekind C and Milinski M. Do three-spined sticklebacks avoid consuming copepods, the first intermediate host of *Schistocephalus solidus*? An experimental analysis of behavioural resistance. *Parasitology*. 1996; 112(4):371–83.

29. Weinreich F, Benesh DP, and Milinski M. Suppression of predation on the intermediate host by two trophically-transmitted parasites when uninfective. *Parasitology*. 2013;140(1):129–35.

30. Hammerschmidt K, Koch K, Milinski M, Chubb JC, and Parker GA. When to go: Optimization of host switching in parasites with complex life cycles. *Evolution*. 2009;63(8):1976–86.

31. Hafer-Hahmann N. Experimental evolution of parasitic host manipulation. *Proc R Soc B: Biol Sci*. 2019; 286(1895):2018–413.

32. Crespi BJ. Vicious circles: Positive feedback in major evolutionary and ecological transitions. *Trends Ecol Evol*. 2004; 19(12):627–33.

33. Brandman O and Meyer T. Feedback loops shape cellular signals in space and time. *Science*. 2008; 322(5900):390–5.

34. Lehtonen J and Kokko H. Positive feedback and alternative stable states in inbreeding, cooperation, sex

roles and other evolutionary processes. *Philos Trans R Soc Lond B Biol Sci*. 2012; 367(1586):211–21.

35. Villalba JJ and Provenza FD. Self-medication and homeostatic behaviour in herbivores: Learning about the benefits of nature's pharmacy. *Animal*. 2007; 1(9): 1360–70.

36. Hite JL, Pfenning AC, and Cressler CE. Starving the enemy? Feeding behavior shapes host–parasite interactions. *Trends Ecol Evol*. 2020; 35(1): 68–80.

37. Provenza FD, Burritt EA, Perevolotsky A, and Silanikove N. Self-regulation of intake of polyethylene glycol by sheep fed diets varying in tannin concentrations. *J Anim Sci*. 2000; 78(5):1206–12.

38. Shurkin J. News feature: Animals that self-medicate. *PNAS*. 2014; 111(49):17339–41.

39. Stockmaier S, Stroeymeyt N, Shattuck EC, Hawley DM, Meyers LA, and Bolnick DI. Infectious diseases and social distancing in nature. *Science*. 2021; 371(6533).

40. Alexander RD. The evolution of social behavior. *Annual Review of Ecology and Systematics*. 1974;5:325–83.

41. Hart BL. Biological basis of the behavior of sick animals. *Neurosci Biobehav Rev*. 1988; 12(2):123–37.

42. Lopes PC, Block P, and König B. Infection-induced behavioural changes reduce connectivity and the potential for disease spread in wild mice contact networks. *Sci Rep*. 2016;6:31790.

43. Ripperger SP, Stockmaier S, and Carter GG. Tracking sickness effects on social encounters via continuous proximity sensing in wild vampire bats. *Behav Ecol*. 2020; 31(6):1296–302.

44. Van Kerckhove K, Hens N, Edmunds WJ, and Eames KTD. The impact of illness on social networks: Implications for transmission and control of influenza. *Am J Epidemiol*. 2013; 178(11):1655–62.

45. Worsley-Tonks KEL and Ezenwa VO. Anthelmintic treatment affects behavioural time allocation in a free-ranging ungulate. *Anim Behav*. 2015; 108:47–54.

46. Stroeymeyt N, Grasse AV, Crespi A, Mersch DP, Cremer S, and Keller L. Social network plasticity decreases disease transmission in a eusocial insect. *Science*. 2018; 362(6417):941–5.

47. Ashby B and Boots M. Coevolution of parasite virulence and host mating strategies. *PNAS*. 2015; 112(43): 13290–5.

48. Owen-Ashley NT, Turner M, Hahn TP, and Wingfield JC. Hormonal, behavioral, and thermoregulatory responses to bacterial lipopolysaccharide in captive and free-living white-crowned sparrows (*Zonotrichia leucophrys gambelii*). *Horm Behav*. 2006; 49(1): 15–29.

49. Lopes PC, Adelman J, Wingfield JC, and Bentley GE. Social context modulates sickness behavior. *Behav Ecol Sociobiol*. 2012; 66(10):1421–8.

50. Hamilton DG, Jones ME, Cameron EZ, Kerlin DH, McCallum H, Storfer A, *et al*. Infectious disease and sickness behaviour: Tumour progression affects interaction patterns and social network structure in wild Tasmanian devils. *Proc R Soc B: Biol Sci*. 2020; 287(1940):2020–454.

51. Owen-Ashley NT and Wingfield JC. Seasonal modulation of sickness behavior in free-living northwestern song sparrows (*Melospiza melodia morphna*). *J Exp Biol*. 2006; 209(16):3062–70.

52. Altizer S, Nunn CL, Thrall PH, Gittleman JL, Antonovics J, Cunningham AA, *et al*. Social organization and parasite risk in mammals: Integrating theory and empirical studies. *Annu Rev Ecol Evol Syst*. 2003; 34(1):517–47.

53. Ezenwa VO, Ghai RR, McKay AF, and Williams AE. Group living and pathogen infection revisited. *Curr Opin Behav Sci*. 2016; 12:66–72.

54. Best A, White A, and Boots M. The coevolutionary implications of host tolerance. *Evolution*. 2014; 68(5):1426–35.

55. Pirrie A, Chapman H, and Ashby B. Parasite-mediated sexual selection: To mate or not to mate? in Ezenwa VO, Altizer S, and Hall RJ (eds), *Animal Behavior and Parasitism*. Oxford: Oxford University Press, 2022.

56. Knell RJ. Sexually transmitted disease and parasite-mediated sexual selection. *Evolution*. 1999; 53(3): 957–61.

57. Sheldon BC. Sexually transmitted disease in birds: Occurrence and evolutionary significance. *Philos Trans R Soc Lond B Biol Sci*. 1993; 339(1290): 491–7.

58. Beltran-Bech S and Richard F-J. Impact of infection on mate choice. *Anim Behav*. 2014; 90:159–70.

59. Lockhart AB, Thrall PH, and Antonovics J. Sexually transmitted diseases in animals: Ecological and evolutionary implications. *Biol Rev Camb Philos Soc*. 1996; 71(3):415–71.

60. Moller A. A fungus infecting domestic flies manipulates sexual behaviour of its host. *Behav Ecol Sociobiol*. 1993; 33(6):403–7.

61. Ashby B. Antagonistic coevolution between hosts and sexually transmitted infections. *Evolution*. 2020; 74(1):43–56.

62. Jennions MD and Petrie M. Variation in mate choice and mating preferences: A review of causes and consequences. *Biol Rev Camb Philos Soc*. 1997; 72(2): 283–327.

63. Lopes PC, French SS, Woodhams DC, and Binning SA. Infection avoidance behaviors across vertebrates. In

Ezenwa VO, Altizer S, and Hall RJ (eds), *Animal Behavior and Parasitism*. Oxford: Oxford University Press, 2022.

64. Poirotte C and Charpentier MJE. Inter-individual variation in parasite avoidance behaviors and its epidemiological, ecological, and evolutionary consequences. In Ezenwa VO, Altizer S, and Hall RJ (eds), *Animal Behavior and Parasitism*. Oxford: Oxford University Press, 2022.

65. Amoroso CR and Antonovics J. Evolution of behavioural resistance in host–pathogen systems. *Biol Lett*. 2020; 16(9):20200508.

66. Amoroso CR. Integrating concepts of physiological and behavioral resistance to parasites. *Front Ecol Evol*. 2021; 9:635607.

67. Best A, White A, and Boots M. The implications of coevolutionary dynamics to host–parasite interactions. *Am Journal*. 2009; 173:779–91.

68. Mooring MS, McKenzie AA, and Hart BL. Grooming in impala: Role of oral grooming in removal of ticks and effects of ticks in increasing grooming rate. *Physiol Behav*. 1996; 59(4–5):965–71.

69. Villa SM, Campbell HE, Bush SE, and Clayton DH. Does antiparasite behavior improve with experience? An experimental test of the priming hypothesis. *Behav Ecol*. 2016; 27(4):1167–71.

70. Hart BL. Behavioural defense against parasites: Interaction with parasite invasiveness. *Parasitology*. 1994;109 Suppl:S139–51.

71. Burkett-Cadena ND. Morphological adaptations of parasitic arthropods. In Mullen GR, and Durden LA (eds), *Medical and Veterinary Entomology*, 3rd edn. Cambridge, MA: Academic Press, 2019, 17–22.

72. Bush SE, Villa SM, Altuna JC, Johnson KP, Shapiro MD, and Clayton DH. Host defense triggers rapid adaptive radiation in experimentally evolving parasites. *Evol Lett*. 2019; 3(2):120–8.

73. Hart BL and Hart LA. Reciprocal allogrooming in impala, *Aepyceros melampus. Anim Behav*. 1992; 44(6):1073–83.

74. Clayton DH, Lee PLM, Tompkins DM, and Brodie ED III. Reciprocal natural selection on host–parasite phenotypes. *Am Journal*. 1999; 154:261–70.

75. Spiegel O, Anglister N, and Crafton MM. Movement data provides insight into feedbacks and heterogeneities in host–parasite interactions. In Ezenwa VO, Altizer S, and Hall RJ (eds), *Animal Behavior and Parasitism*. Oxford: Oxford University Press, 2022.

76. Freeland WJ. Primate social groups as biological islands. *Ecology*. 1979; 60(4):719–28.

77. Rao S, Schieber AMP, O'Connor CP, Leblanc M, Michel D, and Ayres JS. Pathogen-mediated inhibition of anorexia promotes host survival and transmission. *Cell*. 2017;168(3):503–16.e12.

78. Milutinović B, Stock M, Grasse AV, Naderlinger E, Hilbe C, and Cremer S. Social immunity modulates competition between coinfecting pathogens. *Ecol Lett*. 2020; 23(3):565–74.

79. Ingwell LL, Eigenbrode SD, and Bosque-Pérez NA. Plant viruses alter insect behavior to enhance their spread. *Sci Rep*. 2012; 2:578.

80. Hamede RK, McCallum H, and Jones M. Biting injuries and transmission of Tasmanian devil facial tumour disease. *J Anim Ecol*. 2013; 82(1):182–90.

# PART II

## Social Behavior

# CHAPTER 3

# Parasites in a social world: Lessons from primates

Baptiste Sadoughi, Simone Anzà, Charlotte Defolie, Virgile Manin, Nadine Müller-Klein, Tatiana Murillo, Markus Ulrich, and Doris Wu

## 3.1 Introduction

Across the animal kingdom, species display marked variation in **sociality** (see Box 3.1 for glossary). Some **solitary** animals rarely interact with conspecifics (except during reproduction), while other animals are social—from facultative, short-term associations, and loose aggregations based on shared needs, to living in permanent groups with differentiated social relationships (1), (2). The transition from solitary to social living (3) has resulted in far-reaching consequences—both costly and beneficial—for hosts and their parasites.

The repeated emergence of sociality suggests that the benefits of **group living** (e.g., increased vigilance and protection against out-group threats and predators) outweigh the costs (e.g., intragroup competition, infanticide). However, the risk of exposure to parasites or pathogens (these terms are used interchangeably to mean all disease-causing organisms) is considered a looming threat for group members, limiting close or frequent contact among individuals (4), (5). While a positive correlation between group size and infection risk was initially postulated as a major cost of group living several decades ago (6), (7), a growing body of research suggests that links between sociality and infectious diseases are more complex, with mixed outcomes for the transmission of and susceptibility to pathogens (8), (9). Consequently, studying the parasite-related costs and benefits of sociality is an increasingly active focal area in behavioral and evolutionary ecology.

Sociality impacts parasite transmission at multiple levels of interaction—from inter-specific communities, to the group level, and down to the individual—via two distinct mechanisms: exposure and susceptibility (10). While exposure, especially to directly or environmentally transmitted pathogens, depends on direct contact, shared space, or resource use (4), (10), (11), susceptibility depends on individual genetics, physiology, and immunocompetence. Although environmental factors (e.g., rainfall, humidity) and individual immune and genetic profiles affect parasite transmission and susceptibility, this chapter focuses primarily on the sociality–parasite link. Non-human primates (hereafter primates) provide excellent study systems to investigate links between sociality and parasites given their exceptional diversity and remarkable within-species and inter-individual variation in social systems and social behaviors, respectively (Box 3.2). The positive link between sociality and fitness in wild primates (12) also provides real-world study systems to explore these mechanisms. Additionally, wild primates are well-studied, with detailed data on individual behaviors and life histories across a range of long-term study populations worldwide (Figure 3.1). Primates also harbor diverse parasite communities (i.e., micro and macro-parasites: bacteria, viruses, protozoa, fungi, helminths, and arthropods) with various transmission routes (13). Finally, given a shared evolutionary history with humans, primate species are a key model for establishing bridges between field

Baptiste Sadoughi et al., *Parasites in a social world*. In: *Animal Behavior and Parasitism*. Edited by Vanessa O. Ezenwa, Sonia Altizer, and Richard J. Hall, Oxford University Press. © Oxford University Press (2022).
DOI: 10.1093/oso/9780192895561.003.0003

---

### Box 3.1  Glossary

**Despotic:** opposite of tolerant. Despotic dominance translates into higher aggression and lower rates of affiliation compared to more tolerant relationships, with low (sometimes absent) rates of peaceful post-conflict interactions. Dominance is clearly established.

**Grooming:** behavior by which animals clean or maintain their bodies or appearance. Allogrooming is the cleaning of a conspecific partner's skin or fur. In primates, allogrooming serves both hygienic (ectoparasite removal) and social (strengthens bonds between partners) functions.

**Modular social network:** a network divided into subgroups interacting more within than between subgroups. Network modularity increases as more highly cohesive subgroups are formed.

**Multilevel societies:** subgroups of animals from the same species formed at three or four levels. The first level of organization is the one-male unit (semi-permanent reproductive units consisting of one leader male, sometimes a follower male, multiple females, and their offspring). One-male units can associate in teams (second level), with sometimes solitary males or all-male units (units composed exclusively of males). Teams associate in bands (third level) which associate in herds (fourth level). Also see Box 3.2.

**Multi-male–multi-female groups:** groups composed of multiple adult males and adult females, and their offspring.

**One-male–multi-female groups:** groups composed of one adult male, multiple adult females, and their offspring.

**One-female–multi-male groups:** groups composed of one adult female, multiple adult males, and their offspring.

**Pair-living:** groups composed of one adult male and female and their offspring.

**Self-medication:** utilization of plant or animal parts containing secondary compounds or other non-nutritional substances to prevent, combat, or control disease.

**Social buffering:** refers to availability of social support, assumed to mediate the negative relationship between perceived stress and health. Candidate physiological systems underlying social buffering effects include neural (e.g., prefrontal cortex, limbic system), endocrine (e.g., hypothalamic–pituitary–adrenal axis, oxytocin), and immune functions.

**Social immunization:** transfer of low doses of an infectious agent or pathogen during social interactions which activates the immune system, decreases susceptibility, and limits reinfection risk. Behavioral or chemical cues of sickness emitted by infected conspecifics may also elicit preventive activation of the immune system.

**Social network:** a social structure described by nodes (i.e., individuals) and the ties (i.e., connections) between these nodes. In primate studies, connections are often based on interactions (e.g., proximity, mating, grooming, or aggressive interactions). In an epidemiological network, parasites can spread along ties between nodes, representing individuals, groups, or even communities.

**Sociality:** tendency of groups and individuals to develop social bonds and live in communities.

**Solitary vs. group living:** solitary animals spend a majority of their lives alone, with possible exceptions for mating and raising young. Conversely, group living is defined as individuals of the same species (conspecifics) maintaining spatial proximity to one another over time via social mechanisms.

**Spillover:** an event characterized by a pathogen spreading from a reservoir population with high pathogen prevalence to a novel recipient host population.

**Tolerant:** opposite of despotic. Tolerant dominance translates into more symmetrical relationships among dyads, less severe aggressive interactions, and higher rates of affiliation and post-conflict interactions.

---

research and biomedical research on physiology and disease, evolutionary medicine, and public health policies.

This chapter highlights significant contributions from primate research in advancing our understanding of the links between sociality and parasitism. First, we address mechanisms at the group level, which primarily influence exposure to parasites. Next, we move to a finer scale of sociality by discussing how positive or negative interactions alter susceptibility. We also explore how individual behaviors may further mitigate parasite risk. Finally, we identify promising research tools and questions that can help advance our ability to

**Figure 3.1** Several representatives of the Order Primate. From left to right, top to bottom. Red-fronted lemur (*Eulemur rufifrons*), credit Tatiana Murillo; sifakas (*Propithecus verreauxi*), credit Hasina Malalaharivony; chimpanzees (*Pan troglodytes*), credit Liran Samuni, Taï chimpanzee project; grey mouse lemurs (*Microcebus murinus*), credit Johanna Henke-von der Malsburg; coppery titi-monkeys (*Plecturocebus cupreus*), credit Sofya Dolotovskaya; Assamese macaques (*Macaca assamensis*), credit Simone Anzà; baboons (*Papio anubis*), credit Doris Wu.

manage the parasite-related costs and benefits of sociality in primates.

## 3.2 Group level effects of sociality on parasitism

Socio-ecological models of primate sociality provide an integrated picture of the influence of food distribution, predation pressure, female–female tolerance, and male competition for monopolization and access to fertile females in shaping group living (1). These spatiotemporal associations and interactions between individuals, in turn, impact parasite transmission (15). At the group level, parasite-related costs and benefits of sociality are mediated by two important demographic features: group size and **social network** organization (15)–(18).

### 3.2.1 Group size and parasitism

The first readily and easily accessible demographic parameter of any study population is group size. In mammals, strong evidence indicates that the intensity of infection with directly and environmentally transmitted parasites (e.g., fleas, helminths) increases with host group size (16), (19), (20). This is attributed to more frequent direct (body-to-body) and indirect (via environmental contamination with feces) contacts in larger groups promoting the transmission of infectious organisms. In primates, both empirical data and computer simulations lend support for a positive association between parasitism—measured either as parasite load or prevalence—and host group size (4), (11). For example, a long-term study of yellow baboons (*Papio cynocephalus*) found that larger groups had higher counts of helminth eggs (21). A comparative study of two sympatric mouse lemur species (*Microcebus* spp.), a taxon displaying high variation in social organization with social units ranging from solitary to **multi-male–multi-female** sleeping associations, found that species with larger nest associations had increased lice prevalence (22). With lice utilizing host–body contact for transmission, temporary associations at nests created parasite-related costs for larger sleeping groups.

Nevertheless, increasing group size has not been invariably linked to increased parasitism, especially in primate studies (15), (23). First, group size can actually reduce parasitism through an encounter-dilution effect (24) for mobile parasites exhibiting a constant attack rate or targeting one host at a time. As the same number of parasites is distributed among more available hosts, greater host densities drive a lower per-individual infection risk (24). For example, in chimpanzees (*Pan troglodytes*), sleeping in groups may minimize individual exposure to biting insects (25). Second, group size interacts with other ecological variables (specifically predation

pressure and food distribution) that may alter the host's ability to cope with parasitism. Helminth infections in a population of wild ungulates resulted in anorexia, but only for parasitized individuals living in smaller groups, suggesting that individuals living in larger groups cope better with infection (26). Larger social groups may secure access to higher quality food patches and reduce time invested in anti-predator vigilance per individual, allowing members to allocate more energy towards reproduction and immune function. It would be compelling to investigate to what extent the benefits of a larger group size could offset competition within groups (1)—particularly as primates are exposed to seasonal changes in food availability and exhibit diverse feeding regimes from grazing to fruit-based diets with meat supplementation. Finally, although early case studies found that larger groups may harbor more diverse parasite communities (7), several meta-analyses in primates and other vertebrates showed no consistent relationship between host group size and parasite richness (13), (19), (23). This suggests that

---

### Box 3.2   Diversity of primate social systems

In mammals, only 23% of all species live in groups, with 9% being **pair-living** (14). Primates deviate from this general pattern, with ~66% of genera living in permanent mixed-sex groups (1), and 29% of species being pair-living (14). Primate social systems are characterized by three complementary but distinct components found in multiple combinations: social organization, social structure, and mating system (1) (Figure 1).

Social organization refers to the number and composition, as well as cohesion, of a social unit. Five types of social units have been described in primates, with varied sex-age composition and levels of relatedness. There are **solitary** species (e.g., Bornean orangutans *Pongo pygmaeus*), pair-living species (e.g., coppery titi monkeys *Plecturocebus cupreus*); **one-male–multi-female groups** (e.g., patas monkeys *Erythrocebus patas*); **one-female–multi-male groups** (e.g., facultative in many *Callitrichidae*); and **multi-male–multi-female groups** (e.g., Assamese macaques *Macaca assamensis*, red-fronted lemurs *Eulemur rufifrons*).

These units can remain cohesive (most social primates), split into subgroups with changing size and composition over time (fission–fusion societies, e.g., chimpanzees *Pan troglodytes*), or be organized around nested levels forming **multilevel societies** (e.g., geladas *Theropithecus gelada*). Social structure refers to the distribution, quality, and dynamics of relationships between group members. Social relationships may form preferentially between certain partners, with dominance interactions that vary from **despotic** (e.g., chimpanzees, rhesus macaques *M. mulatta*) to **tolerant** (e.g., red-fronted lemurs, tonkean macaques *M. tonkeana*); that are stable or unstable; and inherited or contested. Primates exhibit marked flexibility in social structure, showing variation both within and between species.

Mating system refers to sexual interactions. In primates, these interactions range from monogamous (e.g., coppery titi monkey) to polygynous (e.g., geladas), polyandrous (e.g., many *Callitrichidae*), and polygamous (e.g., chimpanzees).

**Box 3.2** *Continued*

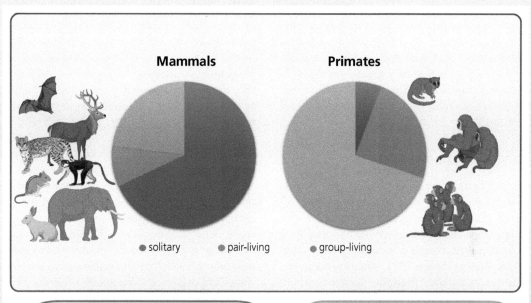

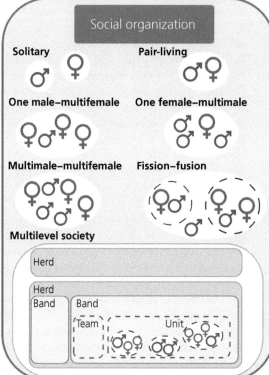

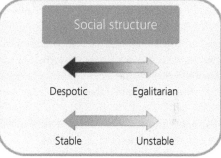

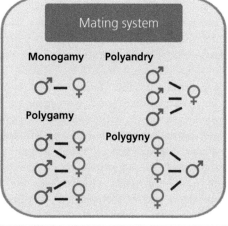

**Box 3.2, Figure 1** The components and diversity of social systems in primates. Figure created using BioRender.

species-specific movement patterns, contact rates (21), habitat use (27), or density (13) can prevail over group size in predicting parasite richness.

Finally, including group size as the sole proxy of sociality assumes that individuals in groups interact with each other at a constant rate, disregarding social dynamics and finer scale social interactions (i.e., the social structure of a group, see Box 3.2). Therefore, group size is often insufficient to explain the link between sociality and parasite infections in full (23), (28).

## 3.2.2 Social network properties and parasitism

Parasite transmission is also influenced by substructure within and between social groups. For instance, the fragmentation of a population into subpopulations with limited movement of individuals among them creates a natural barrier against the spread of pathogens (29). Within groups, animal networks are often **modular**, with interactions unevenly distributed across group members and tending to occur preferentially within subgroups (23), (28), (30). Modularity has a non-linear effect on disease transmission (31), (32). The cost of increased disease transmission within subgroups first offsets the benefit of decreased transmission between subgroups. Above a certain threshold, highly modular social networks limit infection to a few subgroups (31) (Figure 3.2A). Parasites are "trapped" within subgroups resulting in smaller, or at least delayed, disease outbreaks depending on parasite transmissibility (15), (30), (33), (34). Interestingly, comparative studies based on empirical data suggest that modularity increases with group size in primates (15), (30), (31) (but see (32)). The positive relationships between group size and parasitism, group size and modularity, and negative relationship between modularity and parasitism suggest a possible inverted-U relationship between group size and parasitism in real world networks (31), with parasite spread being highest at intermediate group sizes. These results highlight the necessity of considering social structure (see Box 3.2) to understand and predict parasite transmission dynamics in social species.

Individuals are exposed to heterogeneous parasite risks depending on network structure and their respective positions in groups (9), (18) (see also Mistrick *et al.* (35) in this volume for an extensive review of animal networks and pathogen transmission). In primates, higher infection risk of helminth, protozoan, and possibly also viral (36) parasites, is associated with shared habitat use (11), (37), increasing body contact, higher numbers of **grooming** partners, and centrality in grooming networks (28), (37)–(39). Consistent with patterns described in primates, a recent meta-analysis found that more central individuals in a network face higher parasitism (18), although results also indicate wide heterogeneity in the strength and direction of associations. These differences may be expected based on parasite transmission modes and social behaviors measured (38). Indeed, hygienic behaviors bringing individuals into close contact may help fight certain infectious agents, while also increasing exposure to others. For example, being central in a grooming network could lower lice infestation via removal from the partners' fur (40), while simultaneously increasing the risk of gastrointestinal parasites as nematode eggs are ingested during grooming bouts (37), (38). To date, support for such theoretical predictions is ambiguous (38), (40), (41), stressing the necessity to incorporate other factors, such as parasite lifecycle relative to how exposure is measured, when applying network analyses to empirical systems. Furthermore, predictions about individual parasitism likely depend on interactions between group-level organization, an individual's position within the network, and other traits. To illustrate, a study on captive rhesus macaques (*Macaca mulatta*) found links between individual network position and *Shigella* infection in two groups, but not in a third, due to differences in sex–age composition (41). The spread of directly transmitted parasites may also be hampered when the progressive acquisition of resistance among group members increases the chance that transmission chains will be broken before reaching susceptible individuals. Central individuals in the group, often considered to be "superspreaders," may conversely be exceptionally efficient at slowing disease

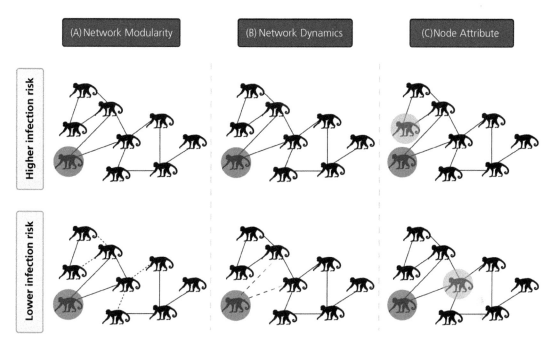

**Figure 3.2** Social network structure influences infection risk in primates. All networks shown have the same number of nodes (individual primates) and edges (connection between the nodes). The spread of infectious diseases from an infected node (red circle) will depend on network properties, such as (A) modularity, which is the extent to which a network is divided into subgroups that interact more within than between subgroups. The lower network has both strong (full link) and weak (dashed link) connections, resulting in higher modularity than in the corresponding upper network. If infection leads to withdrawal from the group or avoidance of group members (B), network dynamics will further decrease infection risk (lower network, dashed lines vs. upper network full lines). Finally, the resistance of a network to disease spread may result from the combined effect of (C) a node's position and attributes. Immune nodes (green circle) with a central position (lower network) may be more efficient at slowing the spread of diseases than more peripheral ones (upper network). Figure created using BioRender.

spread once they have acquired immunity (9), (42) (Figure 3.2C).

In summary, group-level costs and benefits of sociality on parasitism are better understood in the context of multivariate factors (5), (27). Since group size oversimplifies sociodynamics, focusing on group structure and dynamics, rather than group size alone, can offer a more nuanced approach for understanding the vulnerability of social groups to parasites (17), (23), (38) and the costs and benefits of group living (15), (33). Nevertheless, the evidence presented so far in this chapter focuses on the impact of *quantitative* social measures, derived from the number of partners or the number or frequency of contacts, on parasitism. As shown in the next section, the *quality* of social interactions (whether positive or negative) also influences exposure and susceptibility to parasites.

## 3.3 Social interactions, susceptibility, and exposure to parasites

Social interactions inside a group are not random but tend to follow a set of rules with preferential interactions defining a social structure. Competitive and affiliative interactions are part of social living and influence susceptibility and exposure to parasites.

### 3.3.1 Competition and dominance

As individuals compete for access to food and mating partners, both within and between social groups, success depends on competitive abilities and/or alliances. Depending on the degree of tolerance, individuals face different social and environmental adversities. Both social and environmental stressors activate the

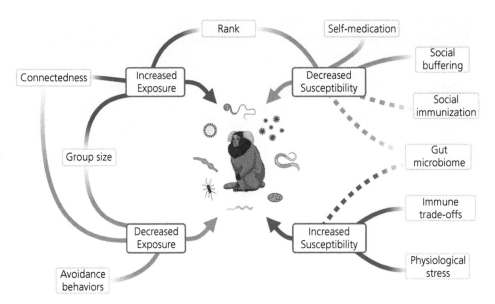

**Figure 3.3** Primate sociality: parasite-related costs and benefits. Illustrated summary of the evidence linking sociality and parasitism in primates. Correlates of sociality increase (red arrows) or decrease (green arrows) parasitism by modulating exposure and susceptibility to parasites. Correlates of sociality with established links to parasitism in primates are depicted with full arrows. Key social drivers of parasitism in other taxa of special interest for primates are depicted with dashed arrows. Evidence gathered from primates highlights that exposure and susceptibility are not determined at one level of sociality but by the interplay of group and individual attributes. Individuals modulate parasitic risks exerted by *group size* and *connectedness* with *avoidance behaviors*. Understanding the consequence of social attributes, such as *rank* on infection risk, requires accounting for increased exposure, engagement in social interactions inducing *physiological stress*, and access to support from close partners providing *social buffering*. Ultimately, the consequences of infection depend on individual susceptibility, shaped by *immune trade-offs*, protection acquired from *social immunization* or modulated by physiological systems such as the *gut microbiome* or the use of *self-medication*. Figure created using BioRender.

hypothalamic–pituitary–adrenal (HPA) axis, triggering a neuroendocrine cascade producing glucocorticoids. Glucocorticoids have immunomodulatory effects, especially if chronically elevated, and can result in reduced immunocompetence and increased parasite susceptibility (43)–(45). For example, in wild olive baboons (*P. anubis*), females harassed more often by aggressive males were also more immunocompromised (46). The effects of such adversity on health and parasitism may vary according to the intensity of competition and inequality between group members. This inequality within groups is often studied through the lens of dominance hierarchies and individual rank (5), (44).

In **despotic** hierarchies, dominant positions are obtained and maintained through physical aggression or threats, leading to more skewed access to resources. Several studies on wild and captive primates show a relationship between dominance

rank and specific physiological profiles related to stress and immune function (44), (47)–(49). For example, being low-ranking in a despotic hierarchy can be associated with higher glucocorticoids levels (44), fewer circulating lymphocytes (44), (46), and higher susceptibility to viral infections (43). Moreover, there is increasing evidence of rank-related effects on immune function directly translating into increased susceptibility to specific pathogens. Using experimental manipulation of social rank in female rhesus macaques, Snyder-Mackler *et al.* (48) found various immune-gene expression patterns causally linked to dominance rank, with high-ranking individuals expressing a more antiviral immune phenotype, and low-ranking individuals showing a pro-inflammatory immune profile more reactive against bacterial infections.

Besides providing evidence of a link between low rank and pro-inflammatory responses known to be associated with several diseases (43), (50), the work from macaques suggests that dominance rank may prime the immune system towards specific types of parasites, possibly as a result of dominance-associated biases in exposure. In partial support of this theory, results from a recent meta-analysis on the relationship between dominance rank and parasitism across several vertebrate clades showed that high-ranking males display greater parasitism than lower-ranking males, and revealed a similar, although non-significant, trend in females (restricted to primates only) (51). The two best supporting explanations for this pattern are that high rank increases parasite exposure via priority access to food and trade-offs between reproduction and immune investment. However, drawing further conclusions about how infection risk for different parasites—and more generally, how parasite exposure and susceptibility—differ by rank, requires studying more diverse social and mating systems (51). For example, in primates, the relationship between dominance, health, and parasitism has focused on strongly despotic systems, with a dearth of data on more **tolerant** dominance styles. In the latter, networks of interactions tend to be less modular; individuals' positions in the dominance hierarchy do not necessarily predict access to food resources, interactions with social partners, or risk of injuries, and offspring may be handled by both relatives and non-relatives, possibly increasing exposure. Conversely, individuals in a despotic system exposed to fierce competition may be more inclined to conceal symptoms, which may exacerbate pathogen spread. Finally, social instability in a group can further complicate the picture: during socially unstable times, high-ranking individuals can temporarily experience higher psychological and physical stress, display impaired HPA-axis activity (44), and develop stress-related diseases faster than low-ranking individuals (44). When hierarchies stabilize, being low-ranking is once again associated with greater physiological indices of stress (44).

### 3.3.2 Social support and social learning

Relationships within groups also include a range of positive interactions between group members. Close preferential relationships provide individuals with social support (e.g., tolerance at feeding sites, coalition partners during agonistic interactions) that may mitigate infection through several mechanisms. First, social support reduces susceptibility to pathogens on a physiological level by buffering social stress. Past experimental work found that social support enhances immune function in captive primates (52), decreases the probability of acquiring influenza in humans (53), and reduces glucocorticoids levels associated with negative social experiences in wild primates (54), (55). All these mechanisms could also explain the relationship between social support and lower mortality risk in humans (56). The protective effect of social support against infection may be particularly valuable to those with heightened stress. Both rhesus macaques with uncertain status in their group and human patients reporting high levels of social tension in their daily lives showed a substantial decrease in infection risk attributed to social support (41), (53). As the definition of social support varies between studies (12), the specific components of sociality relevant to the buffering effect against disease need further characterization (52). For example in Barbary macaques (*M. sylvanus*), strong opposite-sex bonds but not same-sex bonds were found to have a protective effect against contracting gastro-intestinal nematodes (37), although the mechanisms explaining this difference remain unclear. Refining the concept of close friends in primatology to include reciprocity, predictability, and stability of interactions over time will help disentangle the different aspects of social support that contribute to **social buffering** effects.

Another form of social support occurs through (social) learning of anti-parasite behaviors. Some primates actively fight parasitism using natural resources—a phenomenon called **self-medication** (57). First observed in wild African great apes, growing evidence has revealed that self-medication is widespread across the primate order (57). A well-established example of self-medication in great apes is leaf-swallowing to fight internal parasites through properties of anthelmintic phytochemicals and trapping worms in leaf folds (57). To fight external parasites, wedge-capped capuchins (*Cebus olivaceus* (58)) and red-fronted lemurs (*Eulemur rufifrons* (59)) perform fur rubbing, using toxic secretions of millipedes as a repellant and even sharing millipedes with social partners. In these situations, knowledgeable group members may serve as role models for social learning. For example, young chimpanzees acquire knowledge of the curative properties of toxic plants by watching adults (60). In contrast, when adults do not tolerate close proximity during feeding time, as in western lowland gorillas (*Gorilla gorilla gorilla*), young individuals learn curative behaviors by observing kin of similar age (60).

In summary, sociality is best understood if we go beyond group size and organization and also acknowledge the diversity of inter-individual relationships between group members. Primates navigate complex social worlds by mitigating the costs of competition and forming social bonds that in turn, have far-reaching consequences for parasitism and health.

## 3.4 Individual behaviors and parasite risk

Prior sections of this chapter describe how group composition and structure shape exposure and susceptibility to parasites at higher levels of sociality. The focus was on understanding how social dynamics within and between groups, as well as competitive and affiliative interactions, may enhance or impair exposure and susceptibility to parasites in primates. However, despite the overarching effects of such constraints at the species or group level, individuals can still adjust their behaviors in response to parasitism. Thus, to capture in full how sociality interacts with parasitism we must also consider social behaviors that flexibly respond to parasitism (e.g., avoidance behaviors, **social immunization**; Figure 3.3) and how larger scales of social organization shape these behaviors themselves.

### 3.4.1 Avoidance behavior

Depending on the mating system, risk of parasite exposure can be heightened in one sex (e.g., polygyny and polyandry) or equally distributed among males and females (e.g., polygamy and monogamy) (61). In primates, there is evidence of a link between copulation rate and prevalence of sexually transmitted diseases (61): in a group of olive baboons (polygamous), females were found to avoid mating with males showing signs of infection with a sexually transmitted disease similar to syphilis (*Treponema pallidum pertenue*) (62). Infected females also accepted fewer copulations, and all females mated with fewer partners (compared to olive baboons from uninfected populations), despite the large pool of males available (62). Thus, when faced with a heightened risk of infection, sexual behaviors were modified, lowering exposure (Figure 3.3), and reducing mating success for infected individuals, with major implications for disease transmission. Sexual behaviors, which are rarely considered in primate social network studies, could be used to investigate how flexibility of sexual networks may limit pathogen spread.

Beyond sexual interactions, individuals appear to modulate their overall social connectedness in response to disease outbreaks, or in the presence of infected individuals, by adjusting their interactions to minimize risk (36), (37), (63)–(65) (Figure 3.2B). As suggested by Butler and Behringer (29), evidence of social distancing in species as phylogenetically distant as honey bees and chimpanzees, points towards multiple independent emergence events of a similar behavioral strategy (29). Although sick individuals may decrease their social interactions as a result of lethargy and fever resulting from infection, social distancing implies the expression of specific behaviors that reduce the transmission of pathogens by increasing spatial distance among members of the group (29). For example, mandrills (*Mandrillus sphinx*) interact with healthy social partners but avoid conspecifics infected with fecally transmitted gastro-intestinal parasites (63). However, avoidance mechanisms may not always be effective if, for example, individuals disregard cues of infection or try to conceal sickness. Mandrills also do not avoid sick kin, suggesting that kinship interferes with avoidance (66). However, testing whether social distancing is effective in the wild remains challenging since conducting controlled experiments is nearly impossible. Sanctuaries with semi-free ranging populations may provide a semi-natural setting where such studies can be done.

### 3.4.2 Social immunization and the gut microbiome

While the benefits associated with avoiding infected individuals are well-known, the potential benefits resulting from gradual exposure to parasites in building immunity remain mostly untested (8). A lack of immune challenges and acquired immunity during early life can significantly affect resilience to infection with aging (67). Therefore, the possibility that sharing parasitic agents (e.g., from parents to offspring or between individuals during play) may contribute to acquired immunity and partially offset certain future costs associated with socially transmitted parasites needs to be considered. This phenomenon of social immunization (Figure 3.3) through low dose exposure (68) has not yet been investigated in primates, although primate–helminth systems, characterized by high species-specificity (69), offer an exciting potential model. In addition to providing gradual exposure to parasites, social interactions through physical contact, close proximity, and movement between groups also facilitates the acquisition of beneficial microbes composing the gut microbiome (70). This community of bacteria, eukaryotes, archaea, fungi, and viruses inhabiting the gastro-intestinal tract of primates plays an essential role in food digestion and the production of metabolites and vitamins, prevents colonization by opportunistic pathogens, participates in the development of the hosts' immune response, and the metabolism of toxic compounds—all of which help reduce host susceptibility to infection (70), (71). Recent evidence suggests that the gut microbiome may affect host social behavior to promote its own transmission by affecting an individual's olfactory signaling and influencing conspecific recognition and bonding or group-specific scent marks (71). These findings raise new questions about how microbial commensals influence social networks and the evolution of sociality (70), (71). Finally, evidence that early-life microbiome composition influences parasite susceptibility in adult frogs (72) provides a starting point for testing how an individual's past and present social life interact to shape its infection status (Figure 3.3).

Freeland first depicted primate groups as homogeneous biological islands (7) whose parasite diversity is mainly governed by group size and ecology (e.g., diet, habitat use). However, building on these concepts has made room for more fluid and fine-tuned connections between the different levels of sociality. Importantly, the original emphasis on the costs of parasite transmission shaping sociality has been enriched by a better understanding of within-group drivers of parasitism. By including more complex levels of sociality, we expand our understanding on how social networks and individual behaviors are linked to overall health (Figure 3.3). A growing body of research has also allowed us to draw a more nuanced picture of assumed benefits of hygienic behaviors (e.g., grooming), while documenting more cases of rare behaviors (e.g., self-medication). Several bold concepts, including social immunization described in other animal phyla and postulated in primates, remain largely untested; while a recent focus on microbiomes has added a new dimension in understanding links between sociality and parasitism.

## 3.5 Future directions and conclusions

### 3.5.1 Important directions for future research on sociality and parasitism in primates

Social network analysis has greatly contributed towards our understanding of how social structure relates to parasite risk. However, most studies to date rely on networks in which all connections between individuals represent the same type of interaction (using a single behavior or combining multiple behaviors to produce a single aggregate measure). Such static networks conceal important social dynamics; in contrast, multilayer networks can incorporate multiple sets of relationships, with each layer representing a distinct form of connection (e.g., aggression vs. affiliation) (73). Static networks also fail to capture spatiotemporally dynamic environments that can destabilize or alter network ties (Box 3.3). Furthermore, given the interdependencies between hosts and parasites, multilayer networks allow researchers to address questions related to temporally dynamic factors, such as testing the resilience of networks to novel pathogens in comparison to endemic pathogens (5), (9), (74). In addition, multilayer networks can be used not to only compare similar sets of individuals (within species), but also interconnected systems. For example, Gomez *et al.* (20) modeled pathogen transmission on networks where nodes represented species rather than individuals, and found that primate species infected with parasites infecting

---

**Box 3.3  Case study: Dynamic vs. static social networks in models of parasite transmission: predicting *Cryptosporidium* spread in wild lemurs**

**Social networks** are used to assess how pathogens spread among a population, group, or between individuals. However, most network models are static and discount temporal fluctuations, such as seasonal changes. This tends to result in networks with high density, as ties accumulate over time, but in which tie strength, averaged over the study period, likely underestimates the maximal strength between nodes. Springer *et al.* (34) predicted that short-term changes in the distribution of network ties would influence pathogen spread and outbreak size, which could only be successfully modeled by dynamic networks. They created an epidemiological model in which individuals were either susceptible, exposed, infected, recovered, or deceased to test how the pathogen *Cryptosporidium* spread between adjacent groups of wild Verreaux's sifakas (*Propithecus verreauxi*) in two three-month seasons. In their model, individuals could become infected by body contact with an infectious conspecific (direct contact transmission) or by ranging over a contaminated area (environmental transmission). The probability of becoming infected in the modeled network was estimated using empirical data from behavioral observations and GPS tracking. The study explored three different transmission routes of the parasite: (a) both environmental and direct contact, (b) direct contact only, and (c) environmental only. For each season, static versions of networks for intergroup body contact and ranging overlap were created. Additionally, dynamic versions of the model were developed by updating networks of intergroup body contacts and ranging overlap every two weeks (Figure 1).

The dynamic and static models converged in predicting larger outbreaks when taking into account both social and environmental transmission. However, the static network model predicted a smaller outbreak size than the dynamic model in the dry season, due to the rapid increase in intergroup range overlap at the end of the dry season that was only adequately captured by the dynamic model. As a consequence, conclusions about the influence of seasonality on outbreak size (whether larger in the wet or dry season) entirely depended on the choice of network (i.e., dynamic vs.

static). By comparing outbreak sizes across seasons in both dynamic and static networks, it becomes apparent that static models may simplify and limit certain predictions regarding disease dynamics.

To understand further which aspects of network dynamics may influence outbreak size, the authors generated random networks on four social groups with different patterns of intergroup connections. Model predictions on pathogen spread were compared between a set of dynamic networks (updated every two weeks) and the corresponding static networks (generated over the cumulated period considered), for varying probability of infection and recovery. Dynamic networks with a low probability of infection and slow recovery produced larger mean outbreak sizes, especially when the strength of interactions between nodes varied greatly over time. Thus, short-term strong connections detected by dynamic models have a larger influence on outbreak size than averaged connections from static networks—particularly in cases where transmission is low and with long recovery periods.

The importance of short-term bond formation in facilitating parasite transmission raises the possibility that the disappearance of short-term bonds might, conversely, be a major mechanism reducing transmission in social groups. Low probability of infection may provide social individuals with an opportunity to adapt their behavior to reduce infection risk before the pathogens have spread through most of the network. Considering these types of host feedback mechanisms between behavior and parasite transmission is critical for accurately predicting pathogen spread (76). For example, a decrease in ranging behavior during the clinical phase of the infection, or transient social distancing between two closely bonded partners, may not be captured by static networks averaged over months of observation, and yet these factors strongly influence disease spread. Implications of these results go beyond the study of protozoan parasites in primates and influence predictions about the spread of other diseases (e.g., bacteria, viruses) with low transmissibility, such as tuberculosis or latent viral infections.

**Box 3.3** *Continued*

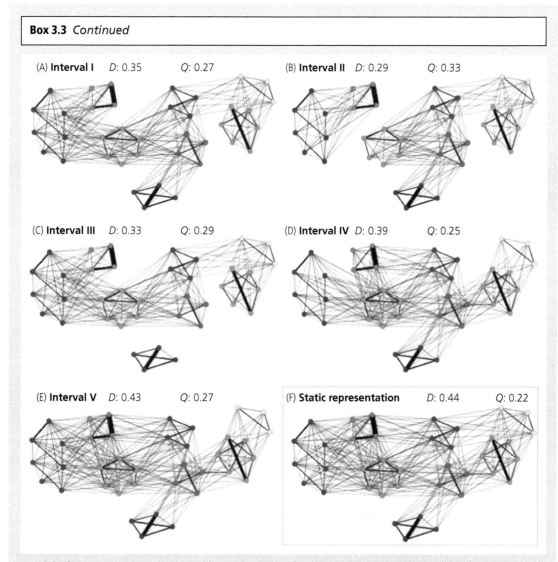

**Box 3.3, Figure 1** Dynamic (A–E) and static (F) networks calculated from behavioral observations of eight groups of Verreaux's sifakas during the dry season. Tie strength is proportional to body contact rates calculated over bi-weekly intervals for the dynamic model, or the entire dry season for the static model. D = density, Q = modularity. Modified with permission from Springer *et al.* (34)

many other primate species were also more likely to harbor pathogens similar to those identified as emerging diseases in humans.

At the individual level, physiological and social mediators of susceptibility to infection are still poorly understood. Although the role of glucocorticoids in response to stress, social isolation, and social buffering (50) makes them ideal candidates linking sociality and parasitism, recent meta-analyses suggest that parasitism itself causes an elevation in glucocorticoids, rather than vice versa (45), (75). The search for physiological mediators between social adversity and susceptibility to parasitism continues. Other possible suspects tying sociality and parasite transmission also require more exploration. This could include flexible behavioral defenses to avoid parasite infections, differences in cognitive capacities to learn anti-parasitic behaviors, and the importance of social interactions in the acquisition of beneficial microorganisms to build up the immune response and compete against pathogens.

### 3.5.2 The relevance of studying host sociality and parasitism in primates

Over the past few decades, the field of primate disease ecology has made great strides towards understanding how group size and composition, social structure, and mating systems relate to parasitism. The level of detail gathered on the interactions and relationships between individually identified primates and their parasites has allowed the field to go beyond the analysis of ecological parameters influencing exposure and susceptibility to identify specific aspects of social life linked to parasitism. In addition, results from empirical studies have contributed to a broader effort of revisiting the links between group living and parasitism by documenting the benefits of sociality against parasitism (8).

Although there is growing understanding of the influence of social behavior on parasite risk, it remains unclear how, and to which degree, parasite infections alter the social behaviors of both infected hosts and their uninfected conspecifics. A perfect example is the global social distancing by humans in response to COVID-19 (76), (77), which illustrates how host social behavior can both respond to parasitism and influence it. Such bidirectional relationships between host social behavior and parasites have important implications for pathogen emergence, spread, and evolution, and ultimately for the evolution of animal sociality (76), (77). Importantly, wild primates, with long-term data collection ongoing in several species worldwide, offer the possibility to explore questions that incorporate the complexities of host social behavior and the role it plays in the transmission of infectious diseases. Such studies in primates have increasing relevance for both public health and conservation.

In terms of public health, studies linking socio-ecological mechanisms involved in parasite transmission in primates may provide information on "best pathogen candidates" for transmission and dissemination in humans (78). Primates are the source of two of the deadliest modern day epidemics in human, HIV-1 (the virus responsible for AIDS) stemming from chimpanzees lentivirus (79), and *Plasmodium falciparum* (which causes malaria) from a strain infecting gorillas (80). Yet, much uncertainty remains around the mechanisms that facilitated zoonotic transmission (81). Primates also represent sentinels for monitoring diseases like anthrax (82) and Ebola virus outbreaks (83) that threaten both wildlife and humans. Although **spillover events** are rare, episodic transmission of extremely severe infections (e.g., Ebola (83), cercopithecine herpesvirus B, (78)) is of serious concern at the growing human–primate interface (78), (84).

Likewise, a better understanding of species' social structure and response to infections can help inform conservation measures (20), (42). An illustration is given by efforts to control tuberculosis in badger populations in the United Kingdom. Culling interventions targeting males disturbed territorial defense behaviors and increased migrations between groups, resulting in greater disease spread (85), (86). Similarly, detailed understanding of the influence of sex, rank, or age on social position gathered from behavioral studies will be critical to implement conservation strategies in primates (42) as 60% of primates species are threatened with extinction (84). The increase in infectious diseases in primates is considered a consequence of deforestation, agriculture expansion, habitat fragmentation, mining, hunting, and climate change, all of which are associated with declines among primate populations (84), (85). Understanding the anthropogenic impact on the socio-ecological systems of primates and the consequences in those affected populations can contribute to an inclusive one health approach by reducing the infection risk for all primate species, including humans (78), (85). Only through long-term monitoring, health surveillance systems for researchers, their study species, as well as sentinel populations of neighboring primates can we detect emerging diseases and study the impact of pathogens on populations (78), (79).

## Acknowledgements

This contribution is the result of collaborative work between past and current PhD students of the DFG-funded Sociality and Health in Primates (SoHaPi) Research Group (DFG FOR 2136) with the encouragement of Prof. Dr. Peter Kappeler. All authors contributed equally to this work. We would like to thank Prof. Dr. Peter Kappeler, Dr. Oliver Schülke, Dr. Jan Gogarten, Dr. Claudia Fichtel, and Prof. Dr. Julia Ostner for their insightful feedback on the manuscript, as well as all members of the SoHaPi Research Group for their support, and the German research foundation (DFG) for funding (grants LE 1813/10-1; OS 201/6-1; FI 929/7-1, KR 3834/5-1; FI 929/7-2, DA 374/312; SCHU 1554/6-1; OS 201/6-2). Finally, we are grateful to the editors, Prof. Vanessa O. Ezenwa, Prof. Sonia Altizer, and Dr. Richard J. Hall for their invitation to contribute to this volume, and to Prof. Ezenwa and three anonymous reviewers for their insightful feedback on earlier versions of the manuscript.

## References

1. Silk JB and Kappeler PM. Sociality in primates in Rubenstein DR and Abbot P (eds), *Comparative Social Evolution.* Cambridge: Cambridge University Press, 2017, 253–83.
2. Smith JE, Lacey EA, and Hayes LD. Sociality in non-primate mammals in Rubenstein DR and Abbot P (eds), *Comparative Social Evolution.* Cambridge: Cambridge University Press, 2017, 284–319.
3. Krause J and Ruxton GD. *Living in Groups.* Oxford: Oxford University Press, 2002.
4. Altizer S, Nunn CL, Thrall PH, Gittleman JL, Antonovics J, Cunningham AA, *et al.* Social organization and parasite risk in mammals: Integrating theory and empirical studies. *Annu Rev Ecol Evol Syst.* 2003;34 517–47.
5. Kappeler PM, Cremer S, and Nunn CL. Sociality and health: Impacts of sociality on disease susceptibility and transmission in animal and human societies. *Philos Trans R Soc Lond B Biol Sci.* 2015;370: 140: 20140116.
6. Alexander RD. The evolution of social behavior. *Annu Rev Ecol Syst.* 1974;5 325–83.
7. Freeland WJ. Primate social groups as biological islands. *Ecol.* 1979;60:719–28.
8. Ezenwa VO, Ghai RR, McKay AF, and Williams AE. Group living and pathogen infection revisited. *Curr Opin Behav Sci.* 2-16;12:66–72.
9. Rushmore J, Bisanzio D, and Gillespie TR. Making new connections: Insights from primate–parasite networks. *Trends Parasitol.* 2017;33:547–60.
10. Hawley DM, Etienne RS, Ezenwa VO, and Jolles AE. Does animal behavior underlie covariation between hosts' exposure to infectious agents and susceptibility to infection? Implications for disease dynamics. *Integr Comp Biol.* 2011;51:528–39.
11. Nunn CL, Thrall PH, Leendertz FH, and Boesch C. The spread of fecally transmitted parasites in socially-structured populations. *PLoS One* 2011;6: e21677.
12. Ostner J and Schülke O. Linking sociality to fitness in primates: A call for mechanisms in Naguib M, Barrett L, Healy SD, Podos J, Simmons LW, and Zuk M (eds), *Advances in the Study of Behavior*, vol. 50. Cambridge, MA: Academic Press, 2018, 127–75.
13. Nunn CL, Altizer S, Jones KE, and Sechrest W. Comparative tests of parasite species richness in primates. *Am Nat.* 2003;162:597–614.
14. Lukas D and Clutton-Brock TH. The evolution of social monogamy in mammals. *Science* 2013;341:526–30.
15. Griffin RH and Nunn CL. Community structure and the spread of infectious disease in primate social networks. *Evol Ecol.* 2012;26:779–800.
16. Côté IM and Poulin R. Parasitism and group size in social animals: A meta-analysis. *Behav Ecol.* 1995;6:159–65.
17. Pastor-Satorras R, Castellano C, Van Mieghem P, and Vespignani A. Epidemic processes in complex networks. *Rev Mod Phys.* 2015;87:925–79.
18. Briard L and Ezenwa VO. Parasitism and host social behaviour: A meta-analysis of insights derived from social network analysis. *Anim Behav.* 2021;172:171–82.
19. Patterson JEH and Ruckstuhl KE. Parasite infection and host group size: A meta-analytical review. *Parasitol.* 2013;140 803–13.
20. Gómez JM, Nunn CL, and Verdú M. Centrality in primate–parasite networks reveals the potential for the transmission of emerging infectious diseases to humans. *Proc Natl Acad Sci.* 2013;110:7738–41.
21. Habig B, Jansen DAWAM, Akinyi MY, Gesquiere LR, Alberts SC, and Archie EA. Multi-scale predictors of parasite risk in wild male savanna baboons (*Papio cynocephalus*). *Behav Ecol Sociobiol.* 2019;73:134.
22. Klein A, Zimmermann E, Radespiel U, Schaarschmidt F, Springer A, and Strube C. Ectoparasite communities of small-bodied Malagasy primates: Seasonal and socioecological influences on tick, mite and lice infestation of *Microcebus murinus* and *M. ravelobensis* in northwestern Madagascar. *Parasit Vectors* 2018;11:459.
23. Rifkin JL, Nunn CL, and Garamszegi LZ. Do animals living in larger groups experience greater parasitism? A meta-analysis. *Am Nat.* 2012;180:70–82.

24. Buck JC and Lutterschmidt WI. Parasite abundance decreases with host density: Evidence of the encounter-dilution effect for a parasite with a complex life cycle. *Hydrobiologia* 2017;784:201–10.

25. Samson DR, Louden LA, Gerstner K, Wylie S, Lake B, White BJ, et al. Chimpanzee (*Pan troglodytes schweinfurthii*) group sleep and pathogen-vector avoidance: Experimental support for the encounter-dilution effect. *Int J Primatol.* 2019;40:47–59.

26. Ezenwa VO and Worsley-Tonks KEL. Social living simultaneously increases infection risk and decreases the cost of infection. *Proc R Soc B Biol Sci.* 2018;285:2018–142.

27. Ezenwa V. Host social behavior and parasitic infection: A multifactorial approach. *Behav Ecol.* 2004;15:446–54.

28. Rimbach R, Bisanzio D, Galvis N, Link A, Di Fiore A, and Gillespie TR. Brown spider monkeys (*Ateles hybridus*): A model for differentiating the role of social networks and physical contact on parasite transmission dynamics. *Philos Trans R Soc Lond B Biol Sci.* 2015;370 20140110.

29. Butler MJ IV and Behringer DC. Behavioral immunity and social distancing in the wild: The same as in humans? *BioScience* 2021; 71: 571–80.

30. Nunn CL, Jordán F, McCabe CM, Verdolin JL, and Fewell JH. Infectious disease and group size: More than just a numbers game. *Philos Trans R Soc Lond B Biol Sci.* 2015;370 20140111.

31. Romano V, Shen M, Pansanel J, MacIntosh, AJJ, and Sueur C. Social transmission in networks: Global efficiency peaks with intermediate levels of modularity. *Behav Ecol Sociobiol.* 2018;72:154.

32. Sah P, Leu ST, Cross PC, Hudson PJ, and Bansal S. Unraveling the disease consequences and mechanisms of modular structure in animal social networks. *Proc Natl Acad Sci.* 2017;114 4165–70.

33. Snaith TV, Chapman CA, Rothman JM, and Wasserman MD. Bigger groups have fewer parasites and similar cortisol levels: A multi-group analysis in red colobus monkeys. *Am J Primatol.* 2008;70:1072–80.

34. Springer A, Kappeler PM, and Nunn CL. Dynamic vs. static social networks in models of parasite transmission: Predicting *Cryptosporidium* spread in wild lemurs. *J Anim Ecol.* 2017;86:419–33.

35. Mistrick J, Gilbertson MLJ, White LA, and Craft ME. Constructing animal networks for parasite transmission inference in Ezenwa VO, Altizer S, and Hall RJ (eds), *Animal Behavior and Parasitism.* Oxford: Oxford University Press, 2022.

36. Patrono LV, Pléh K, Samuni L, Ulrich M, Röthemeier C, Sachse A, et al. Monkeypox virus emergence in wild chimpanzees reveals distinct clinical outcomes and viral diversity. *Nat Microbiol.* 2020;5:955–65.

37. Müller-Klein N, Heistermann M, Strube C, Franz M, Schülke O, and Ostner J. Exposure and susceptibility drive reinfection with gastrointestinal parasites in a social primate. *Funct Ecol.* 2019;33:1088–98.

38. MacIntosh A J J, Jacobs A, Garcia C, Shimizu K, Mouri K, Huffman MA, et al. Monkeys in the middle: Parasite transmission through the social network of a wild primate. *PLoS One* 2012;7:e51144.

39. Friant S, Ziegler TE, and Goldberg TL. Primate reinfection with gastrointestinal parasites: Behavioural and physiological predictors of parasite acquisition. *Anim Behav.* 2016;117:105–13.

40. Duboscq J, Romano V, Sueur C. and MacIntosh AJJ. Network centrality and seasonality interact to predict lice load in a social primate. *Sci Rep.* 2016;6:22095.

41. Balasubramaniam K, Beisner B, Vandeleest J, Atwill E, and McCowan B. Social buffering and contact transmission: Network connections have beneficial and detrimental effects on *Shigella* infection risk among captive rhesus macaques. *PeerJ* 2016;4: e2630.

42. Rushmore J, Caillaud D, Hall RJ, Stumpf RM, Meyers LA, and Altizer S. Network-based vaccination improves prospects for disease control in wild chimpanzees. *J R Soc Interface* 2014;11: 20140349.

43. Cohen S, Janicki-Deverts D, Doyle WJ, Miller GE, Frank E, Rabin BS, et al. Chronic stress, glucocorticoid receptor resistance, inflammation, and disease risk. *Proc Natl Acad Sci.* 2012;109:5995–9.

44. Sapolsky RM. The influence of social hierarchy on primate health. *Science* 2005;308:648–52.

45. Defolie, C, Merkling T, and Fichtel C. Patterns and variation in the mammal parasite–glucocorticoid relationship. *Biol Rev.* 2020;95:74–93.

46. Sapolsky RM. Endocrinology alfresco: Psychoendocrine studies of wild baboons. *Recent Prog Horm Res.* 1993;48:437–68.

47. Capitanio J P and Cole SW. Social instability and immunity in rhesus monkeys: The role of the sympathetic nervous system. *Philos Trans R Soc Lond B Biol Sci.* 2015;370: 20140104.

48. Snyder-Mackler N, Sanz J, Kohn JN, Brinkworth JF, Morrow S, Shaver AO, et al. Social status alters immune regulation and response to infection in macaques. *Science.* 2016;354: 1041–5.

49. Snyder-Mackler N, Sanz J, Kohn JN, Voyles T, Pique-Regi R, Wilson ME, et al. Social status alters chromatin accessibility and the gene regulatory response to glucocorticoid stimulation in rhesus macaques. *Proc Natl Acad Sci.* 2019;116:1219–28.

50. Hostinar CE, Sullivan RM, and Gunnar MR. Psychobiological mechanisms underlying the social buffering of the HPA axis: A review of animal models and human

studies across development. *Psychol Bull.* 2014;140: 256–82.

51. Habig B, Doellman MM, Woods K, Olansen J, and Archie EA. Social status and parasitism in male and female vertebrates: A meta-analysis. *Sci Rep.* 2018;8:3629.

52. Cohen S, Kaplan JR, Cunnick JE, Manuck SB, and Rabin BS. Chronic social stress, affiliation, and cellular immune response in nonhuman primates. *Psychol Sci.* 1992;3:301–5.

53. Cohen S, Janicki-Deverts D, Turner RB, and Doyle WJ. Does hugging provide stress-buffering social support? A study of susceptibility to upper respiratory infection and illness. *Psychol Sci.* 2015;26:135–47.

54. Young C, Majolo B, Heistermann M, Schülke O, and Ostner J. Responses to social and environmental stress are attenuated by strong male bonds in wild macaques. *Proc Natl Acad Sci.* 2014;111:18195–200.

55. Wittig RM, Crockford C, Weltring A, Langergraber KE, Deschner T, and Zuberbühler K. Social support reduces stress hormone levels in wild chimpanzees across stressful events and everyday affiliations. *Nat Commun.* 2016;7:13361.

56. Holt-Lunstad J, Smith TB, Baker M, Harris T, and Stephenson D. Loneliness and social isolation as risk factors for mortality: A meta-analytic review. *Perspect Psychol Sci J Assoc Psychol Sci.* 2015;10:227–37.

57. Huffman MA. Chimpanzee self-medication: A historical perspective of the key findings in Hosaka K, Zamma K, Nakamura M, and Itoh, N (eds), *Mahale Chimpanzees: 50 Years of Research.*Cambridge: Cambridge University Press, 2015, 340–53.

58. Valderrama X, Robinson JG, Attygalle AB, and Eisner T. Seasonal anointment with millipedes in a wild primate: A chemical defense against insects? *J Chem Ecol* 2000;26:2781–90.

59. Peckre LR, Defolie C, Kappeler PM, and Fichtel C. Potential self-medication using millipede secretions in red-fronted lemurs: Combining anointment and ingestion for a joint action against gastrointestinal parasites? *J Primatol.* 2018;59:483–94.

60. Masi S, Gustafsson E, Saint Jalme M, Narat V, Todd A, Bomsel MC, et al. Unusual feeding behavior in wild great apes, a window to understand origins of self-medication in humans: Role of sociality and physiology on learning process. *Physiol Behav.* 2012;105 337–49.

61. Nunn CL, Scully EJ, Kutsukake N, Ostner J, Schülke O, and Thrall PH. Mating competition, promiscuity, and life history traits as predictors of sexually transmitted disease risk in primates. *Int J Primatol.* 2014;35 764–86.

62. Paciência FMD, Rushmore J, Chuma IS, Lipenda IF, Caillaud D, Knauf S, et al. Mating avoidance in female olive baboons (*Papio anubis*) infected by Treponema pallidum. *Sci Adv.* 2019;5:eaaw9724.

63. Poirotte C, Massol F, Herbert A, Willaume E, Bomo PM, Kappeler PM, et al. Mandrills use olfaction to socially avoid parasitized conspecifics. *Sci. Adv.* 2017;3:e1601721.

64. Müller-Klein N, Heistermann M, Strube C, Morbach ZM, Lilie N, Franz M, et al. Physiological and social consequences of gastrointestinal nematode infection in a nonhuman primate. *Behav Ecol.* 2019;30:322–35.

65. Poirotte C and Charpentier MJE. Inter-individual variation in parasite avoidance behaviors and its epidemiological, ecological, and evolutionary consequences. In Ezenwa VO, Altizer S, and Hall RJ (eds), *Animal Behavior and Parasitism.* Oxford: Oxford University Press, 2020.

66. Poirotte C and Charpentier MJE. Unconditional care from close maternal kin in the face of parasites. *Biol Lett.* 2020;16:20190869.

67. De Nys HM, Löhrich T, Wu D, Calvignac-Spencer S, and Leendertz FH. Wild African great apes as natural hosts of malaria parasites: Current knowledge and research perspectives. *Primate Biol.* 2017;4:47–59.

68. Konrad M, Vyleta ML, Theis FJ, Stock M, Tragust S, Klatt M, et al. Social transfer of pathogenic fungus promotes active immunisation in ant colonies. *PLoS Biol.* 2012;10:e1001300.

69. Pedersen AB, Altizer S, Poss M, Cunningham AA, and Nunn CL. Patterns of host specificity and transmission among parasites of wild primates. *Int J Parasitol.* 2005;35:647–57.

70. Sarkar A, Harty S, Johnson KVA, Moeller AH, Archie EA, Schell LD, et al. Microbial transmission in animal social networks and the social microbiome. *Nat Ecol Evol.* 2020;4:1020–35.

71. Sherwin E, Bordenstein SR, Quinn JL, Dinan TG, and Cryan JF. Microbiota and the social brain. *Science* 2019;366: eaar2016.

72. Knutie SA, Wilkinson CL, Kohl KD, and Rohr JR. Early-life disruption of amphibian microbiota decreases later-life resistance to parasites. *Nat Commun.* 2017;8:86.

73. Finn KR, Silk MJ, Porter MA, and Pinter-Wollman N. The use of multilayer network analysis in animal behaviour. *Anim Behav.* 2019;149:7–22.

74. Wilson SN, Sindi SS, Brooks HZ, Hohn ME, Price CR, Radunskaya AE, et al. How emergent social patterns in allogrooming combat parasitic infections. *Front Ecol Evol.* 2020;8: 54.

75. O'Dwyer K, Dargent F, Forbes MR, and Koprivnikar J. Parasite infection leads to widespread glucocorticoid hormone increases in vertebrate hosts: A meta-analysis. *J Anim Ecol.* 2020;89:519–29.

76. Ezenwa VO, Archie EA, Craft ME, Hawley DM, Martin LB, Moore J, *et al.* Host behaviour–parasite feedback: An essential link between animal behaviour and disease ecology. *Proc R Soc B Biol Sci.* 2016;283: 20153078.

77. Hawley DM and Ezenwa VO. Parasites, host behavior and their feedbacks. In Ezenwa VO, Altizer S, and Hall RJ (eds), *Animal Behavior and Parasitism.* Oxford: Oxford University Press, 2020.

78. Devaux CA, Mediannikov O, Medkour H, and Raoult D. Infectious disease risk across the growing human–non human primate interface: A review of the evidence. *Front Public Health* 2019;7: 305.

79. Sharp PM, Rayner JC, and Hahn BH. Great apes and zoonoses. *Science* 2013;340:284–6.

80. Liu W, Li Y, Learn GH, Rudicall RS, Robertson JD, Keele BF, *et al.* Origin of the human malaria parasite *Plasmodium falciparum* in gorillas. *Nature* 2010;467: 420–5.

81. Rupp S, Ambata P, Narat V, and Giles-Vernick T. Beyond the cut hunter: A historical epidemiology of HIV beginnings in Central Africa. *EcoHealth* 2016;13:661–71.

82. Hoffmann C, Zimmermann F, Biek R, Kuehl H, Nowak K, Mundry R, *et al.* Persistent anthrax as a major driver of wildlife mortality in a tropical rainforest. *Nature* 2017;548:82–6.

83. Leroy EM, Rouquet P, Formenty P, Souquière S, Kilbourne A, Froment JM, *et al.* Multiple Ebola virus transmission events and rapid decline of Central African wildlife. *Science* 2004;303:387–90.

84. Estrada A, Garber PA, Rylands AB, Roos C, Fernandez-Duque E, Di Fiore A, *et al.* Impending extinction crisis of the world's primates: Why primates matter. *Sci Adv.* 2017;3:e1600946.

85. Herrera J and Nunn CL. Behavioural ecology and infectious disease: Implications for conservation of biodiversity. *Philos Trans R Soc Lond B Biol Sci.* 2019;374:20180054.

86. McDonald RA, Delahay RJ, Carter SP, Smith GC, and Cheeseman CL. Perturbing implications of wildlife ecology for disease control. *Trends Ecol Evol.* 2008;23:53–6.

# CHAPTER 4

# Constructing animal networks for parasite transmission inference

Janine Mistrick, Marie L.J. Gilbertson, Lauren A. White, and Meggan E. Craft

## 4.1 Introduction

Transmission is the process governing how parasites (transmissible infectious agents including viruses, bacteria, protists, prions, and macroparasites such as worms and arthropods) infect new hosts and spread through populations (1). Simple models of infectious disease dynamics, such as the compartmental frameworks pioneered in early epidemic modeling (2), assume all individuals in a population contribute equally to transmission. However, observations of outbreak dynamics suggest that heterogeneity in transmission is the norm, not the exception, for humans (3), livestock (4), and wildlife (5).

Dynamical disease models consider transmission as the product of two interacting components: (i) the probability of contact between individuals in a population and (ii) the conditional probability of transmission given contact (6). However, transmission events are challenging to document, particularly in wildlife species. Instead, it is often more practical to estimate contacts between individuals and use contacts as a proxy for transmission (7). In disease ecology, "contact" is often used very generally, even up to and including indirect contact via the environment. To acknowledge that not all opportunities for transmission involve a direct, physical association, we use the term "interaction" to refer to opportunities for transmission inclusive of all modes of parasite transmission. These range along a continuum from direct interactions ("same place, same time") such as physical contact or close proximity to indirect interactions ("same place, different time") such as sequential exposure to a shared environment (8).

Networks provide a way to visualize the direction, strength, and structure of interactions within a population (Figure 4.1). The structure or topology (the arrangement of edges between nodes) of the network can serve as a powerful tool to make inferences about parasite transmission by, for example, identifying highly connected individuals or individuals connecting otherwise segregated populations. Importantly, the transmission-relevant host interactions that determine where edges appear in the network ("edge formation") will inform both (i) the thresholds of space and time used to define edges from spatiotemporal data and (ii) the technology used to record spatiotemporal data in animal populations, and these choices can ultimately alter network topology (Figure 4.2) (9), (10). In this chapter, we discuss considerations for defining edges and collecting data to construct networks, synthesize the application of networks in the context of studying parasite–host behavior feedback in animal systems, and identify future directions to improve network studies of parasite transmission.

## 4.2 Constructing networks

### 4.2.1 Node identity

Networks can be built based on the interactions of individuals or groups of individuals (e.g., packs,

Janine Mistrick et al., *Constructing animal networks for parasite transmission inference*. In: *Animal Behavior and Parasitism*. Edited by Vanessa O. Ezenwa, Sonia Altizer, and Richard J. Hall, Oxford University Press. © Oxford University Press (2022). DOI: 10.1093/oso/9780192895561.003.0004

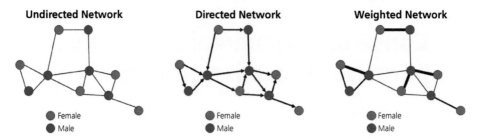

**Figure 4.1** Options for visualization of networks. In networks, individuals can be represented as points (nodes) and lines between nodes can signify interactions (edges). Attributes of individuals may be represented by node size, shape, or color. Edges in a network may be directed to show actions performed by one individual on another, and edges may also be weighted to indicate the strength of the interaction between two nodes.

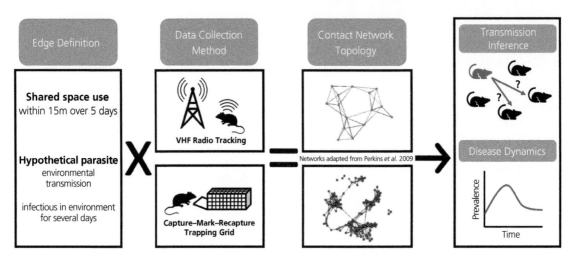

**Figure 4.2** Edge definition and data collection method can impact network topology and transmission inference. Researchers can choose how to define edges and collect network data when constructing networks. These choices can impact the resulting network topology. To illustrate, two empirical networks adapted from Perkins et al. 2009 (10) are shown. Using a common edge definition of shared space use, interaction data on wild mice was collected via both VHF (very high frequency) radio tracking and capture–mark–recapture (CMR) trapping grids, and networks were constructed for each method. Significant differences in network topology arose between the two data collection methods, likely as a result of the resolution with which the method could estimate interactions. These differing network topologies would have downstream effects on transmission outcomes inferred from the networks. Networks adapted from Perkins et al., 2009. Reprinted by permission of John Wiley & Sons, Inc, © 2009.

herds) and nodes may represent a single species or individuals of different species (see section 4.2.6). The social system of a species (encompassing interactions, cohabitation, and mating between individuals) will often inform whether nodes are individuals or groups and how interactions are defined (11). For instance, in group-living species, membership in a herd or pack is often assumed to connect an individual to every other individual in the group (i.e., "gambit of the group" (12)) and thus researchers may be most interested in how groups interact to spread parasites. In contrast, solitary or territorial animals that actively avoid conspecifics

require nodes that represent individuals. It is also important to consider that missing nodes may influence network topology. As such, key goals when defining node identity should be to identify individuals accurately and repeatedly observe a representative proportion of the population.

### 4.2.2 Biological relevance

Edges in a network are only meaningful for understanding transmission if they are defined in a way that is biologically relevant to transmission of the parasite being studied (13). Consider, for example,

**Table 4.1** Examples of biologically relevant edge definitions

| Study | Host–parasite pairing | Social system | Transmission mode | Edge definition | Method/Technology |
|-------|----------------------|---------------|-------------------|-----------------|-------------------|
| Clay et al. 2009 (77) | Deer mouse—Sin Nombre hantavirus | Social, one mature male and several mature females (and offspring) share territory; males may fight with other males | Direct contact (likely aggressive behaviors) | **Direct interaction:** individual with powder color matching that of a marked male | Live trapping capture–mark–recapture grid and powder marking of five individuals to document their subsequent interactions upon next capture |
| Bull et al. 2012 (64) | Australian sleepy lizard—Salmonella | Multi-season monogamous pair bonding; some overlap of home range edges, core area with reduced overlap to same sex neighbors | Unknown, assumed environmental (fecal–oral) | **Spatiotemporal overlap** of GPS locations Synchronized GPS locations of two lizards within 2 meters (m) of each other (conservatively included locations within 14 m to account for varying GPS precision) | GPS loggers, locations recorded every ten minutes |
| Wilber et al. 2019 (15) | Domestic cattle, white-tailed deer, raccoons, opossums—Bovine tuberculosis (bTB) | Herds of cattle on farms; Individual wildlife | Direct and environmental | **Strict spatiotemporal overlap:** proximity logger readings between unique animal pairs within a three-month season (contact rates vary with season) **Shared space use:** animal pair contacts same stationary logger within thirty days (avg. infectious period of bTB in environment) | Proximity logging collars on animals and stationary loggers on cattle-related resources Loggers detect an interaction at a mean distance of 0.88 m—60 seconds of separation before recording a new event |

a sexually transmitted parasite in a solitary, territorial species. Home range overlap estimated from Global Positioning System (GPS) data aggregated over a year would suggest many edges between animals, but these are unlikely to represent the rare, male–female interactions necessary for transmission of the hypothetical parasite. When edges are poorly defined relative to the host–parasite biology, it can lead to problems for transmission inference (7). In Table 4.1, we highlight several examples of edges that are well-defined for parasites of different transmission modes.

It can be challenging to determine how edges in a network should be defined. There may be uncertainty associated with the exact parameters for transmission of a parasite (14) or parasites may have multiple modes of transmission (e.g., direct contact and environmental transmission (15)). Laboratory trials can be a useful tool to suggest a particular dose-response threshold or length of environmental persistence of an infectious agent to inform relevant interactions for transmission. However, laboratory trials are not feasible for many host–parasite pairings and a controlled lab setting is likely an imperfect approximation of what occurs in nature (16). It may be more appropriate to "ground truth" a chosen edge definition with pilot studies in the field or using analytical approaches (this topic will receive further attention in later sections of this chapter: sections 4.2.4; 4.2.6; 4.3.1; 4.6).

### 4.2.3 Defining edges—Direct observation of interactions

Network edges representing direct interactions to estimate transmission of directly transmitted parasites are best defined by the occurrence of a transmission-relevant behavior between two animals (e.g., biting or sexual intercourse). Behavioral observations, focal follows, or video monitoring represent a gold standard for defining edges as they provide information on what animals are doing during interactions and can help identify the interactions most correlated to transmission of various parasites. (17) For instance, Leu *et al.* (18) constructed networks based on sexual interactions, skin contact, synchronous breathing, and social association using behavioral observations

of wild bottlenose dolphins to assess the potential transmission risk of parasites spread by different routes (sexual transmission, physical contact, and aerosol). While powerful for inferring transmission, behavioral observations remain fairly uncommon for constructing networks in disease ecology as they are time intensive, require specialized training in standardized methods, and require host species that can be readily observed. In this chapter, we will focus on methods for estimating interactions between animals, assuming that host behaviors cannot always be observed directly.

### 4.2.4 Defining edges—Estimating interactions

When host behaviors cannot be easily observed, edges in a network are often defined by some degree of host overlap in space and time (spatiotemporal overlap). Spatiotemporal overlap can estimate transmission-relevant interactions for different transmission modes based on how edges are defined, ranging from strict spatiotemporal overlap (direct interactions, e.g., sexual transmission) to asynchronous shared space use (environmental exposure, e.g., fecal–oral transmission). Asynchronous shared space use assumes that spatial overlap correlates with interaction frequency (19) and is often estimated via home range overlap (20). For edges defined by shared space use, a transmission-relevant interaction occurs when two animals use the same space asynchronously, but within a time window relevant to the environmental persistence of the parasite. This could range from days to weeks (e.g., bovine tuberculosis (21)) to months to years (e.g., chronic wasting disease (22)). To estimate transmission opportunities for directly transmitted parasites, edges are better defined by strict spatiotemporal overlap using narrow distance and time thresholds (e.g., fractions of a meter to several meters and seconds to hours, respectively). This can be estimated via visual observations of group membership (23), proximity-logging collars (24), or GPS collars (22).

However, estimating edges using spatiotemporal overlap—even strict overlap—may not provide enough information to confirm an interaction has occurred or to know if the interaction was sufficient for transmission (25). For instance, a direct

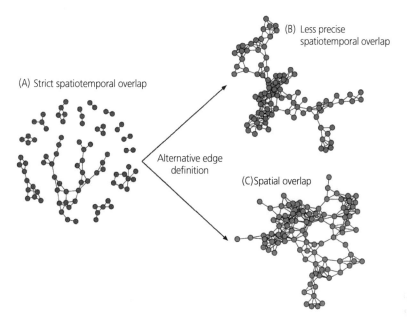

**Figure 4.3** Theoretical example of how edge definition can affect network topology. Gilbertson *et al*. 2020 (9) simulated movements of a host population to generate a "complete" network, and then generated "sample" networks with subsets of the movement data using alternative thresholds to define edges. The "complete" network (A) shows edges as defined by strict spatiotemporal overlap. The same data is shown in two "sample" networks where edges are defined by (B) spatiotemporal overlap with a less precise (i.e., larger) distance threshold or (C) purely spatial home range overlap ignoring temporal elements (i.e., an infinite temporal threshold). The less precise thresholds in sample networks B and C produced more connected networks compared to the complete network (A) while the number and location of connections also led to different network topologies between networks B and C.

interaction may be estimated from two animals with GPS locations within 2 meters of each other at the same time but this neither accounts for attraction or avoidance behavior nor distinguishes if an interaction occurs. As such, grooming may be indistinguishable from sexual intercourse but only the latter is relevant for sexual transmission. Some technologies can help address these uncertainties by approximating behavior states (e.g., accelerometers, see section 4.6.3) or by augmenting interaction data with information about the duration of an interaction (e.g., proximity loggers).

Missing data is a persistent concern in network studies, though intentional study design (e.g., edge definition) and sensitivity analyses can help to limit and quantify the effects on study outcomes. For example, missing edges may influence network topology and could potentially lead to individuals not appearing in the network, both of which would affect transmission inference. Simulation models can be a powerful approach to incorporate

sensitivity analyses into network studies to assess how robust network topology (and thus transmission inference) is to even small changes in how edges are defined (Figure 4.3). These approaches highlight an important intersection between host social system, parasite transmission mode, and transmission inference. Some host social systems, such as highly territorial species which tend to have fewer associations between individuals, or parasite transmission modes, such as those transmitted by direct interactions, are likely more sensitive to changes in how edges are defined as they require more strict definitions (i.e., narrow time and distance thresholds of spatiotemporal overlap) to represent network topology accurately (9).

### 4.2.5 Static versus dynamic networks

Temporal variation in factors such as climate, host social behavior, and host space use can influence transmission dynamics, both for direct and environmentally transmitted parasites (26). This variation should be considered when defining temporal thresholds for edge formation and when aggregating interactions in a population for network construction. For example, climate can seasonally alter survival of parasites transmitted via the environment (27). This may necessitate different edge definitions over time, with longer temporal thresholds in some parts of the year (long parasite survival) than others (shorter parasite survival). Further, temporal or seasonal variation in host social behavior or space use (e.g., fission–fusion dynamics, host breeding season, etc.) can cause heterogeneity in interactions to vary over time (27). For example, in a fission–fusion species like African buffalo, associations may form, dissolve, and shift from month to month (28). A static network aggregating interactions over that period would suggest that all edges are relevant to transmission and would fail to represent the importance of lasting versus ephemeral edges. Temporal dynamics can be managed by using multiple static networks which aggregate interactions over shorter periods, or by using dynamic networks which "rewire" edges over time. However, the increased complexity of dynamic networks requires increased data requirements to collect, store, and process data—especially continuous data—and greater computational complexity to conduct statistical analyses. As such, researchers should only include the temporal complexity necessary to address their research question.

### 4.2.6 Multilayer networks

Network approaches can be extended beyond a single definition for nodes and for edge formation by constructing layered networks to compare different edge definitions or to investigate cross-species interactions in multi-host systems (29). Multilayer networks can represent the same individuals in a series of networks that differ based on how edges are defined. This provides a framework to test which interactions best approximate transmission opportunities (30) or to investigate host–parasite systems with multiple routes of transmission (22). Multilayer networks can also be applied to parasite transmission in multi-host systems. In this context, individuals of different species are represented in

distinct layers and edges may be formed within a layer (same-species interaction) or between layers (cross-species interaction). These approaches have been used to investigate the potential for cross-species transmission via spatial overlap (31) and via direct interactions using proximity loggers (32). Multilayer networks are a new and developing area of research, promising additional realism in the context of ecological interactions, but with the trade-off of complexity in terms of data requirements and network construction which may not be necessary to address all research questions.

## 4.3 Methods of data collection

The choice of how to collect network data requires careful consideration due to the effects these decisions can have on network topology and transmission inference (Figure 4.2). Different methods and technologies have different capabilities for capturing data on animal locations in space and time (particularly with respect to direct versus environmental transmission) and vary in the resolution of data collected. As such, the research question, host–parasite system, and edge definition together should determine the technology used to estimate interactions, with full recognition that the method of data collection may limit the use of the data for questions beyond the scope of the original study.

### 4.3.1 Choice of method

There are many options of data collection methods to record spatial and temporal data on wildlife. These can be generally classified as place-based, technology in the environment recording individuals at specific locations, or individual-based, technology attached to an animal recording their movement or interactions. (33) We present descriptions of a variety of methods that have been commonly used to collect data to construct animal networks (Table 4.2), but this list is by no means exhaustive (see also (34), (35), Section 4.6, and Chapter 6 of this volume (36)). Specific considerations are highlighted in Table 4.2, but we emphasize that these methods range in cost and researcher effort to implement, and some are better suited for certain edge definitions or parasite

**Table 4.2** Data collection methods

| Method/Technology | Explanation | Edge definitions | Host features | Examples | Considerations |
|---|---|---|---|---|---|
| | | **Place-based Monitoring** | | | |
| | | **General considerations:** Technology is generally less costly | | | |
| Trapping grid | Live-capture traps in fixed locations, regular trapping to detect animal locations; often combined with PIT tags to identify individuals | Shared space use; less precise spatiotemporal overlap | Small-bodied and "trap-happy" animals; effective for systems where aggregation naturally occurs (e.g., around food sources) | Rodents (10); brushtail possums (21) | Consider trap density relative to population size (10); added resources may affect interaction patterns |
| Passive Integrated Transponder (PIT) tag | Unique identifier inserted under the skin or attached via leg band; stationary receivers record individuals at a location | Shared space use via visits to common locations (e.g., trap, feeder); may be extended to more strict spatiotemporal overlap | Small-bodied animals *Proximity loggers on animals and stationary devices often used in a similar fashion with larger animals | Finches (78) *Cattle and deer (15) | Assumes host congregation at stationary receivers is congruent with other interaction behavior |
| Visual observations (including gambit of the group, GoG) | Identify individuals by unique markings or VHF/GPS collar; record spatial location, individual associations, or group members | Direct interactions and associated behaviors; strict spatiotemporal overlap; group membership | Uniquely identifiable individuals; most effective with group-living species (GoG can be used to track fission-fusion dynamics) | Meerkats (17); chimpanzees (51) GoG: African buffalo (28) | Time-intensive to collect; difficult to monitor many individuals at once; direct observations may not be feasible for many species |

*Continued*

**Table 4.2** *Continued*

**Individual-based Monitoring**

**General considerations:** Requires entire population to be tagged for optimal resolution; difficulty extrapolating locations between fixes; technology is generally more costly

| Method/Technology | Explanation | Edge definitions | Host features | | Examples | Considerations |
|---|---|---|---|---|---|---|
| Proximity loggers | Devices attached to animals record other devices within a researcher-defined distance | Strict spatiotemporal overlap | Generally larger-bodied hosts *logger size/weight may be prohibitive for small-bodied animals | | Tasmanian devils (24); badgers and cows (32) | Interaction distances set at discrete threshold (e.g., 0.5 m, 2 m, etc.) |
| VHF monitoring | Radio transmitter emits a signal which can be picked up by a manual device (or stationary tower) | Shared space use; strict spatiotemporal overlap | Range of host body sizes | Works well for solitary individuals or long-distance movement | Rodents (10); bobcats and puma (31) | Requires very high frequency of data collection for high temporal resolution |
| GPS monitoring | GPS device broadcasts animal location at specified time points; can provide high-resolution information on animal location and movement | Shared space use; strict spatiotemporal overlap | Generally, larger-bodied hosts *collar size/weight may be prohibitive for some smaller-bodied animals | | Deer (22); lizards (64) | Trade-off temporal resolution vs longevity of sampling. Must consider localization error (error in exact location estimation); can be biased by habitat type, animal behavior |

transmission modes. Moreover, regardless of the data collection method, monitoring a representative sample is key to identifying meaningful patterns that are generalizable to the population of interest. For instance, behavioral heterogeneity may make some individuals more likely to be observed by certain methods which can bias network topology.

Choosing the data collection method that best estimates transmission-relevant interactions may require conducting pilot studies with different technologies or pairing studies of free-ranging wildlife with observations of captive populations (e.g., enclosures with controlled populations). For example, Lavelle *et al.* (37) fitted a population of white-tailed deer with collars equipped with cameras, proximity loggers, and GPS devices to compare interaction rates as measured by the three devices. They found that cameras and GPS underrepresented interactions among deer and only proximity loggers provided interaction rate estimates that were representative of actual rates. Thus, using only one method of data collection and failing to investigate other options or consider potential limitations could result in over- or under-estimation of transmission-relevant interactions.

### 4.3.2 Implementation of the technology

Once the data collection method is chosen, there are additional considerations about how the technology will be implemented: specifically, the temporal resolution (how frequently data are collected) and longevity of sampling (the length of time over which data are collected). For individual-based monitoring in particular, there are trade-offs to balance data resolution and longevity; frequently recording data points provides high spatiotemporal resolution but often comes at the cost of a shorter battery life. However, if the sampling effort is inadequate (e.g., too coarse such that data points are too far apart in time), this can result in failure to detect interactions and, consequently, yield inaccurate inference about transmission (9). We recommend that decisions about implementation of data collection methods be motivated by (i) the host species, including the frequency of transmission-relevant interactions and how seasonality may impact these; (ii) the parasite of interest, including the infectious

period or environmental persistence of infectious agents; and (iii) the technology itself, including the lifetime of monitoring equipment or the resolution at which individual data points can be recorded. Considering all these factors together will help align the appropriate choice of method and the degree of sampling effort necessary to observe transmission-relevant interactions.

## 4.4 Network analysis and application

Methods for analyzing networks can be grouped into three classes: social network analysis, statistical modeling, and simulation modeling. However, these designations are not mutually exclusive and a network analysis pipeline may include several approaches used in parallel. We will briefly discuss network analysis under each of these three classes and illustrate how networks can be used to identify key individuals or potential pathways of transmission, concluding with applications of network data for management of parasite transmission in animal populations. For a deeper dive into the varied uses of network data to explore parasite transmission in wildlife populations, we recommend the review by White *et al.* (11)

### 4.4.1 Social network analysis

Social network analysis (SNA) is used to quantify aspects of network structure at both the node (individual) and network (population) levels. For those new to SNA, several "how-to" guides provide a general introduction (38), (39). A vast number of network metrics (also called network statistics) have been developed to examine both heterogeneity and patterns in node interactions as well as network structure. Sosa *et al.* (40) provide an overview of the most commonly used social network measures including their uses and interpretations. When SNA is applied to parasite transmission, the goal is often to compare node position in a network to node-level characteristics or disease status (41). Studies may seek to correlate characteristics of highly connected individuals to form generalized conclusions about host factors influencing transmission potential. For example, individuals with high betweenness (i.e., frequently lying on the shortest path between other nodes) play an important role in connecting others in their network and may act as superspreaders. In a study comparing social networks and *E. coli* transmission networks in giraffes, individuals that were highly connected or occupied "bottleneck" positions in the social network tended to occupy the same positions in the transmission network, suggesting that an individual's social association patterns could be used to inform their transmission potential (30).

## 4.4.2 Statistical modeling

Statistical modeling can be applied to networks to conduct hypothesis tests or compare networks, however these analyses require careful consideration due to the non-independence of network data (42), (43). Statistical modeling approaches such as exponential random graph models (ERGMs) (44) or stochastic actor-oriented models (45) enable researchers to investigate factors influencing patterns of node interactions in the network. These approaches can help test hypotheses about what factors drive observed network topology; for example, do edges tend to be formed between same-aged individuals. Statistical network models have been widely used in human social and disease research (46), but tools like ERGMs or probabilistic network modeling (47) are relatively new to animal disease ecology. We suggest the review by Silk *et al.* (48) as a starting point to learn more about these approaches.

## 4.4.3 Simulation modeling

Simulation models ultimately link networks to epidemiology to investigate how parasite transmission might play out in a network of a given structure. Generally, simulation modeling also involves social network analysis and statistical modeling to (i) take empirical network data and identify important factors governing edge formation and network structure, (ii) generate new iterations of the network that reflect these same patterns, and (iii) simulate outbreaks on the new networks to understand how network structure affects epidemiological processes. Simulation modeling can be used retrospectively to investigate the network connections that may explain observed transmission dynamics. For instance, simulation modeling of Serengeti

lion networks during the 1994 canine distemper outbreak was used to suggest the critical role of spillover events from other carnivore hosts in maintaining the epidemic in the lion population (49). Simulation models can also be used proactively to predict transmission dynamics or investigate questions of how seasonal or sickness-induced host behavior may affect outbreak dynamics (50).

## 4.4.4 Network applications for disease management

If animals in a population exhibit heterogeneity in the number of transmission-relevant interactions, networks can be important for decision-making in disease management. Network data suggesting heterogeneity in interactions can be used to inform the scope or timing of management efforts (via vaccination, culling, or isolation) by identifying highly connected individuals or seasonal variation in network connections (51). Simulations on empirically derived networks can also be used to compare the outcomes of different management strategies in ways that are not feasible or ethical in empirical contexts. Robinson *et al.* (52) used network simulations based on empirical networks of interactions to compare two candidate vaccination campaigns in endangered Hawaiian monk seals: (i) target the most connected individuals or (ii) vaccinate any available animal. They found that vaccinating any available animal more quickly led to protective levels of immunity than waiting to vaccinate highly connected seals but required more vaccines to achieve the same decrease in outbreak size. As such, social network analysis and simulation modeling approaches can be powerful tools to combine empirical data, hypothetical scenarios, and epidemiological modeling to devise effective management actions to control infectious disease outbreaks in animal populations.

## 4.5 Synthesis: Networks in the context of parasite–host behavior feedback

Thus far, we have emphasized the use of animal networks for understanding how an individual's position in a network or the overall network topology affects transmission inference. Less often is the reverse relationship considered: how parasite infection influences host behavior and thus network position or overall network topology. Assuming that interaction behavior and parasite infection are independent ignores the possibility of parasite–host behavior feedback and limits our understanding of disease dynamics. As such, incorporating feedback loops into disease ecology, and particularly the study of animal networks, has been highlighted as an important gap in order to more fully understand transmission dynamics ((41), (53), (54); see also Chapter 2 of this volume (55)).

The ways in which data are collected and networks are constructed can often make it difficult to identify and effectively study this feedback in free-ranging wildlife systems. First, networks have primarily been constructed using the interactions of uninfected animals (34). However, empirical studies indicate that sickness behavior can have important—and system-specific—consequences for network connections and parasite transmission, further emphasizing the importance of recording observations on infected animals. For example, some infected animals may interact less or be avoided by uninfected conspecifics, decreasing their network connections (56; see also Chapter 14 of this volume (57)) while in other cases, the reverse is true and sickness-induced lethargy can result in an increase in host interactions (58). Second, studies of wildlife disease often operate with a cross-sectional view of infection status and interactions, requiring researchers to extrapolate the dynamic processes that created these patterns (59). Cross-sectional studies are able to quantify how infected hosts are connected in the network but lack the ability to disentangle correlation and causation.

Addressing these limitations to study parasite–host behavior feedback robustly requires longitudinal monitoring to observe individuals that become infected and to document any associated changes in interaction patterns. Collecting such data via observational studies of wildlife populations is no easy task. Instead, field experiments manipulating infection status (e.g., deworming via anthelmintic) can be a powerful means to explicitly examine how parasite infection affects host interactions. The number of studies explicitly manipulating parasite infection in wildlife to investigate the effects on network structure remains limited (11), but

anthelmintics have been used to study the influence of helminth infection on host population cycles (60) and bacterial infection dynamics (61), providing informative precedents for parasite removal experiments in wildlife.

Even when experiments can be implemented in wildlife systems, persistent issues of small sample sizes, confounding factors, and limited replication may hamper the ability to test the impact of parasite infection on host interactions in a robust manner. In these situations, simulation approaches informed by empirical data can be used to examine the effects of parasite-induced host behavioral change on transmission dynamics at a higher resolution (Box 4.1) (11), (54). Building these models requires knowledge about how infection is likely to alter behavior. In wildlife—parasite systems that are difficult to observe, adequate information to inform models may be limited. Nonetheless, even if a range of possible scenarios can be tested, simulations are a powerful approach to model the population-level effects of infection-induced behavioral change and test the implications of varying behavioral change via sensitivity analyses.

## 4.6  Future directions

This section introduces some alternative approaches to investigate animal networks and parasite transmission. It is worth noting that pursuing future research in these directions does not preclude research using the methods already introduced (see section 4.3). We encourage additional field and experimental studies, for example, using anthelmintics to study parasite—host behavior feedback. Nonetheless, the following are promising avenues to track transmission more precisely, improving our ability to study parasite–host behavior feedback in animal populations.

### 4.6.1 Transmission networks from parasite genomic sequencing

One potential approach to circumvent the challenges of estimating interactions is to use transmission networks based on parasite genomes to improve our understanding of social or interaction-based networks and transmission (63). For example,

### Box 4.1  Parasite–host behavior feedback

Reynolds *et al.* (50) used a simulation approach to model the feedback effects of parasite-induced host behavioral change on rabies transmission in wild raccoons. Different rabies-induced behaviors ("furious" or "dumb") have been observed in wild raccoons which may differentially affect transmission (62). Reynolds *et al.* (50) used interaction data from proximity loggers in a healthy population of wild raccoons to simulate a population with a comparable number and durations of edges. They then investigated a series of outbreak scenarios with rabies-induced behavioral change (Figure 1A–D) where infected racoons displaying "dumb" behavior were simulated via a decrease in number of interactions while "furious" behavior corresponded to an increase.

In general, furious behavior led to a faster spread of rabies and increased final outbreak size while dumb behavior had the opposite effect (Figure 1). When a combination of dumb, furious, and normal behavior was simulated, there was little difference in speed of spread or final outbreak size compared to simulations with no behavioral change (50). This study illustrates how simulations can help identify circumstances under which parasite-induced host behavioral change is likely to be important to transmission outcomes. We recommend researchers conduct such sensitivity analyses, as the insights provided can be helpful in developing new hypotheses and/or identifying the most important data to collect to reduce uncertainty about the role parasite–host behavior feedback plays in shaping transmission dynamics.

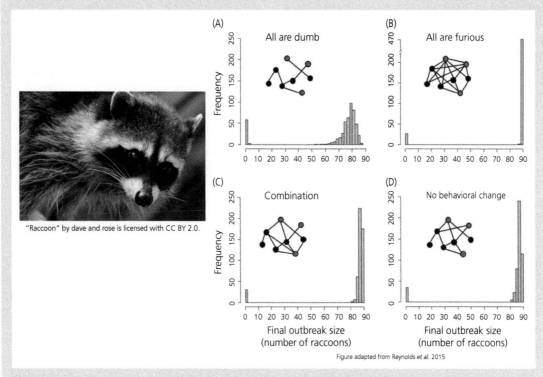

"Raccoon" by dave and rose is licensed with CC BY 2.0.

Figure adapted from Reynolds *et al.* 2015

**Box 4.1, Figure 1** Final rabies outbreak sizes under simulation conditions for a population of suburban raccoons. Panel (A) all infected animals are dumb, (B) all are furious, note y-axis break, (C) one-third are furious, dumb, normal, (D) all are normal. Toy networks illustrate the modeled change to the number of interactions made by infected individuals (represented by red nodes). Photo reproduced under Creative Commons Attribution 2.0 International (CC-BY 2.0) license. Figure adapted from Reynolds *et al.*, 2015. Reprinted by permission of John Wiley & Sons, Inc, © 2015.

several studies have found that genetic relationships among commensal bacterial and viral agents can illuminate social relationships between hosts (30), (64). Such findings, however, have not been consistent across host species and infectious agents (e.g., (65)), so further work is necessary to determine under what conditions infectious agents—including commensal agents—may successfully act as proxies of social or transmission-relevant interactions.

As genomic sequencing technology becomes cheaper and more accessible, advances in software tools can now allow researchers to infer transmission networks directly from high resolution parasite genomes (66). Some transmission network inference approaches can even provide an estimate of the time of infection (67). This is likely to be a coarse or uncertain estimate, and would require parasites with rapid mutation rates, but could provide a temporal window before and after which to look for evidence of behavioral change in studies of parasite-host behavior feedback. Though promising, these transmission network approaches require intensive population sampling, and large genome sizes can introduce computational challenges (but see (67)). Nonetheless, the integration of molecular epidemiology and network analysis is a rapidly developing topic in disease ecology and promises exciting insights into the dynamics of transmission.

### 4.6.2 Intensive non-invasive sampling

A key challenge in studies of parasite–host behavior feedback is determining the time point of infection to improve our understanding of the causal relationships between infection and behavior. Diagnostic field sampling can be challenging in wildlife, especially in cryptic or secretive species, species living in inhospitable habitats, or species of conservation concern where field capture and sampling is particularly risky (35). Advances in non-invasive sampling (e.g., feces (68), saliva (69), carrion- and feces-consuming flies (70)) may help address some of these issues by allowing detection of infection without invasive handling. Longitudinal, repeated non-invasive sampling would increase the probability of detecting infection state conversion and a temporal window for the time of infection. If coupled with direct observations or a single capture event to deploy individual-based monitoring technology, researchers could then analyze behavioral, movement, or interaction data with this known time of infection information, and thereby test for evidence of infection-induced behavioral change in studies of parasite—host behavior feedback.

### 4.6.3 Movement ecology and disease

Increasingly, researchers are calling for integration of methods and theories from movement and disease ecology (see also Chapter 6 of this volume (36), (71), (72)), and these ideas are relevant to studying parasite–host behavior feedback processes. For example, linking movement and disease may help determine if movement patterns are associated with a diseased or infectious state. However, such research questions may require intensive and/or invasive methods in at least a subset of the population either to detect the onset of infection (e.g., body temperature monitoring to detect febrile states as an indicator of infection (73)) or to characterize animal activity acutely (e.g., biological sensor tags to detect locomotion and energetic expenditure or physiological responses like heart rate or ventilation (74)). Intensive non-invasive sampling may be particularly relevant for addressing such issues (see section 4.6.2). In addition, improved statistical approaches or approximations may be particularly relevant for detecting a relationship between disease and changes in movement behavior (e.g., continuous time movement models (75), and step selection functions (76)). Advances in technology can also provide increased resolution of individual-based movement monitoring; for example, applying accelerometers, GPS collars, and proximity logging devices together may be able to improve interrogation of the links between interactions, movement, and behavioral states, albeit with increased financial and data processing costs to implement. Linking movement and disease ecology is likely to be a fruitful avenue for the better understanding of transmission dynamics and parasite–host behavior feedback.

## 4.7 Conclusions

How edges are defined and interactions estimated from spatiotemporal data has the power to affect network topology. Therefore, the process of collecting data and constructing networks must be rooted in an understanding of the biology of the host–parasite system and align with the relevant mode of parasite transmission and the host interactions necessary to facilitate it. Informed choices will best enable meaningful inference about parasite transmission, though it is important to acknowledge that networks of interactions are often, at best, approximations of true transmission networks. And yet, improved technology and modeling frameworks are closing the gap between inferring transmission and directly documenting transmission. Continued advancements in network approaches, transmission inference, infection detection, and movement analyses will open doors to understand transmission processes more fully as a product of feedback between animal behavior and parasite infection, providing powerful insight into generalizable processes driving transmission at an individual and a population level.

## Acknowledgements

The authors would like to thank the editors for the opportunity to contribute to this book.

JM was supported by the National Science Foundation Graduate Research Fellowship (CON-75851, project 00074041). Any opinions, findings, and conclusions or recommendations expressed in this material are those of the authors and do not necessarily reflect the views of the National Science Foundation. MLJG was supported by the Office of the Director, National Institutes of Health (NIH T32OD010993), the University of Minnesota Informatics Institute MnDRIVE program, and the Van Sloun Foundation. The content is solely the responsibility of the authors and does not necessarily represent the official views of the National Institutes of Health. LAW was supported by the National Socio Environmental Synthesis Center (SESYNC) under funding received from the NSF DBI 1639145. M.E.C. was funded by National Science Foundation (DEB-1654609 and 2030509).

## References

1. Thomas JC, Thomas JC, and Weber DJ. *Epidemiologic Methods for the Study of Infectious Diseases*. New York, NY: Oxford University Press, 2001.
2. Anderson RM and May RM. Population biology of infectious diseases: Part I. *Nature*. 1979 Aug;280(5721):361–7.
3. Meyers LA, Pourbohloul B, Newman MEJ, Skowronski DM, and Brunham RC. Network theory and SARS: Predicting outbreak diversity. *J Theor Biol*. 2005 Jan 7;232(1):71–81.
4. Chase-Topping M, Gally D, Low C, Matthews L, and Woolhouse M. Super-shedding and the link between human infection and livestock carriage of *Escherichia coli* O157. *Nat Rev Microbiol*. 2008 Dec;6(12):904–12.
5. VanderWaal KL and Ezenwa VO. Heterogeneity in pathogen transmission: Mechanisms and methodology. *Funct Ecol*. 2016 Oct;30(10):1606–22.
6. White LA, Forester JD, and Craft ME. Covariation between the physiological and behavioral components of pathogen transmission: Host heterogeneity determines epidemic outcomes. *Oikos*. 2018;127(4):538–52.
7. Craft ME. Infectious disease transmission and contact networks in wildlife and livestock. *Philos Trans R Soc B Biol Sci*. 2015 May 26;370(1669):20140107.
8. Richardson TO and Gorochowski TE. Beyond contact-based transmission networks: The role of spatial coincidence. *J R Soc Interface*. 2015 Oct 6;12(111):20150705.
9. Gilbertson MLJ, White LA, and Craft ME. Trade-offs with telemetry-derived contact networks for infectious disease studies in wildlife. *Methods Ecol Evol*. 2021; 12(1): 76–87.
10. Perkins SE, Cagnacci F, Stradiotto A, Arnoldi D, and Hudson PJ. Comparison of social networks derived from ecological data: Implications for inferring infectious disease dynamics. *J Anim Ecol*. 2009 Sep;78(5):1015–22.
11. White LA, Forester JD, and Craft ME. Using contact networks to explore mechanisms of parasite transmission in wildlife. *Biol Rev*. 2017;92(1):389–409.
12. Franks DW, Ruxton GD, and James R. Sampling animal association networks with the gambit of the group. *Behav Ecol Sociobiol*. 2010 Jan 1;64(3):493–503.
13. Grear DA, Luong LT, and Hudson PJ. Network transmission inference: Host behavior and parasite life cycle make social networks meaningful in disease ecology. *Ecol Appl*. 2013;23(8):1906–14.

14. Huyvaert KP, Russell RE, Patyk KA, Craft ME, Cross PC, Garner MG, et al. Challenges and opportunities developing mathematical models of shared pathogens of domestic and wild animals. *Vet Sci.* 2018 Dec; 5(4):92.

15. Wilber MQ, Pepin KM, Campa H, Hygnstrom SE, Lavelle MJ, Xifara T, et al. Modelling multi-species and multi-mode contact networks: Implications for persistence of bovine tuberculosis at the wildlife–livestock interface. *J Appl Ecol.* 2019;56(6):1471–81.

16. Plowright RK, Sokolow SH, Gorman ME, Daszak P, and Foley JE. Causal inference in disease ecology: Investigating ecological drivers of disease emergence. *Front Ecol Environ.* 2008;6(8):420–9.

17. Drewe JA. Who infects whom? Social networks and tuberculosis transmission in wild meerkats. *Proc R Soc B: Biol Sci.* 2010 Feb 22;277(1681):633–42.

18. Leu ST, Sah P, Krzyszczyk E, Jacoby A-M, Mann J, and Bansal S. Sex, synchrony, and skin contact: Integrating multiple behaviors to assess pathogen transmission risk. *Behav Ecol.* 2020 Jun 19;31(3):651–60.

19. Robert K, Garant D, and Pelletier F. Keep in touch: Does spatial overlap correlate with contact rate frequency? *J Wildl Manag.* 2012;76(8):1670–5.

20. Fieberg J and Kochanny CO. Quantifying home-range overlap: The importance of the utilization distribution. *J Wildl Manag.* 2005;69(4):1346–59.

21. Porphyre T, Stevenson M, Jackson R, and McKenzie J. Influence of contact heterogeneity on TB reproduction ratio R0 in a free-living brushtail possum *Trichosurus vulpecula* population. *Vet Res.* 2008 May 1;39(3):1.

22. Schauber EM, Nielsen CK, Kjær LJ, Anderson CW, and Storm DJ. Social affiliation and contact patterns among white-tailed deer in disparate landscapes: Implications for disease transmission. *J Mammal.* 2015 Feb 15;96(1):16–28.

23. Sundaresan SR, Fischhoff IR, Dushoff J, and Rubenstein DI. Network metrics reveal differences in social organization between two fission–fusion species, Grevy's zebra and onager. *Oecologia.* 2007 Feb 1;151(1):140–9.

24. Hamede RK, Bashford J, McCallum H, and Jones M. Contact networks in a wild Tasmanian devil (*Sarcophilus harrisii*) population: Using social network analysis to reveal seasonal variability in social behaviour and its implications for transmission of devil facial tumour disease. *Ecol Lett.* 2009;12(11):1147–57.

25. Silbernagel ER, Skelton NK, Waldner CL, and Bollinger TK. Interaction among deer in a chronic wasting disease endemic zone. *J Wildl Manag.* 2011;75(6):1453–61.

26. Altizer S, Dobson A, Hosseini P, Hudson P, Pascual M, and Rohani P. Seasonality and the dynamics of infectious diseases. *Ecol Lett.* 2006;9(4):467–84.

27. Langwig KE, Frick WF, Reynolds R, Parise KL, Drees KP, Hoyt JR, et al. Host and pathogen ecology drive the seasonal dynamics of a fungal disease, white-nose syndrome. *Proc R Soc B: Biol Sci.* 2015 Jan 22;282(1799):20142335.

28. Cross PC, Lloyd-Smith JO, Bowers JA, Hay CT, Hofmeyr M, and Getz WM. Integrating association data and disease dynamics in a social ungulate: Bovine tuberculosis in African buffalo in the Kruger National Park. *Ann Zool Fenn.* 2004;41(6):879–92.

29. Kinsley AC, Rossi G, Silk MJ, and VanderWaal K. Multilayer and multiplex networks: An introduction to their use in veterinary epidemiology. *Front Vet Sci* 2020;7:596.

30. VanderWaal KL, Atwill ER, Isbell LA, and McCowan B. Linking social and pathogen transmission networks using microbial genetics in giraffe (*Giraffa camelopardalis*). *J Anim Ecol.* 2014;83(2):406–14.

31. Lewis JS, Logan KA, Alldredge MW, Theobald DM, VandeWoude S, and Crooks KR. Contact networks reveal potential for interspecific interactions of sympatric wild felids driven by space use. *Ecosphere.* 2017;8(3):e01707.

32. Silk MJ, Drewe JA, Delahay RJ, Weber N, Steward LC, Wilson-Aggarwal J, et al. Quantifying direct and indirect contacts for the potential transmission of infection between species using a multilayer contact network. *Behaviour.* 2018 Jan 1;155(7–9):731–57.

33. Smouse PE, Focardi S, Moorcroft PR, Kie JG, Forester JD, and Morales JM. Stochastic modelling of animal movement. *Philos Trans R Soc B: Biol Sci.* 2010 Jul 27;365(1550):2201–11.

34. Craft ME and Caillaud D. Network models: An underutilized tool in wildlife epidemiology? *Interdisciplinary Perspectives on Infectious Diseases.* 2011;e676949.

35. Krause J, Krause S, Arlinghaus R, Psorakis I, Roberts S, and Rutz C. Reality mining of animal social systems. *Trends Ecol Evol.* 2013 Sep 1;28(9):541–51.

36. Spiegel O, Anglister N, and Crafton MM. Movement data provides insights into feedbacks and heterogeneities in host–parasite interactions. In Ezenwa VO, Altizer S, and Hall RJ (eds), *Animal Behavior and Parasitism.* Oxford: Oxford University Press, 2022. DOI: 10.1093/os/9780192895561.003.0006.

37. Lavelle MJ, Fischer JW, Phillips GE, Hildreth AM, Campbell TA, Hewitt DG, et al. Assessing risk of disease transmission: Direct implications for an indirect science. *BioScience.* 2014 Jun 1;64(6):524–30.

38. Croft DP, James R, and Krause J. *Exploring Animal Social Networks*. Princeton, NJ: Princeton University Press, 2008, 203.

39. Farine DR and Whitehead H. Constructing, conducting and interpreting animal social network analysis. J Anim Ecol. 2015;84(5):1144–63.

40. Sosa S, Sueur C, and Puga-Gonzalez I. Network measures in animal social network analysis: Their strengths, limits, interpretations and uses. *Methods Ecol Evol*. 2021;12(1):10–21.

41. Godfrey SS. Networks and the ecology of parasite transmission: A framework for wildlife parasitology. Int *J Parasitol Parasites Wildl*. 2013 Dec 1;2:235–45.

42. Croft DP, Madden JR, Franks DW, and James R. Hypothesis testing in animal social networks. *Trends Ecol Evol*. 2011 Oct 1;26(10):502–7.

43. James R, Croft DP, and Krause J. Potential banana skins in animal social network analysis. *Behav Ecol Sociobiol*. 2009 May 1;63(7):989–97.

44. Silk MJ and Fisher DN. Understanding animal social structure: Exponential random graph models in animal behaviour research. *Anim Behav*. 2017 Oct 1;132:137–46.

45. Fisher DN, Ilany A, Silk MJ, and Tregenza T. Analysing animal social network dynamics: The potential of stochastic actor-oriented models. *J Anim Ecol*. 2017;86(2):202–12.

46. Goodreau SM, Kitts JA, and Morris M. Birds of a feather, or friend of a friend? Using exponential random graph models to investigate adolescent social networks. *Demography*. 2009 Feb;46(1):103–25.

47. Yang A, Schlichting P, Wight B, Anderson WM, Chinn SM, Wilber MQ, *et al.* Effects of social structure and management on risk of disease establishment in wild pigs. *J Anim Ecol*. 2021;90(4):820–33.

48. Silk MJ, Croft DP, Delahay RJ, Hodgson DJ, Weber N, Boots M, *et al.* The application of statistical network models in disease research. *Methods Ecol Evol*. 2017;8(9):1026–41.

49. Craft ME, Volz E, Packer C, and Meyers LA. Distinguishing epidemic waves from disease spillover in a wildlife population. *Proc R Soc B: Biol Sci*. 2009 May 22;276(1663):1777–85.

50. Reynolds JJH, Hirsch BT, Gehrt SD, and Craft ME. Raccoon contact networks predict seasonal susceptibility to rabies outbreaks and limitations of vaccination. *J Anim Ecol*. 2015;84(6):1720–31.

51. Rushmore J, Caillaud D, Matamba L, Stumpf RM, Borgatti SP, and Altizer S. Social network analysis of wild chimpanzees provides insights for predicting infectious disease risk. *J Anim Ecol*. 2013;82(5):976–86.

52. Robinson SJ, Barbieri MM, Murphy S, Baker JD, Harting AL, Craft ME, *et al.* Model recommendations meet management reality: Implementation and evaluation of a network-informed vaccination effort for endangered Hawaiian monk seals. *Proc R Soc B: Biol Sci*. 2018 Jan10;285(1870).

53. Ezenwa VO, Archie EA, Craft ME, Hawley DM, Martin LB, Moore J, *et al.* Host behaviour–parasite feedback: An essential link between animal behaviour and disease ecology. *Proc R Soc B: Biol Sci*. 2016 Apr 13;283(1828):20153078.

54. Hawley DM and Altizer SM. Disease ecology meets ecological immunology: Understanding the links between organismal immunity and infection dynamics in natural populations. *Funct Ecol*. 2011;25(1):48–60.

55. Hawley DM and Ezenwa VO. Parasites, host behavior and their feedbacks. In Ezenwa VO, Altizer S, and Hall RJ (eds), *Animal Behavior and Parasitism*. Oxford: Oxford University Press, 2022. DOI: 10.1093/oso/9780192895561.003.0002.

56. Croft DP, Edenbrow M, Darden SK, Ramnarine IW, van Oosterhout C, and Cable J. Effect of gyrodactylid ectoparasites on host behaviour and social network structure in guppies *Poecilia reticulata*. *Behav Ecol Sociobiol*. 2011 Dec 1;65(12):2219–27.

57. Lopes PC, French SS, Woodhams DC, and Binning SA. Infection avoidance behaviors across vertebrate taxa: Patterns, processes and future directions. In Ezenwa VO, Altizer S, and Hall RJ (eds), *Animal Behavior and Parasitism*. Oxford: Oxford University Press, 2022. DOI: 10.1093/oso/9780192895561.003.0014.

58. Franz M, Kramer-Schadt S, Greenwood AD, and Courtiol A. Sickness-induced lethargy can increase host contact rates and pathogen spread in water-limited landscapes. *Funct Ecol*. 2018;32(9):2194–204.

59. Heisey DM, Joly DO, and Messier F. The fitting of general force-of-infection models to wildlife disease prevalence data. *Ecology*. 2006 Sep 1;87(9):2356–65.

60. Pedersen AB and Fenton A. The role of antiparasite treatment experiments in assessing the impact of parasites on wildlife. *Trends Parasitol*. 2015 May 1;31(5):200–11.

61. Ezenwa VO and Jolles AE. Opposite effects of anthelmintic treatment on microbial infection at individual versus population scales. *Science*. 2015 Jan 9;347(6218):175–7.

62. Rosatte R, Sobey K, Donovan D, Bruce L, Allan M, Silver A, *et al.* Behavior, movements, and demographics of rabid raccoons in Ontario, Canada: Management implications. *J Wildl Dis*. 2006 Jul 1;42(3):589–605.

63. Gilbertson MLJ, Fountain-Jones NM, and Craft ME. Incorporating genomic methods into contact networks to reveal new insights into animal behaviour and infectious disease dynamics. *Behaviour*. 2018 Jan 1;155(7–9):759–91.

64. Bull CM, Godfrey SS, and Gordon DM. Social networks and the spread of Salmonella in a sleepy lizard population. *Mol Ecol*. 2012;21(17):4386–92.

65. Blyton MDJ, Banks SC, Peakall R, and Gordon DM. High temporal variability in commensal *Escherichia coli* strain communities of a herbivorous marsupial. *Environ Microbiol*. 2013 Aug;15(8):2162–72.

66. Hall MD, Woolhouse MEJ, and Rambaut A. Using genomics data to reconstruct transmission trees during disease outbreaks. *Rev Sci Tech Int Off Epizoot*. 2016 Apr;35(1):287–96.

67. Didelot X, Fraser C, Gardy J, and Colijn C. Genomic infectious disease epidemiology in partially sampled and ongoing outbreaks. *Mol Biol Evol*. 2017 Apr 1;34(4):997–1007.

68. Cristescu RH, Miller RL, Schultz AJ, Hulse L, Jaccoud D, Johnston S, *et al.* Developing noninvasive methodologies to assess koala population health through detecting Chlamydia from scats. *Mol Ecol Resour*. 2019 Jul 1;19(4):957–69.

69. Evans TS, Gilardi KVK, Barry PA, Ssebide BJ, Kinani JF, Nizeyimana F, *et al.* Detection of viruses using discarded plants from wild mountain gorillas and golden monkeys. *Am J Primatol*. 2016;78(11):1222–34.

70. Hoffmann C, Stockhausen M, Merkel K, Calvignac-Spencer S, and Leendertz FH. Assessing the feasibility of fly based surveillance of wildlife infectious diseases. *Sci Rep*. 2016 Nov 30;6(1):37952.

71. Dougherty ER, Seidel DP, Carlson CJ, Spiegel O, and Getz WM. Going through the motions: Incorporating movement analyses into disease research. *Ecol Lett*. 2018;21(4):588–604.

72. White LA, Forester JD, and Craft ME. Dynamic, spatial models of parasite transmission in wildlife: Their structure, applications and remaining challenges. *J Anim Ecol*. 2018;87(3):559–80.

73. Timsit E, Assié S, Quiniou R, Seegers H, and Bareille N. Early detection of bovine respiratory disease in young bulls using reticulo-rumen temperature boluses. *Vet J*. 2011 Oct 1;190(1):136–42.

74. Wilson ADM, Wikelski M, Wilson RP, and Cooke SJ. Utility of biological sensor tags in animal conservation. *Conserv Biol*. 2015;29(4):1065–75.

75. Hooten MB and Johnson DS. Basis function models for animal movement. *J Am Stat Assoc*. 2017 Apr 3;112(518):578–89.

76. Thurfjell H, Ciuti S, and Boyce MS. Applications of step-selection functions in ecology and conservation. *Mov Ecol*. 2014 Feb 7;2(1):4.

77. Clay CA, Lehmer EM, Previtali A, St. Jeor S, and Dearing MD. Contact heterogeneity in deer mice: Implications for Sin Nombre virus transmission. *Proc R Soc B: Biol Sci*. 2009 Apr 7;276(1660):1305–12.

78. Farine DR, Firth JA, Aplin LM, Crates RA, Culina A, Garroway CJ, *et al.* The role of social and ecological processes in structuring animal populations: A case study from automated tracking of wild birds. *R Soc Open Sci*. 2015;2(4):150057.

# CHAPTER 5

# Collective behavior and parasite transmission

Carl N. Keiser

## 5.1 Introduction

The fields of animal behavior and infectious diseases are both typified by multiscale research perspectives, combining research on individuals, social groups, populations, and communities. For example, the collective movements of fish schools and bird flocks are emergent, self-organizing properties of the behaviors of individual group members. Individual group members behave following sets of rules which vary depending on their own traits and the actions of individuals around them. Likewise, the dynamics of an infectious disease outbreak are emergent properties of individual-level host–parasite interactions. Individual hosts become infected based on their own behavioral and immunological traits, but transmission potential depends jointly on the actions of interacting individuals. One cannot truly understand the dynamics at one scale (e.g., individual) without incorporating information from the other (e.g., group, population).

How, then, do these multi-tier systems of **collective behavior** (see Box 5.1 for glossary) and disease interact? The relationship between individual behavior and infection risk is well-studied (1), (2), as is the effect of parasites on individual behavior traits (3). Less understood, however, are the ways in which the collective behavior of social groups influences the transmission dynamics of parasites. It is important to understand links between collective behavior and parasitism because sociality is such a pervasive and important phenomenon

characterizing animal life. Tight-knit social groups incur unique benefits (**social immunity**, resource acquisition) and costs (parasite exposure), all of which have the potential to impact parasite transmission. Here, I use examples from diverse animal societies to review (i) how individual and group behavior underlies the relationship between collective behavior and parasite transmission; (ii) mechanisms by which groups can modulate the benefits of collective behaviors and minimize the risk of parasitism; and close by (iii) highlighting some contemporary technological and conceptual advances pushing the frontiers of research in collective behavior and disease.

## 5.2 Individual differences and collective outcomes in behavior and disease

Individuals differ from each other in countless ways: body size, body condition, hunger, behavioral phenotypes, experience, parasite load, etc. How, then, do groups of animals produce concerted group-level outcomes? For an anthropocentric yet familiar example: how does a group composed of friends with different tastes, hunger states, and local experiences effectively choose a restaurant? Although this conundrum often results in long-winded discussions rarely concluding in a decision preferred by all members (4), the vast majority of animal societies arrive at collective decisions based on sets of individual-level rules. For example, studies on fish shoals and bird flocks suggest that individuals are attracted to others at long distances,

Carl N. Keiser, *Collective behavior and parasite transmission*. In: *Animal Behavior and Parasitism*. Edited by Vanessa O. Ezenwa, Sonia Altizer, and Richard J. Hall, Oxford University Press. © Oxford University Press (2022). DOI: 10.1093/oso/9780192895561.003.0005

## Box 5.1 Glossary

**Box 5.1, Figure 1** Examples of collective behavior in animal groups with varying social systems. Photo sources: Wikimedia Commons. Clockwise from top left: locust swarm by Iwoelbern, social spiders by Wynand Uys, termite mound by Bernard Gagnon, wildebeest migration by T.R. Shankar Raman, shoaling fish by lifelish.

**What is collective behavior?** Through direct or indirect interactions between individuals, groups of animals produce coordinated actions at the group level. Collective behaviors are unachievable by individuals alone, and thus many collective behaviors represent emergent properties of social groups. As pictured earlier, collective behavior can emerge in locomotion, decision-making, group foraging and defense, construction of built environments, and more.

**Allogrooming:** Unidirectional and/or reciprocal grooming between individuals, including the removal of ectoparasites and debris.

**Collective personality:** Temporally consistent differences between groups of individuals in the execution of collective behavior.

**Infection–information trade-off:** Social interactions are the basis of beneficial information spread and harmful parasite transmission. A trade-off emerges if these two processes are regulated via the same social interactions.

**Keystone individual:** Akin to the keystone species concept of community ecology, keystone individuals are those that exert an inordinate influence over dynamics at higher levels of biological organization (social groups, communities, etc.).

**Social context:** Synonymous with "social environment," conditions of the social group in which an individual resides (e.g., group size, sex ratio, group phenotypic composition), which often influence how individuals behave.

**Social immunity:** Antiparasitic defenses mounted by groups to protect individuals against disease and therefore protect the group from the loss of individuals or the transmission of parasites. Some researchers use the term social immunity exclusively in the context of collective behaviors in eusocial insects and highly complex primate societies, whereas others refer to social immunity more broadly as any immune response or antiparasitic defense that benefits others.

**Social fulcrum hypothesis:** In cases where trade-offs exist between the execution of a collective behavior (e.g., foraging, exploration) and the transmission of parasites, herein I propose the hypothesis that groups can resolve this trade-off by shifting the relative composition of different phenotypes within the group (see section 5.4 for more details).

**Social heterosis:** Benefits shared by group-mates via representation of diverse genotypes or phenotypes relative to monotypic groups.

repulsed by individuals at close distances, and align with group-mates depending on orientation and speed (5), (6). Most importantly, perhaps, are the rules regarding the source of social information: focal individuals respond to either every individual within a certain *distance* (7), a certain *number* of individuals (8), or only their single *nearest neighbor* (5). However, as we understand more about behavioral variation among individuals, contemporary studies aim to generate frameworks for how individual heterogeneity influences the mechanisms by which collective behaviors are organized (9)–(11).

Interestingly the ways in which individual behaviors contribute to the execution of collective behaviors can vary in response to infection. In shoaling fishes, individuals taking leading positions have a disproportionate influence over navigational decision-making (12). It has been hypothesized that individuals assume leading positions when most in need of finding a particular resource or locality ("lead according to need," (13)), as nutritionally deprived fish often assume leading positions, acquire more food, and then retreat to posterior positions (14), (15). However, evidence also suggests that more experienced (16) or risk-tolerant fish (17) repeatedly take leading positions. Interestingly, Killifish infected with the trematode *Crassiphiala bulboglossa* and sticklebacks infected with the microsporidian *Glugea anomala* are both more likely to take leading positions in their shoals (18), (19) (Figure 5.1). The foraging benefits of a leading position may help ameliorate the deleterious effects of parasite infection, or perhaps leaders are simply more likely to encounter parasites. Experimental infections in laboratory shoals can help differentiate these competing hypotheses.

Just as individual-level differences in behavior show broad explanatory power for various topics in behavioral ecology, individual variation in disease susceptibility and transmission potential have been at the forefront of infectious disease research for decades (e.g., (20)). The influx of studies focusing on consistent individual differences in behavior (i.e., animal personalities, behavioral syndromes (21), (22)) over the past two decades has laid the framework for uniting animal behavior and research on host heterogeneity in wildlife diseases with multiscale perspectives. Differences among individuals on axes of behavioral variation like activity, aggressiveness, and sociability that underlie ecological outcomes like exploration of new environments (23) and predator–prey dynamics (24) similarly influence host–parasite dynamics (1), (25). Although most personality studies on wild-caught animals are correlative, studies have demonstrated that more exploratory (26) or active (27) individuals are more likely to encounter and acquire parasites. Of course, correlations between behavior and immune

**Figure 5.1** Example depicting one relationship between parasitism and collective behavior. Killifish in shoals where the majority of individuals are infected show a phalanx-like shoal formation relative to the direction of travel (↑) compared to the processional shoal of majority uninfected groups. Adapted from Ward *et al*. (2002) (19).

traits likely play a role in behavior-parasitism relationships (2), (28), like the positive relationship between boldness and immunity in crickets (29).

The most extreme examples of individual variation are those where one or a few individuals exert an inordinately large influence over group outcomes, termed **keystone individuals** (30). For example, group exploration in guppies is not driven by social conformity of the entire group, but rather by the behavior of the least active member of the shoal (31). Keystone individuals may be beneficial to the groups in which they reside, like a small subset of "elite" workers in *Temnothorax* spp. that contribute to the majority of work in colony tasks (32). However, keystone individuals may also be detrimental to their groups, like disease superspreaders, where a large number of infections are caused by a small subset of infectious individuals (20). Martin *et al.* (33) hypothesized that correlated suites of behavioral and physiological traits associated with parasite exposure, susceptibility, and transmission have the potential to generate "extremely competent" individuals that have disproportionate impacts on disease spread, for example by generating an excess of infections. The degree to which underlying traits predict the influence of keystone individuals on both collective behavior and disease dynamics is an overlooked but important phenomenon that requires deeper study.

Some notable case studies have identified traits which may underlie keystone individuals' joint influence over collective behaviors and disease. Sapolsky and Share (34) describe in detail a case where a number of highly aggressive male baboons, *Papio anubis*, ate contaminated meat from a garbage dump and died of bovine tuberculosis infections. The death of highly aggressive males left behind groups with altered sex ratios and a greater representation of non-aggressive survivors that persisted for over a decade. Here, aggressiveness *per se* (not dominance rank or age) predicted which individuals fed at the high-risk resource patch and subsequently engaged in more dominance interactions between infected and susceptible baboons (34). In the social spider *Stegodyphus dumicola*, the presence of highly risk-tolerant or "bold" individuals in a colony is associated with groups attacking prey more rapidly (35). When these potentially influential individuals are exposed to a cocktail of harmful cuticular bacteria, collective prey capture is dampened even in the absence of spider mortality (36). Thus, collective behavior is dampened not because of the loss of participating individuals, but potentially because keystone individuals behaved differently after bacterial exposure and lost their influence over group-mates (e.g., (37)). Using simulations parameterized with data from laboratory colonies of *S. dumicola*, Pinter-Wollman *et al.* (38) suggest that keystone individuals might alter group-mates' behavior via their interaction patterns which lead to trade-offs between disease risk and cooperative prey attack.

## 5.3 Collective behavior and group infection dynamics

Just as behavioral traits like sexual promiscuity, sociality, or territory size can be important predictors of parasite risk for individuals, the collective traits of social groups can similarly predict the likelihood of disease outbreaks therein. Group-level traits like density are often predictors of parasite transmission (39), and "crowding" behavior in beetle larvae (*Plagiodera versicolora*) is positively associated with group-level mortality from parasitoids (40). However, the effects of individual traits on individual infection risk will not always scale linearly to group behavior and group transmission dynamics. For example, *Poecilia reticulata* guppies experience female-biased parasite intensity, but no differences in disease dynamics are found between single-sex and mixed-sex groups (41). This is because the expression of individual behaviors, and their associated relationships with infection risk, change depending on **social context** (42). For example, glucocorticoids have immunosuppressive effects on low-ranking but not high-ranking baboons (43), and zebra finches respond to immune challenge by reduced activity level, but only in isolation; birds in a social setting do not exhibit this change in behavior (44). Therefore, studying group-level traits (group size, group composition, collective behaviors) and how individual traits contribute to them is important in predicting group-level disease dynamics.

In some cases, collective behaviors and infectious disease dynamics may be driven by the same underlying organizational forces. Modular social network structures, where social interactions are more likely to occur within subgroups rather than between subgroups, are hypothesized to promote the evolution of cooperation (45), (46), and the severity of disease outbreaks decline at higher levels of network modularity (47). Similarly, the spatial and temporal separation of interactions between nurses (specializing on brood care) and foragers (specializing on food collection) in eusocial insect colonies may serve as a means to organize division of labor (48) and to restrict parasite transmission (49).

Particularly interesting cases are those where a collective behavior meant to decrease infection risk can also serve as a route for parasite transmission. **Allogrooming**, where individuals clean debris and ectoparasites off group-mates, results from emergent, self-organizing social behaviors that reduce parasite burden in many animal groups (50). *Formica* ants increase allogrooming when exposed to entomopathogenic fungal spores (51) and allogrooming decreases tick load in baboons (52). This collective behavior serves, in addition to social bonding in some mammals, to reduce parasite burden yet also facilitates additional routes for parasite transmission (53). However, in some animal societies where allogrooming and dominance are positively correlated (e.g., meerkats (54)) certain allogrooming strategies (e.g., indiscriminate allogrooming versus preferential allogrooming based on social connectedness) are ineffective in parasite removal and may rather serve to reinforce social bonding (50), (55). In contrast, in other societies, like baboons, allogrooming is both effective at parasite removal and positively correlated with dominance (52).

Allogrooming represents a promising phenomenon with which to test hypotheses linking the execution of collective behaviors with potential antiparasitic defenses across species with different systems of social organization. For example, in which social systems do the consequences of collective behaviors like allogrooming on group cohesion and parasitism align versus misalign? Are these contrasts driven solely by social organization, or under higher parasite intensity will these outcomes become aligned? Allogrooming is a major component of behavioral disease defenses in eusocial insects, referred to collectively as **social immunity**,

and how colonies execute various components of social immunity to optimize collective behavior and disease defenses is another interesting avenue of research (Box 5.2).

Extended phenotypes of animal collective behavior, like the built environment of social groups, can also influence infectious disease dynamics (56). Nest architecture has been a focus of evolutionary parasitology in eusocial insects, as the physical structures around individuals directly constrain the frequency, duration, and location of social interactions (57)–(59). Some features of nest architectures, like subdivision into separate chambers (57) or small nest entrances (60), have been hypothesized to reduce parasite transmission, but some experimental studies have found no effect of colony architecture on disease prevalence (e.g., (61)). The built environment represents an interesting phenomenon for the study of collective behavior and parasite transmission because the disease consequences of the collective behaviors involved in constructing that environment are temporally offset. That is, a social group builds a nest whose features will influence future social interaction patterns therein, and those patterns may differentially impact parasite transmission within groups (62).

## 5.4 Performing collective behaviors while minimizing parasite transmission

Although collective behavior and parasite transmission are often studied separately, these group-level outcomes may be interrelated in some cases. The collective outcomes of social groups are driven, in part, by the composition of individual phenotypes within the group. Groups of social spiders containing more risk-prone or bold individuals attack prey more quickly (74), and guppy shoals containing both bold and shy individuals experience increased collective foraging success compared to monotypic groups (75). However, the optimal mixture of phenotypes for one task may be detrimental in other contexts (Figure 5.2C). For example, the **infection–information trade-off** describes the trade-off between the sharing of beneficial information and the transmission of parasites via social interactions (76). Evans *et al.* (77) suggest that modular social network structures and long-term social bonds can promote the sharing of information

## Box 5.2  Social immunity as a collective behavioral syndrome

Despite living in dense groups of highly related individuals, eusocial insects are remarkably proficient at reducing the risk of disease outbreaks within the colony (63). Eusocial insects mitigate outbreak risk via collective behaviors referred to as **social immunity** (64); a set of colony-level protections against parasites performed by workers either individually (e.g., corpse removal; antimicrobial secretions), in dyadic interactions (e.g., allogrooming), or collectively (e.g., social fever; altered social network structure) (65), (66). A few studies have quantified aspects of social immunity over time and identified consistent among-colony differences in the expression of social immunity under identical environmental conditions (i.e., social immunity as "**collective personalities**" (121)). For example, honey bees exhibit among-colony differences in hygienic behaviors which are repeatable across years (67), and defensive resistance against *Varroa* mites is a repeatable and heritable trait at the colony level (68). Social immunity traits also correlate with other collective behaviors, like a negative relationship between corpse-removal behavior and nest relocation in *Temnothorax* ants (69) and a positive relationship between corpse-removal and foriging activity in honey bees (70). Therefore, factors of social immunity may similarly be related to one another in collective behavioral syndromes (correlations between collective personality traits), as has been found in other collective traits in ant societies (71).

For example, in the hypothetical population depicted in Figure 1, colonies that exhibit more stringent parasite avoidance or selective nest entry also express increased hygienic corpse-removal behaviors, but show decreased time investment in allogrooming. Thus, depending on local parasite pressure, colonies may adjust the relative expression of different social immunity traits as alternative strategies to achieve a necessary level of colony protection. There has not yet been a study which explicitly addressed collective immune syndromes and their survival/fitness consequences in the face of parasitism, though some studies have shown that social immunity is negatively correlated with individual immunity (72). This suggests that colonies may regulate the expression of social immunity based on the physiological immunocompetence of workers therein. Cassidy *et al.* (73) found that *T. curvispinosus* ant colonies containing workers with weaker individual immune defenses showed increased social immunity (faster corpse removal). Future studies should test whether the relative investment in social versus individual immunity, or relationships between social immune traits, change across time and space based on local parasite pressures.

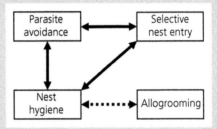

**Box 5.2, Figure 1** Conceptual map depicting a hypothetical collective behavioral syndrome involving four social immunity traits, where arrows represent positive correlations and dashed lines represent negative correlations between traits.

while mitigating parasite transmission. Notably, they predict that behavioral plasticity modifying social interactions in the presence of parasites is "a key mechanism by which this balance between the costs and benefits of being highly socially connected is mediated." (77) Therefore, trade-offs in collective outcomes may be resolved via the adaptive allocation of individual phenotypes within groups. For example, the trade-off between exploiting known resource patches and exploring new patches is mediated in honey bees via the mixture of different "finder" and "refiner" learning phenotypes (78).

Here, I propose the "**social fulcrum hypothesis**," where shifting phenotypic composition is a mechanism by which groups can adaptively modulate the competing outcomes of collective behavior and disease risk (Figure 5.2D). Imagine a case where (i) groups containing more aggressive individuals outperform groups with fewer aggressive individuals in collective foraging, (ii) parasites are transmitted via aggressive interactions, and (iii) the benefits garnered from foraging are outweighed by the costs associated with parasitism. In the presence of parasites, the optimal group composition

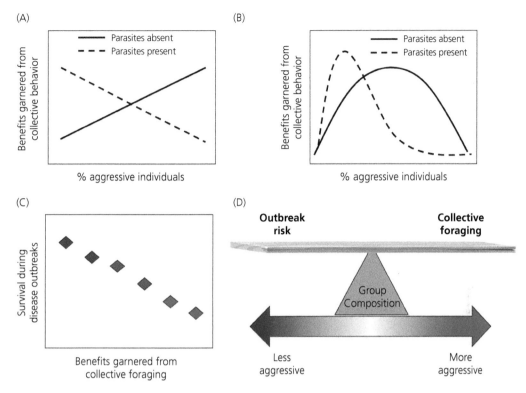

**Figure 5.2** (A, B) Hypothetical relationships between group composition (the percentage of aggressive individuals in a group) and the benefits garnered from the execution of a given collective behavior like cooperative hunting. The optimal group composition may change depending on the presence of transmissible parasites, and these relationships may be (A) linear or (B) nonlinear. (C) Proposed trade-off between the benefits garnered from collective foraging and survival during disease outbreaks. Redder colors represent groups with more aggressive individuals while violet colors represent groups with fewer aggressive individuals. (D) The social fulcrum hypothesis of group composition posits that group phenotypic composition will shift towards the optimum to balance opposing demands of collective behavior and disease risk.

should shift from aggressive-dominated to potentially mixed behavioral composition. Regardless of whether the effect of group composition on collective behavior is linear (Figure 5.2A) or non-linear (Figure 5.2B), shifts in phenotypic composition may materialize via two non-mutually exclusive mechanisms: (i) there may be *local adaptation* in host–parasite coevolution as in geographic mosaic theory (79) where the collective behavior of social groups differs among populations depending on parasite pressure (Figure 5.3A); or (ii) groups may shift group phenotypic composition as a result of prevailing parasite cues (e.g., across a "landscape of disgust"; (80)) (Figure 5.3B). Shifts in group composition could occur via numerical changes in the relative representation of different phenotypes,

as in shifting caste ratios in response to competition in ants (81) and wasps (82) or individual phenotypic plasticity may shift group composition towards the optimum, as in task switching in harvester ants (83) or individualized changes of specialized roles in social groups (84). Alternatively, numerical changes in group composition may occur due to phenotype-biased mortality. The motivation for groups to modulate group composition will also depend on whether collective behaviors are associated with the transmission of *beneficial* microorganisms (85) or anti-parasitic substances (63). Currently, I am unaware of any direct evidence in support of the social fulcrum hypothesis. Rather, several systems provide indirect evidence from separate experiments that group

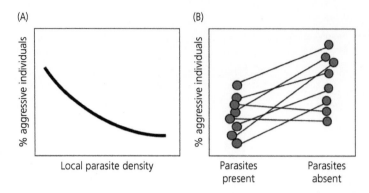

**Figure 5.3** Two ways by which the social fulcrum hypothesis of group composition could manifest. (A) Local adaptation in host–parasite coevolution where the percentage of aggressive individuals in social groups differs among populations, and therefore the execution of collective behaviors also varies, depending on local parasite pressure (B) Active shifts in group composition, and thereby changes in collective behavior, as a result of parasite presence. Variation in group composition in the absence of parasites may be due to prey availability, whereas variation in group composition in the presence of parasites may be due to differences in parasite pressure. Shifts in group composition may be numerical via differential mortality from parasite infection, or individual phenotypic plasticity may adaptively shift group composition.

composition underlies both collective behavior and disease dynamics (Table 5.1). For example, the same group compositions that benefit social spider group foraging (74), (86) also increase the transmission of cuticular bacteria (87). And female-biased groups of *Drosophila melanogaster* exhibit more cohesive aggregation when choosing food patches (124), but experience more severe outbreaks of a fungal pathogen (42). Future studies can utilize these systems to test for direct trade-offs between collective behaviors and disease risk, and whether phenotypic composition varies among groups to resolve these conflicting demands.

## 5.5 Frontiers in collective behavior and disease

Researchers interested in the interface between collective behavior and parasite transmission are fortunate to have multiple fields from which theory and methodology can be drawn. Here I highlight some of the current conceptual advances (e.g., parasite-collective behavior feedbacks) and methodological advances (e.g., multi-level network modeling, animal tracking) in research uniting collective behavior and parasitism, along with some currently unanswered questions to inspire future research.

### 5.5.1 Parasite-collective behavior feedbacks

Ezenwa *et al.* (89) provide a framework for studying feedbacks between host behavior and parasite infection where there is a reciprocal exchange between host behavior and parasitism. Host behavior influences parasite infection risk, parasite infection influences host behavior, and these dynamics do not occur in isolation of each other. The framework of behavior–parasitism feedbacks can similarly be applied to feedbacks between parasite transmission and the execution of collective behaviors. For example, aggregation in locusts can increase the transmission of the microsporidian parasite *Paranosema locustae*, though infection suppresses the hindgut bacteria that produce aggregation pheromones in their locust hosts, thereby preventing swarming behavior (Box 5.3; (88)). Future research should identify systems where there is evidence for reciprocal effects of collective behavior on parasite transmission and parasite presence on collective behavior, and use longitudinal observations to identify whether these effects are indeed linked via a feedback loop. Then, time series analyses can be used to test for temporal associations between collective behavior and parasite prevalence over time or parameterize simulations to generate predictions regarding these potentially reciprocal interactions.

**Table 5.1** Examples of study systems in which group composition effects have been described as predictors of both collective behavior and disease. Each of these systems represent promising cases in which the social fulcrum hypothesis may be tested directly. Group composition in these examples is broken up into mixtures of behavioral phenotypes, group genotypic composition, and demographics like caste ratio, sex ratio, and age class distribution.

| Host species | Group composition | Collective behavior outcome | Disease outcome | Notes |
|---|---|---|---|---|
| Social spider, *Stegodyphus dumicola* | Bold and shy behavioral types | More bold spiders in colony increase attack speed and repeatability of collective foraging (74), (86) | More bold spiders increase bacterial transmission rates (87) | Bacteria used in transmission experiments were naturally occurring and likely non-pathogenic |
| Cavity-nesting ants, *Temnothorax* spp. | Aggressive and non-aggressive behavioral types | More aggressive individuals in colony associated with more rapid nest relocation behavior (112) | Colonies with greater nest relocation tendency have greater immune defenses (69) | An example of a non-contrasting outcome, though positive selection for "all-aggressive" colonies is unlikely due to **social heterosis** (113), (114) |
| Honey bee, *Apis mellifera* | Aggressive and non-aggressive behavioral types | Colonies containing more aggressive bees exhibit more rapid nest defense (115) | Aggressive and mixed-phenotype colonies exhibit enhanced nest hygiene compared to non-aggressive colonies (115) | This was a single study that manipulated worker composition and measured collective behaviors related to colony defense and nest hygiene |
| Guppy, *Poecilia reticulata* | Bold and shy behavioral types | Shoals containing bold individuals discover novel food sources more quickly (116) | Shy fish acquire more ectoparasites within shoals (117) | A potential case where phenotype-biased illness/mortality underlies shifts in group composition |
| Threespine sticklebacks, *Gasterosteus aculeatus* | Sociability and boldness scores | Trait compositions explain the emergent structure and coordination of self-organized shoaling, also driving foraging success (118) | Collective movement in pairs of infected fish is less cohesive and coordinated than non-infected pairs; mixed pairs largely led by the non-infected fish (119) | One piece of evidence missing in this example is whether shoals containing different mixtures of behavioral types are more or less likely to encounter parasites and experience different rates of transmission/infection |
| Honey bee, *Apis mellifera* | Genotypic composition | Genetically diverse colonies show enhanced collective foraging and weight gain compared to uniform colonies (120) | Colonies containing workers from breeding lines predisposed to hygienic behavior perform more hygienic behaviors (121) | Genetic variation may benefit colonies under some circumstances, but if parental genotypes not selected for hygienic behavior, then genetic composition may have maladaptive effects on the distribution of anti-parasite behavior |

*continued*

**Table 5.1** *Continued*

| Host species | Group composition | Collective behavior outcome | Disease outcome | Notes |
|---|---|---|---|---|
| Bumble bee, *Bombus impatiens* | Body size variation | Colony body size variation does not affect colony thermoregulation, where bees fan their wings to cool the nest after heating. | Colonies with similarly sized bees outperform those with a wider range of body sizes in corpse-removal behavior. | This was a single study that manipulated worker body size composition and measured collective behaviors related to colony thermoregulation and nest hygiene (122) |
| Cavity-nesting ants, *Temnothorax curvispinosus* | Queen presence | Queenless colonies discover food more slowly and exhibit less brood-care behavior. | Queenless colonies experience more rapid disease-associated mortality. | Single study; queenless colonies that exhibited more brood care (i.e., behaved more like queenright colonies) experienced less severe outbreaks (123) |
| Guppy, *Poecilia reticulata* | Sex ratio | Male-biased sex ratios reduce group cohesion (males disperse; (124)) | Despite female-biased parasite intensity, no difference in disease dynamics between single-sex and mixed-sex groups (41) | Groups may benefit from female-biased sex ratios, but this depends on whether the benefits of group cohesion are sex-biased |
| Fruit fly, *Drosophila melanogaster* | Sex ratio | Female-biased groups exhibit more cohesive aggregation when choosing food patches (125) | Female-biased social groups experience more severe outbreaks of a fungal pathogen (42) | Promising system to test the social fulcrum hypothesis, as one metric of group composition (sex ratio) affects collective behavior and disease risk, with a potential trad-eoff in group outcomes |

**Box 5.3  Parasite-collective behavior feedbacks**

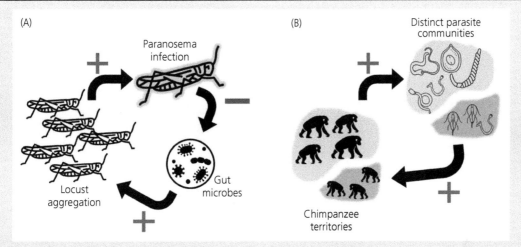

(A)    (B)

**Box 5.3, Figure 1** Potential positive (A) and negative (B) feedbacks between collective behavior and parasitism. Icons obtained from the Noun Project. Grasshopper by Yu Luck; chimpanzee by Abby; parasites by Olena Panasovska.

Ezenwa *et al*. (89) provide a framework for studying feedback between host behavior and parasite infection. This framework can similarly be applied to feedbacks between parasite transmission and the execution of collective behaviors. Figure 1A gives an example of a negative feedback between parasite transmission and collective behavior. The microsporidian parasite *Paranosema locustae* can be transmitted horizontally in aggregations of their locust host (90). Microsporidian infection suppresses the hindgut bacteria responsible for producing aggregation pheromones, thereby reducing swarming tendencies generated through aggregation and reducing horizontal transmission. Figure 1B shows an example of a positive feedback between parasite transmission and collective behavior. Chimpanzees maintain territories via aggressive interactions at territory borders. Within territories, chimpanzees accumulate locally abundant parasites (91) and parasite load increases with levels of aggression/dominance via immunosuppression from testosterone and cortisol (92). This establishes a potential positive feedback between group-level aggression and parasite transmission.

### 5.5.2  Multi-level network modeling

The analysis of social networks has been an invaluable tool in the study of social behavior and infectious disease epidemiology (93). One criticism of traditional social network analysis is that a connection (edge) between two individuals (nodes) is often meant to represent a single action, like grooming, information transfer, bodily contact, or mere proximity. However, this singular view of social behavior ignores the multifaceted nature of social interactions. The advent of multi-level network modeling integrates multiple network topologies atop one another either simultaneously or over time to address interrelatedness between different contexts of social interactions (94). For example, one could use multi-level network models to ask how different types of social interaction (e.g., allogrooming versus cooperative hunting) underlie the transmission of different pathogens, and how those transmission dynamics may interact (95). This powerful analytical tool could be applied to any system, from social insects (96) to bird flocks (97) and mass migrations (98).

### 5.5.3 High-resolution tracking

Advances in real-time tracking technology have facilitated leaps in progress in studying collective behaviors. As questions about living systems become grander, so do the tools required to address them and the resulting scale and dimensionality of datasets (99). For example, automated tracking via miniature QR codes attached to *L. niger* ants demonstrated that colony social networks change after parasite introduction (49). By tracking juvenile white storks using high-resolution GPS and accelerometers, Flack *et al.* (100) discovered that clear leaders and followers emerge in flocking dynamics, where leaders discover thermal uplifts and are followed by followers. In this example, leaders spend more time in thermals, requiring less flapping during flight, and migrate further annually. The migration of juvenile white storks has been the source of introduced avian diseases in the past, like West Nile Virus in Israel in 1998 (101) and Newcastle Disease Virus in Germany in 1992–1993 (102). Given that some sub-genotypes of Newcastle Disease Virus exhibit significant panzootic potential (103), this study system has the potential to link individual traits and collective flocking behavior with migration outcomes to predict and prevent disease outbreaks (104), (105).

### 5.5.4 Collective behavior of clones

Given how important inbred genotype lines in model systems (*Drosophila* flies and *C. elegans* nematodes) have been to understanding the intrinsic and extrinsic factors of behavior, clonal animals are a growing tool for the study of social behavior. Bierbach *et al.* (106) used tightly controlled ontogenetic experiments with the clonal Amazon molly to identify the degree to which behavioral individuality emerges despite no differences in genotype, no social input, and near-identical rearing conditions. They discovered substantial individual variation among genetically identical siblings isolated directly after birth into standardized environments, and the degree of variation did not change depending on varying levels of social exposure. Laskowski *et al.* (107) highlight the untapped potential of clonal vertebrates in experimental animal behavior, and how systems like the Amazon Molly may be

predisposed for incorporating parasite transmission dynamics due to a history of studies on the major histocompatibility complex and its regulation on genotypic immune variation (108), (109). In some cases, research on vertebrate models is not feasible, and fruit flies represent an amazing system for studying collective behavior (e.g., aggregation) and disease dynamics in single and mixed-genotype populations (110). Using inbred laboratory lines of *Drosophila melanogaster*, one can generate replicate social groups of specific genotype combinations over a relatively short time span (111), allowing for large-scale testing of hypotheses combining collective behavior and disease outcomes.

## 5.6 Conclusions

The transition from solitary life to social living was one of the major biological transitions in evolution and as such, social animals are amazingly successful with huge impacts on ecological communities. Quantifying the dynamics which influence parasite transmission in social groups is complicated because the collective behaviors of animal societies incur unique benefits and costs associated with parasitism. The goals of this chapter were threefold: (i) describe notable phenomena and trends in research focusing on animal collective behavior and its disease outcomes, (ii) propose a mechanism (the social fulcrum hypothesis) by which selection may operate on group traits and their joint outcomes on collective behavior and parasitism, and (iii) describe burgeoning research frontiers in which the dynamics of collective behavior and parasitism can be tested. With advances in tracking technologies and the computational power that facilitates analyses of tracking data, identifying the costs and benefits of complex collective behaviors for social groups becomes more feasible. Experiments designed to test hypotheses explicitly regarding potential trade-offs in collective outcomes, and how they are resolved, are likely to expand our foundational understanding of how disease dynamics play out in animal societies.

## Acknowledgements

I thank the editors and three anonymous reviewers for comments on earlier versions of this chapter.

Thanks also to the Perspectives in Ecological Research (PEERS) seminar group at the University of Florida for helpful discussion on the social fulcrum hypothesis.

# References

1. Barber I and Dingemanse NJ. Parasitism and the evolutionary ecology of animal personality. *Phil Trans Roy Soc B*. 2010;365(1560):4077–88.
2. Hawley DM, Etienne RS, Ezenwa VO, and Jolles AE. Does animal behavior underlie covariation between hosts' exposure to infectious agents and susceptibility to infection? Implications for disease dynamics. *Integr Comp Biol*. 2011:icr062.
3. Poulin R. Parasite manipulation of host personality and behavioural syndromes. *J Exp Bio*. 2013;216(1):18–26.
4. Conradt L. Group decisions: How (not) to choose a restaurant with friends. *Curr Bio*. 2008;18(24):R1139–R40.
5. Herbert-Read JE, Perna A, Mann RP, Schaerf TM, Sumpter DJ, and Ward AJW. Inferring the rules of interaction of shoaling fish. *PNAS*. 2011;108(46):18726–31.
6. Lukeman R Li Y-X, and Edelstein-Keshet L. Inferring individual rules from collective behavior. *PNAS*. 2010;107(28):12576–80.
7. Couzin ID, Krause J, James R, Ruxton GD, and Franks NR. Collective memory and spatial sorting in animal groups. *J Theoretical Bio*. 2002;218(1):1–12.
8. Ballerini M, Cabibbo N, Candelier R, Cavagna A, Cisbani E, Giardina I, et al. Interaction ruling animal collective behavior depends on topological rather than metric distance: Evidence from a field study. *PNAS*. 2008;105(4):1232–37.
9. Jolles JW, King AJ, and Killen SS. The role of individual heterogeneity in collective animal behaviour. *Trends Ecol Evol*. 2020;35(3):278–91.
10. Farine DR Montiglio P-O, and Spiegel O. From individuals to groups and back: The evolutionary implications of group phenotypic composition. *Trends Ecol Evol*. 2015;30(10):609–21.
11. LeBoeuf AC and Grozinger CM. Me and we: The interplay between individual and group behavioral variation in social collectives. *Curr Opinion Insect Sci*. 2014;5:16–24.
12. Bumann D and Krause J. Front individuals lead in shoals of three-spined sticklebacks (*Gasterosteus aculeatus*) and juvenile roach *(Rutilus rutilus)*. *Behaviour*. 1993;125(3-4):189–98.
13. Conradt L, Krause J, Couzin ID, and Roper TJ. "Leading according to need" in self-organizing groups. *Am Nat*. 2009;173(3):304–12.
14. McLean S, Persson A, Norin T, and Killen SS. Metabolic costs of feeding predictively alter the spatial distribution of individuals in fish schools. *Curr Bio*. 2018;28(7):1144–49.e4.
15. Hansen MJ, Schaerf TM, Krause J, and Ward AJW. Crimson spotted rainbowfish (*Melanotaenia duboulayi*) change their spatial position according to nutritional requirement. *PLoS One*. 2016;11(2):e0148334.
16. Reebs SG. Can a minority of informed leaders determine the foraging movements of a fish shoal? *Anim Behav*. 2000;59(2):403–9.
17. Reebs SG and Leblond C. Individual leadership and boldness in shoals of golden shiners (*Notemigonus crysoleucas*). *Behaviour*. 2006;143(10):1263.
18. Ward AJW, Duff AJ, Krause J, and Barber IJ. Shoaling behaviour of sticklebacks infected with the microsporidian parasite, *Glugea anomala*. *Env Bio Fishes*. 2005;72(2):155–60.
19. Ward AJW, Hoare DJ, Couzin ID, Broom M, and Krause J. The effects of parasitism and body length on positioning within wild fish shoals. *J Anim Ecol*. 2002:10–14.
20. Lloyd-Smith JO, Schreiber SJ, Kopp PE, and Getz W. Superspreading and the effect of individual variation on disease emergence. *Nature*. 2005;438(7066):355–9.
21. Sih A and Bell AM. Insights for behavioral ecology from behavioral syndromes. *Advances in the Study of Behavior*. 2008;38:227–81.
22. Bell AM, Hankison SJ, and Laskowski KL. The repeatability of behaviour: A meta-analysis. *Anim Behav*. 2009;77(4):771–83.
23. Cote J, Clobert J, Brodin T, Fogarty S, and Sih A. Personality-dependent dispersal: Characterization, ontogeny and consequences for spatially structured populations. *Phil Trans Roy Soc B*. 2010;365(1560):4065–76.
24. Toscano BJ, Gownaris NJ, Heerhartz SM, and Monaco CJ. Personality, foraging behavior and specialization: Integrating behavioral and food web ecology at the individual level. *Oecologia*. 2016:182:55–69.
25. Kortet R, Hedrick AV, and Vainikka A. Parasitism, predation and the evolution of animal personalities. *Eco Lett*. 2010;13(12):1449–58.
26. Gyuris E, Hankó JF, Feró O, and Barta Z. Personality and ectoparasitic mites (*Hemipteroseius adleri*) in firebugs (*Pyrrhocoris apterus*). *Behav Proc*. 2016;122:67–74.
27. Paquette C, Garant D, Savage J, Réale D, and Bergeron PJ. Individual and environmental determinants of Cuterebra bot fly parasitism in the eastern chipmunk (*Tamias striatus*). *Oecologia*. 2020;193(2):359–70.

28. Demas GE and Carlton ED. Ecoimmunology for psychoneuroimmunologists: Considering context in neuroendocrine–immune–behavior interactions. *Brain Behav Immunity.* 2015;44:9–16.

29. Niemelä PT, Dingemanse NJ, Alioravainen N, Vainikka A, and Kortet RJ. Personality pace-of-life hypothesis: Testing genetic associations among personality and life history. *Behav Ecol.* 2013;24(4):935–41.

30. Modlmeier AP, Keiser CN, Watters JV, Sih A, and Pruitt JN. The keystone individual concept: An ecological and evolutionary overview. *Anim Behav.* 2014;89:53–62.

31. Brown C and Irving E. Individual personality traits influence group exploration in a feral guppy population. *Behav Ecol.* 2014;25(1):95–101.

32. Pinter-Wollman N, Hubler J, Holley J-A, Franks NR, and Dornhaus A. How is activity distributed among and within tasks in *Temnothorax* ants? *Behav Ecol and Sociobiol.* 2012;66(10):1407–20.

33. Martin LB, Addison B, Bean AG, Buchanan KL, Crino OL, Eastwood JR, et al. Extreme competence: Keystone hosts of infections. *Trends Ecol Evol.* 2019;34(4):303–14.

34. Sapolsky RM and Share LJ. A pacific culture among wild baboons: Its emergence and transmission. *PLoS Biol.* 2004;2(4):e106.

35. Hunt ER, Mi B, Geremew R, Fernandez C, Wong BM, Pruitt JN, et al. Resting networks and personality predict attack speed in social spiders. *Behav Ecol Sociobiol.* 2019;73(7):1–12.

36. Keiser CN, Wright CM, and Pruitt JN. Increased bacterial load can reduce or negate the effects of keystone individuals on group collective behaviour. *Anim Behav.* 2016;114:211–18.

37. Beros S, Jongepier E, Hagemeier F, and Foitzik SJ. The parasite's long arm: A tapeworm parasite induces behavioural changes in uninfected group members of its social host. *Proc R Soc B: Biol Sci.* 2015;282(1819):20151473.

38. Pinter-Wollman N, Keiser CN, Wollman R, and Pruitt JN. The effect of keystone individuals on collective outcomes can be mediated through interactions or behavioral persistence. *The Am Nat.* 2016;188(2):240–52.

39. Brown CR and Brown MB. Ectoparasitism as a cost of coloniality in cliff swallows (*Hirundo pyrrhonota*). *Ecology.* 1986;67(5):1206–18.

40. McCauley DE. Intrademic group selection imposed by a parasitoid–host interaction. *Am Nat.* 1994;144(1):1–13.

41. Tadiri C, Scott M, and Fussmann GJ. Impact of host sex and group composition on parasite dynamics in experimental populations. *Parasitol.* 2016;143(4):523–31.

42. Keiser CN, Rudolf VHW, Sartain E, Every ER, and Saltz JB. Social context alters host behavior and infection risk. *Behav Ecol.* 2018;29(4):869–75.

43. Archie EA, Altmann J, and Alberts SC. Social status predicts wound healing in wild baboons. *PNAS.* 2012;109(23):9017–22.

44. Lopes PC, Adelman J, Wingfield JC, and Bentley GE. Social context modulates sickness behavior. *Behav Ecol Socio.* 2012;66(10):1421–8.

45. Marcoux M and Lusseau DJ. Network modularity promotes cooperation. *J Theoretical Bio.* 2013;324:103–8.

46. Gianetto DA and Heydari BJ Sr. Network modularity is essential for evolution of cooperation under uncertainty. *Sci Rep.* 2015;5:9340.

47. Sah P, Leu ST, Cross PC, Hudson PJ, and Bansal S. Unraveling the disease consequences and mechanisms of modular structure in animal social networks. *PNAS.* 2017:201613616.

48. Mersch DP, Crespi A, and Keller LJ. Tracking individuals shows spatial fidelity is a key regulator of ant social organization. *Science.* 2013;340(6136):1090–93.

49. Stroeymeyt N, Grasse AV, Crespi A, Mersch DP, Cremer S, and Keller L. Social network plasticity decreases disease transmission in a eusocial insect. *Science.* 2018;362(6417):941–5.

50. Wilson SN, Sindi SS, Brooks HZ, Hohn ME, Price CR, Radunskaya AE, et al. How emergent social patterns in allogrooming combat parasitic infections. *Front Ecol Evol.* 2020;8:54.

51. Reber A, Purcell J, Buechel SD, Buri P, and Chapuisat M. The expression and impact of antifungal grooming in ants. *J Evol Biol.* 2011;24:954–64.

52. Akinyi MY, Tung J, Jeneby M, Patel NB, Altmann J, and Alberts SC. Role of grooming in reducing tick load in wild baboons (*Papio cynocephalus*). *Anim Behav.* 2013;85(3):559–68.

53. Nowak MA and May RM. Superinfection and the evolution of parasite virulence. *Proc R Soc B: Biol Sci.* 1994;255(1342):81–9.

54. Kutsukake N and Clutton-Brock TH. Social functions of allogrooming in cooperatively breeding meerkats. *Anim Behav.* 2006;72(5):1059–68.

55. Madden JR and Clutton-Brock TH. Manipulating grooming by decreasing ectoparasite load causes unpredicted changes in antagonism. *Proc R Soc B: Biol Sci.* 2009;276(1660):1263–8.

56. Pinter-Wollman N, Jelić A, and Wells NMJ. The impact of the built environment on health behaviours and disease transmission in social systems. *Proc R Soc B: Biol Sci.* 2018;373(1753):20170245.

57. Pie MR, Rosengaus RB, and Traniello JF. Nest architecture, activity pattern, worker density and the

dynamics of disease transmission in social insects. *J Theoretical Biol*. 2004;226(1):45–51.

58. Fefferman NH, Traniello JF, Rosengaus RB, and Calleri DV. Disease prevention and resistance in social insects: Modeling the survival consequences of immunity, hygienic behavior, and colony organization. *Behav Ecol Sociobiol*. 2007;61(4):565–77.

59. Pinter-Wollman N, Fiore SM, and Theraulaz GJ. The impact of architecture on collective behaviour. Nature Eco Evo. 2017;1(5):1–2.

60. Drum NH and Rothenbuhler WC. Differences in non-stinging aggressive responses of worker honeybees to diseased and healthy bees in May and July. *J Apicultural Res*. 1985;24(3):184–7.

61. Loreto RG and Hughes DP. Disease in the society: Infectious cadavers result in collapse of ant sub-colonies. *PloS One*. 2016;11(8):e0160820.

62. Pinter-Wollman N, Penn A, Theraulaz G, and Fiore SM. Interdisciplinary approaches for uncovering the impacts of architecture on collective behaviour. *Phil Trans Roy Soc B* 2018: 373:20170232.

63. Hughes WO, Eilenberg J, and Boomsma JJ. Trade-offs in group living: Transmission and disease resistance in leaf-cutting ants. *Proc R Soc B: Biol Sci*. 2002;269(1502):1811–19.

64. Cremer S, Armitage SAO, and Schmid-Hempel P. Social immunity. *Curr Biol*. 2007;17:R693–R702.

65. Cremer SJ. Social immunity in insects. *Curr Biol*. 2019;29(11):R458–R63.

66. Pull CD and McMahon DP. Superorganism immunity: A major transition in immune system evolution. *Front Ecol Evol*. 2020;8:186.

67. Facchini E, Bijma P, Pagnacco G, Rizzi R, and Brascamp EW. Hygienic behaviour in honeybees: A comparison of two recording methods and estimation of genetic parameters. *Apidologie*. 2019;50(2):163–72.

68. Villa JD, Danka RG, and Harris JW. Repeatability of measurements of removal of mite-infested brood to assess varroa sensitive hygiene. *J Apicultural Res*. 2017;56(5):631–4.

69. Scharf I, Modlmeier AP, Fries S, Tirard C, and Foitzik S. Characterizing the collective personality of ant societies: Aggressive colonies do not abandon their home. *PLoS One*. 2012;7(3):e33314.

70. Wray MK, Mattila HR, and Seeley TD. Collective personalities in honeybee colonies are linked to colony fitness. *Anim Behav*. 2011;81(3):559–68.

71. Bengston S and Dornhaus AJ. Be meek or be bold? A colony-level behavioural syndrome in ants. *Proc R Soc B: Biol Sci*. 2014;281(1791):20140518.

72. Cotter SC, Littlefair JE, Grantham PJ, and Kilner RM. A direct physiological trade-off between personal and social immunity. *J Anim Ecol*. 2013;82(4):846–53.

73. Keiser CN and Pruitt JN. Personality composition is more important than group size in determining collective foraging behaviour in the wild. *Proc R Soc B: Biol Sci*. 2014;281(1796):20141424.

74. Dyer JR, Croft DP, Morrell LJ, and Krause J. Shoal composition determines foraging success in the guppy. *Behav Ecol*. 2009;20(1):165–71.

75. Romano V, Sueur C, and MacIntosh AJ. The trade-off between information and pathogen transmission in animal societies. *Oikos*. 2021;00:1–11.

76. Evans JC, Silk MJ, Boogert NJ, and Hodgson DJ. Infected or informed? Social structure and the simultaneous transmission of information and infectious disease. *Oikos*. 2020. 129:1271–88.

77. Lemanski NJ, Cook CN, Ozturk C, Smith BH, and Pinter-Wollman N. The effect of individual learning on collective foraging in honey bees in differently structured landscapes. *Anim Behav*. 2021;179: 113–23.

78. Nuismer SL. Parasite local adaptation in a geographic mosaic. *Evolution*. 2006;60(1):24–30.

79. Weinstein SB, Buck JC, and Young HS. A landscape of disgust. *Science*. 2018;359(6381):1213–14.

80. Passera L, Roncin E, Kaufmann B, and Keller LJN. Increased soldier production in ant colonies exposed to intraspecific competition. *Nature*. 1996;379(6566): 630–1.

81. Harvey JA, Corley LS, and Strand MR. Competition induces adaptive shifts in caste ratios of a polyembryonic wasp. *Nature*. 2000;406(6792):183–6.

82. Gordon DM. Dynamics of task switching in harvester ants. *Anim Behav*. 1989;38(2):194–204.

83. Réale D and Dingemanse NJ. Personality and individual social specialisation in Székely T, Moore AJ, and Komdeur J (eds), *Social Behaviour: Genes, Ecology and Evolution*. Cambridge: Cambridge University Press, 2010, 417–41.

84. Keiser CN, Wantman T, Rebollar EA, and Harris RN. Tadpole body size and behaviour alter the social acquisition of a defensive bacterial symbiont. *Roy Soc Open Sci*. 2019;6(9):191080.

85. Lichtenstein JL, Wright CM, and Pruitt JN. Repeatability of between-group differences in collective foraging is shaped by group composition in social spiders. *J Arachnology*. 2019;47(2):276–79.

86. Keiser CN, Howell KA, Pinter-Wollman N, and Pruitt JN. Personality composition alters the transmission of cuticular bacteria in social groups. *Bio Lett*. 2016;12(7).

87. Shi W, Guo Y, Xu C, Tan S, Miao J, Feng Y, *et al*. Unveiling the mechanism by which microsporidian parasites prevent locust swarm behavior. *PNAS*. 2014;111(4):1343–8.

88. Ezenwa VO, Archie EA, Craft ME, Hawley DM, Martin LB, Moore J, *et al*. Host behaviour–parasite feedback: An essential link between animal behaviour and disease ecology. *Proc R Soc B: Biol Sci*. 2016. 283:20153078.

89. Wang-Peng S, Zheng X, Jia WT, Li AM, Camara I, Chen HX, *et al*. Horizontal transmission of *Paranosema locustae* (Microsporidia) in grasshopper populations via predatory natural enemies. *Pest Management Sci*. 2018;74(11):2589–93.

90. Nunn CL and Dokey AT-W. Ranging patterns and parasitism in primates. *Bio Lett*. 2006;2(3):351–4.

91. Muehlenbein MP and Watts DP. The costs of dominance: Testosterone, cortisol and intestinal parasites in wild male chimpanzees. *Biopsych Social Med*. 2010;4(1):1–12.

92. White LA, Forester JD, and Craft ME. Using contact networks to explore mechanisms of parasite transmission in wildlife. *Biol Rev*. 2017;92(1):389–409.

93. Silk MJ, Finn KR, Porter MA, and Pinter-Wollman N. Can multilayer networks advance animal behavior research? *Trends Ecol Evol*. 2018;33(6): 376–8.

94. Finn KR, Silk MJ, Porter MA, and Pinter-Wollman N. The use of multilayer network analysis in animal behaviour. *Anim Behav*. 2019;149:7–22.

95. Lemanski NJ, Cook CN, Smith BH, and Pinter-Wollman N. A multiscale review of behavioral variation in collective foraging behavior in honey bees. *Insects*. 2019;10(11):370.

96. Papageorgiou D, Christensen C, Gall GE, Klarevas-Irby JA, Nyaguthii B, Couzin ID, *et al*. The multilevel society of a small-brained bird. *Curr Biol*. 2019;29(21):R1120–R1.

97. Torney CJ, Hopcraft JGC, Morrison TA, Couzin ID, and Levin SA. From single steps to mass migration: The problem of scale in the movement ecology of the Serengeti wildebeest. *Phil Trans Roy Soc B*. 2018;373(1746):20170012.

98. Graving JM and Couzin ID. VAE-SNE: A deep generative model for simultaneous dimensionality reduction and clustering. *BioRxiv*. 2020. Available from: <https://doi.org/10.1101/2020.07.17.207993>.

99. Flack A, Nagy M, Fiedler W, Couzin ID, and Wikelski M. From local collective behavior to global migratory patterns in white storks. *Science*. 2018;360(6391): 911–14.

100. Malkinson M, Banet C, Weisman Y, Pokamunski S, King R, Drouet M-T, *et al*. Introduction of West Nile virus in the Middle East by migrating white storks. *Emerg Infect Dis*. 2002;8(4):392–7.

101. Kaleta EF and Kummerfeld N. Isolation of herpesvirus and Newcastle disease virus from White Storks (*Ciconia ciconia*) maintained at four rehabilitation centres in northern Germany during 1983 to 2001 and failure to detect antibodies against avian influenza A viruses of subtypes H5 and H7 in these birds. *Avian Path*. 2012;41(4):383–9.

102. Miller PJ, Haddas R, Simanov L, Lublin A, Rehmani SF, Wajid A, *et al*. Identification of new sub-genotypes of virulent Newcastle disease virus with potential panzootic features. *Infect Genet Evol*. 2015;29:216–29.

103. Fritzsche McKay A, and Hoye BJ. Are migratory animals superspreaders of infection? *Int Comp Biol*. 2016;56(2):260–7.

104. Altizer S, Bartel R, and Han BA. Animal migration and infectious disease risk. *Science*. 2011;331(6015): 296–302.

105. Bierbach D, Laskowski KL, and Wolf M. Behavioural individuality in clonal fish arises despite near-identical rearing conditions. *Nature Commun*. 2017;8(1):15361.

106. Laskowski KL, Doran C, Bierbach D, Krause J, and Wolf M. Naturally clonal vertebrates are an untapped resource in ecology and evolution research. *Nature Ecol Evol*. 2019;3(2):161–9.

107. Warren WC, García-Pérez R, Xu S, Lampert KP, Chalopin D, Stöck M, *et al*. Clonal polymorphism and high heterozygosity in the celibate genome of the Amazon molly. *Nature Ecol Evol*. 2018;2(4):669–79.

108. Schaschl H, Tobler M, Plath M, Penn DJ, and Schlupp I. Polymorphic MHC loci in an asexual fish, the amazon molly (*Poecilia formosa; Poeciliidae*). *Mol Ecol*. 2008;17(24):5220–30.

109. White LA, Siva-Jothy JA, Craft ME, and Vale PF. Genotype and sex-based host variation in behaviour and susceptibility drives population disease dynamics. *Proc R Soc B: Biol Sci*. 2020;287(1938):20201653.

110. Wice EW and Saltz JB. Selection on heritable social network positions is context-dependent in Drosophila melanogaster. *Nature Commun*. 2021;12(1):3357.

111. Modlmeier AP, Keiser CN, Shearer TA, and Pruitt JN. Species-specific influence of group composition on collective behaviors in ants. *Behav Ecol Sociobiol*. 2014;68(12):1929–37.

112. Modlmeier AP, Liebmann JE, and Foitzik S. Diverse societies are more productive: A lesson from ants. *Proc R Soc B: Biol Sci*. 2012;279(1736):2142–50.

113. Modlmeier AP and Foitzik S. Productivity increases with variation in aggression among group members in Temnothorax ants. *Behav Ecol*. 2011;22(5):1026–32.

114. Paleolog J. Behavioural characteristics of honey bee (*Apis mellifera*) colonies containing mix of workers of divergent behavioural traits. *Anim Sci Pap*. 2009;27(3):237–48.

115. Dyer JRG, Croft DP, Morrell LJ, and Krause J. Shoal composition determines foraging success in the guppy. *Behav Ecol*. 2008;20(1):165–71.

116. Richards EL. Foraging, personality and parasites: Investigations into the behavioural ecology of fishes. PhD dissertation, Cardiff University (United Kingdom). 2010.

117. Jolles JW, Boogert NJ, Sridhar VH, Couzin ID, and Manica A. Consistent individual differences drive collective behavior and group functioning of schooling fish. *Curr Biol*. 2017;27(18):2862–68. e7.

118. Jolles JW, Mazué GPF, Davidson J, Behrmann-Godel J, Couzin ID. Schistocephalus parasite infection alters sticklebacks' movement ability and thereby shapes social interactions. *Sci Rep*. 2020;10(1):12282.

119. Mattila HR and Seeley TD. Genetic diversity in honey bee colonies enhances productivity and fitness. *Science*. 2007;317(5836):362–46.

120. Arathi HS and Spivak M. Influence of colony genotypic composition on the performance of hygienic behaviour in the honeybee, Apis mellifera L. *Anim Behav*. 2001;62(1):57–66.

121. Jandt JM and Dornhaus A. Bumblebee response thresholds and body size: Does worker diversity increase colony performance? *Anim Behav*. 2014;87: 97–106.

122. Keiser CN, Vojvodic S, Butler IO, Sartain E, Rudolf VHW, and Saltz JB. Queen presence mediates the relationship between collective behaviour and disease susceptibility in ant colonies. *J Anim Ecol*. 2018;87(2): 379–87.

123. Croft DP, Albanese B, Arrowsmith BJ, Botham M, Webster M, and Krause J. Sex-biased movement in the guppy (*Poecilia reticulata*). *Oecologia*. 2003;137(1): 62–8.

124. Lihoreau M, Clarke IM, Buhl J, Sumpter DJT, and Simpson SJ. Collective selection of food patches in Drosophila. *J Exp Biolfri*. 2016;219(5): 668–75.

# Movement Behavior

CHAPTER 6

# Movement data provides insight into feedbacks and heterogeneities in host–parasite interactions

Orr Spiegel, Nili Anglister, and Miranda M. Crafton

## 6.1 Introduction

### 6.1.1 Considering movement in the feedback between parasite dynamics and host behavior

In a world facing challenges of the COVID-19 pandemic, the importance of understanding wildlife disease dynamics and their zoonotic potential are perhaps more apparent than ever. Indeed, scholars have long realized that parasites and pathogens (here used as synonyms for simplicity) serve as substantial ecological and evolutionary drivers of natural ecosystems (1), (2). Over large spatiotemporal scales, parasites affect their hosts in a number of ways, shaping their distributions in both time and space and even their social structure. As a result, parasites can become a strong factor affecting both the natural and sexual selection experienced by hosts (3), (4). At local scales, parasites and pathogens influence host fitness, habitat preference, social interactions, and behavior (5)–(7). The influence of parasites on host behavior can be mediated by directly depleting the host's internal resources and thus altering its physiological state (8–10); or indirectly by modifying its social environment (e.g., repulsion by conspecifics) (4), (5), (9), (11). Furthermore, in specific cases, parasites may directly manipulate host behavior in favor of their own fitness (12)–(14). The behavior of both the host and the parasite interactively affects the chances of parasite encounter, infection, survival, and the transmission

of the parasites to new hosts. Accordingly, this behavioral feedback between host and parasite is attracting growing attention, aimed at identifying both its underlying mechanisms and its diverse ecological impacts (4), (15). Here we focus on the role of animal movement in such feedbacks, and how tracking data can provide unique insights when exploring parasite–host relationships.

Movement (spatial displacement of the whole organism) is an essential feature of life, and one of the most central aspects of behavior (16), (17). For mobile species, most (if not all) behaviors include decisions on where, when, and how to move, including foraging, predator avoidance, and mate finding. Even apparently sessile creatures typically move during some phases of their lives before settling (e.g., as seed, or larva). For example, coral and barnacles rely on movement for important dispersal processes, either via floating or rafting on debris. An organism's movement is the product of both its internal state (*sensu* Nathan *et al.* (16)); namely the explanation for the motivation for moving) and its environment (e.g., where are resources, barriers and competitors located?). The accumulation of an individual's local movements subsequently generates both its broader space use pattern and lifetime track, affecting social interactions, foraging, survival, and fitness. The interactive feedback between host behavior and parasite dynamics is also readily expressed in movements: movement

Orr Spiegel, Nili Anglister, and Miranda M. Crafton, *Movement data provides insight into feedbacks and heterogeneities in host–parasite interactions.*
In: *Animal Behavior and Parasitism.* Edited by Vanessa O. Ezenwa, Sonia Altizer, and Richard J. Hall, Oxford University Press.
© Oxford University Press (2022). DOI: 10.1093/oso/9780192895561.003.0006

decisions affect parasite encounter rates and the potential for subsequent effective transmission (18), (19). In parallel, parasites can alter host movements either through morbidity (lethargy caused by parasitic load or immune response) or by manipulation of behavior (e.g., hairworms (Nematomorpha) causing crickets to jump into water where the worms reproduce (13), (14), (20). These effects, in turn, may alter the space use patterns of infected hosts by changing their movement rates, timing, structure, and destinations. The feedback on host–parasite dynamics can be positive, facilitating transmission through manipulation and aggregations, or suppressive due to decreased movements and social contacts (7), (9), (21). This inherent dependency between host movement and parasites suggests that collecting movement data before, during, or after infection can contribute to our understanding of host–parasite dynamics. For instance, collecting movement data can reveal the magnitude of reduction in activity and range by infected individuals, or parasite avoidance behavior by others (15), (22). These behavioral changes may ultimately prevent the spread of a predicted outbreak. Similarly, accounting for the influence of parasites on host movement can facilitate predictions on animal space use patterns, such as predicting migration, dispersal, and colonization rates (7), (9).

In this chapter, we argue that incorporating movement ecology approaches and tracking data into host–parasite studies is instrumental for developing a better understanding of these interactions in natural conditions. To present our argument, and outline some of the potential benefits of this integration, we have structured our chapter into five main sections. First, we briefly summarize the basics of movement ecology (section 6.1.2). Next, we discuss the potential effects of host movements on host–parasite interactions (e.g., through the spatial scale of transmission, Section 6.2). Then, to demonstrate the interactive aspects of this phenomenon we explain how parasites, in turn, affect host movement (e.g., by affecting host locomotion capacity, section 6.3). After establishing the necessity of movement data for studying this behavioral feedback, we describe some of the tools and techniques for animal tracking and collecting this data (section 6.4). Finally, we highlight a few points

that can facilitate integration among studies of ecology, parasitology, and animal behavior, and parameterization of relevant models (section 6.5) (see Figure 6.1).

### 6.1.2 The basics of movement ecology

Researchers in the field of movement ecology focus on measuring animal spatial positions over time, describing the emerging movement patterns, identifying the factors that shape these patterns, and examining their consequences for ecological and evolutionary processes. Fine scale movement decisions of individuals can scale up to shape space use patterns at population and community levels. Hence studies may span over a range of spatiotemporal scales, reaching large ecological processes such as nutrient flow across ecosystems and disease ecology research (19), (23), (24). Accordingly, studies typically follow either a Lagrangian approach, where subsets of individuals are tracked over time and space at fine scales, or a Eulerian approach, where population-level fluxes of individuals to/from specific sites or habitats are quantified. While the latter approach is naturally connected to classical compartmental models studying disease/parasite dynamics in populations such as the SEIR models (Susceptible, Exposed, Infected, Removed/Recovered), the Lagrangian approach can be advantageous for parameterizing individual-based models (IBMs) that might often offer valuable insight into the factors governing variation and structure in host–parasite transmission rates (19).

The last two decades have seen a growing number of studies addressing movement-related ecological questions (16), (17). This growing interest is driven in part by the ecological importance of movement, and in part by the technological developments that have facilitated bio-logging of a growing number of species. While later in the chapter (section 6.4) we provide a detailed guide on how to choose the proper tracking methods, Table 6.1 outlines a basic overview of important techniques and considerations required to collect movement data. Regardless of the differences among the various methods in their resolution and accuracy, telemetry data fundamentally contain a series of consecutive relocations of an individual over time. An

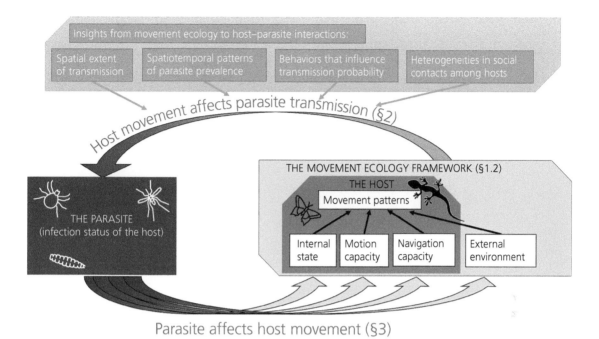

**Figure 6.1**  A graphical representation of the proposed links between movement ecology and host-parasite interactions. Using the terminology of the movement ecology framework (summarized in section 6.1.2) we argue that host movement affects parasite transmission through a few alternative pathways (detailed in section 6.2), eventually leading to population-level outcomes, such as parasite gene flow. Parasite infection, in turn, may also alter host movement through several alternative mechanisms (section 6.3); either affecting the host itself (namely its internal state, locomotion, or navigation capacities), or the behavior of its conspecifics (thus modifying the host's external environment). This feedback eventually shapes system dynamics.

individual's trajectory is then summarized by a set of statistical currencies such as step length, relative turning angle, and directional persistence. Temporal changes in the distribution of these currencies allow one to cluster movement segments into canonical activity modes and to detect behavioral changes in movement, such as transitions from resting periods to local foraging and fast commutes (16), (25). In the context of host–parasite feedbacks, these approaches can allow, for instance, comparing healthy and sick individuals or inferring disease status from changes in movement. Such tools and data underscore the potential of understanding the patterns and processes shaping animal movement, and have led to a wealth of both analytical methods and conceptual innovations.

The movement ecology paradigm has been suggested as an overarching framework for these diverse studies (16). The paradigm aims to integrate the study of movement across disciplines,

postulating four main factors driving movement patterns, each of which can be related to host–parasite ecology (Figure 6.1): **(i) The internal state**. *Why move?* This is a summary of the fundamental motivation to move, essentially the behavioral decisions generating movements and the underlying selective forces that drive these decisions. Note that following this definition, hereafter internal state refers directly to the motivation to move and not to the ability to move. Two common motivations relevant for parasites include resource acquisition and enemy avoidance. For instance, fish infected by larval tapeworms have higher oxygen demands causing them to approach the water surface (8), and grazing mammals exhibit a behavior known as "gadding" once bitten by parasitic flies, where they run away to avoid further bites (26). **(ii) Motion capacity**. *How to move?* This component refers to the diverse means by which animals move. For instance, pelicans can fly, walk, swim,

**Table 6.1** An overview of some of the commonly used tracking methods.

| Method | Output | Quality of data | Quantity of data | Cost | Host–parasite feedback |
|---|---|---|---|---|---|
| Marking—bird ringing, ear tagging etc. | Mark—resight or recapture data. Can measure influx of individuals to certain sites and temporal activity, or general spatial extent by an individual. | High (individuals are identified accurately in space) | Low (sparse observations on few individuals, many are lost) | Low for equipment High for personnel effort | Often for Eulerian approach. Mostly for population-based models (SEIR). Recapture enables investigation of parasite load in different spatiotemporal points. Yet, movement between resights is largely unknown. Limited by re-trapping\resight effort and extent. |
| Radio Frequency Identification (RFID) | Mark—resight or recapture data. Can measure temporal activity in certain sites. Very local geographical scale. Local social networks. | Medium (presence of marked individual is known only near the readers and not accurately within range) | Medium (can generate high temporal resolution of presence data only) | Low for tags High for RFID readers | Recaptures enable investigation of parasite loads through time. Co-occurrence of individuals can predict parasite transmission in hotspots (feeding areas, water holes, roosts). Yet, movement beyond readers' local detection is largely unknown. |
| Passive Integrated Transponders (PIT) | | | | | |
| Geo-locators | Daily location data. Can generate approximate trajectories. | Low (tens of km accuracy) | Medium (daily data across long periods) | Medium for tags | Limited to migration and population connectivity applications. Rarely integrated in disease ecology but see for instance (79). |

| | | | | |
|---|---|---|---|---|
| **Radio transmitters** | Location, trajectories, home range, small scale social networks. Geographical scale usually local, can be mobile. | Medium (location data is not very accurate, and depends on tracking) | Medium (limited by tracking effort per tag, unless automatic stations exist) | Medium for tag, High for personnel effort and radio receiver | Lagrangian approach—Routine recaptures enable investigation of parasite loads. Co-occurrence of individuals and overlapping home-ranges can predict parasite transmission. |
| **GPS tags** | Location, trajectories, home range, Social networks, Large Geographical scale. | Very high (accurate locations and uniform sampling) | High (varies by data retrieval method, manual vs. GSM/satellite/VHF communications) | High for tags with retrieval communication | Can be used for individual based models. Different host behavior (such as exploratory behavior) can affect parasite transmission (30, 40). Spatially explicit network analyses of shared space use provide another way movement data can be used to reveal dependencies between host and parasite behavioral feedbacks (52). |
| **Atlas tags** | Location, trajectories, home range, Social networks. Local geographical scale | High (accurate locations and uniform sampling but limited geographical extent) | Very High | Low for Tags, High for required infrastructure and personnel | |

and dive, depending on the context and motivation. Parasites can affect host movement ability though its morphology, physiology, or energetics. For example, the trematode *Riberia ondatrae* interferes with limb development in the Pacific chorus frog (*Pseuacris regilla*) causing shorter jumping distances and slower swimming (27). Typically, a sick individual moves less due to limited available energy, and heavily tick-infested lizards run slower due to loss of vigor (7), (20). **(iii) Navigational capacity.** *Where to move?* For self-propelling organisms, movement decisions are based on their current knowledge of their environment. To maximize fitness, animals collect knowledge (or utilize existing), and process and store spatial information for determining their destination. Parasites may manipulate behavior through neuromodulation (e.g., hormones), as discussed previously in the example of the nematodes altering crickets' behavior (13). **(iv) External conditions.** All of the above-mentioned factors interact with the environment, where physical conditions, resources, predators, and conspecifics shape the movement paths (trajectories) of animals. Interactions among these factors (many of which are relevant for parasite transmission) create movement patterns that are extraordinarily diverse and dynamic. For instance, moving further away from infested habitats may help minimize the costs of infection, in the context of migratory behavior, this phenomenon has been coined "migratory escape."(9)

Untangling these different factors affecting movement, and the fine resolution changes in its patterns, can help measure, describe, and predict heterogeneity in transmission between hosts, and in addition provide a new lens for looking for answers to a wide variety of research questions. Movement data also offer opportunities to pose novel questions relating to the proximate reasons for observed social contacts (e.g., what proportion of social encounters occur from socially-independent reasons such as visiting a waterhole (28)); or on the effects of contact on transmission (e.g., how does duration and proximity of contact affect the transmission probability?); and the impact of infection on movement (e.g., does infection decrease or increase daily movement ranges in a particular system?). Next, we describe five central aspects through which tracking

data can inform our understanding of host–parasite dynamics, potentially improving epidemiological models by better representing behavioral, individual, and spatial heterogeneity (Figure 6.1). We begin by simple coarse scale data on location and home range, and gradually continue towards fine-scale resolution tracking and identification of social encounters.

## 6.2 The relevance of movement data for host–parasite systems

### 6.2.1 The value of temporal longitudinal data for host–parasite studies

The first benefit arising from collecting movement data is the ability to follow the same individual, regardless of the specific features of its movement. Longitudinal data (from the same individual) on parasite load and behaviors can allow scholars to document potential changes in these parameters and evaluate the social context. While re-sights and recaptures of the same free-ranging individual are not logistically feasible in many systems, the use of tracking devices can relax this constraint and facilitate repeat visits to a focal individual. Thus, following movement-ecology approaches (i.e., attaching tracking devices such as radio transmitters or Global Positioning System (GPS) tags) might be beneficial, even if the tracking data itself is not the core objective of a study. In Siberian chipmunks and other small mammals, capture–recapture methods have been used to study personality, space-use, and tick load (29); however this method is more difficult to use for species with large home ranges or seasonal migrations. In a recent example, Payne *et al.* (30) studied differences between individuals in parasite load in a wild population of sleepy lizards (*Tiliqua rugosa*). GPS tracking across eight years allowed routine recapture and ectoparasite load assessment (two tick species). The authors examined multiple factors that may explain differences in parasite load and found that lizards exhibited consistent individual differences both within and across years. They also show that both the spatial location (proximity to a road that served as preferred habitat with higher lizard density), and lizard traits (morphology and personality) affected observed loads. While the tracking data were only tangentially used in this particular study, such insights would have been impossible without the ability to relocate the same individual host.

Consistency in parasite loads and in transmission-relevant behaviors can indicate that parasitism constitutes a phenotypic trait of the host, and even a heritable tendency in some instances (10). Various aspects of the host phenotype can influence its role as a spreader or receiver of pathogens and parasites. Predictors include sex, age, body condition (e.g., malnutrition), immune system performance, as well as behavior. Individual hosts often differ consistently in their behavior across time and context, sociability, and space use, potentially affecting the number of contacts, the probability of transmission, and their susceptibility to infection (31), (32). At the extreme, some individuals can act as superspreaders that contribute disproportionately to disease transmission, either because they shed higher doses of parasites, carry more infectious strains, or have higher numbers of contacts compared to other hosts (10), (31), (33). The notorious "Typhoid Mary," who was a highly mobile asymptomatic carrier of *Salmonella enterica* Typhi (Typhus), is a textbook example of the devastating potential of a single superspreader (34). Due to her mobility and extensive contacts while working as a cook, she infected 53 members of 7 different families in New York in the early twentieth century. Indeed, recently there is a growing awareness of the role of consistent differences in individual movement and their potential to amplify the heterogeneity among hosts in their ability to spread or receive parasites (31), (33), (35). These findings indicate the potential of temporal longitudinal data to promote our understanding of parasite-host dynamics, and the value of tracking free-ranging animals.

## 6.2.2 Tracking data reveal the spatial extent of transmission potential

Even when described at a coarse spatiotemporal scale, space use data may still provide important details on host home or dispersal ranges and the spatial extent of their movements. Until recently, tracking devices provided locations at low resolution and accuracy (e.g., a daily location with

errors at the scale of hundreds of meters). These datasets can only provide a general description of space use patterns, home ranges, and habitat preference. Nevertheless, these basic features of movement can inform the potential spatial scales of parasite transmission and the diversity of parasites encountered (36), (37). In many host–parasite systems, parasites do not "quest" (i.e., actively search for hosts), at least not at a scale that is meaningful with respect to the scale of the host movement (e.g., ticks and sleepy lizards (30)). Hence the scale of host movement, home range, and dispersal ability can affect the transmission potential and destination of parasites (36), (38). Home range overlap among neighbors sets the template for transmission events, either through direct interactions among adjacent individuals (e.g., in giraffes (39)) or through indirect transmission of parasites with a latent phase in the environment between successive hosts (19). For instance, Boyer *et al.* (29) report that more exploratory chipmunks (here those with larger home ranges) had more ticks, demonstrating how variation in life-history strategies can alter parasite loads and transmission trajectories. The intensity of spatial usage also can show a strong positive association with the prevalence of parasite transmission, as shown in models of gastrointestinal parasites in socially structured mammal populations (40). Further, movement data can be used to track changes in space use patterns of hosts and parasites spread across seasons and environmental gradients. For example, Farnsworth *et al.* (18) studied Chronic Wasting Disease (CWD), a prion disease carried by mule deer (*Odocoileus hemionus*). They found that deer winter ranges were much smaller and rarely overlapped, whereas summer ranges had substantial overlap among population units (clusters of individuals). They concluded that summer ranging behavior was likely responsible for the spread of CWD among subpopulations, whereas winter ranging behavior had the potential to amplify CWD prevalence within a subpopulation if an infected individual was present. These empirical findings corroborate theoretical predictions on the importance of host movement in structured populations (3), (33).

On a broader spatial scale, host ranges can affect parasite connectivity and population structure. For example, migrating species or dispersing individuals can connect otherwise isolated populations and contribute to the long-range transport of parasites and create gene flow of the parasites between distant locations (41), (42). This unique ability is illustrated, for instance, by the transmission of haematozoan parasites by migratory waterfowl (43). In addition to the connectivity of parasites across extended scales, migratory species may affect haematozoan parasite community composition; an increase in the migration distance of waterfowl species was found to be positively related to parasite diversity. While the importance of migratory species to host–parasite dynamics is covered elsewhere (see Chapter 7 in this book, (44)), the same principles also hold for non-migratory species, where the scale of host movement determines the scale of connectivity of the parasite (36), (38). Occasional long-range forays beyond the animal's typical home range may be critical for parasite transmission events. While tracking studies report such forays (sometimes reaching hundreds of kilometers beyond the home range, e.g., (45)), we are unaware of models directly addressing the importance of this particular behavior for parasite structure. Yet, such behaviors are unlikely to be observed without tracking devices, highlighting the relevance of host movement scale (even if limited to coarse movement data) for parasite transmission.

### 6.2.3 Details on the spatiotemporal patterns of hosts and their effect on parasite prevalence

Another layer of realism in which movement data can contribute to our understanding of host–parasite interactions is through the consideration of spatial heterogeneity in animal movement due to habitat preference and the shared space-use among individuals. These insights do not require high-resolution movement data, and even intermediate resolutions are suitable for inferring habitat preference and shared space use. Movement data may help identify behavioral drivers (e.g., resource distribution and its seasonal changes), and thus could be used to mechanistically model encounter probabilities or factors contributing to shared space use (19), (46). Investigating movement patterns that shape emergent shared space use is therefore essential for predicting parasite

hotspots, especially for less mobile parasites (32), (33). Hotspots of parasite prevalence can emerge at sites or habitat types where hosts aggregate in high densities or preferably spend more time, if they offer favorable conditions for parasite survival and/or transmission. The relative importance of host co-occurrence (encounter) and time-lagged shared space use depends on the life history of the parasite and the suitability of the environment for their off-host survival. For instance, shared roosting sites are often infection hotspots for animals such as Bechstein's bats and sleepy lizards; the use of these sites predicts individual parasite loads (46), (47). Local infectious zones of anthrax (*Bacillus anthracis*) and other agents emerge from the aggregation patterns of their host around (often ephemeral) key resources (21). Water sources in arid conditions, for instance, or attractive food sources, may lead to environmental hotspots of infection, which are analogous to super spreaders. Such aggregated resources may enhance spatial overlap among hosts in specific times (e.g., dry season), both of which can increase transmission (21). Indeed, microhabitats surrounding shared resources were found to present areas of increased parasite transmission risk and contact rates, among sick, lethargic animals (37). Furthermore, given the poor condition of infected (or sick) individuals, they are often expected to remain closer to key resources, thus enhancing spatial aggregation at these transmission hotspots (48).

### 6.2.4 Identifying behaviors that influence transmission probability

Compartmental SEIR models often assume an average transmission probability across all interactions, yet the behaviors of infected or susceptible hosts can substantially alter these probabilities. Infected hosts may modify their social activities, habitat preference, and other aspects of their movement and behavior, thus changing their effectiveness in transmission, and the dynamics of parasite spread. For example, red-capped mangabeys (*Cercocebus torquatus*) treated with anti-parasitic drugs spent more time foraging and less time resting and vigilant compared with when they were infected (49). Group cohesion also changed following treatment, suggesting parasite-induced social avoidance. Similarly, Passive Integrated Transponder- (PIT) tagged bats were found to avoid roosts (hotspots) with high infestation of bat fly puparia (47). Beyond self-avoidance by infested individuals, behavior can modulate transmission through social distancing by group members. Many parasites cause conspicuous changes in host appearance or behavior and studies have demonstrated avoidance of individuals that display physical characteristics of infection ((15), see section 6.4.4). For example, infected sticklebacks displaying visible lesions and infected tissues triggered conspecifics to avoid shoaling with them (50). In some cases, cannibalism of infected young is also reported (11). While many of these behavioral changes might reduce the probability of transmission due to the degraded condition of the infected hosts, some behavioral changes may actually enhance transmission probability, especially if driven by adaptive (for the parasite) host manipulation (14). as will be discussed in detail in section 6.3.2. Movement data, if collected at sufficient resolution with respect to the relevant behavioral changes, can add another layer of complexity and realism and lead to qualitatively different results compared to population-scale research (19).

### 6.2.5 Describing heterogeneities in social contacts among hosts

Finally, as discussed in section 6.2.1, heterogeneity in the rate and nature of social contacts is central for the dynamics of directly transmitted parasites, exactly as the details of variation in shared space use matter for indirectly transmitted ones ((32), see also Chapter 4, (51)). The growing popularity of social network analyses highlight the need for fine-resolution datasets of social encounters, or time lagged interactions of asynchronous shared space use (46). Yet, obtaining such datasets for free-ranging populations is challenging. Direct observations of social interactions are only possible for a small subset of group-living species like monkeys or large mammals, often members of human-habituated troupes (6), (28). Alternatively, these datasets can be obtained for a much broader diversity of species by attaching proximity loggers or GPS

tags (30), (39). These networks, often referred to as spatial-proximity social networks, infer encounters from the frequency or intensity of co-occurrences. Typically, tracking resolution and accuracy for these datasets needs to be sensitive enough to detect relatively short encounters, and quantify their variation in length, spatial proximity, or frequency. Hence, such data collection is only applicable with cutting-edge and often costly devices (see Table 6.1 and section 6.4).

Spatially explicit network analyses of shared space use provides another way movement data can be used to reveal dependencies between host and parasite behavioral feedbacks (52), (53). with accumulating examples from a wide range of taxa such as giraffes and sleepy lizards (39, 54); Figure 6.2). For instance, in studies conducted on Bovine tuberculosis (bTB) in badgers, researchers found statistically significant relationships between network position and bTB infection status. Social behavior was related to spatial behavior; time spent resting at outlying setts was negatively related to time spent with badgers from their own group but positively related to time spent with members of other social groups (55). Similarly, contact networks among Tasmanian devils (*Sarcophilus harrisii*) differ among sexes and seasons, predicting the spread of a transmissible disease (56). Thus, tracking data underlie many of the modern applications of social networks. Addressing movement-related constraints and considerations is essential in the construction of models and hypothesis testing using these datasets and the potential contribution of movement behavior to parasite transmission (28), (52).

## 6.3 How do parasites affect host movement?

Until now, we have talked about how collecting movement data of individuals (normally the hosts) contributes to our knowledge of transmission dynamics. Now we turn to discuss the other direction of this interaction, focusing on how movement data can be used to evaluate the effect of parasites on their hosts' movement. As mentioned earlier, parasites can alter host movement across different scales, and sometimes in opposing directions. For

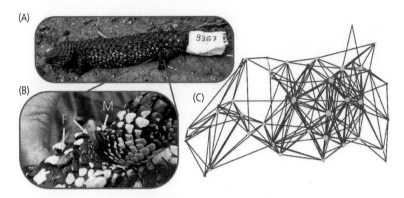

**Figure 6.2** An example of a proximity-based social network in the context of parasite transmission. (A) GPS-tagged sleepy lizards (*Tiliqua rugosa*) were tracked at Bundey Bore, Australia, to study parasite transmission. (B) A close-up of three *Bothriocroton hydrosauri* ticks (two engorged females, one male) near the armpit of a lizard (C) A graph of the lizard social network based on co-occurrences. Nodes (circles) depict individuals and edges (lines) depict contacts between individuals. Node size represents its degree (number of contacts), and node color represents sex: females in pink ($n$ = 30) and males in light blue ($n$ = 29). Both direct contact and time-lagged (shared space-use) networks affect the transmission of diseases and parasites (46, 54). Photo credit: Orr Spiegel.

example, if infection increases local movement and leads to host aggregation, there may be less of a need for individuals to disperse to seek conspecifics. Identifying the particular pathway of influence can provide a mechanistic understanding of the interaction and can facilitate future predictions on the outcome (and sometimes interventions). For instance, if the parasite decreases travel distances it can reduce spatial scale of transmission, enhancing local outbreaks, but if it results in social distancing it may suppress outbreaks. In the next section, we discuss several non-mutually exclusive pathways for possible parasite effects, classifying them according to the main components of the movement ecology paradigm (namely: the locomotion capacity, internal state, and navigation capacities, and external environment that includes both the social and ecological contexts) as introduced in section 6.2. Adhering to this terminology is useful for connecting to the broad literature on the topic.

### 6.3.1 Parasites affect motion capacity of their hosts

The influence of parasites on animal movement can often be seen through alterations in both host physiology and physical condition, often resulting in reduction of an individual's locomotion and fitness (7), (27). For example, a study on ectoparasites of coral reef fish showed that fish parasitized by isopods exhibited higher metabolic rates and lower swimming speeds owing to reduced streamlining, and this effect disappeared when isopods were removed (7). Various examples, such as mange-infected wolves or tick-infected lizards, demonstrate how the poor body condition of an infected host can result in reduced energy reserves, vigor, and locomotion capacity (20), (57). Perhaps one of the most direct examples of the effects of parasitism on host motion capacity can be seen in Pacific chorus frogs infected by trematodes, resulting in severe limb malformations and thus decreasing the ability of a range of motions such as jumping and swimming distance, swimming duration, and burst speed (27) (Figure 6.3). Notably, naturally occurring parasite loads may vary among host species, individuals, and time, influencing the magnitude of parasite effect on locomotion. Commonly observed low infestation rates may thus fail to influence host movement (58). See also Box 6.1 for an empirical example where *Mycoplasma* is associated with reduced movement in vultures.

# The Effects of Parasitism on Host Movement

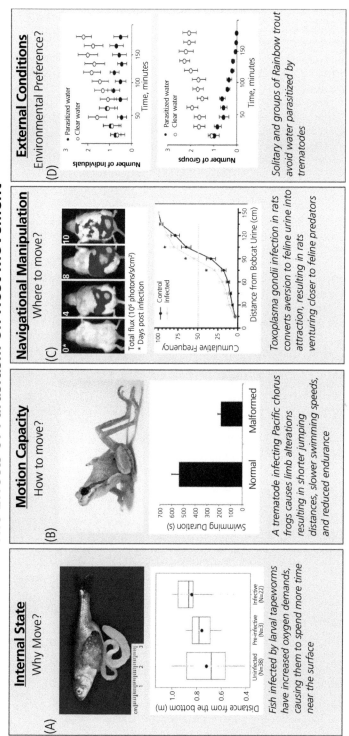

**Figure 6.3** Demonstrating possible effects of parasites on their host movement through the movement ecology paradigm. (A) Internal state: Why does an individual move (77)? (B) Motion capacity: How to move (27)? (C) Navigational manipulation: Where to move (12)? (D) External conditions: modulating the host environment (64).

**(A) Internal State**
Why Move?

Fish infected by larval tapeworms have increased oxygen demands, causing them to spend more time near the surface

**(B) Motion Capacity**
How to move?

A trematode infecting Pacific chorus frogs causes limb alterations resulting in shorter jumping distances, slower swimming speeds, and reduced endurance

**(C) Navigational Manipulation**
Where to move?

Toxoplasma gondii infection in rats converts aversion to feline urine into attraction, resulting in rats venturing closer to feline predators

**(D) External Conditions**
Environmental Preference?

Solitary and groups of Rainbow trout avoid water parasitized by trematodes

## Box 6.1   A case study of the influence of parasites on animal movement: Griffon vultures infected by *Mycoplasma spp.* move less

By documenting intraspecific differences in behavior (e.g., movement), we can improve our ability to identify threats to endangered wildlife species. Towards this end, we have explored the relationship between bacterial infection by *Mycoplasma spp.* and the movements of Griffon vultures (*Gyps fulvus*) in Israel. This vulture species is an obligate scavenger that is locally critically endangered and is the focus of an intensive management program (80).

*Mycoplasma* is a bacterial genus that includes commensal and pathogenic species infecting various raptors. Pathogenic species can cause clinical or subclinical infections that may interfere with reproductive success. However, little is known about the prevalence of *Mycoplasma* in captive and wild raptors, nor its significance to their health. To investigate their potential effect on Griffon vultures, we combined GPS-tracking and clinical sampling (Mycoplasma and other bacteria) by choanal swabs and universal PCR. During the non-breeding period (September–November, 2018–2020), ~100 griffons were captured during population monitoring, of which 44 were negative and 56 were positive for Mycoplasma. Eighty individuals had known origin and age: 48 were wild-born in Israel, 32 were released from management restocking; 37, 33, and 10 were adults (> 4 years (y)), immatures (1–4 y), and juveniles (< 1 y), respectively (Anglister *et al*. unpublished data).

Using a subset of 43 vultures fitted with "Ornitela" Ornitrack 50 3G transmitters (Ornitela, Ltd, Lithuania), and focusing on the first two weeks, two days post-capture, we calculated the daily maximal displacement of vultures. We then modeled the effect of Mycoplasma status, age group, and origin on this movement index, using general linear mixed models with a gamma distribution and a log-link function and considering individuals as random factors. Models were ranked with Akaike's Information Criteria.

Among the eight models considered, the top four models (cumulative weight of 99%) included Mycoplasma status (infected or not). Fixed effects included in the top model (weight 74%) were the cumulative effects of the Mycoplasma status and individual origin. Effect sizes from the top model indicate that non-infected vultures had a mean maximal daily displacement of 16.5 kilometers (km), compared to infected individuals that flew only 10.3 km daily. Vultures originating from the breeding program also flew shorter distances. While these results suggest that infection with Mycoplasma causes a decrease in flight distance, we are unable to exclude an unknown causative factor that could both increase Mycoplasma infection and decrease movement. Future study will explore the clinical and reproductive significance of the different species of *Mycoplasma* spp. in this threatened Israeli population.

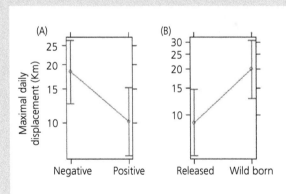 

**Box 6.1, Figure 1**  The relationship between (A) *Mycoplasma* spp. and (B) individual origin (breeding program or wild-born) and griffon vulture (*Gyps fulvus*) movement. Effect size for fixed factors from the top-ranked model included their additive effect. Photo credit: Tovale Solomon.

### 6.3.2 Parasites affect the internal state of their hosts

Beyond their effect on host *ability* to move (i.e., motion capacity), parasites can dramatically influence their hosts' motivation and reasoning for moving (i.e., internal state). Changes in host motivation can occur either through avoidance of infectious hotspots by uninfected individuals (e.g., landscape of disgust (11)), or through resource consumption (by the parasites) that forces hosts to respond to these losses. For example, tapeworms that parasitize certain fish species deplete their oxygen levels drastically, which results in the fish spending more time near the surface ((8), Figure 6.3). By diverting resources and increasing host energetic demands, parasites can dramatically influence the ways in which their hosts allocate energy to movement (7). On the one hand they can lead to reduced movement due to shortage in available energy, and on the other to lead to an increase in movement due to higher motivation and risk taking for obtaining sufficient energy for existence. For instance, coyotes infected with mange (caused by the mite *Sarcoptes scabei*) shifted their home ranges to more anthropogenically disturbed areas where they had easier access to food despite the higher apparent threat (48). These non-mutually exclusive dual effects of (parasite-driven) resource shortage demonstrate the potential insight from movement data to explain variation across host–parasite systems.

While many recent advances in telemetry and biologging have made large-scale movement tracking easier than ever (17), the effects of animal movement and behavior at a micro-movement level has recently begun to draw attention. Analyses of these relationships can be used to quantify the effect of parasites on the host "internal state," such as behavior and energy expenditure (59), (60). Undeniably, changes in the motivation are not always empirically distinguishable from other factors (such as capacity or information) and unambiguously explaining a particular movement is a general challenge in the study of wildlife, where context, perception, internal states, and particular environmental cues all determine an animal's response, but are often unobserved. Nevertheless, striving towards a mechanistic understanding can improve our ability to predict host parasite dynamics. Similarly, a prominent example of changing the internal state is

the ability of some parasites to manipulate behaviors and decisions made by the host (13), (14). Under parasite manipulation, hosts substantially change their priorities, risk-taking behavior, habitat preference, and motivations, but these changes do not necessarily coincide with resource shortage; instead, they are selected for, ultimately increasing the chances of parasite transmission to its subsequent host (see Chapter 12, (61)). Future research into this topic can improve our ability to identify and characterize changes in the internal state of the host.

### 6.3.3 Parasites affect host navigation capacity (cognitive functions)

Parasites may affect host movement through their navigation capacity by influencing their ability to acquire information on their environment (e.g., interfering with sensory modalities), or to process such information when deciding on their movement goals. Parasites may even modify neural pathways, by either up- or down-regulation, ultimately altering the host's behavior and fitness. These impacts are usually considered as examples of host manipulation and are arguably among the most sophisticated ways for parasites to influence host movement and fitness. For example, the *Cordyceps* fungi produce chemicals that alter the navigational sense of their arthropod host, which causes infected hosts to climb to the upper part of a plant before death; this syndrome is named "summit disease" (62). Crickets can fall victim to hairworms which increase water-seeking behavior, causing the host to commit suicide in water, and various parasitoid wasps (e.g., emerald wasp) can turn cockroaches, spiders, or caterpillars into zombies by inserting a venom that either paralyzes their hosts or even allows the wasps to steer them towards a preferred laying site (62). *Toxoplasma gondii* infection in rats (*Rattus norvegicus*) alters the rat's perception of cat predation risk and increases activity levels that expose them to cats (12) (Figure 6.3C). Admittedly, classification of the influence of parasites on movement into one of the two pathways ("internal state" versus "navigation capacity") is sometimes challenging. In other words, it is not always possible to determine if the animal "wanted" to behave in a specific manner (i.e., an "internal state" alteration) or if it was confused (interference in its "navigation capacity").

### 6.3.4 Can parasites affect the external environment of their hosts?

How can internal parasites, or even external ones, affect the environment of their immediate hosts? We are unaware of examples of a change to the actual habitat by parasites. Hence, the main influence of a parasite on a host's surroundings, when it exists, is likely to be through indirect effects that alter its social environment, namely through social avoidance by conspecifics (15). Avoidance of infected hosts or habitats is only beneficial for uninfected hosts when the cost of avoidance is lower than infection itself. Moreover, avoidance is only possible if parasites are associated with certain obvious cues, such as physical marks on infested individuals (e.g., mange) or changes in their behavior (i.e., symptoms) or if parasites prevail in certain habitats. Conspecifics can avoid infections by habitat choice and migration (poor-condition individuals may be incapable of migrating), or by avoiding highly infested individuals from this habitat (11), (47), (63). Spatial avoidance through detection of infective stages of the parasites in the water has been studied in detail in many fish species. For example, one study by Mikheev *et al.* (64). found that both individuals and groups of juvenile rainbow trout (*Oncorhynchus mykiss*) avoided areas of water where trematodes were added (Figure 6.3). While evidence for more direct effects of parasites on their host movement through alternation of the external environment are rare (like ecosystem engineering), future studies can further investigate this pathway.

## 6.4 How to collect movement data for studying host–parasite interactions

### 6.4.1 General guidelines for choosing a tracking approach

After establishing the relevance of movement data for researchers studying host–parasite interactions, and outlining the potential insights these data can provide, we turn to describe some of the common tracking methods and the guiding principles that can facilitate their prioritization. Broadly speaking, the field of movement ecology focuses on measuring animal spatial positions over time, describing the emerging movement patterns, identifying the factors that shape these patterns, and examining their consequences. Nowadays a wide variety of methods and devices are available for collection of movement data (Figure 6.4), focusing here largely on the Lagrangian approach of following known individuals. Selection of these devices is dependent on several criteria.

First, the research question will dictate the type of data needed to address the focal aspects of the host–parasite interaction (namely the temporal resolution and spatial accuracy). For instance, questions on the potential spatial extent of host transmission, or on the population connectivity for either the host or the parasite, are likely to involve extended tracking periods, which in turn compromises accuracy and resolution. In contrast, questions related to transmission networks or the nature of contacts require accurate and high-resolution data from multiple individuals simultaneously, often compromising duration and focusing on a small spatial extent.

Second, the appropriate approach depends on the properties of the habitat and the characteristics of the focal species. Aquatic environments often prevent acoustic tracking since electromagnetic radiation is ineffective in water, whereas mountainous terrain or dense jungles might limit the usefulness of tags that depend on satellite reception. Body size is an obvious limitation for some devices, and the broad space use pattern can also dictate methodology. Some methods require the tagged individual to be recaptured (e.g., geo-locators), while others rely on observer proximity or a base station for direct triangulation or data download, thus constraining the effectiveness of the device for creatures that roam over vast areas (e.g., migrating herbivores (65)).

Third, a chosen tracking method reflects the trade-offs among data *quantity*, data *quality*, and the *cost* of data collection. Methods can vary by orders of magnitude in each of these three axes. *Quantity* can be simplified to the number of fixes (location identifications) obtained per individual over the tracking period. For example, bird-band recovery may merely provide two fixes (locations of banding and recapture), while radio tagging provides tens to hundreds of points per individual, and GPS tags hundreds to thousands. The *quality* of fixes includes their spatial accuracy (i.e., size of position

**Figure 6.4** Examples of tracking methods. (A) A rock pigeon (*Columba livia domestica)* with colored rings for identification and a GPS-ACC tag. (B) Barn owl (*Tyto alba*) tagged with an ATLAS radio-tag. (C) Two Nubian ibex (*Capra nubiana*) with ear tags and GPS collars. (D) PIT tag in the ventral side of a salmon fry. (E) Searching for radio telemetry transmitter of vultures in the Negev desert, Israel. Photo credit: Orr Spiegel, Jonathan Tichon, and Anna Steel.

error, ideally approaching a couple of meters). It also depends on temporal precision (i.e., uniformity of sampling interval) that can facilitate analysis of the collected trajectories. While geo-locators provide position errors of tens of kilometers, GPS, or ATLAS tag (66) positions tend to have errors of less than 3-5 meters. *Costs* incorporate the price of the animal-borne unit (e.g., the radio tag), the tracking equipment (e.g., the antennae or base station), and the required personnel. Tags, for instance, can range from a few US dollars for a simple PIT tag, to $4000 for a sophisticated solar-powered GPS unit and data logger with remote communication ability (see Table 6.1 for a few haphazardly chosen examples). Yet, it should be noted that these substantial differences can be masked by the high variation in personnel effort needed to support the data collection. While some GPS devices require no additional effort after deployment, since they can directly send data via cellular/satellite communication, cheaper radio telemetry tags often require intensive tracking effort after deployment.

Finally, even within a given type of technology, there is often an inherent trade-off between tracking duration and sampling intensity, with their product being limited by the body size of the focal organism and the battery capacity it can carry. This trade-off is particularly pronounced when using GPS technology where power constraints dictate the overall number of fixes it can provide. Dynamic sampling regimes allow researchers to alleviate this trade-off and optimize data collection by using duty cycles (e.g., avoid sampling during daytime for nocturnal creatures), geo-fencing (i.e., enhance sampling rates when a tracked individual is within a polygon of interest), and/or intensive sampling bursts. For diurnal, medium-to-large species living in open habitats (i.e., mostly birds), solar panels can circumvent this problem by obtaining a very large number of fixes, almost irrespective of battery size.

### 6.4.2 Incorporating complementary sensors

While abilities to collect longitudinal data on hosts and their movement are imperative for the understanding of parasite ecology, the technological revolution enhancing these abilities is not limited to movement alone. Many tags serve as platforms for additional independent sensors that facilitate the collection of complementary datasets on the focal individual and its environment (17), (59), (65). These sensors can assess the physiological condition of the individual, its motivation, its vocal communication with others, and the social context. For example, accelerometer data can be used to recognize feeding events, and proximity sensors can log social encounters with other tagged individuals. While some of these aspects can be inferred from movement data alone (e.g., a stop at a feeding station, or apparent nearby stops of different individuals), having independent data provides a direct and robust evaluation, and reduces interpretation errors in correlative findings (67). For instance, imagine a bat switching from fast and directional flight to slow and tortuous flight. This change can be explained by a shift from a commuting flight to a local foraging search, or entering a cluttered environment (68). Having independent information on bat feeding events (e.g., from acoustic recordings of echolocation feeding "buzz") can allow a researcher to differentiate between these alternatives. The distinction between these alternatives can be valuable for understanding transmission rates (e.g., if parasites are consumed with prey) and infection status of focal individuals (e.g., if infected individuals lose appetite).

Sensors that provide data on the internal state of an individual include loggers of body temperature, heart rate, blood pressure, and neural activity (69). Accelerometers also belong to this group; these sensors measure the forces experienced by the tag due to gravity and the motion of the animal, allowing one to infer an individual's energy expenditure, behavior (e.g., walking, flying, standing, etc.) and micro-movements (59), (60). For instance, a study done on Griffon vultures used accelerometer data to identify feeding events and assess changes in flight characteristics as a function of time since the last feeding event (deprivation periods), demonstrating a trade-off between increasing motivation to find food and the risk of starvation (70). In another study, domestic sheep were treated for gastrointestinal nematodes and their activity patterns (measured with accelerometers) changed, demonstrating the effectiveness of these sensors in quantifying parasite load and influence (71). Other sensors are useful for characterizing the immediate environment and social context of the focal individual. These include sensors that measure physical conditions (such as

light level, diving depth, external temperature, etc.) or tags that collect data on the biological context (presence of resources or conspecifics). Animal-borne cameras, audio recordings (especially in bats that rely on echolocation), and proximity sensors are all developing technologies that are becoming increasingly popular and can allow us to address questions regarding social interactions and parasite transmission. Future studies can elaborate on the use of these sensors for identifying the effect of parasites on host condition, internal state, and locomotion capacity. For instance, this can be achieved through *in situ* experiments treating free-ranging tagged animals for parasites and quantifying the changes in activity budget and movement periods.

## 6.5 Future directions and concluding remarks

Traditionally, parasite and pathogen dynamics have been modeled using compartmental models (e.g., SEIR models). However, to remain tractable, SEIR models often (over)simplify the reality of disease dynamics. Hence, while these models provide valuable insight into general patterns, they often fail in generating accurate predictions due to unaccounted heterogeneity in the transmission process (especially in encounter rates (section 6.2.1). and transmission probabilities (section 6.2.4)). While SEIR models continue to serve as a fundamental tool in modeling disease and parasite dynamics, new modeling approaches are becoming popular alternatives. These include models of metapopulation, lattice or cellular automata, individual-based and continuous-space models, and social network analyses (72). These models often generate better predictions for the behaviors of a particular system, and allow testing of realistic scenarios (such as the expected effect of habitat fragmentation on transmission rate; (23)). Nevertheless, these models are less transferable across systems, and require more intensive system-specific parameterization. Many of these parameters (e.g., variation among hosts in contact rates, or in the influence of infection on their behavior), directly rely on tracking data. For instance, the growing popularity of social network analysis for investigating parasite spread, is driven, in part, by the ability to automate simultaneous tracking of multiple individuals (53). We suggest that this trend towards more sophisticated ("data hungry") models will be accelerated thanks to the synergetic effect of a few methodological and conceptual factors. First, growing accessibility of computational power allows researchers across the world to readily run such demanding simulation models. Second, technological innovations facilitate better (and cheaper) tracking of an ever-growing array of species and individuals of hosts in various systems. To date, most parasites are still too small to be tracked directly, but this gap might be soon within reach. For instance, by experimentally infesting GPS-tracked sleepy lizards with known genetic lines of *B. hydrosauri* ticks, researchers are able to track their transmission individually among hosts and empirically test model predictions. Third, this wealth of data also boosts conceptual developments and analytical methods that promote understanding of ecological complexities. For instance, social network studies now highlight the importance of dissecting the contributions of social preference and ecological constraints (e.g., spatial configuration of resources) for shaping network structures and derived transmission dynamics (28), (52). Finally, knowledge dissemination through the scientific community is much easier nowadays, giving rise to a diverse array of user-friendly modeling platforms (e.g., Numerus epidemic simulator; (73)), online courses, and shared codes.

The fields of movement ecology, parasite ecology, and behavioral ecology have evolved rapidly over the last two decades, each embracing new concepts and tools (e.g., the notions of "animal personalities," or "host manipulation"). Yet, these parallel developments remain poorly connected, slowing progress and fruitful fusions (19). For instance, intraspecific consistency in movement (that draws ideas from both behavioral and movement ecology) and its consequences for ecological dynamics (including disease) is only now becoming a new research frontier (35). The growing attention in the ecological literature to the behavioral feedback between host and parasite highlights the potential benefit from multidisciplinary studies and the importance of collecting longitudinal data on hosts before, during, and after infection.

In addition to insights into the various mechanisms of local interactions on larger and longer spatiotemporal scales it is the host movement (occasionally rare dispersal and long-range forays) that shapes the population of parasites (41), (74). For instance, a study on ectoparasites (lice and fleas) of shearwaters (42). found no genetic structure in parasite populations. The study suggests that an underestimation of the magnitude of movement of lice between hosts (host movement and parasite movement) may explain this finding, highlighting how movement data can also help at these broader scales. Ongoing global changes alter the relative distributions of hosts and parasite species, often creating mismatches, or introducing new parasites that drive species to extinctions (e.g., Hawaiian honeycreepers that became locally extinct at low elevations due to introduced avian malaria (75)). Insights into the drivers of these spatial dynamics are particularly needed to avoid such detrimental effects. Local anthropogenic effects such as land use change and fragmentation may promote outbreaks of pathogens in wildlife, such as avian malaria and avian influenza (76), (77). Fragmentation can also enhance the risk of zoonotic outbreaks due to stronger overlap between humans and wildlife and their parasites increasing. For example, in Malaysian Borneo, increases in human infections by the zoonotic malaria, *Plasmodium knowlesi*, were found to be connected to increasing contact between people and macaques due to deforestation (77). A mechanistic understanding of transmission processes is mandatory for efficient interventions and healthy ecosystem functioning, and integrating movement data into studies of host–parasite interaction can help achieve this goal in many cases.

## Acknowledgments

We are grateful to the editors of this book (VO Ezenwa, SM Altizer, and RJ Hall) for inviting us to take part in this exciting initiative. We also appreciate careful and constructive feedback from RJ Hall and two anonymous reviewers. This research was supported by the NSF-BSF grant to OS (IOS# 2015662–2019082), and by the Israeli Science Foundation (#396/20). The Faculty of Life Science at Tel Aviv university supported MC's stipend and Ramat Hanadiv supported NA.

# References

1. Dobson AP and Hudson PJ. Parasites, disease and the structure of ecological communities. *Trends Ecol Evol.* 1986 Jul 1;1(1):11–5.

2. Price PW. Parasite mediation in ecological interactions. *Annu Rev Ecol Syst Vol 17.* 1986;17(1986):487–505.

3. Altizer S, Nunn CL, Thrall PH, Gittleman JL, Antonovics J, Cunningham AA, et al. Social organization and parasite risk in mammals: Integrating theory and empirical studies. *Annu Rev Ecol Evol Syst.* 2003;34:517–47.

4. Beldomenico PM and Begon M. Disease spread, susceptibility and infection intensity: Vicious circles? *Trends Ecol Evol.* 2010;25(1):21–7.

5. Ezenwa VO, Archie EA, Craft ME, Hawley DM, Martin LB, Moore J, et al. Host behaviour–parasite feedback: An essential link between animal behaviour and disease ecology. *Proc R Soc B: Biol Sci.* 2016;283(1828).

6. Vanderwaal KL, Obanda V, Omondi GP, McCowan B, Wang H, Fushing H, et al. The "strength of weak ties" and helminth parasitism in giraffe social networks. *Behav Ecol.* 2016;00:arw035.

7. Binning SA, Shaw AK, and Roche DG. Parasites and host performance: Incorporating infection into our understanding of animal movement. *Integr Comp Biol.* 2017;57(2):267–80.

8. Giles N. A comparison of the behavioural responses of parasitized and non-parasitized three-spined sticklebacks, *Gasterosteus aculeatus* L., to progressive hypoxia. *J Fish Biol.* 1987;30:631–8.

9. Altizer S, Bartel R, and Han BA. Animal migration and infectious disease risk. *Science (80-).* 2011 Jan 21;331(6015):296–302.

10. McDonald JL, Robertson A, and Silk MJ. Wildlife disease ecology from the individual to the population: Insights from a long-term study of a naturally infected European badger population. *Journal of Animal Ecology.* 2018;87: 101–12.

11. Weinstein SB, Buck JC, and Young HS. A landscape of disgust. *Science.* 2018 Mar 16;359(6381):1213–14.

12. Vyas A, Kim S-K, Giacomini N, Boothroyd JC, and Sapolsky RM. Behavioral changes induced by Toxoplasma infection of rodents are highly specific to aversion of cat odors. *PNAS.* 2007 Apr 10;104(15):6442–7.

13. Ponton F, Otálora-Luna F, Lefvre T, Guerin PM, Lebarbenchon C, Duneau D, et al. Water-seeking behavior in worm-infected crickets and reversibility of parasitic manipulation. *Behav Ecol.* 2011;22(2):392–400.

14. Poulin R and Maure F. Host manipulation by parasites: A look back before moving forward. *Trends Parasitol.* 2015;31(11):563–70.

15. Stockmaier S, Stroeymeyt N, Shattuck EC, Hawley DM, Meyers LA, and Bolnick DI. Infectious diseases and social distancing in nature. *Science.* 2021 Mar 5;371(6533):eabc8881.

16. Nathan R, Getz WM, Revilla E, Holyoak M, Kadmon R, Saltz D, et al. A movement ecology paradigm for unifying organismal movement research. *PNAS.* 2008;105(49):19052–9.

17. Kays R, Crofoot MC, Jetz W, and Wikelski M. Terrestrial animal tracking as an eye on life and planet. *Science.* 2015 Jun 12;348(6240):aaa2478–aaa2478.

18. Farnsworth ML, Hoeting JA, Hobbs NT, and Miller MW. Linking chronic wasting disease to mule deer movement scales: A hierarchical Bayesian approach. *Ecol Appl.* 2006;16(3):1026–36.

19. Dougherty ER, Seidel DP, Carlson CJ, Spiegel O, and Getz WM. Going through the motions: Incorporating movement analyses into disease research. *Ecol Lett.* 2018;21(4):588–604.

20. Main AR and Bull CM. The impact of tick parasites on the behaviour of the lizard *Tiliqua rugosa*. *Oecologia.* 2000;122(4):574–81.

21. Blackburn JK, Ganz HH, Ponciano JM, Turner WC, Ryan SJ, Kamath P, et al. Modeling R0 for pathogens with environmental transmission: Animal movements, pathogen populations, and local infectious zones. *Int J Environ Res Public Heal.* 2019;16:954.

22. Ripperger SP, Stockmaier S, and Carter GG. Sickness behaviour reduces network centrality in wild vampire bats. *bioRxiv.* 2020;2020.03.30.015545.

23. Tracey JA, Bevins SN, Vandewoude S, and Crooks KR. An agent-based movement model to assess the impact of landscape fragmentation on disease transmission. *Ecosphere.* 2014;5(9):1–24.

24. Mcinturf AG, Pollack L, Yang LH, and Spiegel O. Vectors with autonomy: What distinguishes animal-mediated nutrient transport from abiotic vectors? *Biol Rev.* 2019;94(5):1761–73.

25. Gurarie E, Bracis C, Delgado M, Meckley TD, Kojola I, and Wagner CM. What is the animal doing? Tools for exploring behavioural structure in animal movements. *Journal of Animal Ecology.* 2016;85: 69–84.

26. Thieltges DW, and Poulin R. Parasites and pathogens: Avoidance. *Encycl Life Sci.* 2008;1–8.

27. Goodman BA and Johnson PTJ. Disease and the extended phenotype: Parasites control host performance and survival through induced changes in body plan. *PLoS One.* 2011;6(5).

28. Spiegel O, Leu ST, Sih A, and Bull CM. Socially interacting or indifferent neighbours? Randomization of movement paths to tease apart social preference and spatial constraints. *Methods Ecol Evol.* 2016;7(8):971–9.

29. Boyer N, Réale D, Marmet J, Pisanu B, and Chapuis JL. Personality, space use and tick load in an introduced population of Siberian chipmunks *Tamias sibiricus*. *J Anim Ecol*. 2010;79(3):538–47.

30. Payne E, Sinn DL, Spiegel O, Leu ST, Wohlfeil C, and Godfrey SS, *et al*. Consistent individual differences in ecto-parasitism of a long-lived lizard host. *Oikos*. 2020;(March):1–11.

31. Lloyd-Smith JO, Schreiber SJ, Kopp PE, and Getz WM. Superspreading and the effect of individual variation on disease emergence. *Nature*. 2005;438(7066):355–9.

32. Sih A, Spiegel O, Godfrey S, Leu S, and Bull CM. Integrating social networks, animal personalities, movement ecology and parasites: A framework with examples from a lizard. *Anim Behav*. 2018;136(2018): 195–205.

33. Cross PC, Lloyd-Smith JO, Johnson PLF, and Getz WM. Duelling timescales of host movement and disease recovery determine invasion of disease in structured populations. *Ecol Lett*. 2005;8(6):587–95.

34. Paull SH, Song S, McClure KM, Sackett LC, Kilpatrick AM, and Johnson PTJ. From superspreaders to disease hotspots: Linking transmission across hosts and space. *Front Ecol Environ*. 2012;10(2):75–82.

35. Spiegel O, Leu ST, Bull CM, and Sih A. What's your move? Movement as a link between personality and spatial dynamics in animal populations. *Ecol Lett*. 2017;20(1):3–18.

36. Bordes F, Morand S, Kelt DA, and Van Vuren DH. Home range and parasite diversity in mammals. *Am Nat*. 2009;173(4):467–74.

37. Benavides J, Walsh PD, Meyers LA, Raymond M, and Caillaud D. Transmission of infectious diseases en route to habitat hotspots. *PLoS One*. 2012;7(2):1–9.

38. Pérez-Tris J and Bensch S. Dispersal increases local transmission of avian malarial parasites. *Ecol Lett*. 2005;8(8):838–45.

39. VanderWaal KL, Atwill ER, Isbell LA, and McCowan B. Linking social and pathogen transmission networks using microbial genetics in giraffe (*Giraffa camelopardalis*). *Anim Ecol*. 2014 Mar;83(2): 406–14.

40. Nunn CL, Thrall PH, Leenndertz FH, and Boesch C. The spread of fecally transmitted parasites in socially-structured populations. *PLoS One*. 2011;6(6):21677.

41. Clayton DH, Bush SE, and Johnson KP. Ecology of congruence: Past meets present. *Syst Biol*. 2004;53(1): 165–73.

42. Gómez-Díaz E, González-Solís J, Peinado MA, and Page RDM. Lack of host-dependent genetic structure in ectoparasites of Calonectris shearwaters. *Mol Ecol*. 2007;16(24):5204–15.

43. Figuerola J and Green AJ. Haematozoan parasites and migratory behaviour in waterfowl. *Evol Ecol*. 2000;14(2):143–53.

44. Hall RJ, Altizer S, Peacock SJ, and Shaw AK. Animal migration and infection dynamics: Recent advances and future frontiers. In Ezenwa VO, Altizer S, and Hall RJ (eds), *Animal Behavior and Parasitism*. Oxford: Oxford University Press, 2022. DOI: 10.1093/oso/9780192895561.003.0006.

45. Spiegel O, Leu ST, Sih A, Godfrey SS, and Bull CM. When the going gets tough: Behavioural type-dependent space use in the sleepy lizard changes as the season dries. *Proc R Soc B: Biol Sci*. 2015;282(1819).

46. Leu ST, Kappeler PM, and Bull CM. Refuge sharing network predicts ectoparasite load in a lizard. *Behav Ecol Sociobiol*. 2010 Sep 21;64(9):1495–503.

47. Reckardt K and Kerth G. Roost selection and roost switching of female Bechstein's bats (*Myotis bechsteinii*) as a strategy of parasite avoidance. *Oecologia*. 2007;154:581–8.

48. Murray M, Edwards MA, Abercrombie B, and St. Clair CC. Poor health is associated with use of anthropogenic resources in an urban carnivore. *Proc R Soc B: Biol Sci*. 2015;282(1806).

49. Friant S, Ziegler TE, and Goldberg TL. Changes in physiological stress and behaviour in semi-free-ranging red-capped mangabeys (*Cercocebus torquatus*) following antiparasitic treatment. *Proc R Soc B: Biol Sci*. 2016 Jul 27;283(1835):20161201.

50. Ward AJ., Duff AJ, Krause J, and Barber I. Shoaling behaviour of sticklebacks infected with the microsporidian parasite, *Glugea anomala*. *Environ Biol Fishes*. 2005;(72):155–60.

51. Mistrick J, Gilbertson MLJ, White LA, and Craft ME. Constructing animal networks for parasite transmission inference. In Ezenwa VO, Altizer S, and Hall RJ (eds), *Animal Behavior and Parasitism*. Oxford: Oxford University Press, 2022. DOI: 10.1093/oso/9780192895561.003.0004.

52. Albery GF, Kirkpatrick L, Firth JA, and Bansal S. Unifying spatial and social network analysis in disease ecology. *J Anim Ecol*. 2021;90(1):45–61.

53. Smith JE and Pinter-Wollman N. Observing the unwatchable: Integrating automated sensing, naturalistic observations and animal social network analysis in the age of big data. *J Anim Ecol*. 2021 Jan 20;90(1): 62–75.

54. Bull CM, Godfrey SS, and Gordon DM. Social networks and the spread of Salmonella in a sleepy lizard population. *Mol Ecol*. 2012 Sep;21(17):4386–92.

55. Weber N, Carter SP, Dall SRX, Delahay RJ, McDonald JL, Bearhop S, *et al*. Badger social networks

correlate with tuberculosis infection. *Curr Biol.* 2013;23(20):R915–6.

56. Hamede RK, Bashford J, McCallum H, and Jones M. Contact networks in a wild Tasmanian devil (*Sarcophilus harrisii*) population: Using social network analysis to reveal seasonal variability in social behaviour and its implications for transmission of devil facial tumour disease. *Ecol Lett.* 2009;12(11):1147–57.

57. Cross PC, Almberg ES, Haase CG, Hudson PJ, Maloney SK, Metz MC, *et al.* Energetic costs of mange in wolves estimated from infrared thermography. *Ecology.* 2016;97(8):1938–48.

58. Taggart PL, Leu ST, Spiegel O, Godfrey SS, Sih A, and Bull CM. Endure your parasites: Sleepy lizard (*Tiliqua rugosa*) movement is not affected by their ectoparasites. *Can J Zool.* 2018 Dec;96(12):1309–16.

59. Nathan R, Spiegel O, Fortmann-Roe S, Harel R, Wikelski M, and Getz WM. Using triaxial acceleration data to identify behavioral modes of free-ranging animals: General concepts and tools illustrated for griffon vultures. *J Exp Biol.* 2012;215(6):986–96.

60. Brown DD, Kays R, Wikelski M, Wilson R, and Klimley AP. Observing the unwatchable through acceleration logging of animal behavior. *Anim Biotelemetry.* 2013;1(1):1–16.

61. Godfrey SS and Poulin, R. Host manipulation by parasites: From individual to collective behavior. In Ezenwa VO, Altizer S, and Hall RJ (eds), *Animal Behavior and Parasitism*. Oxford: Oxford University Press, 2022. DOI: 10.1093/oso/9780192895561.003.0012.

62. Libersat F, Delago A, and Gal R. Manipulation of host behavior by parasitic insects and insect parasites. *Annu Rev Entomol.* 2009;54:189–207.

63. Poulin R and FitzGerald GJ. Risk of parasitism and microhabitat selection in juvenile sticklebacks. *Can J Zool.* 1989 Jan 1;67(1):14–8.

64. Mikheev VN, Pasternak AF, and Taskinen JVT. Grouping facilitates avoidance of parasites by fish. *Parasit Vectors.* 2013;6(1):301.

65. Wilmers CC, Nickel B, Bryce CM, Smith JA, Wheat RE, and Yovovich V. The golden age of bio-logging: How animal-borne sensors are advancing the frontiers of ecology. *Ecology.* 2015;96(7):1741–53.

66. Toledo S, Shohami D, Schiffner I, Lourie E, Orchan Y, Bartan Y, *et al.* Cognitive map-based navigation in wild bats revealed by a new high-throughput tracking system. *Science.* 2020;369(6500):188–93.

67. Spiegel O, Harel R, Centeno-Cuadros A, Hatzofe O, Getz WM, and Nathan R. Moving beyond curve fitting: Using complementary data to assess alternative explanations for long movements of three vulture species. *Am Nat.* 2015 Feb;185(2):E44–54.

68. Grodzinski U, Spiegel O, Korine C, and Holderied MW. Context-dependent flight speed: Evidence for energetically optimal flight speed in the bat *Pipistrellus kuhlii*? *J Anim Ecol.* 2009;78(3):540–8.

69. López-López P. Individual-Based Tracking Systems in Ornithology: Welcome to the era of big data. *Ardeola.* 2016;63(1).

70. Spiegel O, Harel R, Getz WM, and Nathan R. Mixed strategies of griffon vultures' (*Gyps fulvus*) response to food deprivation lead to a hump-shaped movement pattern. *Mov Ecol.* 2013;1(1):1–12.

71. Burgunder, J, Petrželková, KJ, Modrý, D, Kato, A, and MacIntosh, AJJ. Fractal measures in activity patterns: Do gastrointestinal parasites affect the complexity of sheep behaviour? *Appl. Anim. Behav. Sci.* 2018;205: 44–53.

72. White LA, Forester JD, and Craft ME. Dynamic, spatial models of parasite transmission in wildlife: Their structure, applications and remaining challenges. *J Anim Ecol.* 2018;87:559–80.

73. Getz WM, Salter R, Muellerklein O, Yoon HS, and Tallam K. Modeling epidemics: A primer and Numerus Model Builder implementation. *Epidemics.* 2018;25(July):9–19.

74. Mazé-Guilmo E, Blanchet S, McCoy KD, and Loot G. Host dispersal as the driver of parasite genetic structure: A paradigm lost? *Ecol Lett.* 2016;19(3):336–47.

75. Benning TL, LaPointe D, Atkinson CT, and Vitousek PM. Interactions of climate change with biological invasions and land use in the Hawaiian Islands: Modeling the fate of endemic birds using a geographic information system. *PNAS.* 2002 Oct 29;99(22): 14246–9.

76. Gilbert M, Prosser DJ, Zhang G, Artois J, Dhingra MS, Tildesley M, *et al.* Could changes in the agricultural landscape of Northeastern China have influenced the long-distance transmission of highly pathogenic avian influenza H5Nx viruses? *Front Vet Sci.* 2017;4(DEC): 1–8.

77. Stark DJ, Fornace KM, Brock PM, Abidin TR, Gilhooly L, Jalius C, *et al.* Long-tailed macaque response to deforestation in a *Plasmodium knowlesi*-Endemic Area. *Ecohealth.* 2019;16(4):638–46.

78. Gabagambi NP, Salvanes AGV, Midtøy F, and Skorping A. The tapeworm *Ligula intestinalis* alters the behavior of the fish intermediate host *Engraulicypris sardella*, but only after it has become infective to the final host. *Behav Processes.* 2019;158(July 2018):47–52.

79. Hahn S, Briedis M, Barboutis C, Schmid R, Schulze M, Seifert N, *et al.* Spatially different annual cycles but similar haemosporidian infections in distant populations of collared sand martins. *BMC Zool.* 2021;6(1): 1–11.

80. Choresh Y, Burg D, Trop T, Izhaki I, and Malkinson D. Long-term griffon vulture population dynamics at Gamla Nature Reserve. *J Wildl Manage.* 2019;83(1): 135–44.

CHAPTER 7

# Animal migration and infection dynamics: Recent advances and future frontiers

Richard J. Hall, Sonia Altizer, Stephanie J. Peacock, and Allison K. Shaw

## 7.1 Introduction

Animal migration, where individuals or groups of animals move in broadly predictable directions at predictable times in response to changing abiotic or biotic conditions, is a common movement behavior in birds, insects, mammals, fish, and other taxa (1). By connecting distant locations and diverse habitat types, animal migrations provide crucial ecosystem services, including the transfer of nutrients or dispersion of less-mobile organisms (2). Beyond their ecological importance, long-distance animal migrations are of cultural importance to people, marking the changing of the seasons, providing psychological benefits associated with nature-based recreation, and economic benefits from hunting, ecotourism, or seasonal pest control (3). During the past 50 years, researchers have uncovered many of the drivers of migratory behavior and the ecological consequences of these movements for migrants and the species with which they interact (4).

One interaction that intimately shapes, and is shaped by, migratory behavior is parasitism (here defined broadly to include infection by microparasites such as viruses and bacteria, as well as infestation by macroparasites such as helminths). Recognition that migration can influence parasitism first appeared in the scientific literature in the early twentieth century, when it was noted that migratory birds were more commonly infected with blood parasites than non-migrants (5). Early studies also

proposed mechanisms by which parasites persist in migratory hosts, including dormancy of avian malaria during migration (6), which reduces costs of infection until infections reactivate in the summer, when biting vectors emerge and a pulse of host offspring facilitates transmission. Later studies suggested a key role of migratory behavior in shaping parasite transmission, including allowing hosts to avoid or escape from habitats where parasite transmission stages accumulate, as noted for reindeer (*Rangifer tarandus*) and their warble fly parasites (7).

More recently, scientists have sought to identify general patterns and mechanistic support for how migratory behavior can shape parasite transmission, and how parasites influence migratory propensity over ecological and evolutionary timescales, through experimental manipulations (8)–(10), cross-taxon comparative analyses (11)–(14), and mathematical modeling (15)–(17). Notably, the majority of these studies focus on unidirectional effects of migration on parasitism, with fewer studies considering how parasites (or risk of parasite exposure) shape migration. In doing so, these studies miss the potential importance of bidirectional interactions between migration and parasitism that give rise to feedbacks that can crucially impact animal behavior and parasitism ((18); also see Chapter 2 of this book (19)). Further, in studying migration as a population-level phenomenon, the effects

Richard J. Hall et al., *Animal migration and infection dynamics*. In: *Animal Behavior and Parasitism*. Edited by Vanessa O. Ezenwa, Sonia Altizer, and Richard J. Hall, Oxford University Press. © Oxford University Press (2022). DOI: 10.1093/oso/9780192895561.003.0007 of individual host variation in whether, when, and where they migrate, and their ability to acquire, transmit, and survive or clear infection, are overlooked, in spite of their potential importance in explaining population-level outcomes such as partial migration (20) and superspreading of infection (21). The importance of understanding migratory host–parasite relationships is heightened given the pace of environmental change in the Anthropocene, which simultaneously threatens the persistence of long-distance migration (22) and brings migrants into close contact with humans and domesticated animals, potentially resulting in parasite sharing and spread of pathogens of conservation or public health concern (23).

In this chapter, we first provide an overview of theoretical and empirical support for proposed mechanisms by which migration influences parasitism and vice versa. Next, we highlight how studying bidirectional interactions between migration and parasitism that lead to feedbacks could serve as a unifying framework for explaining diverse migratory behaviors and infection outcomes at the individual and population level. We then synthesize recent advances in understanding how migratory host–parasite interactions respond to global change in the context of migration–parasite feedbacks. We conclude by highlighting three key themes that represent emerging research frontiers in migration–parasitism research: migration–parasitism feedbacks; individual (co)variation in migration and infection status; and global change impacts on migratory host-parasite dynamics in the context of global health; and outline some promising future directions to advance our holistic understanding of migration–parasitism interactions in the Anthropocene.

## 7.2 Effects of migration on parasitism

Empirical studies have sought to quantify relationships between host migratory behavior and metrics of parasitism, including parasite diversity (number of parasite strains, species, or taxa), infection prevalence (the fraction of a population infected), or intensity (parasite burden within hosts and its distribution among hosts). Evidence of positive, negative, or no relationships between migration and parasitism have been documented, prompting

the development of hypotheses potentially explaining these diverse outcomes (Box 7.1). Most of the proposed mechanisms are unidirectional, and focus on how aspects of host migratory strategy (broadly defined here to include the probability of migration, distance travelled, route choice, stopover duration and frequency, and timing of movement) influence observed infection patterns via host exposure and susceptibility to, or transmission and clearance of, infection. Fewer mechanisms focus on how parasite infection influences migratory behavior, and fewer still explicitly explore the role of feedbacks between migratory strategy and parasitism. By varying traits associated with host migration (e.g., migration distance and phenology) and parasites (e.g., transmission mode and virulence), mathematical models can illuminate the conditions under which migration alters infection dynamics, and when migratory strategies may be shaped by parasitism, on ecological and evolutionary timescales. In the next section we synthesize empirical support for hypothesized mechanisms linking migration and parasitism, and theoretical work that explores the ecological and evolutionary consequences of these linkages for host–parasite dynamics, and further highlight some potential consequences of migration–parasite feedbacks.

### 7.2.1 Migratory strategy and exposure to diverse parasites

One way in which migratory behavior influences parasitism is by increasing the pool of parasites to which animals are exposed. Comparative studies of resident and migratory birds, ungulates, and fishes have found evidence that migrants harbor more parasite species than residents (32). The most intuitive explanation is that migrants encounter more parasites than residents because they typically have larger ranges and sample diverse habitat types with distinct parasite communities (**environmental sampling**, Table 7.1). However, the extent to which migrations elevate exposure to parasites is contingent on traits of the migrants, including their movement mode and how frequently they sample environments along their migratory routes. Animals that move primarily by walking or swimming, and that feed continuously along their route, face more frequent opportunities for exposure to

## Box 7.1  Overview of migration–parasite interactions

Many aspects of parasite transmission can be influenced by host migration (Figure 1A), including exposure to infected hosts or habitats, attributes of host condition and immunity that influence host susceptibility to infection or parasite development and replication, and clearance of infection through recovery or host mortality. In turn, infection (or risk of infection) can influence whether, when, and how (i.e., which routes and stopover sites are used) animals migrate, and their ability to successfully complete migration (Figure 1C). The primary coupling mechanisms proposed in past work by which migration influences infection, or infection influences migration (Figure 1B) are described in Table 7.1. The outcomes of interactions between these mechanisms, where migration influences parasitism, which in turn influences migration, could drive eco-evolutionary feedbacks that shape outcomes for host migratory strategy and infection prevalence.

Four key examples represent diverse host and parasite taxa in which migration has been demonstrated to influence patterns of infection both within and among migratory populations. First, monarch butterflies (*Danaus plexippus*) are famous for their fall migration in eastern North America,

traveling from as far north as southern Canada to wintering sites in central Mexico. Monarchs are commonly infected with a debilitating protozoan parasite *Ophryocystis elektroscirrha*, henceforth OE, which is transmitted when spores scattered onto eggs and milkweed by adult monarchs are ingested by young caterpillars (24) (Figure 1B, lower left). Parasitized monarchs emerge as adults covered with parasite spores on their bodies. Infections lower adult monarch lifespan, size, and flight performance (8). Work across multiple populations showed that OE prevalence decreases with migratory propensity and annual distance flown (25), (26), with the lowest infection prevalence occurring among eastern North American monarchs, and among individual monarchs that migrated the farthest distances. Field and experimental studies provided evidence for both migratory culling, where heavily infected individuals are less likely to survive strenuous migrations, and migratory escape, where successive breeding generations reduce their exposure by vacating sites where OE spores accumulate on their host plants, in lowering infection prevalence in migratory populations (8), (27).

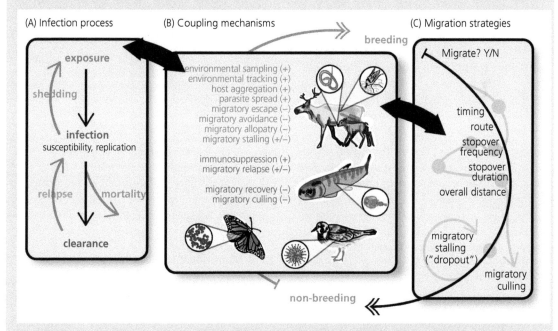

**Box 7.1, Figure 1**  Mechanisms coupling migratory behavior and infection processes. (A) Infection processes relating to parasite exposure (green), susceptibility, replication, and shedding (red), and clearance of (or mortality from) infection (blue) can be influenced by, or influence, aspects of host migration strategy (C), including whether, when and how (far) to migrate via unidirectional coupling mechanisms (B) which are described in Table 7.1 of the main text. Interactions between infection and migration could further result in migration-parasitism feedbacks. Images denote some well-studied systems in which one or more of these mechanisms has been empirically demonstrated. Artwork by S. Peacock.

**Box 7.1** *Continued*

Caribou (*Rangifer tarandus*) occur in the Arctic tundra, where they have a circumpolar distribution. Herds migrate hundreds to thousands of kilometers between calving grounds, summer feeding grounds, and wintering grounds. These movements have been posited to facilitate migratory escape from warble flies (*Hypoderma tarandi*, (7)), which negatively influence caribou body condition, and helminths (*Ostertagia* spp., (28)), which can reduce body condition and pregnancy rates (Figure 1B, top right). In Norwegian caribou (or reindeer), herds that undertook post-calving migrations showed lower larval abundance of warble flies than herds that remained close to calving grounds all summer; additionally, distance migrated was negatively associated with warble fly abundance. Together, these observations suggested that post-calving migration could be a behavioral adaptation to reduce or escape warble fly infection (7). In North American caribou, delays due to development of environmental stages of *Ostertagia* helminths mean that caribou movement away from areas where parasites have been shed in feces allows hosts to escape re-exposure (28).

Pink salmon (*Oncorhynchus gorbuscha*) and other salmonids migrate from freshwater spawning streams in western North America to the Pacific Ocean. These salmon are often infested by sea lice (*Lepeophtheirus salmonis*), which attach to host surface tissue to feed, resulting in host stress and increased susceptibility to viral or bacterial coinfection (Figure 1B, middle right). Juvenile salmon face a much higher mortality risk from sea louse infestation than adults.

Since freshwater rapidly kills sea lice, season-specific and stage-specific use of freshwater habitats can reduce infection transmission and prevalence. First, migration of adults to freshwater can promote migratory recovery from infection prior to spawning (29). Second, juvenile salmon hatch in freshwater and mature in near-shore environments where exposure to adult salmon—and their parasites—is naturally low. By reducing spatiotemporal overlap with infectious adults that spend most of their lives in the open ocean, separation of infection-vulnerable and infection-tolerant life history stages reduces parasite transmission, a phenomenon known as migratory allopatry (9).

Last, ruddy turnstones (*Arenaria interpres*) are globally distributed shorebirds that breed in the Arctic tundra and migrate to temperate or tropical latitudes to overwinter, and are one of several waterbird species infected with LPAIV (Figure 1B, lower right). Turnstones and other shorebirds aggregate reliably at spring stopover sites, and time their arrival to coincide with the mass spawning of horseshoe crabs, an important nutritional resource for attaining breeding condition. LPAIV is shed in bird feces where it can persist in water and is ingested by feeding waterbirds. The high density of migratory shorebirds that coincide with LPAIV-infected resident and overwintering ducks can increase LPAIV transmission within and among species (i.e., host aggregation), and facilitate virus dispersal along flyways and to breeding grounds (pathogen dispersal) (30), (31).

environmental parasite stages during transit than aerial migrants that make few or no stopovers (33), especially if migrants track environmental conditions that favor survival of infected hosts, external parasite stages, and disease vectors (**environmental tracking**, Table 7.1). Additionally, migrants that aggregate at stopover sites shared by multiple species can face elevated exposure to multi-host pathogens (**host aggregation**, Table 7.1).

It is worth noting that observed parasite richness measured in hosts is a subset of the parasites that migrants encounter, and further depends on the ability of parasites to successfully persist through the annual cycle and (re)infect migrants in subsequent years. Persistence of parasites in migrants may depend critically on how

movements shape transmission opportunities, how physiological changes in hosts related to migration influence host susceptibility and parasite fitness, and parasite traits that favor their ability to exploit migrants as hosts and as vectors for dispersal.

## 7.2.2 Migration and parasite avoidance

Empirical studies in birds, mammals, fish, and insects have shown that for a focal parasite, migratory animals can have lower infection prevalence or intensity than their non-migratory counterparts (Box 7.1). Together, these results suggest that migratory strategy could allow migratory individuals or populations to reduce their risk of exposure to parasites. Three related avoidance

**Table 7.1** Glossary of mechanisms proposed for how migration influences parasitism and vice versa, via changes to parasite exposure, within-host infection processes, and infection clearance, along with empirical examples for each mechanism. The + or − indicate whether the mechanism increases or decreases parasitism in migrants (parasite richness and/or infection prevalence or intensity of a focal parasite).

| Mechanism | Definition | Example |
|---|---|---|
| Environmental sampling (+) | By using many geographic regions and habitats, migrants are exposed to a greater diversity of parasites than residents | Waterfowl and avian malaria (11) |
| Environmental tracking (+) | Moving within a favorable environmental window increases survival of infected hosts and parasites | Ungulates, diverse parasites (14) |
| Host aggregation (+) | Concentrations of animals at stopover sites facilitates transmission | Waterfowl and avian influenza (31) |
| Parasite spread (+) | Migrants are responsible for transporting parasites over large geographic distances | Waterfowl and avian influenza (54) |
| Migratory escape (−) | Seasonally vacating parasite-contaminated habitats reduces transmission opportunities | Caribou and warble flies (7) |
| Migratory avoidance (−) | Avoidance of routes or stopover sites with high exposure risk | Shorebirds and avian malaria (35) |
| Migratory allopatry (−) | Stage-specific migratory behavior separates vulnerable individuals from infected hosts | Salmon and sea lice (9) |
| Migratory dropout and stalling (+/−) | Costs of infection cause infected animals to abandon migration or migrate more slowly. If slow-migrating animals encounter more parasites, feedbacks between migration speed and infection intensity arise (Figure 7.1D) | Bewick's swan and avian influenza (50) |
| Immuno-suppression (+) | Energetic costs associated with migration preparation or onset reduce immune function, increasing susceptibility and/or shedding | Migratory thrushes (immune measures only; (39)) |
| Migratory relapse (+/−) | Migration preparation or initiation reactivates latent infections, either increasing transmission and prevalence, or exacerbating migratory culling to lower prevalence (Box 7.2). | Redwings and *Borrelia* infections (41) |
| Migratory recovery (−) | Moving to habitats that improve host condition, or reduce parasite replication or survival, reduces infection prevalence | Spiny toads and chytrid fungus (46) |
| Migratory culling (−) | Infected animals are less likely to survive migrations, reducing infection prevalence | Monarchs and protozoan OE (8) |

mechanisms have been proposed: periodically vacating used habitats with high exposure risk (**migratory escape**); avoiding routes or sites with high exposure risk (**migratory avoidance**); and separation of vulnerable hosts from potentially infectious conspecifics (**migratory allopatry**) (Table 7.1) (9), (34), (35). Of these mechanisms, migratory escape has been invoked most frequently in support of empirical patterns observed across host and parasite taxa (26), (27), (36).

The above studies highlight a unidirectional effect of migration on lowering infection prevalence. If, additionally, hosts can detect and behaviorally remove themselves from high-risk habitats or host aggregations, and/or migratory strategies that reduce exposure risk are evolutionarily favored, feedbacks between infection avoidance and migratory strategy could shape the dynamics of migration and parasitism (see also Chapter 14 of this book, (37)). Mathematical modeling has shed light on the ecological and evolutionary consequences of

migratory escape for host–parasite dynamics. For example, one study (15) showed that for infected populations, migrating further than is optimal for a disease-free population could reduce both infection prevalence and the negative impacts of the pathogen on host population size; escape was especially beneficial for pathogens that are highly transmissible and of intermediate virulence. Subsequent studies highlighted that the benefits of escape were influenced by parasite transmission mode; escape was more beneficial for avoiding directly transmitted parasites with frequency-dependent rather than density-dependent transmission (38). For hosts that escape environmentally acquired macroparasites by moving into parasite-free habitats, reductions in parasite burdens occurred more quickly with higher within-host parasite mortality, lower parasite transmission rates, and faster host migration speeds (17). Evolutionary models that explicitly simulate the introduction and fate of novel "mutant" migration strategies from wholly resident populations can provide explicit mechanistic support for the role of parasite avoidance in shaping migration. Such models have demonstrated that infection can select for greater migratory propensity, even when migration increases parasite richness (and infection cost), as long as migration also decreases infection prevalence (i.e., infection risk) (16).

### 7.2.3 Migration and host competence for infection

Preparation for migration, and active migratory movement, can involve dramatic physiological or behavioral changes and impose extreme energetic demands on migrants. Together these physiological changes can influence the ability of a host to acquire and transmit a parasitic infection (referred to here as host competence for infection) by altering susceptibility to infection, internal parasite development and replication, and clearance of infection. Prior to migration, metamorphosis, molt, or shrinking organs to increase fat deposition or locomotive capacity, or costly behavioral shifts such as the onset of migratory restlessness, can come at the expense of components of immune defense (Box 7.2). These can increase susceptibility to parasite infection, or facilitate parasite development and replication, before or during movement (39), (40) (**migratory immunosuppression**, Table 7.1). Additionally, migration preparation or onset can cause latent infections to reactivate (41), which can increase or decrease infection prevalence and alter the timing of peak prevalence (**migratory relapse**, Table 7.1). Alternatively, migratory behavior can be associated with no change, or reduced susceptibility to infection. For example, skylarks (*Alauda arvensis*) are able to maintain costly acute phase immune responses during migration (42). Further, a comparative analysis of migratory and non-migratory birds showed that migrants typically had larger spleens (an organ that produces lymphocytes, a type of white blood cell, and filters blood to destroy antigens), perhaps as an adaptation to enhanced encounter rates with diverse parasites (43).

By spending part of the year in habitats unfavorable to parasite survival (or, analogously, that promote a host's ability to clear infection), changing habitat types can also allow migrants to clear infection (**migratory recovery**, Table 7.1). For example, by abandoning freshwater habitats necessary for zoospore persistence, spiny toads (*Bufo spinosus*) can clear chytrid fungus (*Batrachochytrium dendrobatidi*) infections on their terrestrial postbreeding migrations (46) (Table 7.1). Further, a theoretical model demonstrated that migratory recovery from infection could favor the evolution of migration over residency, even in the absence of other environmental drivers of migration (47), potentially providing mechanistic support for a migration–parasitism feedback that alters migratory behavior and regulates parasitism. Subsequent work showed that recovery from infection was more likely to drive the evolution of migration than escape from areas of high transmission.

Energetic demands of migration can result in disproportionately high mortality of infected migrants, reducing infection prevalence within the migrating host population, a phenomenon known as **migratory culling** (Table 7.1). Mathematical models shed light on when migratory culling is especially likely to operate. First, even non-selective migratory culling (i.e., both uninfected and infected migrants experience high mortality during migration) is effective at reducing infection prevalence (38). The benefits of migratory culling for reducing infection prevalence and intensity increase with parasite virulence (15), (17), and for relapsing infections, culling is enhanced when relapse of virulent pathogens occurs at migration onset (45) (Box 7.2).

## 7.3 Effects of parasite infection on migratory behavior, and infection–migration feedbacks

The majority of work on migration–parasite interactions has focused on how migration can influence infection patterns. Due to challenges of disentangling the multiple (and covarying) causes of migration in comparative studies (48), and critiques of theory being distanced from empirical reality, there is still active debate about the magnitude of the role that parasites play in shaping host migration. Patterns from some field studies indicate that infection can influence migration probability and subsequent performance. In the zooplankton *Daphnia pulicaria*, healthy hosts migrate to lake surfaces at night, while those actively infected with the chytrid

### Box 7.2  Immunity, migratory behavior, and parasite transmission: Case studies from migratory songbirds

Physiological tradeoffs associated with preparation for (and initiation of) migration can influence host immune defense, and in turn, immune defenses mounted in response to infection challenge can influence migration performance and success. Together, interactions between immune defense and migratory performance could drive feedbacks that shape population-level infection outcomes and/or migration strategies. Much of the evidence for unidirectional linkages between immune and migratory performance comes from migratory songbirds, since migration preparation can involve large physiological changes, and nonstop flapping flight sustained over periods of hours to days is energetically costly

and likely to trade off with immunity and infection. Further, in both captive and field settings the migratory state of birds can be assessed and manipulated experimentally, and similarly immune status measured and manipulated with sham or real infections. To date, logistical challenges of tracking birds and their infection status throughout the annual cycle mean that the bidirectional linkages necessary for feedbacks to emerge have not been demonstrated in a single empirical system. However, mathematical models whose assumptions reflect mechanistic linkages between immunity and migration provide one way to assess the consequences of feedbacks for host–parasite dynamics.

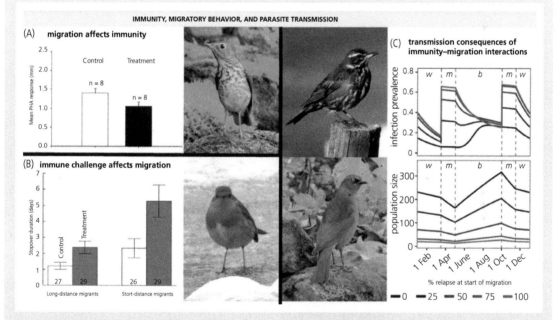

**Box 7.2, Figure 1** (A) Migratory behavior can influence immune defense. Swainson's thrushes (*Catharus ustulatus*) experimentally induced into a migratory state (Treatment, black bar) exhibited a lower Phytohemagglutinin (PHA) response than control birds (white bar) (adapted from (39)). (B) Immune defense can influence migratory behavior. Across six species of European migratory songbirds, immune challenge with a sham infection (gray bar) increased the duration of migratory stopover compared to control birds (white bar), especially for short-distance migrants such as the European robin (*Erithacus rubecula*) (adapted from (44)). (C) Migration is associated with reactivation of *Borrelia* infections in redwings (*Turdus iliacus*, top left), and haemosporidian infections in rusty blackbirds (*Euphagus carolinus*, bottom left). Results of a mathematical model where migration onset influences the fraction of latent infections that relapse (colored lines) and costs of migration influence the survival of relapsing individuals. Example model output shows the dynamics of infection prevalence (top right) and host population size (bottom right), through a calendar year (w=winter, m=migration, b=breeding). When costs of migration and infection are relatively low, relapse results in a switch from a single to a double peak in annual infection prevalence, increases peak infection prevalence, and lowers host population size (adapted from (45)). All photos by R. Hall.

---

**Box 7.2** *Continued*

One way in which migration could increase infection is if energetic costs of undertaking migration result in reduced immune performance (**migratory immunosuppression**, Table 7.1 [main text]). In support of this mechanism, captive Swainson's thrushes (*Catharus ustulatus*) experimentally induced into a migratory state by manipulating photoperiod exhibited a lower phytohemagglutinin (PHA) response (a metric of cellular immunity) than control birds (39) (Figure 1A). Activation of the immune system, or parasite replication in immunocompromised hosts, could also incur energetic costs that reduce migratory performance by reducing the probability of migration or migration speed (migratory stalling), or causing individuals to drop out or die from infection (migratory culling). In support of migratory stalling, immune challenge of six species of wild European migratory songbirds with a sham infection increased the duration of migratory stopover, especially for short-distance migrants such as the European robin (*Erithacus rubecula*) (44) (Figure 1B).

One mechanism by which effects of migration on immunity and infection can lead to feedbacks is if migration reactivates latent infections (**migratory relapse**, Table 7.1), as observed in *Borrelia* bacterial infections in Redwings (*Turdus iliacus*), and hemosporidian infections in Rusty Blackbirds

(*Euphagus carolinus*) (Figure 1C). A mathematical modeling study (45) investigated how migration influenced the probability of infection relapse, how infection influenced the probability of survival during migration, and the subsequent consequences for infection patterns. In the absence of migratory relapse, a single peak in annual infection prevalence occurred late in the breeding season (i.e., when host density was highest). In contrast, infection relapse at the onset of migration led to a double peak in infection prevalence coincident with the start of spring and fall migration. When costs of infection during migration were low, increasing the probability of relapse increased peak infection prevalence, and resulted in stronger regulation of the migratory population by parasites (Figure 1C, upper and lower panels). In this case, infection relapse during migration could be regarded as a beneficial strategy for low-virulence pathogens, or those that exploit hosts whose migrations are less costly (e.g., short distance or using soaring flight). However, if costs of infection (or fighting infection) during active migration were high, migratory relapse enhanced the effect of migratory culling to reduce infection prevalence or even prevent pathogen invasion. Thus, bidirectional linkages between infection relapse and migration costs can result in feedbacks that either enhance or reduce parasitism in migrants.

---

*Polycaryum laeve* do not migrate (49). Bewick's swans (*Cygnus columbianus*) infected with low pathogenic avian influenza virus departed later, and moved less far, than uninfected conspecifics (50). Finally, a cross-taxon meta-analysis indicated that infection status was weakly associated with migratory delay, while infection intensity altered migratory phenology, and reduced migrant endurance, migration speed, and extent (13).

Experimental manipulation of infection to test its effects on migration is a relatively new approach that offers promise for demonstrating this causal link (Box 7.2). Trout artificially infested with sea lice had shorter sea migrations than control fish, and returned prematurely to freshwater, which facilitates delousing (51). Additionally, juvenile salmon experimentally infected with sea lice were more likely to be at the back of schools than uninfected fish (52). In monarch butterflies, flight-mill experiments revealed that *Ophryocystis*

*elektroscirrha* (OE)-infected monarchs have reduced flight capacity relative to uninfected butterflies (8) (Box 7.1), providing mechanistic support for **migratory dropout** (Table 7.1) and culling. Whether and how these impacts of infection on individual movement behavior translate to population-level responses and feedback cycles is the topic of section 7.3.2.

### 7.3.1 Parasite exploitation of migrants as vectors for dispersal

Many parasites rely on migrants for long-distance dispersal in addition to exploiting individual hosts for transmission (**parasite spread**, Table 7.1). Thus, host migratory behavior could, in many cases, shape parasite gene flow, genetic diversity, and the evolution of parasite traits. Given the limited distances over which parasites and their biting vectors can disperse unaided, migratory species are likely to be crucial for transporting parasites over large geographic distances. Indeed, host migratory routes could explain the contemporary distribution of some widespread parasites; for example, genetic analyses of seabird tick diversity in the Cape Verde islands revealed that trans-oceanic movements of seabirds explained tick dispersal to these sites (53). While causal links between parasite distribution and host migration remain rare, there are several instances where migrants are implicated in the long-distance movement of emerging or range-expanding pathogens of human health concern, including inter-continental spread of avian influenza viruses by waterfowl (54). When viewing ticks as both parasites and disease vectors, the transportation of ticks to more northerly latitudes by neotropical migratory songbirds could further expand the range of the bacteria causing Lyme disease (*Borrelia burgdorferi* (55)).

Importantly, parasites infecting migratory species might evolve traits which facilitate their persistence and transmission throughout the annual migratory cycle. In some cases, strenuous migratory journeys could select for lower parasite virulence. In support of this prediction, protozoan parasites in monarchs that were isolated from longer-distance migratory populations caused lower mortality and less severe sub-lethal effects than parasites from other populations (56). Host migratory phenology could also select on the timing of parasite replication within hosts to increase spatiotemporal overlap in the production of infectious stages and the availability of susceptible hosts. As one example, the timing of trematode maturation within snail intermediate hosts coincides with snail migration to breeding sites where definitive (avian) hosts are also present (57). In other work, a phylogenetic analysis of avian blood-borne parasites shows a switch from seasonal within-host replication and transmission (summer only) to year-round transmission, which would facilitate transport to new areas via host migration (58). Furthermore, blood-borne parasites infecting migratory birds adjust resource allocation seasonally, between sexual reproduction in summer (when hosts are in areas with abundant insect vectors) and clonal reproduction in winter (59).

For parasites that rely on migrants for their persistence and spread, it is likely that migration selects for additional parasite traits that favor parasite survival in migratory hosts, their onward transmission, and establishment in novel habitats. While comparative analyses within and among host species with varying degrees of migration have been used to understand how migration influences parasite diversity and prevalence (48), similar synthetic analyses of parasite traits in relation to migration have not been conducted. Analyses that compare key parasite traits (e.g., transmissibility, virulence, transmission mode, and host generalism) among migratory and non-migratory hosts could yield exciting insights as to how migration shapes parasite life histories. Surprisingly, there is a dearth of theory for how parasites evolve in response to host migration, and this is a key area ripe for development.

### 7.3.2 A framework for studying migration–infection feedbacks

Examples from section 7.2 hinted at how interactions between migration and (i) parasite exposure, (ii) host susceptibility and parasite replication, and (iii) recovery or mortality from infection can result in feedbacks that shape host migration strategy and parasite infection prevalence. Foundational theory exploring the consequences of these feedbacks assumed that all individuals in the population attempt to undertake the same migration strategy *irrespective of their infection status*, and that differential survival of infected and uninfected hosts leads to a fixed, directional change in (obligate) migratory strategy and infection prevalence over multiple generations (Figure 7.1A). However, infection status and intensity can also influence whether, when, how far, and how fast an animal migrates. Within-population variation associated with infection status could therefore generate a rich array of outcomes of migration–parasitism feedbacks, including partial migration (where a fraction of the population migrates in a given year) or reinforce individual (co)variation between infection status and migratory strategy. Mathematical models and mechanistic hypotheses accounting for this variation are in their infancy, and a great need remains to

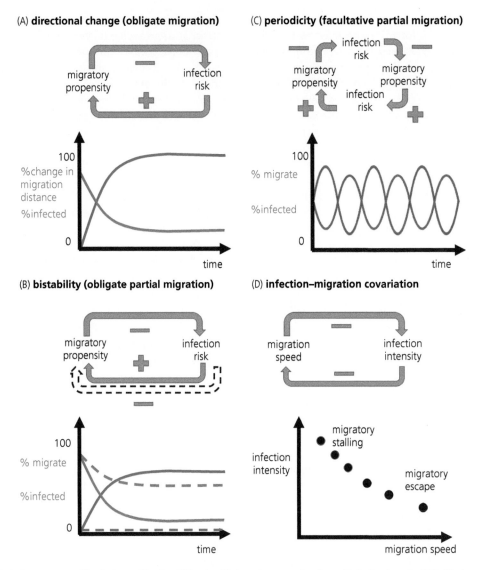

**Figure 7.1** Consequences of feedbacks resulting from bidirectional linkages between migration and infection. In each of (A)–(D), the gray arrows indicate the effect of migratory propensity on infection risk, and vice versa, with the + or − sign indicating a respective increase or decrease in the response variable when the predictor variable increases. In A–C, the lower graphs indicate population level responses in migratory propensity (blue) and infection prevalence (red), with the dashed lines in (B) indicating the responses of residents (dashed lines) compared to migrants (solid lines). In (D), dots represent individual variation of infection intensity and migration speed resulting from feedbacks. (A) Directional change in migratory propensity (here representing distance migrated) and infection prevalence. (B) Bistable strategy (obligate partial migration). (C) Cyclic changes in migratory propensity (facultative partial migration) and infection prevalence. (D) Reinforcement of individual covariation in infection intensity and migration performance (here migration speed).

examine within-population variation in both migratory propensity and parasite infection.

One potential consequence of feedbacks is the emergence of bistable migration strategies, where a fixed fraction of the population migrates,

and infection prevalence in residents is higher than that in migrants (obligate partial migration, Figure 7.1B). Mechanistically, partial migration lowers population-level infection prevalence of density-dependent transmitted parasites relative to historically resident populations, by effectively splitting the population, reducing local density and thus infection risk (60). Partial migration can also emerge as a conditional strategy, either where only infected individuals migrate (when migratory recovery occurs (61)) or where only individuals with lower infection intensity migrate (when migratory escape but not recovery occurs (62)). Alternatively, for historically fully migratory populations, pathogens can result in the maintenance of partial migration if residents tolerate infection better by forgoing energetic costs associated with migration, and if infection causes former migrants to forgo migration. Maintenance of transmission by residents can, in turn, erode benefits of migratory escape for migrants, especially if migratory culling operates (63). As a third possibility, if animals adjust their annual migration probability in relation to their current infection risk, this could result in cycles of high migratory propensity and low infection prevalence, followed by low migratory propensity and high infection prevalence (facultative partial migration, Figure 7.1C). While facultative partial migration in response to interannual variation in condition is common in ungulates (64), to date, the potential for infection–migration feedbacks to drive alternating cycles of high prevalence and high fraction migrating has neither been demonstrated empirically nor explored with transmission models.

Finally, if migratory performance covaries with infection intensity, feedbacks could reinforce individual differences in migration strategy and infection status (Figure 7.1D), with consequences for population-level infection prevalence. In a model of avian influenza transmission in mallards, infection-induced delays to migration reduced overall infection prevalence and tended to delay the peak in infection prevalence until later in the migration (65). Slower migrations arising from macroparasite burden also tended to reduce transmission among some migrating hosts during the annual cycle, as faster hosts "outran" their infected counterparts. However, hosts that were left behind suffered a "double whammy" of increased exposure and reduced capacity for migratory escape, a feedback that may lead to **migratory stalling** (17), (66) (Table 7.1).

## 7.4 Predicting migratory host–pathogen dynamics under global change

Humans are causing widespread environmental changes in climate and habitats, with cascading effects on wildlife health. For migratory wildlife, annual movements themselves are already changing in response to global change (67), (68). In many systems, the abundance of migrants, and even migratory behavior itself, are in decline (22). Novel environmental conditions and emerging diseases can exacerbate pathogen-related threats to migrants. For example, saiga antelope have been endangered by poaching and habitat loss for decades, such that recent outbreaks of hemorrhagic septicemia triggered by extreme environmental conditions (69) resulted in a 62% decline in the global saiga population. Similarly, wild salmon (*Oncorhynchus* spp.) populations have declined throughout the Pacific due to overfishing and loss of spawning habitat, and the transmission of pathogens from novel domesticated hosts in open-net salmon farms is a serious concern (70). Because global change can alter migratory strategy in ways that impact infection outcomes and simultaneously change infection risk in ways that influence migratory performance, we argue that knowledge of migration–infection linkages, and the feedbacks they produce, is crucial for predicting the net consequences of climatic and habitat changes on migratory host–pathogen dynamics. In section 7.4.1 we harness our mechanistic understanding of migration–infection linkages and summarize recent literature on migration and infection in changing environments to explore how three prominent aspects of global change (climate change, habitat disruption, and anthropogenic food subsidies) influence migration and infection outcomes.

### 7.4.1 Climate change impacts on host–parasite phenology and spatial distributions

For pathogens with multiple hosts or intermediate hosts (including arthropod vectors), the response of migration timing to warming-induced shifts in resource phenology can alter host–parasite temporal overlap, and therefore the efficacy of migratory escape. For example, a mathematical model of low pathogenic avian influenza virus (LPAIV) (30) showed that observed advances in the timing of ruddy turnstone migration can increase the prevalence of LPAIV due to increased overlap with alternate host species (resident ducks) at a time when infection prevalence in ducks is at its peak (Figure 7.2). Alternatively, when migration timing does not track advances in breeding site phenology, but emergence of arthropod disease vectors at migrant breeding sites shifts earlier, reduced temporal overlap of hosts and vectors could decrease transmission and prevalence ("migratory host–pathogen mismatch" (71)).

Non-linear responses of parasite development and mortality to climate warming can alter both

the timing of transmission and infection intensity in migrants. For example, arctic warming has increased developmental rates of *Ostertagia gruehneri*, a gastrointestinal helminth that infects migratory barren-ground caribou (72) (Figure 7.2). Previously, the parasite had a two-year lifecycle: caribou would migrate away from summer breeding grounds before the parasite larvae had time to develop to the infectious stage, and so larvae would remain in the environment until the return of their hosts the following spring. Warmer temperatures allow the parasite to complete its life cycle within a single season, but suffer higher mortality of larvae during peak summer temperatures. Theory predicts that the combined effects of increasing

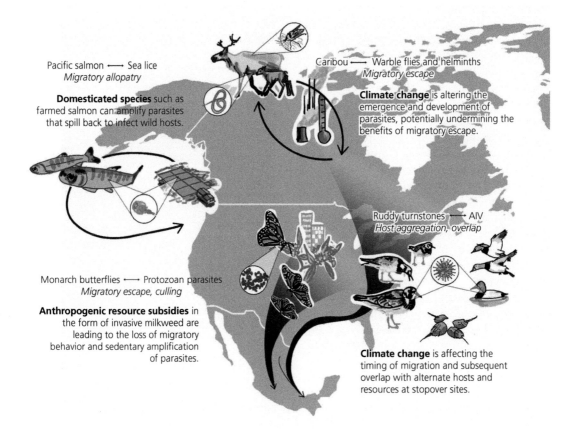

**Figure 7.2** Impacts of global change on migratory host–parasite interactions. Global change is influencing migratory host–parasite systems in a myriad of ways, including through climate-change impacts on parasite lifecycles and on host migration timing and extent, increased overlap with domesticated animals, and anthropogenic resource subsidies promoting residency in some migratory hosts. The map highlights how and where four well-studied migratory host–parasite systems are responding to these threats. Artwork by S. Peacock.

temperatures on development and mortality of *Ostertagia* leads to two transmission peaks, in the spring (from last year's larvae) and in the fall (from larvae of the same year) (73). At present, most models of environmentally transmitted macroparasites and migratory hosts have yet to include the impact of temperature on parasite phenology and survival. There is an opportunity here to integrate spatial models of migration with temperature-dependent parameters (e.g., via the metabolic theory of ecology (73)) to improve our understanding of and predict the impact of changing climate on host–pathogen dynamics, and how this alters patterns of individual covariation between infection speed and intensity driven by migratory stalling.

Range shifts (e.g., poleward or elevational shifts in breeding habitat) could also increase or decrease host–parasite overlap, and the consequences for wildlife health will depend on the initial ranges and the thermal responses of host and parasites (74), (75). If climate change leads to milder winter conditions, some animals will migrate shorter distances, or forego migration altogether to remain in temperate breeding grounds (this loss of migratory behavior can also result from anthropogenic resource subsidies). This could undermine the disease-related benefits of migratory escape and migratory culling, leading to "sedentary amplification" of disease (76) (Box 7.3).

### 7.4.2 Habitat disruption, host aggregation, and parasite sharing with domestic animals

Human-driven habitat disruptions are changing the availability and connectivity of habitat for migratory animals and their pathogens. Impacts on stopover habitat in particular can increase food stress, which might increase transmission via migratory immunocompromise and relapse, or lower transmission through exacerbating migratory culling. Habitat loss can also crowd animals into remaining stopover sites, potentially increasing transmission (via migratory aggregation), as seen in the case of migratory shorebirds and their avian influenza viruses (77). Sharing habitats with domesticated animals can further

---

**Box 7.3 Shifts towards residency and consequences for parasitism: Monarchs and their protozoan parasites as a case study**

Two decades of work focused on monarch butterflies and a debilitating protozoan parasite, OE (Figure 1A), showed how anthropogenic resource subsidy can alter host migratory behavior (Figure 1B) and the spread of infectious disease. Like many other migrants, monarchs have experienced population declines coincident with the loss of breeding and overwintering habitats (83). Nature enthusiasts eager to create monarch habitat in their yards are encouraged to provide milkweed host plants for monarchs, but the most popular milkweed in gardens is a non-native species, tropical milkweed (*Asclepias curassavica*; Figure 1A), which is attractive, easy to grow, and commonly sold by nurseries. Unlike most native milkweeds that enter dormancy in the autumn, tropical milkweed persists throughout the year in mild climates, and continues to flower into the fall months. In the southern United States, especially along the Gulf Coast, newly formed resident monarch populations have become common (76), enabled by the year-round breeding habitat afforded by *A. curassavica*. Exposure to tropical

milkweed in the autumn was recently shown to induce some migrants to break reproductive diapause (a pre-migratory physiological state associated with the delay of reproductive development) and to stop migrating (84).

A study involving citizen scientists (76) showed that resident monarchs at winter-breeding sites in the southern United States were far more likely to be infected compared to migratory monarchs (sampled in both their seasonal breeding and wintering ranges; Figure 1C). Other work showed that in Texas, where monarch flyways overlap with resident breeding sites, parasite-contaminated tropical milkweeds and infected monarchs could act as a source of infection for migratory monarchs during autumn and spring (84). Using chemical markers to identify monarchs sampled in Texas as migrants versus residents (e.g., Figure 1D), this same study showed that migrants captured at resident breeding sites were more likely to be infected, and also showed more evidence for reproductive development (compared to migrants captured at fall stopover sites without tropical

**Box 7.3** *Continued*

milkweed)—suggestive of migratory dropout of infected monarchs (Figures 1E, 1F). Collectively, this work on the monarch–protozoan interaction suggests that human-driven changes to the seasonal availability of milkweed host plants in areas with mild winter climates has reduced monarch migration and health. In response to these findings, many gardeners and monarch enthusiasts are now planting native milkweeds, and either avoiding tropical milkweeds, or cutting them back during the fall and winter, to make them unavailable during the critical time when migratory monarchs are traveling south to winter in Mexico.

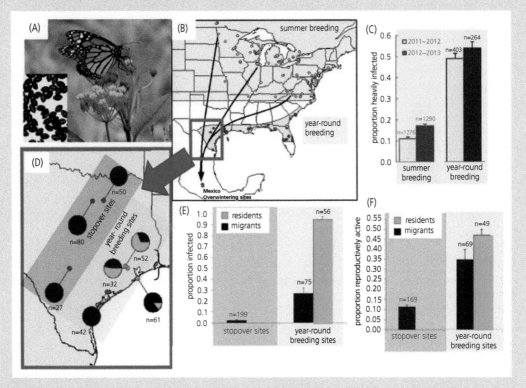

**Box 7.3, Figure 1** Resource subsidy and monarch-protozoan interactions. (A) Monarch butterfly nectaring on tropical milkweed, and *Ophryocystis elektroscirrha* (OE) parasite spores (inset); photos by R. Hall and S. Altizer respectively. (B) Map showing the migratory range of eastern North American monarchs. Yellow denotes the summer breeding range, blue denotes the year-round breeding range supported by tropical milkweed, and the blue rectangle indicates study sites where resident and migratory monarchs potentially overlap in south Texas. (C) OE infection prevalence in migratory monarchs (yellow) and resident monarchs (blue) in two successive years (unshaded and shaded bars); b and c adapted from (76). (D) Location of migratory stopover sites (pink shading) and year-round breeding sites (blue shading). Pie charts indicate the proportion of migrants (black) and residents (turquoise) captured at each site type. (E) Infection prevalence and (F) fraction of reproductively active monarchs at stopover (pink) and year-round breeding sites (blue), according to whether captured monarchs were migrants (black) or residents (turquoise); D–F adapted from (84).

increase migratory species' exposure to novel or virulent pathogens (78), and disrupt natural parasite-avoidance benefits of migration, such as the separation of vulnerable juveniles from infected adults (migratory allopatry; (9)). For example, farmed salmon are a reservoir of sea-louse parasites that can spill back to infect naive juvenile wild salmon, with consequences for the long-term survival of wild salmon populations (70) (Figure 7.2). Management of parasites in domesticated host populations could decrease the impact on migratory wildlife populations, but timing treatments to coincide with wildlife migration patterns is important (79). The migration-infection framework proposed in this chapter could inform understanding of how migration influences where, when, and how long migrants encounter high-risk habitats, and migrants' susceptibility to infection at these overlap points.

### 7.4.3 Responses of migratory species and their parasites to supplemental feeding

Human-driven landscape changes such as urbanization and agricultural expansion can provide novel food resources for wildlife including recreational feeding stations in backyards, or when wildlife access food in dumpsters and crop fields (80). Human-provided food can disrupt animal migrations that have evolved to capitalize on seasonal fluctuations in resources. In response, some animals now migrate shorter distances, or have halted their migrations altogether, forming resident populations that live year round in the same location. For example, many Spanish white storks (*Ciconia ciconia*) now forego their annual migration to Africa and instead subsist on city landfills in Spain year round (81) and monarchs are showing similar reductions in movement tied to supplemental feeding (Box 7.3). Importantly, evolutionary losses of traits important for migration might occur in newly formed resident populations in the absence of migration-driven selection, reinforcing the degradation of long-distance movement (82).

Recent work showed that changes in animal migrations in response to supplemental feeding can alter mechanisms determining the outcomes of host–parasite dynamics (68). Although less mobile hosts might encounter fewer parasite species (i.e., reduced environmental sampling), or have more resources to allocate to fighting infection (i.e., reduced migratory immunosuppression), for many species, shifts towards residency in response to human-provided food are likely to increase

parasite transmission. The most immediate impacts of increased residency will occur through the loss of migratory escape and culling, as well as the loss of migratory recovery and allopatry. For example, the build-up of environmentally transmitted pathogens in even high-quality habitats with stable resources can create ecological traps for animals, as recent modeling work showed (85).

In response to supplemental feeding, newly formed resident populations could reach higher local densities than migrants and, as a result, experience greater contact rates, crowding, and stress. Implications of higher host density around resource subsidies for pathogen transmission have been examined in host–pathogen models (86) and in several empirical systems. For example, elk in Wyoming have formed high-density, sedentary winter populations at sites with human-provided food; this led to higher brucellosis transmission compared with free-ranging elk (87). Whether other newly resident populations supported by human-provided resources experience crowding, stress, and increased pathogen transmission remains an area open for future work. Research is also needed to examine how these changes might affect pathogen virulence evolution as residency combined with host crowding might favor the transmission of more virulent pathogen strains (76).

Importantly, newly formed resident populations could act as sources of infection for remaining migrants (Box 7.3). In waterfowl, for example, residents can act as pathogen reservoirs and expose migrants to infection by avian influenza subtypes (88). Pathogen infections at locations where residents and migrants interact could feed back to accelerate the loss of migratory behavior still further, because a higher prevalence of infection in departing migrants leads to higher mortality of migrants due to migratory culling (63). Empirical studies and mathematical models that incorporate multiple mechanisms by which resource-induced residency influences host immunity, pathogen exposure, and infection impacts are needed to understand whether migration–parasitism feedbacks will reinforce residency and its net infection outcomes, or lead to divergent outcomes (e.g., habitat specialization that leads to isolation of resident and migratory populations and their pathogens).

## 7.5 Conclusions and future research

Dramatic growth in the field of migratory disease ecology has yielded new mechanistic insights on the ecology and evolution of migratory host–parasite dynamics. By synthesizing past research and knowledge gaps on migration–parasitism linkages, three key themes have emerged from this chapter that represent research frontiers ripe for future investigation. First, by focusing on bidirectional interactions, where parasites influence migration and migration in turn influences parasites, we can advance our understanding of how feedbacks influence patterns and process in real-world migratory host–parasite systems. Second, consideration of variation in movement and competence for infection among individuals and populations may reveal a richer range of outcomes for migratory behavior and parasitism than classical population-level approaches that mask this variation. Third, quantifying movement–parasitism linkages and feedbacks can provide a scaffold for predicting and managing the health of migrants and other highly mobile species under diverse forms of global change. In section 7.5.1 we highlight key priorities for future research in these three areas.

### 7.5.1 Ecological and evolutionary feedbacks between migration and parasitism

Studies to date of migratory host–pathogen interactions often focus on unidirectional processes that could contribute to migration–parasitism feedbacks, without fully studying the feedbacks themselves. In this chapter we have outlined various ways in which feedbacks could arise to shape diverse outcomes for host migration and infection prevalence or intensity. Due to challenges in studying migration and infection throughout the annual cycle and across multiple generations, empirical studies have rarely been able to capture the bidirectional relationships between components of migration and infection necessary to generate feedbacks in a single system. However, increasing capacity of lightweight trackers and data loggers to monitor the location and condition of migrants, and innovative experiments that manipulate infection or migratory status in both captive and field animals, show promise for filling these data gaps (see also

Chapter 6 (89)). While we have outlined how feedbacks can shape migration and infection outcomes for hosts, and discussed how migration might select for parasite traits that favor their persistence in migrants, coevolutionary feedbacks between host migration and parasite life history have not been explored. For example, adaptations in parasites that allow them to exploit migrants, such as reduced virulence and seasonal dormancy, could also reinforce selection pressure for hosts to maintain breeding and wintering site fidelity (90). Intriguingly, for the evolutionary ecology of migrant–parasite interactions, theory outpaces empirical work in understanding how parasites might shape migration, while empirical studies outpace theory in understanding how migration shapes parasites. Experimental and theoretical studies in both directions are rare and much needed (91).

### 7.5.2 Embracing variability in animal movement

Theoretical studies of migratory host–parasite dynamics are often conducted at the population scale, but individual heterogeneity in behavior and susceptibility to infection underpin many of the mechanisms driving population-scale patterns in infection. Empirical studies have revealed associations between individual infection status and migration propensity (92), but more empirical studies that experimentally manipulate host infection and look for a response in migratory behavior (36) are needed to understand the causes and consequences of parasites on migration at the individual level. Recent theory has demonstrated how feedbacks can reinforce differences in migration performance and infection intensity (17). Individual-based models that allow for interactions among hosts may provide additional insights into how individual heterogeneity affects outcomes for both hosts and parasites, but to our knowledge such models have not yet been developed. Finally, while most work on migratory–host parasite interactions has treated seasonal migration as a highly repeatable two-way journey, little attention has focused on parasitism in relation to long-distance movements that are less predictable across time and space, and among individuals (e.g., irruptive migrations and seasonal nomadism (14)). Advances in animal tracking technologies, and adopting a movement ecology perspective whereby animal movement is determined both by internal state and the external environment (Chapter 6 (89)), show promise for revealing novel associations and feedbacks between movement and parasitism.

### 7.5.3 Migratory hosts and their parasites under global change: An evolutionary and one health perspective

This chapter highlighted how global change is altering the arena of migratory host–parasite interactions in ways that can enhance or erode migration–parasitism feedbacks. Prior work has highlighted how unidirectional relationships between migration and parasitism are altered under global change and their recent, perhaps transient consequences for migratory propensity and parasite prevalence. Additional study of hosts (and their parasites) that have undergone recent shifts in migratory propensity are needed to understand whether selection across host and parasite generations will maintain and reinforce these behaviors, and how parasite adaptation to migration shifts will shape feedbacks. There is increasing recognition of the interconnectedness of environmental, human, and animal health, and the roles that highly mobile wildlife can play in affecting the health of people (and vice versa). Global change is strengthening these connections as humans encroach upon wildlife habitat, and interactions between domesticated animals, wildlife, and people become more frequent. Migratory wildlife in particular can connect disparate habitats and populations, potentially facilitating the spread of emerging and zoonotic diseases, such as highly pathogenic avian influenza viruses (23). Infectious disease impacts on migratory wildlife can also influence human health indirectly by altering ecosystem services, such as impacting food safety and security among Indigenous communities that rely heavily on migratory animals for subsistence (93). Employing the One Health paradigm (94) to study feedbacks between human behavior, animal behavior, and their shared parasites as an integrated system offers a promising way forward. Adopting this approach may yield novel insights and solutions for maintaining unique and important animal

migrations and their associated contributions to human wellbeing, conserving threatened specialist parasites of declining migratory species, and mitigating threats posed by emerging pathogens.

## Acknowledgements

The authors gratefully acknowledge funding support from the National Science Foundation grants DEB-1518611 and DEB-1754392 (RJH and SA), DEB-1911925 (RJH), DEB-1654609 to AKS; and the Natural Sciences and Engineering Research Council of Canada (NSERC) postdoctoral and Banting fellowships to SJP.

## References

1. Dingle H. *Migration: The Biology of Life on the Move*. New York, NY: Oxford University Press, 2014.
2. Bauer S and Hoye BJ. Migratory animals couple biodiversity and ecosystem functioning worldwide. *Science*. 2014 Apr 4;344(6179).
3. Kendrick A, Lyver PO, and Łutsël K'é Dene First Nation. Denésqliné (Chipewyan) knowledge of barren-ground caribou *(Rangifer tarandus groenlandicus)* movements. *Arctic*. 2005 Jun 1:175–91.
4. Bowlin MS, Bisson IA, Shamoun-Baranes J, Reichard JD, Sapir N, Marra PP, *et al.* Grand challenges in migration biology. *Integr Comp Biol*. 2010 Sep 1;50(3):261–79.
5. Manwell RD and Herman CM. Blood-parasites of birds and their relation to migratory and other habits of the host. *Bird-Banding*. 1935 Oct 1;6(4):130–3.
6. Beaudoin RL, Applegate JE, Davis DE, and McLean RG. A model for the ecology of avian malaria. *J Wildl Dis*. 1971 Jan;7(1):5–13.
7. Folstad I, Nilssen AC, Halvorsen O, and Andersen J. Parasite avoidance: The cause of post-calving migrations in *Rangifer? Can J Zool*. 1991 Sep 1;69(9):2423–9.
8. Bradley CA and Altizer S. Parasites hinder monarch butterfly flight: Implications for disease spread in migratory hosts. *Ecol*. 2005 Mar;8(3):290–300.
9. Krkošek M, Gottesfeld A, Proctor B, Rolston D, Carr-Harris C, and Lewis MA. Effects of host migration, diversity and aquaculture on sea lice threats to Pacific salmon populations. *Proc R Soc B: Biol Sci*. 2007 Dec 22;274(1629):3141–9.
10. Eikenaar C, Isaksson C, and Hegemann A. A hidden cost of migration? Innate immune function versus antioxidant defense. *Ecol Evol*. 2018 Mar;8(5):2721–8.
11. Figuerola J and Green AJ. Haematozoan parasites and migratory behaviour in waterfowl. *Evol Ecol*. 2000 Mar;14(2):143–53.

12. Altizer S, Bartel R, and Han BA. Animal migration and infectious disease risk. *Science*. 2011 Jan 21;331(6015):296–302.

13. Risely A, Klaassen M, and Hoye BJ. Migratory animals feel the cost of getting sick: A meta-analysis across species. *J Anim Ecol*. 2018 Jan;87(1):301–14.

14. Teitelbaum CS, Huang S, Hall RJ, and Altizer S. Migratory behaviour predicts greater parasite diversity in ungulates. *Proc R Soc B: Biol Sci*. 2018 Mar 28;285(1875):20180089.

15. Hall RJ, Altizer S, and Bartel RA. Greater migratory propensity in hosts lowers pathogen transmission and impacts. *J Anim Ecol*. 2014 Sep;83(5):1068–77.

16. Shaw AK, Sherman J, Barker FK, and Zuk M. Metrics matter: The effect of parasite richness, intensity and prevalence on the evolution of host migration. *Proc R Soc B: Biol Sci*. 2018 Nov 21;285(1891):20182147.

17. Peacock SJ, Krkošek M, Lewis MA, and Molnár PK. A unifying framework for the transient parasite dynamics of migratory hosts. *PNAS*. 2020 May 19;117(20):10897–903.

18. Ezenwa VO, Archie EA, Craft ME, Hawley DM, Martin LB, Moore J, *et al.* Host behaviour–parasite feedback: An essential link between animal behaviour and disease ecology. *Proc R Soc B: Biol Sci*. 2016 Apr 13;283(1828):20153078.

19. Hawley DM and Ezenwa VO. Parasites, host behavior and their feedbacks. In Ezenwa VO, Altizer S, and Hall RJ (eds), *Animal Behavior and Parasitism*. Oxford: Oxford University Press, 2022. DOI: 10.1093/oso/9780192895561.003.0002.

20. Chapman BB, Brönmark C, Nilsson JÅ, and Hansson LA. The ecology and evolution of partial migration. *Oikos*. 2011 Dec;120(12):1764–75.

21. Lloyd-Smith JO, Schreiber SJ, Kopp PE, and Getz WM. Superspreading and the effect of individual variation on disease emergence. *Nature*. 2005 Nov;438(7066):355–9.

22. Wilcove DS and Wikelski M. Going, going, gone: Is animal migration disappearing. *PLoS Biology*. 2008 Jul;6(7):e188.

23. Liu J, Xiao H, Lei F, Zhu Q, Qin K, Zhang XW, *et al.* Highly pathogenic H5N1 influenza virus infection in migratory birds. *Science*. 2005 Aug 19;309(5738):1206–.

24. Majewska AA, Sims S, Schneider A, Altizer S, and Hall RJ. Multiple transmission routes sustain high prevalence of a virulent parasite in a butterfly host. *Proc R Soc B: Biol Sci*. 2019 Sep 11;286(1910):20191630.

25. Altizer SM, Oberhauser KS, and Brower LP. Associations between host migration and the prevalence of a protozoan parasite in natural populations of adult monarch butterflies. *Ecol Entomol*. 2000 May;25(2):125–39.

26. Altizer S, Hobson KA, Davis AK, De Roode JC, and Wassenaar LI. Do healthy monarchs migrate farther? Tracking natal origins of parasitized vs. uninfected monarch butterflies overwintering in Mexico. *PloS One*. 2015 Nov 25;10(11):e0141371.

27. Bartel RA, Oberhauser KS, De Roode JC, and Altizer SM. Monarch butterfly migration and parasite transmission in eastern North America. *Ecology*. 2011 Feb;92(2):342–51.

28. Hoar BM, Eberhardt AG, and Kutz SJ. Obligate larval inhibition of *Ostertagia gruehneri* in Rangifer tarandus? Causes and consequences in an Arctic system. *Parasitology*. 2012 Sep;139(10):1339–45.

29. Wright DW, Oppedal F, and Dempster T. Early-stage sea lice recruits on Atlantic salmon are freshwater sensitive. *J Fish Dis*. 2016 Oct;39(10):1179–86.

30. Brown VL and Rohani P. The consequences of climate change at an avian influenza "hotspot." *Biol Lett*. 2012 Dec 23;8(6):1036–9.

31. Hill NJ, Ma EJ, Meixell BW, Lindberg MS, Boyce WM, and Runstadler JA. Transmission of influenza reflects seasonality of wild birds across the annual cycle. *Ecol*. 2016 Aug;19(8):915–25.

32. Poulin R and de Angeli Dutra D. Animal migrations and parasitism: Reciprocal effects within a unified framework. *Biol Rev*. 2021 Mar 4.

33. Daversa DR, Fenton A, Dell AI, Garner TW, and Manica A. Infections on the move: How transient phases of host movement influence disease spread. *Proc R Soc B: Biol Sci*. 2017 Dec 20;284(1869):20171807.

34. Loehle C. Social barriers to pathogen transmission in wild animal populations. *Ecology*. 1995 Mar;76(2):326–35.

35. Mendes L, Piersma T, Lecoq M, Spaans B, and Ricklefs R. Disease-limited distributions? Contrasts in the prevalence of avian malaria in shorebird species using marine and freshwater habitats. *Oikos*. 2005 Apr;109(2):396–404.

36. Halttunen E, Gjelland KØ, Hamel S, Serra-Llinares RM, Nilsen R, Arechavala-Lopez P, *et al.* Sea trout adapt their migratory behaviour in response to high salmon lice concentrations. *J Fish Dis*. 2018 Jun;41(6):953–67.

37. Lopes PC, French SS, Woodhams DC, and Binning SA. Infection avoidance behaviors across vertebrate taxa: Patterns, processes and future directions. In Ezenwa VO, Altizer S, and Hall RJ (eds), *Animal Behavior and Parasitism*. Oxford: Oxford University Press, 2022. DOI: 10.1093/oso/9780192895561.003.0014

38. Johns S and Shaw AK. Theoretical insight into three disease-related benefits of migration. *Popul*. 2016 Jan 1;58(1):213–21.

39. Owen JC and Moore FR. Swainson's thrushes in migratory disposition exhibit reduced immune function. *J Ethol.* 2008 Sep 1;26(3):383–8.

40. Van Dijk JG, Hoye BJ, Verhagen JH, Nolet BA, Fouchier RA, and Klaassen M. Juveniles and migrants as drivers for seasonal epizootics of avian influenza virus. *J Anim Ecol.* 2014 Jan;83(1):266–75.

41. Gylfe Å, Bergström S, Lundstróm J, and Olsen B. Reactivation of *Borrelia* infection in birds. *Nature.* 2000 Feb;403(6771):724–5.

42. Hegemann A, Matson KD, Versteegh MA, and Tieleman BI. Wild skylarks seasonally modulate energy budgets but maintain energetically costly inflammatory immune responses throughout the annual cycle. *PLoS One.* 2012 May 3;7(5):e36358.

43. Møller AP and Erritzøe J. Host immune defence and migration in birds. *Evol Ecol.* 1998 Nov;12(8):945–53.

44. Hegemann A, Alcalde Abril P, Sjöberg S, Muheim R, Alerstam T, Nilsson JÅ, *et al.* A mimicked bacterial infection prolongs stopover duration in songbirds—but more pronounced in short-than long-distance migrants. *J Anim Ecol.* 2018 Nov;87(6): 1698–708.

45. Becker DJ, Ketterson ED, and Hall RJ. Reactivation of latent infections with migration shapes population-level disease dynamics. *Proc R Soc B: Biol Sci.* 2020 Sep 30;287(1935):20201829.

46. Daversa DR, Monsalve-Carcaño C, Carrascal LM, and Bosch J. Seasonal migrations, body temperature fluctuations, and infection dynamics in adult amphibians. *PeerJ.* 2018 May 8;6:e4698.

47. Shaw AK and Binning SA. Migratory recovery from infection as a selective pressure for the evolution of migration. *Am Nat.* 2016 Apr 1;187(4):491–501.

48. Poulin R, Closs GP, Lill AW, Hicks AS, Herrmann KK, and Kelly DW. Migration as an escape from parasitism in New Zealand galaxiid fishes. *Oecologia.* 2012 Aug 1;169(4):955–63.

49. Johnson PT, Stanton DE, Forshay KJ, and Calhoun DM. Vertically challenged: How disease suppresses *Daphnia* vertical migration behavior. *Limnol Oceanogr.* 2018 Mar;63(2):886–96.

50. Van Gils JA, Munster VJ, Radersma R, Liefhebber D, Fouchier RA, and Klaassen M. Hampered foraging and migratory performance in swans infected with low-pathogenic avian influenza A virus. *PloS One.* 2007 Jan 31;2(1):e184.

51. Serra-Llinares RM, Bøhn T, Karlsen Ø, Nilsen R, Freitas C, Albretsen J, *et al.* Impacts of salmon lice on mortality, marine migration distance and premature return in sea trout. *Mar Eco. Prog Ser.* 2020 Feb 6;635:151–68.

52. Krkošek M, Connors BM, Ford H, Peacock S, Mages P, Ford JS, *et al.* Fish farms, parasites, and predators: Implications for salmon population dynamics. *Ecol App.* 2011 Apr;21(3):897–914.

53. Gómez-Díaz E, Morris-Pocock JA, González-Solís J, and McCoy KD. Trans-oceanic host dispersal explains high seabird tick diversity on Cape Verde islands. *Biol Lett.* 2012 Aug 23;8(4):616–9.

54. Global Consortium for H5N8 and Related Influenza Viruses. Role for migratory wild birds in the global spread of avian influenza H5N8. *Science.* 2016 Oct 14;354(6309):213–7.

55. Ogden NH, St-Onge L, Barker IK, Brazeau S, Bigras-Poulin M, Charron DF, *et al.* Risk maps for range expansion of the Lyme disease vector, *Ixodes scapularis,* in Canada now and with climate change. *Int J Health Geogr.* 2008 Dec;7(1):1–5.

56. Altizer SM. Migratory behaviour and host–parasite co-evolution in natural populations of monarch butterflies infected with a protozoan parasite. *Evol Ecol Res.* 2001;3(5):567–81.

57. Born-Torrijos A, Poulin R, Pérez-del-Olmo A, Culurgioni J, Raga JA, and Holzer AS. An optimised multi-host trematode life cycle: Fishery discards enhance trophic parasite transmission to scavenging birds. *Int J Parasitol.* 2016 Oct 1;46(11):745–53.

58. Perez-Rodriguez A, de la Hera I, Bensch S, and Perez-Tris J. Evolution of seasonal transmission patterns in avian blood-borne parasites. *Int J Parasitol.* 2015 Aug 1;45(9–10):605–11.

59. Soares L, Young EI, and Ricklefs RE. Haemosporidian parasites of resident and wintering migratory birds in the Bahamas. *Parasitol Res.* 2020 May;119(5):1563–72.

60. Shaw AK, Craft ME, Zuk M, and Binning SA. Host migration strategy is shaped by forms of parasite transmission and infection cost. *J Anim Ecol.* 2019 Oct;88(10):1601–12.

61. Naven Narayanan, Binning SA, and Shaw AK. Infection state can affect host migratory decisions. *Oikos.* 2020 Oct;129(10):1493–503.

62. Balstad LJ, Binning SA, Craft ME, Zuk M, and Shaw AK. Parasite intensity and the evolution of migratory behavior. *Ecology.* 2021 Feb;102(2):e03229.

63. Brown LM and Hall RJ. Consequences of resource supplementation for disease risk in a partially migratory population. *Philos Trans R Soc Lond B: Biol Sci* 2018 May 5;373(1745):20170095.

64. Berg JE, Hebblewhite M, St Clair CC, and Merrill EH. Prevalence and mechanisms of partial migration in ungulates. *Front Ecol Evol.* 2019 Aug 29;7:325.

65. Galsworthy SJ, Ten Bosch QA, Hoye BJ, Heesterbeek JA, Klaassen M, and Klinkenberg D. Effects of infection-induced migration delays on the epidemiology of avian influenza in wild mallard populations. *PloS One.* 2011 Oct 18;6(10):e26118.

66. Peacock SJ, Bouhours J, Lewis MA, and Molnár PK. Macroparasite dynamics of migratory host populations. *Theor Popul Biol.* 2018 Mar 1;120:29–41.

67. Robinson RA, Crick HQ, Learmonth JA, Maclean IM, Thomas CD, Bairlein F, *et al.* Travelling through a warming world: Climate change and migratory species. *Endanger Species Res.* 2009 May 14;7(2): 87–99.

68. Satterfield DA, Marra PP, Sillett TS, and Altizer S. Responses of migratory species and their pathogens to supplemental feeding. *Philos Trans R Soc Lond B Biol Sci.* 2018 May 5;373(1745):20170094.

69. Kock RA, Orynbayev M, Robinson S, Zuther S, Singh NJ, Beauvais W, *et al.* Saigas on the brink: Multidisciplinary analysis of the factors influencing mass mortality events. *Sci Adv* 2018 Jan 1;4(1):eaao2314.

70. Ford JS and Myers RA. A global assessment of salmon aquaculture impacts on wild salmonids. *PLoS Biol.* 2008 Feb;6(2):e33.

71. Hall RJ, Brown LM, and Altizer S. Modeling vector-borne disease risk in migratory animals under climate change. *Integr Comp Biol.* 2016 Aug 1;56(2):353–64.

72. Hoar BM, Ruckstuhl K, and Kutz S. Development and availability of the free-living stages of *Ostertagia gruehneri,* an abomasal parasite of barren ground caribou *(Rangifer tarandus groenlandicus),* on the Canadian tundra. *Parasitology.* 2012 Jul;139(8):1093–100.

73. Molnár PK, Kutz SJ, Hoar BM, and Dobson AP. Metabolic approaches to understanding climate change impacts on seasonal host-macroparasite dynamics. *Ecol.* 2013 Jan;16(1):9–21.

74. Gehman AL, Hall RJ, and Byers JE. Host and parasite thermal ecology jointly determine the effect of climate warming on epidemic dynamics. *PNAS.* 2018 Jan 23;115(4):744–9.

75. Hurford A, Cobbold CA, and Molnár PK. Skewed temperature dependence affects range and abundance in a warming world. *Proc R Soc B: Biol Sci.* 2019 Aug 14;286(1908):20191157.

76. Satterfield DA, Maerz JC, and Altizer S. Loss of migratory behaviour increases infection risk for a butterfly host. *Pro. Royal Soc B.* 2015 Feb 22;282(1801): 20141734.

77. Poulson RL, Luttrell PM, Slusher MJ, Wilcox BR, Niles LJ, Dey AD, *et al.* Influenza A virus: Sampling of the unique shorebird habitat at Delaware Bay, USA. *R Soc Open Sci.* 2017 Nov 15;4(11):171420.

78. Daszak P. Emerging infectious diseases of wildlife—threats to biodiversity and human health. *Science.* 2000;287(5459):1756.

79. Peacock SJ, Krkošek M, Proboszcz S, Orr C, and Lewis MA. Cessation of a salmon decline with control of parasites. *Ecol App.* 2013 Apr;23(3):606–20.

80. Oro D, Genovart M, Tavecchia G, Fowler MS, and Martínez-Abraín A. Ecological and evolutionary implications of food subsidies from humans. *Ecol.* 2013 Dec;16(12):1501–14.

81. Gilbert NI, Correia RA, Silva JP, Pacheco C, Catry I, Atkinson PW, *et al.* Impacts of landfill use on the movement and behaviour of resident white storks *(Ciconia ciconia)* from a partially migratory population. *Mov.* 2016;4(7).

82. Pérez-Tris J and Tellería JL. Migratory and sedentary blackcaps in sympatric non-breeding grounds: Implications for the evolution of avian migration. *J Anim Ecol.* 2002 Mar;71(2):211–24.

83. Thogmartin WE, Wiederholt R, Oberhauser K, Drum RG, Diffendorfer JE, Altizer S, *et al.* Monarch butterfly population decline in North America: Identifying the threatening processes. *R Soc Open Sci.* 2017 Sep 20;4(9):170760.

84. Satterfield DA, Maerz JC, Hunter MD, Flockhart DT, Hobson KA, Norris DR, *et al.* Migratory monarchs that encounter resident monarchs show life-history differences and higher rates of parasite infection. *Ecol.* 2018 Nov;21(11):1670–80.

85. Leach CB, Webb CT, and Cross PC. When environmentally persistent pathogens transform good habitat into ecological traps. *R Soc Open Sci.* 2016 Mar 23;3(3):160051.

86. Becker DJ and Hall RJ. Too much of a good thing: Resource provisioning alters infectious disease dynamics in wildlife. *Biol Lett.* 2014 Jul 31;10(7):20140309.

87. Cross PC, Cole EK, Dobson AP, Edwards W, Hamlin KL, Luikart G, *et al.* Probable causes of increasing brucellosis in free-ranging elk of the Greater Yellowstone Ecosystem. *Ecol App.* 2010 Jan;20(1): 278–88.

88. Hill NJ, Takekawa JY, Ackerman JT, Hobson KA, Herring G, Cardona CJ, *et al.* Migration strategy affects avian influenza dynamics in mallards *(Anas platyrhynchos). Mol.* 2012 Dec;21(24):5986–99.

89. Spiegel O, Anglister N, and Crafton MM. Movement data provides insights into feedbacks and heterogeneities in host–parasite interactions. In Ezenwa VO, Altizer S, and Hall RJ (eds), *Animal Behavior and Parasitism.* Oxford: Oxford University Press, 2022. DOI: 10.1093/oso/9780192895561.003.0006.

90. Møller AP and Szép T. The role of parasites in ecology and evolution of migration and migratory connectivity. *J Ornithol.* 2011 Sep 1;152(1):141–50.

91. Birnie-Gauvin K, Lennox RJ, Guglielmo CG, Teffer AK, Crossin GT, Norris DR, *et al.* The value of experimental approaches in migration biology. *Physiol Biochem Zool.* 2020 May 1;93(3):210–26.

92. Mysterud A, Qviller L, Meisingset EL, and Viljugrein H. Parasite load and seasonal migration in red deer. *Oecologia*. 2016 Feb 1;180(2): 401–7.

93. Meakin S and Kurvits T Assessing the impacts of climate change on food security in the Canadian Arctic. GRID Arendal, Norw. 2009. Available from <http://www.grida.no/files/publications/foodsec_updt_LA_lo.pdf>.

94. Jenkins EJ, Simon A, Bachand N, and Stephen C. Wildlife parasites in a One Health world. *Trends Parasitol*. 2015 May 1;31(5):174–80.

# CHAPTER 8

# Seasonal human movement and the consequences for infectious disease transmission

Hannah R. Meredith and Amy Wesolowski

## 8.1 Introduction

Variation in environmental factors and host behaviors can result in spatial and temporal fluctuations in the transmission of infectious diseases to humans, including malaria, schistosomiasis, Chagas disease, and leishmaniasis (1)–(4). Seasonal fluctuations in rainfall, humidity, and temperature can affect the pathogen population and transmissibility directly or indirectly, via its intermediate host (1), (2). Host mobility can affect the spatial spread of pathogens, with travelers importing or exporting infections to susceptible individuals in other populations that are receptive to transmission (5)–(7). In turn, seasonal mobility can be affected by pathogen transmission levels, with increased risk of transmission causing either residents and visitors to flee and further spread the pathogen (8), or travel restrictions enacted to control the spread of disease (9). Being able to predict large fluctuations in population sizes can help intervention strategies allocate and plan for additional resources more effectively (10).

Unfortunately, seasonal travel and its interruptions are rarely quantified or modeled, typically because the data requirements to measure these patterns are high. This is particularly true in low- or middle-income settings which often have limited data availability. Information is needed to establish the baseline and identify the deviations in mobility patterns. Furthermore, to define the relationship between mobility and disease spread,

the temporal resolution of the mobility data should match that of the disease transmission patterns of interest. Population-level representative data (e.g., national census) are often not time-resolved (i.e., they are only available for a single time point) and individual-level data (e.g., Global Positioning Systems (GPS), travel surveys) may be limited to a small population or geographic area that may not make it generalizable to the broader population. To improve understanding of the spread of infectious diseases there is a need to identify and characterize the spatially dynamic trends, drivers, and underlying demographics of seasonal human behaviors. This requires mobility data collected at sufficiently high spatial and temporal resolutions over a long period of time so that seasonal and baseline mobility patterns can be defined. To this end, the COVID-19 pandemic and increased importance of capturing time-resolved mobility data across many geographical settings is making some of these possibilities realized. However, it may still be difficult to obtain information on the motivations for travel and/or mobility patterns from certain socio-demographic groups.

This chapter reviews the current state of understanding and implications of seasonal mobility with regards to infectious disease transmission. Section 8.2 describes what drives long- and short-term mobility patterns and their consequences for transmission. It also explores how infection (or infection risk) can influence mobility, potentially

Hannah R. Meredith and Amy Wesolowski, *Seasonal human movement and the consequences for infectious disease transmission*.
In: *Animal Behavior and Parasitism*. Edited by Vanessa O. Ezenwa, Sonia Altizer, and Richard J. Hall, Oxford University Press.
© Oxford University Press (2022). DOI: 10.1093/oso/9780192895561.003.0008

creating feedbacks between seasonal movement and infection dynamics. In section 8.3, we compare the utility and limitations of direct and proximate measures of human mobility. Section 8.4 compares different modeling approaches that incorporate mobility data to predict infection dynamics. Finally, section 8.5 outlines the gaps in this field that need to be addressed to improve estimates of how seasonal mobility patterns impact the spread of pathogens.

## 8.2 Defining seasonal mobility

Variations in human mobility and aggregation patterns are often seasonal, reflecting a range of factors including seasonal work (e.g., agriculture, fishing, forestry, tourism), weather, and societal factors (e.g., school terms, national and religious holidays) (11)–(14). Each driver results in movements that are characterized by demographic (e.g., age, gender, religion, occupation), spatial (e.g., scale and location(s)), and temporal information (e.g., duration, time of year, and frequency). Identifying these characteristics and their associated risks for disease is important for understanding how seasonal mobility contributes to the spread of diseases. Ultimately, this information can be used to determine the appropriate tool(s) necessary to measure these movements and help focus intervention strategies. Here, we describe different kinds of seasonal travel, their motivation, and their relationship to disease transmission.

### 8.2.1 Drivers and disease implications of long-term seasonal mobility patterns

Long-term seasonal shifts in human mobility and aggregation can occur when individuals move for opportunities in agriculture, forestry, fisheries, or tourism and spend the entire season (upwards of months) residing in a new location (14), (15). Depending on if a seasonal worker moves from a low- to high-transmission area or vice versa, they may be associated with an increased risk of being exposed to a pathogen or importing a pathogen for local transmission upon relocation, respectively. In addition to changes in prevalence, seasonal changes in location may be accompanied by changes in host behaviors that could alter the likelihood of being exposed to (or transmitting) pathogens, having access to preventative measures, or receiving treatment. For instance, seasonal laborers may spend more time outside and may not have access to preventative measures, such as bednets, thus increasing their exposure time to different pathogens and subsequently leading to infection (16), (17). If the pathogen is transmitted seasonally, then the risk of exposure may be amplified or reduced if the seasonal travel is in or out of sync with the transmission season (Box 8.1). Despite their increased risk for infection, seasonal laborers may be less likely to seek care due to a lack of sick leave, health insurance, ability to communicate with local healthcare officials, or transportation to a healthcare facility (18), (19). This could result in untreated infections being exported when the workers travel to another position or return home.

### 8.2.2 Drivers and disease implications of short-term seasonal mobility patterns

There may also be seasonal shifts in short-term or daily movements and activities driven by environmental conditions or social factors. For instance, a study in rural Zambia found that people took fewer long-distance trips in the rainy season than the dry season, possibly because roads became impassable (20). At the end of the rainy season, the number of long-distance trips increased, possibly taking travelers into areas of higher malaria transmission, and coincided with the area's seasonal peak in malaria cases. Travel for school terms (particularly boarding schools) or national or religious holidays (13), are two factors that may also drive seasonal shifts in short-term movements. School term schedules have been identified as drivers of disease incidence as they are associated with increases in the density of susceptible populations, contact rates between specific age groups, and the rate of introduction events that can lead to outbreaks (21), (22). In fact, much of our understanding of the relationship between seasonal mobility and disease incidence has been driven by school-based studies (22)–(24). Further, national holidays (religious, political, etc.) are often periods when a large proportion of the population may travel nationally and/or internationally (25), (26). Hence, characterizing shorter, seasonal movements can help determine if changes in mobility are associated with regularly imported infections that could subsequently spread to the surrounding areas.

### 8.2.3 Impact of disease on human mobility

Disease dynamics and human mobility are often connected in a feedback loop whereby disease status and transmission levels can impact human behavior and, consequently, disrupt regular human mobility patterns, which in turn can alter pathogen transmission and spread. Human behavior may voluntarily change due to one's disease status or perceived threat of disease in an area. For instance, decreases in mobility have been observed when individuals are infected by dengue (27), (28). When patients were infected with dengue, they spent more time at home, had fewer contacts, visited

fewer locations, and traveled to locations closer to home compared to when they were healthy (27), (28). These effects may be impacted by the severity of illness and may also vary during the course of an infection, which further generates dynamic and complex interactions between the host and secondary infections (27). Decreases in the number of social contacts outside of the home have also been observed for symptomatic people with influenza-like illness (29). Regardless of disease status, people may also alter their behavior as a disease becomes more prevalent in a particular area. For example, the Ebola outbreak of 2013–2016 elicited fear-related behaviors that caused people to flee from high-incidence communities and introduce the disease to new regions (8). Another way disease impacts human behavior is by triggering institutional recommendations aimed at mitigating transmission through reduced contact rates. For instance, school closures, public gathering bans, and lockdowns have been implemented to try to control outbreaks

---

**Box 8.1 Impact of seasonal human mobility on disease dynamics**

To demonstrate the impact of different seasonal human mobility patterns, we consider how malaria cases can be affected due to different travel schemes between two locations. Here, malaria transmission in humans is modeled using a Susceptible, Exposed, Infected, Recovered, Asymptomatic (SEIRA) model and in mosquitoes using an SEI model (Figure 1A). Mosquito breeding is seasonal, resulting in Peak (April–July) and Off-peak (January–April) malaria season (indicated in Figure 1C). Locations 1 and 2 have different levels of malaria transmission and all population members, regardless of infection status, are assumed to move between populations at the same rate (Figure 1B). The proportion of infections detected in each location is observed over the course of one year (top row) for five different mobility patterns (bottom row) (Figure 1C). In the *Baseline* scenario, people move at a low, constant rate between Location 1 (black) and Location 2 (blue). *Relocate for Off-Peak* or *Peak* are similar to seasonal workers relocating to another place for a period of time coinciding with Off-peak or Peak malaria season before returning home. In these scenarios, baseline movement rates are present year round, except for an initial

week when workers from Location 2 relocate to Location 1 at an increased rate and a later week when workers from Location 1 return to Location 2 at an increased rate. In the *Off-Peak* scenario, there is a small increase in Location 1 infections as members of Location 2 arrive. In the *Peak* scenario, there is a larger increase in Location 1 infections as more members arriving from Location 2 are likely to be infected at that time. *Increase for Off-peak* or *Peak* simulate scenarios similar to when there is a period of sustained increase in daily travel between both locations, such as a winter or summer holiday. To evaluate the impact of different mobility patterns on the spread of malaria, the percent change in the number of cases observed over the year relative to the baseline scenario was calculated (Figure 1D). For Location 1, Peak travel resulted in larger increases in infections than Off-Peak travel and scenarios with longer periods of increased travel rates resulted in more infections than those with relocation. Location 2 saw a decrease in infections, as members traveling to Location 1 were less likely to become infected and import infections with them upon return.

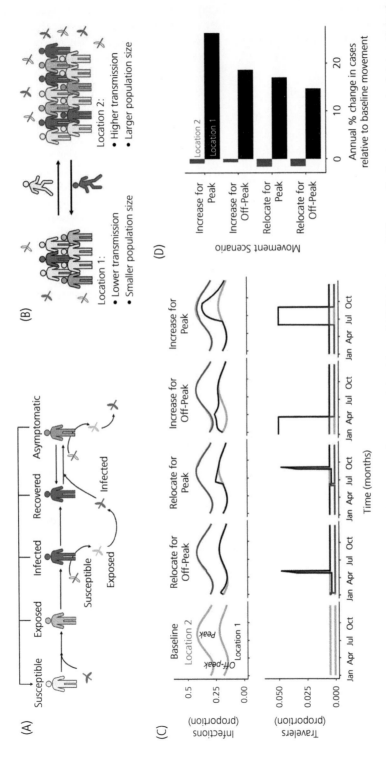

**Box 8.1, Figure 1** Demonstration of how seasonal mobility can impact disease transmission. (A) A model (SEIRA for humans, SEI for mosquitoes) captures seasonal malaria transmission dynamics. (B) The model accounts for the spread of malaria in two locations connected by travelers. (C) Temporal dynamics of the proportion of infected human populations (top row) due to different mobility patterns (bottom row) are considered for location 1 (black) and 2 (blue) and compared to a baseline scenario with constant mobility rates (faded curves repeated in each subplot). See Box text for description of scenarios. (D) The impact of different mobility patterns on resulting cases is quantified and compared for each location.

and pandemics (30)–(32). These factors have further been explored in theoretical models (33). Characterizing the trigger, timing, and extent of changes in seasonal behaviors in response to disease dynamics would help further hone our ability to estimate the spread of disease and efficacy of interventions.

## 8.3. Detecting when and where seasonal travel occurs

Ideally, travel data would capture information on how many people are moving between specific locations as well as trip duration. Further insight into travel patterns could be gleaned if the data included details such as the socio-demographic groups traveling and motivations for travel. Realistically, however, there is often a trade-off between the level of detail available for individual trips and the generalizability of travel patterns across populations, spatial scales, and geographies. As a result, data typically only capture some aspects of travel, such as estimating the number of trips made between a given origin and destination, but without traveler demographic information. Quantifying seasonal travel has the additional requirement that these data are time-resolved (i.e., travel data need to be collected regularly over time) such that baseline rates of travel can be defined and seasonal changes can be measured. These seasonal changes in movement can then be matched to particular events (e.g., holidays, school terms, or environmental factors), thus allowing for drivers to be identified and tested. The requirement for quantifying mobility at multiple time points greatly limits which datasets may be applicable and, consequently, the settings in which seasonal mobility can be measured. The choice of datasets may be further constrained by the seasonal movement's spatial scale (e.g., international, intranational, regional, or city). Here, we outline many of the existing tools and datasets available to quantify human mobility patterns and evaluate their ability to characterize seasonal travel patterns (Table 8.1 and Figure 8.1).

### 8.3.1 Census data

Census data have been used to discern mobility patterns by querying households about changes in residence (34). While this information is better suited for long-term or permanent changes in location, studies suggest it could capture some aspects of shorter-term patterns (35). The low frequency at which census data are collected (often administered every 5 or 10 years) makes it difficult to establish a baseline mobility pattern and detect seasonal differences. Furthermore, the census is typically only collected from households at a permanent address and therefore has a tendency to miss populations who may be absent from or do not have a permanent address (e.g., migrant workers or highly mobile populations) that are relevant in driving some seasonal mobility patterns (36), (37).

### 8.3.2 Surveys

Surveys are widely used tools for capturing information on individual travel history and can be developed to address specific questions of interest. They can be targeted to quantify mobility patterns of particular groups (e.g., by age or income group), capture detailed information on particular travel patterns of interest, and collect additional information (e.g., behaviors during travel or infection). Travel surveys are not limited by geographic or temporal scale so they can capture both intra- and international patterns and detailed information on when and where travel occurred. However, surveys may be limited in scope and size. Although it is possible to ask a wide range of questions about travel, the amount of time and cost to answer these questions to obtain detailed information may limit their scope. Given the time and costs involved in conducting travel surveys, they often capture only a particular subset of the population and surveys may not capture highly mobile individuals. Further, travel surveys are subject to recall bias (which increases as participants are asked to recall trips further in the past) and require skilled interviewers to elicit detailed information (38), (39).

### 8.3.3 Commuting data

Another proxy for mobility is measurements of daily traffic, captured by monitoring commuter traffic or tickets purchased for bus, train, or plane routes (40)–(42). The most commonly used version of these

**Table 8.1** Tools to quantify human travel patterns and their relevance to seasonal patterns

| Data source | Spatial resolution | Temporal resolution | Traveler information | Limitations |
|---|---|---|---|---|
| **Surveys** | | | | |
| Census (34), (35) | —Depends on the country, with a wide range | —Often only measured every 5 or 10+ years | —Basic household member details may be recorded (gender, age, race) —Location of a previous residence | —Does not capture short/seasonal trips, travel motivation —Often excludes hard to reach populations or non-residents |
| Travel history surveys (38), (66) | —Depends on the level of detail asked for. Typically the name of village, town, or district visited | —Typically trip arrival and departure dates or arrival date and duration are recorded | —Good for collecting demographic information and motivation for travel —Can be designed for particular groups of interest | —Only represents a subset of the population —Affected by recall bias and recall period is typically on the order of months to a year |
| **Transportation data** | | | | |
| Bus, train, and commuter data (40)–(42) | —Bus and train routes defined by origin and destination —Road traffic data defined by the road network system | —Departure and arrival times may be aggregated by day or finer time scales —Road traffic may be aggregated by hours or rush hour periods | —Not recorded or available | —Often limited to high income countries and those using commuting systems (e.g., public transportation) —Does not reflect the initial origin and final destination of passengers —In some settings, routine bus or train trips are not documented |
| Flight data (40)–(42) | —Plane routes defined by origin and destination | —Departure and arrival times could be aggregated by day or finer time scales | —Not recorded or available | —Captures a small portion of trips within/between countries —May not reflect the initial origin and final destination of passengers |
| **Technology** | | | | |
| Satellite imagery (14) | —High spatial resolution —Covers a large area | —High temporal resolution (on the order of minutes) | —No information | —Does not quantify trips between locations —Reports relative changes in population size, not counts |
| Call data records (6), (45) | —Capable of high spatial resolution | —Capable of high temporal resolution | —Sometimes recorded, but often unavailable | —User bias (must own a phone) —Spatial resolution depends on cell tower density —In general, limited to intranational trips —Can be challenging to obtain from network providers |
| GPS loggers (20) | —High spatial resolution | —High temporal resolution | —Not directly recorded; however, information could be collected when assigning and/or collecting the logger | —Cost of loggers limits the study size, so the sample size of mobility patterns is limited. —Trade-offs between battery life, data collection frequency, and storage capacity |
| Mobile phone apps (48), (49) | —High spatial resolution | —High temporal resolution | —Demographics are sometimes included, depending on the app and data-use agreement. —Sometimes records trip purpose | —User bias (must own a smartphone, download the app, enable location history setting) —Not all establishments are mapped, thus the app may not identify type of destination motivating travel |

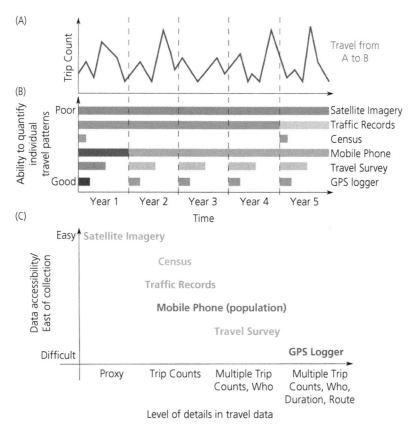

**Figure 8.1** Mobility data sources and their ability to capture different levels of detail for seasonal mobility patterns. (A) Human travel patterns may fluctuate with seasonal differences in travel between locations. A hypothetical pattern of travel between two locations A and B demonstrates how these patterns may vary within and between years. (B) The ability for each data type to quantify individual travel patterns (poor to good) and capture change over time is visualized. While many of these data sources can generate datasets that span multiple years, this is less common. General patterns of dataset duration are indicated by shaded areas, with satellite imagery datasets often spanning the longest period of consistent data collection and GPS loggers often spanning the shortest. These two data sources also represent the spectrum of the ability to quantify individual travel patterns. While there is variability in the resolution of these data (e.g., it is possible to collect survey data over longer time frames than shown), these represent general patterns of how the data have been primarily used to date. (C) Datasets also vary in terms of accessibility for researchers (easy to difficult) and the level of detail collected. Satellite imagery data are freely available whereas travel surveys or GPS logger data would need to be actively collected. The level of detail for each dataset include: generates proxy data (cannot quantify actual trip counts), quantifies trip counts (quantify the number of trips between two locations), quantifies multiple trips per individual (Multiple Trip Counts, Who), and/or characterizes additional factors such as duration or routes of travel (Multiple Trip Counts, Duration, Who, Routes). The amount of information about trips per dataset (duration, route take, frequency) and demography of the traveler also varies by data source. Some datasets, such as mobile phone data, may have subscriber-level granularity (multiple trip counts, who) or only aggregated population-level estimates (trip counts).

data is airline flight routes, which have been extensively applied to measure global mobility patterns (40). If sufficient data are archived, then a baseline and seasonal trends for each mode of transportation could be extracted. However, trip records may not be archived for long-term use and it is challenging to monitor commuting traffic in areas where road networks are not well established (43). While each

bus, train, or plane route is associated with an origin and destination, it is often not possible to determine an individual passenger's true start and end points. Further limitations include the fact that routine trips or trips between rural areas may go undocumented in some settings and that trip records may not be archived for long-term use (43). Finally, these data may only represent a particular demographic of the population. For example, airline data are unlikely to be representative of the many travel patterns in low- or middle-income settings.

### 8.3.4 Satellite imagery

Satellite imagery, particularly light intensity at night, is used as a proxy for movement where images over long periods of time are used to estimate a baseline intensity (as a measure of population size) from which deviations are detected (14). These data are available globally (often freely accessible in near-real time from websites such as the Visible Infrared Imaging Radiometer Suite, <https://ncc.nesdis.noaa.gov/VIIRS/>), at a high spatial and temporal resolution, but do not inherently capture a particular type of travel pattern. Given that they are available globally, they can detect inter- and intra-national changes in population sizes and may also be able to detect movement in difficult-to-reach populations (e.g., migrant populations). While this approach is useful to identify when and where population changes are occurring and can be used to study large areas, it only provides relative changes over time at specific locations and does not provide information on the number of trips between locations. For example, you may observe a large difference in the image intensity of a location which may serve as a proxy for a change in the population at that location. However, you cannot determine where those individuals may have traveled to or from. While you can capture fine temporal and spatial detail, the ability to translate these data to actual travel patterns may be limited. For example, travel that does not increase the number of lights at night or detectable development (e.g., building new houses or infrastructures) may not be captured. Further, light intensity can saturate, which limits the use of these data in areas that are already highly populated.

### 8.3.5 Mobile phone data

As mobile phone ownership increases (67% of the global population owned mobile phones in 2019 (44)), call data records stamped with the time and tower at which a call/text was sent have been used to study intranational mobility patterns from large portions of populations at high spatial and temporal resolution (6), (45). If call data records from a sufficiently long period of time can be acquired, baseline and seasonal trends in the number of people traveling to/from specific locations can be ascertained. For instance, mobile phone data were used to quantify seasonal mobility trends around national and school holidays in Kenya, Namibia, and Pakistan (11). Studies have shown that ownership has penetrated a wide range of socio-economic levels (46), so it is likely that movements by particular demographic groups (such as migrant workers and mobile populations) may be included. However, the mobility patterns detected by call data records are biased by phone ownership, which is often lowest in the youngest and oldest age groups (46), and by the density of cell towers, which is often lower in remote and/or rural areas where mobile populations reside and migrant workers may travel to/from (47). Due to national network providers, call data records are not able to capture cross-border travel, thus limiting their utility in capturing international seasonal travel.

### 8.3.6 GPS data

GPS data can be used to quantify mobility patterns at high spatial and temporal resolution (20). One way of collecting coordinates is through assigning GPS loggers to participants while they go about their daily routines. One such study was able to detect seasonal trends in trip duration (shorter in rainy season and longer in dry season) after collecting GPS coordinates from participants over a six-month period (20). Due to cost, GPS logger studies are typically limited to capturing mobility patterns of a particular subset of the population over a limited time period that is not conducive to detecting multi-year seasonal trends. More recently, mobile phone applications (apps) that record users' locations have begun to serve as GPS coordinate data sources with a wider coverage of the population. For instance, data collected from the Google Location History app was able to discern strong seasonal effects on mobility in different parts of the world due to national holidays, weather patterns, and amount of daylight available at a given time of year (48). Facebook's Data for Good has also been used to develop population density maps and monitor global mobility patterns and could be a viable source for measuring seasonal variations in human mobility and aggregations (49), (50). However, the data collected is likely to be biased by which users enable the app to record locations as well as smartphone ownership, which is increasing globally but is not as pervasive as mobile phones in general (44).

## 8.4. Connecting human mobility patterns with infectious disease dynamics

Ultimately, data that quantifies and characterize seasonal mobility patterns can be integrated into models that predict disease transmission dynamics, optimize healthcare interventions, or identify behaviors associated with disease risk that should be mitigated. Here, we present examples of a range of models relating different types of mobility data to disease dynamics (Table 8.2), and discuss their ability to capture seasonal changes. While there are many other examples of ways mobility data are integrated into infectious disease models, these examples provide a range of commonly used approaches.

### 8.4.1 Risk models

Often, infection dynamics are quantified by summarizing an individual's "risk of infection" based on their mobility patterns. In order to take into account travel and the chance of being infected, the risk of an individual being infected can be estimated over time using a combination of time-resolved infection risk maps and travel patterns. An individual's (i) risk score ($r_i$) can be calculated by summing the product of the risk value ($r_k$) and time spent ($\tau_{ij}$) at each location (j) visited and normalizing by the individual's total participation time ($T_i$)(EQ. 8.1)(51). The longer one spends in a high-risk location, the greater the risk of being infected. The impact of seasonal mobility on risk of infection can be determined by calculating $r_i$ for time periods of interest (e.g., rainy or dry season, school session or holiday).

$$r_i = \frac{\sum_{j-i}^{n} \tau_{ij}{}^*r_k}{T_i} \qquad 8.1$$

This integration structure is flexible. For example, location, dates, and trip durations of destinations

can be characterized by data sources that capture individual travel patterns, such as GPS loggers, travel history surveys, and, depending on privacy laws, mobile phone data (52). These different dimensions of mobility can be integrated into a general measure of time spent at various locations. The risk of infection associated with each location can be extracted from a previously defined infection map at the relevant time of year. Depending on the pathogen of interest, this measure could also vary with time or integrate additional uncertainty in transmission potential (and hence risk of individual infection). The spatial and temporal resolution of risk scores is limited by the dataset with the lowest resolutions.

### 8.4.2 Statistical models

Statistical models fit to mobility data can characterize the relationship between an individual's travel patterns and infection status. At the individual level, models' fit to travel history surveys have determined that changes in mobility patterns (e.g., time spent at home, number of places visited) are often related to being symptomatic (28). These changes are important to consider when incorporating seasonal mobility into disease models: seasonal changes in movement may impact the proportion of a population that is infected and displaying reduced mobility patterns, ultimately impacting the risk of onward transmission and an outbreak (53).

At the population level, regression models have been used to determine the relationship between population fluctuations and incidence outcome (54), (55). Population fluctuations can be quantified by census migration data, call data records (CDRs), and satellite imagery (14), (54), (55). While census migration data collected every 5–10 years can approximate movement over shorter time scales (35), it does not capture seasonal changes in movement. Thus, CDRs and satellite images analyzed at fine temporal and spatial scales would be more suitable inputs for models aiming to characterize the relationship between seasonal movement and disease dynamics. Often standard statistical approaches are used that can also integrate other factors associated with disease risk (e.g., age, demographics).

**Table 8.2** Using different mobility datasets to enhance understanding of infectious disease dynamics. Here are examples for how different mobility data sources have been integrated into models with disease data to inform transmission patterns.

| Data source | Example | Disease data | Mobility data | Model integrating data sources |
|---|---|---|---|---|
| **Surveys** | | | | |
| Census | Census migration data identified areas predicted to act as major exporters/importers of malaria(54) | PAHO incidence data for malaria for Mesoamerica aggregated to the administrative one unit of each country | Census microdata for the first level administrative unit of a household's current and prior residence, if within the same country | A logistic regression model fit to census microdata and scaled by incidence data determined the flow of infected people between areas |
| Travel history surveys | Retrospective movement interviews determined how fever impacted an infected person's mobility(28), (67) | Fever status was recorded for participants in a longitudinal cohort study in Iquitos, Peru | Survey information on locations visited in the past two weeks. Coordinates, frequency of visits per day, and duration of stay are recorded for each location | A range of models relating fever status to changes in mobility patterns (distance traveled, number of locations visited, time spent at home) were fit to travel history interviews |
| **Transportation data** | | | | |
| Bus, train, and commuter data | Commuter data informed models to predict spatial risk of influenza transmission during different seasons(68) | Weekly time series of influenza-like-illnesses were determined from medical claims in the United States and aggregated to cities | County-to-county work commutes documented by the US Census | Semi-parametric mechanistic models that incorporated inter-county movement quantified risk of infection for each city for different seasons |
| Flight data | The Vector-Borne Disease Air Importation Risk tool estimated monthly risk of air travel introducing a disease and/or vector and the risk of onward transmission(69) | Global distribution of vector-borne diseases (i.e., malaria, dengue, yellow fever, chikungunya) and their associated vectors and climate constraints | Origin and destination airports and the number of passengers flying each route per month | An entity-relationship model integrated flight data with global distribution of disease, vector and climate, and travel time to nearest settlement |
| **Technology** | | | | |
| Satellite imagery | Night-time satellite images explained fluctuations in measles and improved vaccination coverage strategies(14), (70) | Local health centers provided daily measles case records, aggregated at the commune level for three cities in Niger | Nighttime satellite images quantified daily fluctuations in brightness as a proxy for changes in population size due to immigration and emigration | An SEIR model with a migration component was fit to migration data, defined by changes in brightness values |
| Call data records | Mobile phone data quantified impact of nation-wide seasonal population fluxes on rubella transmission(55) | Rubella cases reported for each Kenyan province | Mobile phone data quantified daily population flux (relative changes in mobility) over time | A regression model using population flux predicted changes in transmission (fit to incidence data) |
| GPS loggers | GPS loggers identified possible areas of exposure to malaria stratified by demographic group, season, and infection status(51) | Seasonal malaria risk maps developed from a longitudinal study of malaria cases in Zambia | Every 2.5 minutes, the coordinates, date, and time were collected for participants carrying a GPS logger | The risk of malaria infection was calculated for each participant based on how long they spent time in a high- or low-risk setting |
| Mobile phone apps | Facebook data quantified potential of SARS-CoV-2 to spread faster due to changes in mobility patterns when lockdowns were announced(71) | The spread of SARS-CoV-2 is simulated for various urban centers in the US as well as all of Spain, India, France, and Bangladesh | The number of Facebook users moving between predefined tiles is calculated every eight hours | A metapopulation model integrated mobility data to examine potential epidemiological outcomes of lockdown announcement responses |

### 8.4.3 Compartmental models

Compartmental models capture the progression of population members through different stages of disease (see Box 8.1 for an example). These models can be solved numerically to track the number of infected individuals in a population or location through time, or to derive summary population-level metrics of infection risk such as the pathogen basic reproductive number ($R_0$), the number of expected new cases following the introduction of a single index case in a wholly susceptible population. They can account for seasonal mobility patterns by incorporating a time-dependent migrating population that impacts the proportion of the population in each compartment (14). To consider disease transmission across $n$ discrete populations defined by specific locations and/or demographics, a metapopulation model can be used. Each population has its own transmission model (such as an SIR compartmental model for all $n$ populations) and is connected to other populations at a rate often estimated as the rate of travel between locations (53). The connectivity value of two populations can be informed by data such as satellite imagery or mobile phone data. Other population-level sources of data, such as census data, can also help inform the rate of flow of individuals between populations. Depending on the prevalence of disease at the traveler population's origin(s) and destination(s) and the seasonal mobility patterns, the rate at which infection is imported or exported can vary with time.

## 8.5. Identifying gaps and future directions

Ultimately, data that quantify and characterize seasonal mobility patterns can be integrated into models that can predict disease transmission dynamics, predict additional healthcare resources needed, or identify behaviors associated with disease risk that should be mitigated. As outlined earlier, these datasets all include various strengths and weaknesses regarding their ability to improve our understanding of seasonal travel and their impact on disease dynamics. Next we outline directions to move this area of research forward.

### 8.5.1 Improving models of seasonal travel

In combination with mobility data or when mobility data are unavailable, outdated, or incomplete, mobility models have often been used to estimate travel between locations (56). Mobility models have been used extensively to couple locations spatially in transmission modeling frameworks. These mobility models rely on fairly standard assumptions to inform drivers of movement: individuals are more likely to travel to more populated areas and the overall cost to travel is approximated by a measure of distance. However, the standard spatial interaction models of mobility do not vary with time (Box 8.2 reviews commonly used models (57), (58)) and typically estimate a baseline rate of travel between locations. A better understanding of how seasonality impacts these drivers (e.g., seasonally varying population estimates) or costs to travel (e.g., climate factors impacting the ease of travel) could help improve our ability to model seasonal travel. These drivers and costs to travel may also be impacted by trip duration and need to be considered when distinguishing between short- and long-term seasonal mobility patterns (59). To date, there have been limited data to fit these models to, either generally or seasonally. However, as mobile phone data become more accessible, their ability to characterize mobility patterns by space, time, and duration could be leveraged to further improve how these models predict seasonal travel.

### 8.5.2 Identifying who is traveling and motivations for travel

It may also be relevant to characterize who is traveling and their motivations at particular times of the year, as different demographics or occupations may be associated with different seasonal movements and risks of infection (60)–(62). Thus, data sources that provide insight into motivation of travel are necessary to complement the quantified mobility patterns. As mentioned in section 8.3, most existing datasets have a trade-off between the level of individual detail available and the generalizability of travel patterns. Some data sources, such as satellite images, travel network data, and call data records may capture population level movement,

## Box 8.2 Mobility models

Spatial interaction models are often used to estimate trips or connectivity between geographic locations. While mobility models have been reviewed extensively elsewhere (56), here we briefly review the two most common forms: the gravity and radiation models (57), (58). Both models rely on the similar assumptions that the number of trips between locations is determined based on the attractiveness of the destination (often approximated via population size, Figure 1A) and how hard it is to travel to that destination (often approximated via distance, travel time, or intervening opportunities). The output for these models is typically the number of trips between a location; however, they can be used to estimate a rate (i.e., if a model is fit to data for trips taken over a month, then the output will be the number of trips per month) or the probability of travel to a location. Notably, neither model accounts for temporal dynamics in trips.

**Gravity model:** The gravity model assumes that the number of trips between an origin $i$ and destination $j$ ($T_{ij}$) increases as a function of the population size at each location ($N_i$ and $N_j$) and decreases as a function of distance between the two locations ($d_{ij}$) (Figure 1B). The impact of the origin population, destination population, and distance can be weighted by parameters $\alpha$, $\beta$, and $\gamma$, respectively. Often, a scaling factor $\theta$ is included to improve the model's fit to data.

$$T_{ij} = \theta \frac{N_i^\alpha N_j^\beta}{d_{ij}^\gamma} \qquad \text{EQ (1)}$$

**Radiation model:** The radiation model defines the number of trips made between an origin $i$ and destination $j$ ($T_{ij}$) as the fraction of all trips originating at location $i$ ($T_i$). Like the gravity model, the radiation model also assumes that $T_{ij}$ increases as a function of $N_i$ and $N_j$. However, it is different in the way that it accounts for the attraction of any additional, alternative destinations within a circle centered on the origin with a radius equal to $d_{ij}$ ($s_{ij}$) (Figure 1C).

$$T_{ij} = T_i \frac{N_i N_j}{(N_i + s_{ij})(N_i + N_j + s_{ij})} \qquad \text{EQ (2)}$$

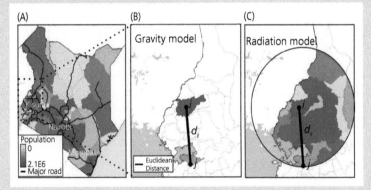

**Box 8.2, Figure 1** Comparing components of the gravity and radiation model. (A) The attractiveness of a location is often modeled as a function of population size. Here, the population size of each county (outlined in gray) is determined for the study area (Kenya). The number of trips between origin county $i$ and destination county $j$ can be estimated by a (B) gravity model or (C) radiation model. (B) The gravity model estimates trip counts as a function of the population sizes at origin $i$ and destination $j$ as well as the distance between the two ($d_{ij}$). (C) The radiation model is based on the population size of the origin and destination, as well as the total population within a circle centered on the origin $i$ with a radius of $d_{ij}$. Figure adapted from (57).

but have no direct link to motivation or demographics (Box 8.1 and Figure 8.1). While census data could provide an indirect assessment of motivations behind short-term movements, it is collected at the household level every 5–10 years and may not capture movements of individuals (e.g., children going to boarding school or a member traveling to work seasonally). Travel surveys are a good tool for collecting information on individuals, their motivation behind travel, and if they have access to preventative measures (62), (63), although recall bias (e.g., when individuals do not accurately remember previous events) may impact the quality of answers and it is difficult to conduct surveys on a wide scale.

To avoid recall bias, mobile phone apps that record locations can provide information on the particular destinations of the phone user which may be used to infer motivation. For instance, Google has generated mobility reports based on the destinations a user visits (64). However, these apps do not typically collect user demographics and may not be able to capture specific destinations as well in low- and middle-income settings and remote/rural areas where many establishments have not yet been mapped or uptake of these services is low. Thus, there is a clear need to relate population-level, generalizable datasets to individual demographics and behaviors.

### 8.5.3 Improving model estimates of transmission

As demonstrated in section 8.4, datasets on mobility patterns, demographics, and transmission are often combined in infectious disease models to identify risk factors, calculate basic reproduction numbers ($R_0$), and determine sources/sinks of travel and pathogens (6), (40), (45), (63). These model predictions are informative for designing intervention strategies, such as determining where preventative measures should be introduced and which groups should be prioritized for vaccination, and predicting health care needs, like ensuring enough resources are available at local health facilities to handle an influx of cases (65). Most models assume mobility, traveler demographics, and infection rates are stable, either for model simplicity or due to lack of data. Consequently, models would not detect the spatio-temporal fluctuations in risk factors, susceptible populations, and exposure rates that would be informative for predicting outbreaks and allocating health care resources. Incorporating time-resolved datasets of both mobility patterns from mobile phone data and demographic patterns from travel surveys into these models could improve their ability to inform public health decisions.

## 8.6. Conclusions

Although seasonal human travel is ubiquitous, there are several challenges involved with quantifying and characterizing these patterns, including procuring detailed temporal data at a high spatial resolution, obtaining representative coverage of the population of interest, and quantifying how these patterns translate to disease transmission or risk. Here, we have outlined several datasets commonly used to quantify human mobility and highlight their ability to inform seasonal estimates of travel and, ultimately, disease transmission. Many of these datasets have not been extensively used for quantifying seasonal changes in mobility and there are clear gaps in our ability to quantify these patterns. Nonetheless, access to relevant data continues to improve which may greatly enhance our understanding of how human behavior impacts the spread of infectious diseases.

In parallel with advances in the study of human mobility in relation to infectious disease transmission and spread, technological advances in animal tracking are also yielding new insights into linkages between movement behavior and infection (see Chapters 4, 6, and 7) (72)–(74)). While these two fields of study share some common approaches to summarizing movement data and modeling infection dynamics, further integration of studies linking human and animal movement to infection risk could yield important insights. For example, this integrated approach could help elucidate the transmission of zoonotic diseases, such as highly pathogenic avian influenza viruses, where movement of wild birds, domestic animals, and humans is crucial for holistic prediction of global infection risk. Additionally, human mobility data have the capacity to detect human behavioral changes in response to infection (or perceived infection risk), as well as how human movement behavior influences local transmission and regional spread. Interpreting these data within the rich ecological literature on drivers and consequences of behavioral responses to infection could yield new insights and mechanistic hypotheses for explaining observed infection patterns. Finally, the capacity to determine how infection influences human mobility and vice versa lays the groundwork for investigating behavior-infection feedback loops (see Chapter 2) (75), which to date have not been well explored in human disease systems.

# References

1. Teklehaimanot HD, Lipsitch M, Teklehaimanot A, and Schwartz J. Weather-based prediction of Plasmodium falciparum malaria in epidemic-prone regions of Ethiopia I. Patterns of lagged weather effects reflect biological mechanisms. *Malar J.* 2004; 3(41).

2. Remais J. Modelling environmentally-mediated infectious diseases of humans: Transmission dynamics of schistosomiasis in China. *Adv Exp Med Biol.* 2010;673:79–98.

3. Bayer AM, Hunter GC, Gilman RH, Cornejo del Carpio JG, Naquira C, Bern C, et al. Chagas Disease, Migration and Community Settlement Patterns in Arequipa, Peru. Gyapong JO, editor. *PLoS Negl Trop Dis.* 2009;3(12):e567.

4. Galgamuwa LS, Dharmaratne SD, and Iddawela D. Leishmaniasis in Sri Lanka: Spatial distribution and seasonal variations from 2009 to 2016. *Parasites Vectors* 2018; 11:1–10.

5. Mott KE, Desjeux P, Moncayo A, Ranque P, and De Raadt P. Parasitic diseases and urban development. *Bull World Health Organ.* 1990; 168:691–8.

6. Wesolowski A, Eagle N, Tatem AJ, Smith DL, Noor AM, Snow RW, et al. Quantifying the impact of human mobility on malaria. *Science.* 2012;338(6104):267–70.

7. Prothero RM. Population movements and problems of malaria eradication in Africa. *Bull World Health Organ.* 1961;24:405–25.

8. Shultz JM, Cooper JL, Baingana F, Oquendo MA, Espinel Z, Althouse BM, et al. The role of fear-related behaviors in the 2013–2016 West Africa Ebola virus disease outbreak. *Curr Psychiatry Rep.* 2016; 18(1):104.

9. Lau H, Khosrawipour V, Kocbach P, Mikolajczyk A, Schubert J, Bania J, et al. The positive impact of lockdown in Wuhan on containing the COVID-19 outbreak in China. *J Travel Med.* 2021;27(3):1–7.

10. Zu Erbach-Schoenberg E, Alegana VA, Sorichetta A, Linard C, Lourenço C, Ruktanonchai NW, et al. Dynamic denominators: The impact of seasonally varying population numbers on disease incidence estimates. *Popul Health Metr.* 2016;14(1).

11. Wesolowski A, Zu Erbach-Schoenberg E, Tatem AJ, Lourenço C, Viboud C, Charu V, et al. Multinational patterns of seasonal asymmetry in human movement influence infectious disease dynamics. *Nat Commun.* 2017;8(1):1–9.

12. Buckee CO, Tatem AJ, Metcalf CJE. Seasonal population movements and the surveillance and control of infectious diseases. *Trends Parasitol.* 2017; 33:10–20.

13. Lessler J, Rodriguez-Barraquer I, Cummings DAT, Garske T, Van Kerkhove M, Mills H, et al. Estimating potential incidence of MERS-CoV associated with Hajj pilgrims to Saudi Arabia, 2014. *PLoS Curr.* 2014;6.

14. Bharti N, Tatem AJ, Ferrari MJ, Grais RF, Djibo A, and Grenfell BT. Explaining seasonal fluctuations of measles in Niger using nighttime lights imagery. *Science.* 2011;334(6061):1424–7.

15. Faulkingham RH and Thorbahn PF. Population dynamics and drought: A village in Niger. *Popul Stud (NY)* 1975;29(3):463–77.

16. Guyant P, Canavati SE, Chea N, Ly P, Whittaker MA, Roca-Feltrer A, et al. Malaria and the mobile and migrant population in Cambodia: A population movement framework to inform strategies for malaria control and elimination. *Malar J.* 2015;14(1):252.

17. Sturrock HJW, Hsiang MS, Cohen JM, Smith DL, Greenhouse B, Bousema T, et al. Targeting asymptomatic malaria infections: Active surveillance in control and elimination. *PLoS Med.* 2013;10(6):e1001467.

18. Hansen E and Donohoe M, Health Issues of migrant and seasonal farmworkers. *J Health Care Poor Underserved.* 2003;14(2):153–64.

19. Benach J, Muntaner C, Delclos C, and Menéndez M, Ronquillo C. Migration and "low-skilled" workers in destination countries. *PLoS Med.* 2011;8(6):e1001043.

20. Searle KM, Lubinda J, Hamapumbu H, Shields TM, Curriero FC, Smith DL, et al. Characterizing and quantifying human movement patterns using GPS data loggers in an area approaching malaria elimination in rural Southern Zambia. *R Soc Open Sci.* 2017;4(5):1–12.

21. Dalziel BD, Bjørnstad ON, van Panhuis WG, Burke DS, Metcalf CJE, and Grenfell BT. Persistent chaos of measles epidemics in the prevaccination United States caused by a small change in seasonal transmission patterns. *PLoS Comput Biol.* 2016;12(2):1004655.

22. Bjørnstad ON, Finkenstädt BF, and Grenfell BT. Dynamics of measles epidemics: Estimating scaling of transmission rates using a time series SIR model. Vol. 72, *Ecological Monographs.* 2002:169–94.

23. Cauchemez S, Ferguson NM, Wachtel C, Tegnell A, Saour G, Duncan B, et al. Closure of schools during an influenza pandemic. *Lancet Infect Dis.* 2009; 9:473–81.

24. London WP and Yorke JA. Recurrent outbreaks of measles, chickenpox and mumps: I. Seasonal variation in contact rates. *Am J Epidemiol.* 1973;98(6):453–68.

25. Chen S, Yang J, Yang W, and Wang C, Bärnighausen T. COVID-19 control in China during mass population movements at New Year. *Lancet* 2020;395(10226): 764–6.

26. Ewing A, Lee EC, Viboud C, and Bansal S. Contact, travel, and transmission: The impact of winter holidays on influenza dynamics in the United States. *J Infect Dis.* 2017;215(5):732–9.

27. Schaber KL, Paz-Soldan VA, Morrison AC, Elson WHD, Rothman AL, Mores CN, et al. Dengue illness impacts daily human mobility patterns in Iquitos, Peru. *PLoS Negl Trop Dis.* 2019;13(9).

28. Perkins TA, Paz-Soldan VA, Stoddard ST, Morrison AC, Forshey BM, Long KC, et al. Calling in sick: Impacts of fever on intra-urban human mobility. *Proc R Soc B: Biol Sci.* 2016;283(1834).

29. Kerckhove K Van, Hens N, Edmunds WJ, and Eames KTD. The impact of illness on social networks: Implications for transmission and control of influenza. *Am J Epidemiol.* 2013;178(11):1655–62.

30. Cauchemez S, Valleron A-J, Boëlle P-Y, Flahault A, and Ferguson NM. Estimating the impact of school closure on influenza transmission from Sentinel data. 2008;452:750–4.

31. Bootsma MCJ and Ferguson NM. The effect of public health measures on the 1918 influenza pandemic in U.S. cities. *PNAS.* 2007 May 1;104(18):7588–93.

32. Pepe E, Bajardi P, Gauvin L, Privitera F, Lake B, Cattuto C, et al. COVID-19 outbreak response: A first assessment of mobility changes in Italy following national lockdown. *medRxiv.* 2020. Available from: <https://doi.org/10.1101/2020.03.22.20039933>.

33. Funk S, Salathé M, and Jansen VAAA. Modelling the influence of human behaviour on the spread of infectious diseases: A review. *J R Soc Interface.* 2010;7(50):1247–56.

34. Bell M and Muhidin S. Cross-national comparison of internal migration. *HDRP* 2009;30.

35. Wesolowski A, Buckee CO, Pindolia DK, Eagle N, Smith DL, Garcia AJ, et al. The use of census migration data to approximate human movement patterns across temporal scales. *PLoS One.* 2013;8(1):e52971.

36. Randall S. Where have all the nomads gone? Fifty years of statistical and demographic invisibilities of African mobile pastoralists. *Pastoralism.* 2015 Dec 1;5(22).

37. Grietens KP, Gryseels C, Dierickx S, Bannister-Tyrrell M, Trienekens S, Uk S, et al. Characterizing types of human mobility to inform differential and targeted malaria elimination strategies in Northeast Cambodia. *Sci Rep.* 2015;5.

38. Wesolowski A, Stresman G, Eagle N, Stevenson J, Owaga C, Marube E, et al. Quantifying travel behavior for infectious disease research: A comparison of data from surveys and mobile phones. *Sci Rep.* 2014;4.

39. Jäckle A, Lynn P, Sinibaldi J, and Tipping S. The effect of interviewer experience, attitudes, personality and skills on respondent co-operation with face-to-face surveys. *Surv Res Methods.* 2013;7(1):1–15.

40. Balcan D, Colizza V, Gonçalves B, Hud H, Ramasco JJ, and Vespignani A. Multiscale mobility networks and the spatial spreading of infectious diseases. *PNAS.* 2009;106(51):21484–9.

41. Tatem AJ. Mapping population and pathogen movements. *Int Health.* 2014 Mar;6(1):5–11.

42. Tizzoni M, Bajardi P, Decuyper A, Kon Kam King G, Schneider CM, Blondel V, et al. On the use of human mobility proxies for modeling epidemics. *PLoS Comput Biol.* 2014;10(7):1003716.

43. Pindolia DK, Garcia AJ, Huang Z, Fik T, Smith DL, and Tatem AJ. Quantifying cross-border movements and migrations for guiding the strategic planning of malaria control and elimination. *Malar J.* 2014;13(1):169.

44. GSM Association. The mobile economy 2020. 2020. Available from: <https://www.gsma.com/mobileeconomy/wp-content/uploads/2020/03/GSMA_MobileEconomy2020_Global.pdf>.

45. Ruktanonchai NW, DeLeenheer P, Tatem AJ, Alegana VA, Caughlin TT, zu Erbach-Schoenberg E, et al. Identifying malaria transmission foci for elimination using human mobility data. Koelle K, editor. *PLoS Comput Biol.* 2016;12(4):e1004846.

46. Wesolowski A, Eagle N, Noor AM, Snow RW, and Buckee CO. The impact of biases in mobile phone ownership on estimates of human mobility. *J R Soc Interface.* 2013;10(81).

47. Buckee CO, Wesolowski A, Eagle NN, Hansen E, and Snow RW. Mobile phones and malaria: Modeling human and parasite travel. *Travel Med Infect Dis.* 2013;11:15–22.

48. Kraemer MUG, Sadilek A, Zhang Q, Marchal NA, Tuli G, Cohn EL, et al. Mapping global variation in human mobility. *Nat Hum Behav.* 2020;4:800–10.

49. Kissler SM, Kishore N, Prabhu M, Goffman D, Beilin Y, Landau R, et al. Reductions in commuting mobility correlate with geographic differences in SARS-CoV-2 prevalence in New York City. *Nat Commun.* 2020;11(1):1–6.

50. Facebook Data for Good. (Cited Feb 24, 2021). Available from: <https://dataforgood.fb.com/>.

51. Hast M, Searle KM, Chaponda M, Lupiya J, Lubinda J, Sikalima J, et al. The use of GPS data loggers to describe the impact of spatio-temporal movement patterns on malaria control in a high-transmission area of northern Zambia. *Int J Health Geogr.* 2019 Dec;18(1).

52. Ray EL, Wattanachit N, Niemi J, Kanji AH, House K, Cramer EY, et al. Ensemble forecasts of coronavirus disease 2019 (COVID-19) in the U.S. *medRxiv.* 2020.

53. Schaber KL, Perkins TA, Lloyd AL, Waller LA, Kitron U, Paz-Soldan VA, et al. Disease-driven reduction in human mobility influences human-mosquito contacts and dengue transmission dynamics. *PLoS Comput Biol.* 2021;17(1):e1008627.

54. Ruktanonchai NW, Bhavnani D, Sorichetta A, Bengtsson L, Carter KH, Córdoba RC, et al. Census-derived migration data as a tool for informing malaria elimination policy. *Malar J.* 2016; 15(1):273.
55. Wesolowski A, Metcalf CJE, Eagle N, Kombich J, Grenfell BT, Bjørnstad ON, et al. Quantifying seasonal population fluxes driving rubella transmission dynamics using mobile phone data. *PNAS.* 2015;112(35):11114–9.
56. Barbosa H, Barthelemy M, Ghoshal G, James CR, Lenormand M, Louail T, et al. Human mobility: Models and applications. *Phys.* 2018; 734:1–74.
57. Wesolowski A, O'Meara WP, Eagle N, Tatem AJ, and Buckee CO. Evaluating spatial interaction models for regional mobility in Sub-Saharan Africa. *PLoS Comput Biol.* 2015;11(7):1004267.
58. Simini F, González MC, Maritan A, and Barabási AL. A universal model for mobility and migration patterns. *Nature.* 2012;484(7392):96–100.
59. Giles JR, zu Erbach-Schoenberg E, Tatem AJ, Gardner L, Bjørnstad ON, Metcalf CJE, et al. The duration of travel impacts the spatial dynamics of infectious diseases. *PNAS.* 2020; 117(36):201922663.
60. Marshall JM, Touré M, Ouédraogo AL, Ndhlovu M, Kiware SS, Rezai A, et al. Key traveller groups of relevance to spatial malaria transmission: A survey of movement patterns in four sub-Saharan African countries. *Malar J.* 2016 Dec 12;15(1):200.
61. Duc Thang N, Erhart A, Speybroeck N, Xuan Hung L, Khanh Thuan L, Trinh Hung C, et al. Malaria in central Vietnam: Analysis of risk factors by multivariate analysis and classification tree models. *Malar J.* 2008.
62. Tatem AJ, Adamo S, Bharti N, Burgert CR, Castro M, Dorelien A, et al. Mapping populations at risk: Improving spatial demographic data for infectious disease modeling and metric derivation. Population Health Metrics. *BioMed Central* 2012; 10:1–14.
63. Lowa M, Sitali L, Siame M, and Musonda P. Human mobility and factors associated with malaria importation in Lusaka district, Zambia: A descriptive cross sectional study. *Malar J.* 2018;17(1):404.
64. Google. COVID-19 Community Mobility Reports. Google 2021. Available from: <https://www.google.com/covid19/mobility/>.
65. Heesterbeek H, Anderson RM, Andreasen V, Bansal S, DeAngelis D, Dye C, et al. Modeling infectious disease dynamics in the complex landscape of global health. *Science.* 2015;347:aaa4339.
66. Pindolia DK, Garcia AJ, Huang Z, Smith DL, Alegana VA, Noor AM, et al. The demographics of human and malaria movement and migration patterns in East Africa. *Malar J.* 2013;12(1):397.
67. Perkins TA, Garcia AJ, Paz-Soldan VA, Stoddard ST, Reiner RC, Vazquez-Prokopec G, et al. Theory and data for simulating fine-scale human movement in an urban environment. *J R Soc Interface.* 2014;11(99).
68. Charu V, Zeger S, Gog J, Bjørnstad ON, Kissler S, Simonsen L, et al. Human mobility and the spatial transmission of influenza in the United States. *PLoS Comput Biol.* 2017; 13(2):e1005382.
69. Huang Z, Das A, Qiu Y, and Tatem AJ. Web-based GIS: The vector-borne disease airline importation risk (VBD-AIR) tool. *Int J Health Geogr.* 2012;11(1):1–14.
70. Bharti N, Djibo A, Tatem AJ, Grenfell BT, and Ferrari MJ. Measuring populations to improve vaccination coverage. *Sci Rep.* 2016 Oct 5;5.
71. Kishore N, Kahn R, Martinez PP, De Salazar PM, Mahmud AS, Buckee CO. Lockdowns result in changes in human mobility which may impact the epidemiologic dynamics of SARS-CoV-2. *Sci Rep.* 2021;11(1):6995.
72. Mistrick J, Gilbertson MLJ, White LA, and Craft ME. Constructing animal networks for parasite transmission inference. In Ezenwa VO, Altizer S, and Hall RJ (eds), *Animal Behavior and Parasitism.* Oxford: Oxford University Press, 2022. DOI: 10.1093/oso/9780192895561.003.0004.
73. Spiegel O, Anglister N, and Crafton MM. Movement data provides insights into feedbacks and heterogeneities in host–parasite interactions. In Ezenwa VO, Altizer S, and Hall RJ (eds), *Animal Behavior and Parasitism.* Oxford: Oxford University Press, 2022. DOI: 10.1093/oso/9780192895561.003.0006.
74. Hall RJ, Altizer S, Peacock SJ, and Shaw AK. Animal migration and infection dynamics: Recent advances and future frontiers. In Ezenwa VO, Altizer S, and Hall RJ (eds), *Animal Behavior and Parasitism.* Oxford: Oxford University Press, 2022. DOI: 10.1093/oso/9780192895561.003.0007.
75. Hawley DM and Ezenwa VO. Parasites, host behavior and their feedbacks. In Ezenwa VO, Altizer S, and Hall RJ (eds), *Animal Behavior and Parasitism.* Oxford: Oxford University Press, 2022. DOI: 10.1093/oso/9780192895561.003.0002.

# Sexual Selection and Mating Behavior

# CHAPTER 9

# Parasite-mediated sexual selection: To mate or not to mate?

Alistair Pirrie, Hettie Chapman, and Ben Ashby

## 9.1 Introduction

Sexual selection is a fundamental evolutionary phenomenon, responsible for the exceptional diversity of secondary sex characteristics found across the natural world, including antlers in deer and ornate plumage in many bird species (Figure 9.1). Many of these traits are costly in the absence of sexual selection, either through metabolic costs (1), an increased risk of predation, or a reduction in foraging capabilities, but may be advantageous if they increase mating opportunities (2). Sexual selection also plays a crucial role in population dynamics (3), the evolution of mating systems (e.g., polygyny, polyandry, monogamy (4)), and sexual conflict (5). Sexual selection occurs when individuals compete for access to mates (intrasexual competition), or when individuals exhibit non-random mate choice (intersexual competition). Both intra- and intersexual selection can occur pre- or post-copulation, for example through rutting or sperm competition (intrasexual) and through lekking displays or cryptic female choice (intersexual).

The importance of sexual selection was first recognized by Darwin (2), yet the notion that parasitism may be a major driver of sexual selection is relatively recent (6), (7). Sparked by the inception of the Hamilton–Zuk "Good Genes Hypothesis" nearly four decades ago, parasite-mediated sexual selection (PMSS; see Box 9.1) has since been a major focus of theoretical and empirical research in

**Figure 9.1** Parasite-mediated sexual selection (PMSS) may explain the evolution of secondary sex characteristics, mate choice, and mating systems. For example, females may choose males based on the condition of showy traits as an indicator of genetic quality (resistance or tolerance genes) or infection status. Here, two male wild turkeys (*Meleagris gallopavo*) display tail fans in contrasting condition. Image credit: Ben Ashby.

Alistair Pirrie, Hettie Chapman, and Ben Ashby, *Parasite-mediated sexual selection: To mate or not to mate?*. In: *Animal Behavior and Parasitism*. Edited by Vanessa O. Ezenwa, Sonia Altizer and Richard J. Hall, Oxford University Press.

evolutionary biology (8). Parasites (in the broadest sense, including bacterial pathogens and viruses) are found throughout the natural world and can impose significant fitness costs on their hosts due to increased mortality or a loss in reproductive output. However, the fitness costs associated with parasitism typically differ from the costs associated with sexual selection in several key aspects. First, unlike metabolic costs, which are typically fixed, infection is a dynamic cost that varies with parasite prevalence and the intensity of infection in the host population, producing a feedback loop between the ecology (or epidemiology) of a parasite and the evolution of the host. Second, parasites are also subject to selection, leading to another feedback loop in the form of reciprocal adaptations and counter-adaptations (antagonistic coevolution). Moreover, parasites typically have short generation times and large population sizes, which facilitate rapid evolution. Finally, sexually transmitted infections (STIs; see Box 9.1), are inextricably linked with host reproduction, and therefore the mating dynamics of the host will play a critical role in the epidemiology of these particular parasites.

> **Box 9.1  Glossary of key terms**
>
> **Parasite-mediated sexual selection (PMSS)**—a mode of sexual selection driven by infectious agents (parasites)
>
> **Sexually transmitted infection (STI)**—an infectious agent (parasite) which is transmitted primarily during mating
>
> **Good Genes Hypothesis (GGH)**—PMSS is driven by genetic quality in the form of resistance or tolerance genes
>
> **Direct fitness effects**—PMSS affects the reproductive success of the parent
>
> **Indirect fitness effects**—PMSS affects the reproductive success of the offspring
>
> **Parental Care Hypothesis** (PCH)—PMSS is driven by assessment of the quality of parental care that could be provided by a prospective mate
>
> **Transmission Avoidance Hypothesis (TAH)**—PMSS is driven by avoiding mating with individuals who present a risk of infection

> **Ordinary infectious disease (OID)**—an infectious agent (parasite) which is not primarily transmitted during mating

In this chapter, we discuss the development and current state of research on PMSS, from the main hypotheses, to empirical data and insights from mathematical modeling, including the critical importance of ecological and coevolutionary feedbacks on PMSS. We also identify directions for future theoretical and empirical research.

## 9.2  Hypotheses for parasite-mediated sexual selection

To date, hypotheses for PMSS have almost entirely focused on female pre-copulatory mate choice and can be broadly defined by whether the fitness effects are conferred indirectly (e.g., good genes, (7)) or directly (e.g., transmission avoidance (9), (10), and resource provisioning/parental care, (11)–(13)) (Figure 9.2). For example, a female may choose to mate with a male who appears to be in good condition because he may have a heritable genetic advantage for resisting or tolerating parasites, or because her risk of contracting an infection appears to be low.

Other mechanisms by which PMSS can occur include mate competition between males and assortative mating. Mate competition between males occurs when access to females is limited and therefore males must compete for access to females. Parasitism can impact a male's ability to compete (14), for example by reducing male body or weapon size resulting in fewer mating opportunities (15), or conversely driving an increase in mating efforts to compensate for their lowered fitness (16). Alternatively, parasitism may lead to segregation, leading to assortative mating in uninfected or infected sub-populations (17), and therefore potentially also assortative mating between resistant and non-resistant sub-populations. For example, Thomas *et al.* (17) found assortative mating in parasitized and non-parasitized sub-populations of lagoon sand shrimp (*Gammarus insensibilis*), as parasitized hosts become positively phototactic, driving them to the surface of the water.

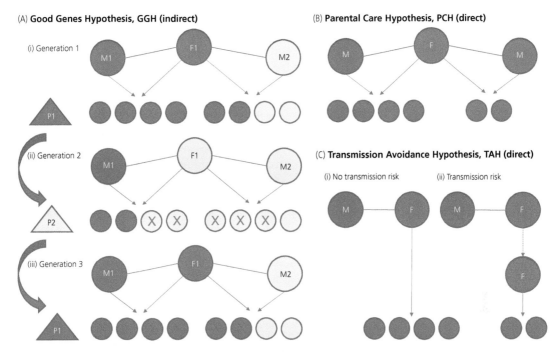

**Figure 9.2** Hypotheses for parasite-mediated sexual selection (PMSS). Parental generation is shown by large circles (males, M; females, F), and offspring by small circles. Lines indicate mating and solid arrows indicate offspring produced. Females choose to mate with males based on the condition of showy or non-showy traits. (A) Good Genes Hypothesis (GGH): individuals are either susceptible to parasite 1 (P1, blue) or to parasite 2 (P2, yellow). Male condition may indicate whether a male is genetically resistant or tolerant to the prevailing parasite (triangle). Differential survival or reproductive success of offspring is imposed by the parasite (crosses indicate death). Genetic variation is maintained by coevolutionary (coadaptive) cycles in genes associated with host resistance and parasite infectivity. (B) Parental Care Hypothesis (PCH): male condition may indicate whether a prospective mate is infected (red) or not (blue), and hence whether they may be a risky partner for shared parental care (the infected male rears fewer surviving offspring). (C) Transmission Avoidance Hypothesis (TAH): male condition indicates infection status and hence the risk of transmission (dotted arrow) to the female. Infection reduces the lifetime reproductive success.

Here, we focus on female pre-copulatory mate choice, following the majority of studies on PMSS. We consider indirect and direct fitness effects separately, but the reasons for mate choice may be unclear, with potential for both indirect and direct fitness effects (see Box 9.1).

## 9.2.1 Indirect fitness effects

The fitness effects of mate choice are indirect when choosiness increases the reproductive success of offspring. Early theories of mate choice assumed that females choose who to mate with based on perceptions of genetic quality, so that offspring can inherit "good genes" (7). If secondary sex characteristics such as bright plumage or other showy traits are honest indicators of male quality, then

female choice may select for exaggerated male traits. However, directional selection imposed by female choice is predicted to erode genetic variation in male traits, eventually rendering female choice obsolete. This is known as the "lek paradox" (13), (18) and was a major stumbling block for good genes hypotheses for most of the twentieth century.

In 1982, Bill Hamilton and Marlene Zuk proposed a parasite-centric "Good Genes Hypothesis" (GGH; sometimes referred to as the Hamilton–Zuk Hypothesis, see Box 9.1) in order to overcome the depletion of male genetic variation in the lek paradox (7). The GGH offers a unique resolution to the lek paradox in the form of fluctuating selection driven by host–parasite coevolution (referred to by the authors as "coadaptive cycles," and in the literature often referred to as "Red Queen Dynamics," eponymously named after the character in

Lewis Carroll's *Through the Looking Glass* due to her insistence that "it takes all the running you can do, to keep in the same place"). Suppose hosts who are unburdened by parasitism develop showier traits (Hamilton and Zuk focused on male and female "brightness," and male song ranked by variety and complexity). Then, if there is heritable (additive) genetic variation in resistance to different parasites, individuals resistant to the most common parasite phenotype will subsequently develop showier traits. Females should therefore choose to mate with the showiest males as this may indicate a strong genetic resistance to the most common parasite present in the population. Crucially, the optimal resistance phenotype may change through time due to counter-adaptations in the parasite and negative frequency-dependent selection (rare advantage) in the host, therefore preventing the erosion of genetic variation and maintaining female choice.

For example, consider a host and parasite population with genetic heritability in resistance where there are two host resistance phenotypes, $H_1$ and $H_2$, which can each be infected by their matching parasite phenotypes $P_1$ and $P_2$. If $H_1$ is initially more common, then $P_1$ will increase in frequency. Males of type $H_2$, being resistant to the most common parasite, will tend to develop showier traits and mate more often due to female choice. Hosts with resistance phenotype $H_2$ will therefore increase in frequency, as will parasite phenotype $P_2$, tracking the host. Now males of type $H_1$ are more resistant, on average, and so will have the showiest traits and higher reproductive success. As the frequency of $H_1$ increases in subsequent generations, so too does the frequency of $P_1$, and the "coadaptive cycle" repeats (Figure 9.2A). Genetic diversity in disease resistance is therefore maintained through fluctuating selection in both the host and parasite, also maintaining diversity in showy traits. Although originally envisaged in terms of resistance genes by Hamilton and Zuk, the GGH may also work if there is heritable variation in host tolerance to different parasites.

The GGH was of great historical importance, sparking a rush to test its predictions empirically (8)–(10), (13), (19)–(22), and generating considerable interest in PMSS. As noted by Hamilton and Zuk (7), their hypothesis hinges on a number of critical assumptions. In particular, key requirements of the theory are that there is heritable variation

in resistance, and that there are cycles in genes governing host resistance and parasite infectivity, the latter of which has yet to be shown (8). A second, potentially more fundamental issue is that secondary sex characteristics may not be a reliable indicator of resistance (10), (23) (but see Chapter 10 in this volume (24) for a discussion on why sexually selected traits might be fundamentally linked to parasite resistance). Alongside these limitations of the GGH are factors that can weaken selection for mate choice for showy traits as an indicator of resistance. For example, aggregation of parasites can weaken selection for secondary sex characteristics as an indicator of resistance, as: (i) a male may be in good condition simply because he has not come into contact with the parasite, and therefore a female may choose to mate with him mistaking his chance disease-free status for resistance; (ii) parasitized males are less likely to be able to compete for partners because of increased morbidity and mortality; (iii) the variability of showy traits may not be stark enough to facilitate mate choice (25). Parasites may also directly influence female choice. Experiments in upland bullies (*Gobiomorphus breviceps*) (26) and guppies (*Poecilia reticulata*) (27) have shown a reduction in female discriminatory behavior between males of differing quality when parasitized. López (27) suggests an energetic cost to females when parasitized, which may limit their ability to conduct choosy behavior, which will tend to weaken selection on mate choice for showy traits.

While the GGH offers a potential resolution to the lek paradox, it may also be resolved by so-called genic capture, where the degree of expression of sexually selected traits is highly dependent on an individual's condition, which is determined by many loci (28). The inaccuracy of selection towards the many loci determining condition results in deleterious mutations maintaining the mutation–selection balance in condition and therefore also variation in condition-dependent traits (28).

### 9.2.2 Direct fitness effects

Fitness effects are direct when choosiness increases reproductive success of the individual expressing mate choice. For example, mate choice may confer direct fitness effects in species with shared or male parental care, a form of resource provisioning which we refer to as the "Parental Care Hypothesis" (PCH; (11)–(13); see Box 9.1). If a female chooses a male who is in poor condition, for example due to infection, then he may die prematurely or be unable to provide sufficient care to offspring (Figure 9.2B). Showy traits are therefore assumed to be indicators of male disease status, rather than indicators of good genes, and so unlike the GGH, the PCH does not rely on heritable variation in resistance or cycling in allele frequencies.

Another direct benefit of choosiness is a reduction in the risk of transmission during mating, which we refer to broadly as the "Transmission Avoidance Hypothesis" (TAH; Figure 9.2C; see Box 9.1) (6), (9), (10), (29), (30). The TAH posits that mate choice is primarily to identify mating partners who represent a low risk of passing on parasites during mating. Individuals may want to avoid infection for several reasons, including: (i) effects of disease on fecundity or survival due to parasite virulence; (ii) effects on reproduction due to metabolic costs associated with mounting an immune response (31); (iii) effects on survival such as a weakened ability to forage or hunt, to evade predators; or (iv) increased susceptibility to other parasites due to being in a weakened state. The primary fitness effect is therefore direct to the individual expressing the choice, although offspring may also benefit if there is a risk of vertical transmission (parent to offspring) or if the chosen partner possesses genes for resistance. The TAH does not preclude indirect fitness effects and is compatible with the GGH and PCH.

The TAH requires the parasite to be transmitted during mating (note that a parasite may be transmitted during mating due to close contact, but may not be a conventional STI), and for females to be able to reliably detect signs of infection and choose a less risky (healthy) partner accordingly. In principle, disease status may be detectable through visual, behavioral, or olfactory cues (32), (33). For example, pre-copulatory inspections may reveal ectoparasites (20) or lesions (19), (22). Showy traits are not necessary for the TAH to operate provided females are able to judge male condition accurately (30). However, showy traits may be more reliable indicators of disease status—for instance, if infection limits the development of showy traits or the ability to compete for access to females—and therefore females may evolve preferences for more exaggerated male traits.

### 9.2.3 Comparison of hypotheses

While there are important differences between the various PMSS hypotheses—for instance, indirect fitness effects are likely to induce weaker selection than direct fitness effects (10)—there are also considerable overlaps (Figure 9.3). In all cases, females must be free to choose who to mate with and male condition must be a reliable indicator of quality or risk. If females are unable to freely choose their mate(s), for instance due to male coercion or highly skewed sex ratios, then mate choice cannot operate effectively. Similarly, mate choice will be ineffective if male condition is an unreliable indicator of genetic quality (GGH), parental quality (PCH), or transmission risk (TAH).

What constitutes a good choice according to one hypothesis may be a poor choice under the assumptions of another (Table 9.1). For example, suppose a male who is in good condition is chosen by a female as her mate. If he has not been infected then this may be indicative of possessing resistance genes or simply that he has not been exposed to infection. According to the GGH, the first scenario would represent a good choice, but the second scenario would be a poor choice. Yet according to the PCH or TAH, mating with this male would be advantageous regardless of whether he possesses resistance genes or not (Table 9.1). Recovery and tolerance may complicate the decision process, depending on the long-term impact on male condition. If a male was previously (but no longer) infected and remains in poor condition, rejection would be the correct choice according to the GGH but the wrong choice according to the TAH, and potentially according to the PCH depending on the impact on parental care. If a male is infected but tolerates his parasites and remains in good condition, then he would be a poor choice according to the TAH but may be a good choice according to the GGH in terms of tolerance rather than resistance genes.

Beyond the common assumptions of free female choice and reliable indicators, the hypotheses diverge (Figure 9.3). For example, the TAH requires transmission to be primarily associated with mating, whereas the PCH and GGH do not.

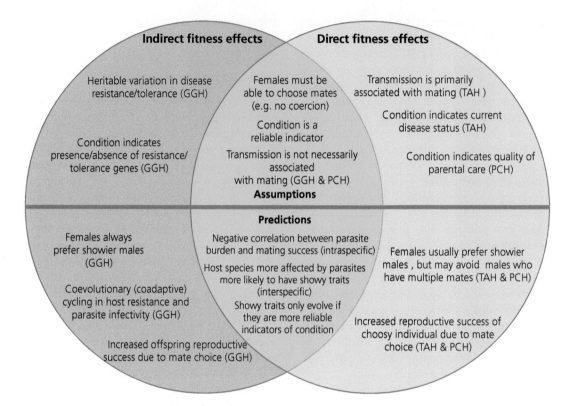

**Figure 9.3** Assumptions and predictions of hypotheses based on indirect fitness effects (Good Genes Hypothesis, GGH) and direct fitness effects (Parental Care Hypothesis, PCH; Transmission Avoidance Hypothesis, TAH).

**Table 9.1** Summary of whether acceptance of a mate is a good or poor choice given variation in condition and infection status, according to the Good Genes Hypothesis (GGH), Parental Care Hypothesis (PCH), or Transmission Avoidance Hypothesis (TAH).

| | Male in good condition (accept) | | | Male in poor condition (reject) | | |
|---|---|---|---|---|---|---|
| | GGH | PCH | TAH | GGH | PCH | TAH |
| Not currently infected (resistant) | Good choice | Good choice | Good choice | Poor choice | Good choice for avoiding transmission to offspring | Poor choice |
| Not currently infected (lack of exposure, not resistant/tolerant) | Poor choice | Good choice | Good choice | Good choice | Good choice for avoiding transmission to offspring | Poor choice |
| Currently infected | Poor choice for resistance, but may be good choice if male is tolerant to infection | Poor choice if transmission risk to offspring or reduced parental ability | Poor choice | Good choice | Good choice | Good choice |
| Previously infected | Poor choice for resistance, but may be good choice if male is tolerant to infection | Good choice | Good choice | Good choice | Poor choice for avoiding transmission to offspring, good choice if parental care compromised | Poor choice |

Condition must indicate current disease status according to the TAH but indicates the presence of resistance/tolerance genes according to the GGH, and the quality of parental care according to the PCH. As with the assumptions, some predictions are common to all of the PMSS hypotheses. First, at the intraspecific level, uninfected males should typically have greater mating success than infected males. Second, at the interspecific level, species more affected by parasites (e.g., higher virulence) are more likely to evolve showy traits. Third, showy traits are only predicted to evolve provided they are more reliable indicators than non-showy traits. At this point, the hypotheses diverge, with the GGH predicting that females always prefer showier males, whereas the TAH and PCH imply that females may prefer less showy (or less popular) males (4), (34). The GGH also predicts that mate choice will lead to an increase in offspring reproductive success, whereas the TAH and PCH predict there to be an increase in reproductive success of the choosy individual, but not necessarily for the offspring. While it is possible that mate choice may result in an increase in offspring reproductive success, this is a secondary effect mediated through the choosy parent.

Given the overlap between the proposed mechanisms for PMSS, it is possible that fitness effects of mate choice may be both direct and indirect. For example, a male who is resistant to infection will represent a low risk of transmission to a female and her offspring and may be a better candidate for shared parental care. Thus, even if the identification of resistance genes is the primary driver of PMSS, as envisaged by the GGH, females who choose males who are in good condition are less likely to be exposed to infection. It may therefore be difficult to distinguish between the hypotheses and identify which, if any, is the major driver of PMSS.

### 9.2.4 Proposed mechanisms for links between sexually selected traits and immune genes

We briefly discuss two hypotheses for mechanistic links between sexually selected traits and immunity genes: the Immunocompetence Handicap Hypothesis (ICHH), and major histocompatibility complex (MHC)-mediated mate choice. These are discussed in more detail in Chapters 10 (24) and 11 (35) of this volume, respectively.

The ICHH proposes that there is a trade-off between showy traits and immunocompetence (a special case of the handicap principle) (36). For example, males may need to allocate resources to either immunity or showy traits, as seen in the development of carotenoid-based yellow plumage in greenfinches (*Carduelis chloris*) (37), great tits (*Parus major*) (38), and blue tits (*Cyanistes caeruleus*) (39). Carotenoids are used in both the development of yellow plumage, and in the host's immune system (37)–(39), and females prefer yellow plumage since it is an accurate indicator of host health. However, carotenoid-based showy traits do not necessarily act as a good indicator of parasite load in other systems; for example, guppies (*Poecilia reticulata*) with low carotenoid diets harbor fewer parasites (*Gyrodactylus turnbulli*) than guppies with high carotenoid diets, seemingly because guppies with high carotenoid diets are bigger and can support more parasites (40). Alternatively, immunosuppressive hormones such as testosterone may be required to develop showy traits, in which case showier males may be less immunocompetent (41). In either case, to remain attractive to choosy females, showy males must be uninfected, possess resistance genes (to avoid becoming infected in the first place), or have surplus energy/resources (to maintain condition).

The MHC is a highly polymorphic region of the vertebrate genome involved in pathogen recognition that can also influence the odor of an individual (42), (43), providing a mechanism by which individuals could in principle discriminate between prospective mates to obtain complementary MHC alleles for offspring (42), (43). For example, in three-spined sticklebacks (*Gasterosteus aculeatus*), MHC-driven sexual selection seems to cause mate choice for MHC alleles which confer the greatest resistance to offspring (42). However, MHC-mediated mate choice does not always imply PMSS. In mice, mate choice appears to be driven by inbreeding avoidance, with relatedness indicated by similarity of MHC alleles (44). MHC-driven sexual selection for optimal resistance alleles against parasites in the population may therefore indicate PMSS, but mate choice for partners with dissimilar MHC alleles is not necessarily due to PMSS. Whether MHC-driven sexual selection can be considered a mechanism for PMSS will depend on the specific system and will need to be determined experimentally. Furthermore, mate choice in some species has been described as occurring in two stages: attraction of potential mates via MHC-related odors, and then selection of a partner due to the condition of their showy traits, but their relative importance is not clear (43).

## 9.3 Empirical evidence of parasite-mediated sexual selection

Empirical evidence for PMSS is mixed, resulting in a lack of resolution over the dominant PMSS mechanisms. Empirical tests specific to the Hamilton–Zuk GGH have proved particularly difficult as the genetic basis for resistance and infectivity—if it exists—is often unknown. Moreover, demonstrating coadaptive cycles, which may take place over many generations, is extremely challenging (38). While studies have shown coadaptive parasite-mediated selection in other contexts (45), no studies have yet shown coadaptive cycles in the context of sexual selection. Studies have however shown heritable resistance linked to showy traits. For example, the offspring of brighter male sticklebacks (*Gasterosteus aculeatus*) are more resistant to infection (46). Most empirical studies have instead explored the relationship between: (i) host infection status and showiness (intraspecific); (ii) disease incidence and showiness (interspecific); and (iii) host infection status and reproductive success. There appears to be reasonable support for negative correlations between parasite load or health status and the showiness or condition of male traits at the intraspecific level, with examples found in several avian species. This includes the aforementioned examples of plumage yellowness, where, in general, brighter individuals have fewer blood parasites (37)–(39). The only exception to this is in older male great tits, where unparasitized individuals have less bright plumage, but parasitized individuals have a lower chance of mortality (38). Plumage brightness therefore seems to indicate good health rather than low blood parasite levels *per se* (38). Coinfections by multiple blood parasites have also been found to

result in more dull plumage, suggesting a potential compound effect (39). However, there does not appear to be a negative correlation between ectoparasite load (*Myrsidea ptilonorhynchi*) and plumage brightness in satin bowerbirds (*Ptilonorhynchus violaceus*) (20). Furthermore, there is a positive correlation between nematode load (*Dictyocaulus* sp., *Elaphostrongylus cervi*, and *Varestrongylus sagittatus*) and antler development in red deer (*Cervus elaphus*) which appears to contradict the intraspecific predictions of PMSS (47). However, a negative correlation between parasite load and showy traits is not predicted by the GGH, as individuals with the most parasites do not necessarily experience the most severe health effects (e.g., they may have higher tolerance) (48).

Correlations between disease incidence and the showiness of traits at the interspecific or population level are also ambiguous. Hamilton and Zuk (7) found a generally positive correlation between male brightness or song complexity in passerine bird species and disease incidence of five genera of protozoa and one nematode, yet the findings are less supportive when viewed across taxa (22) and when confounding factors are taken into account (49). In a larger and more detailed analysis across a wide range of bird species, Garamszegi and Møller (49) did not detect a strong interspecific relationship between prevalence of blood parasites and sexual traits but did find a difference between zero and non-zero parasite prevalence and male trait expression. A potential reason for these mixed results is that ecological and coevolutionary feedbacks exist between hosts and parasites, which could in principle cause positive, negative, or non-monotonic relationships between disease incidence and the showiness of traits. For example, selection for showy traits might be weak when disease incidence is low (low risk of infection) or when disease incidence is high (unable to avoid infection).

Empirical evidence for relationships between host infection status and reproductive success is also mixed. Evidence in support comes primarily from birds, including satin bowerbirds (*P. violaceus*) and red jungle fowl (*Gallus gallus*), which appear to choose mates based on infection status (19), (51). Similarly, male mating success in Lawes' parotia

(*Parotia lawesii*) negatively correlates with blood parasite load (52), and female rock doves (*Columba livia*) have been shown to prefer males not infected with lice (21). However, evidence from other taxa also exists, especially in fish. For example, female guppies (*Poecilia reticulata*) prefer males with relatively few nematodes (*Camallanus cotti*) (53), and female sticklebacks (*Gasterosteus aculeatus*) make mate choice decisions based on the intensity of red carotenoid-based coloration, which is negatively affected by parasitism (54). There is also evidence of a negative correlation between male tapeworm load (*Hymenolepis diminuta*) and the number of larvae produced by female grain beetles (*Tenebrio molitor*), with females preferring the odor of uninfected males (55). Since tapeworms cannot be transmitted during mating, this suggests that females choose to avoid the fecundity costs of mating with parasitized males. In contrast, studies of STIs in milkweed and eucalypt leaf beetles (*Labidomera clivicollis* and *Chrysophtharta agricola*) have failed to show evidence of discrimination between infected and uninfected individuals (16), (56). Similarly, studies of two spot ladybirds (*Adalia bipunctata*) have failed to reveal differences in mating preferences based on the presence of a sexually transmitted mite (*Coccipolipus hippodamiae*) (57). Moreover, wild populations do not appear to have evolved different mating rates to those that are mite free (58). Together, these studies may indicate that beetle STIs have evolved to be cryptic due to strong selection imposed by the host (59).

## 9.4 Are some parasites more likely to influence sexual selection than others?

Hosts can be infected by many different parasites, with various life-history traits and life cycles, modes of transmission, and varying degrees and modes of virulence. For example, parasites may be transmitted through social contact (e.g., respiratory viruses such as common colds or influenza), via sexual contact (e.g., the Human Immunodeficiency Virus (HIV), or the bacterium *Chlamydia trachomatis*, which causes chlamydia infections), from the environment (e.g., *Vibrio* cholerae, the causative agent of cholera), or by vectors (e.g., the malaria parasite *Plasmodium falciparum*). Moreover, while some parasites may have a high mortality rate (e.g., Ebola

virus in humans), others may cause varying levels of morbidity (e.g., the bacterium *Treponema pallidum*, which causes syphilis) or may sterilize their hosts (e.g., the bacterium *Chlamydia trachomatis*, causing chlamydia). Not all parasites will influence sexual selection, but are parasites with certain characteristics more likely to be involved in sexual selection than others?

As there are often significant costs associated with sexually selected traits (e.g., foraging costs (60), increased risk of mortality due to predation or injury (60), missed mating opportunities for choosy females (19)), it is likely that only parasites which impose a significant risk to host health will play an important role in sexual selection. Hence parasites that are sufficiently rare or are not very virulent are unlikely to influence host sexual selection. The mode of transmission is also likely to be critical, depending on the mechanism(s) driving PMSS. For example, the TAH is more likely to apply when there is a significantly higher risk of infection during mating compared to other sources of exposure (to reap greater benefits of transmission avoidance), whereas sexual transmission is not necessary (and may even make it harder) for the GGH or PCH to operate. Sexually transmitted infections (STIs) are therefore closely associated with the TAH, although close contact during mating may also increase the risk of contracting parasites that are not exclusively sexually transmitted (e.g., respiratory viruses, ectoparasites). For simplicity, we will refer to infections that are primarily transmitted through sexual contact as STIs, and all others as ordinary infectious diseases (OIDs; see Box 9.1).

STIs tend to differ from OIDs in several important characteristics. In general, STIs tend to be less likely to cause mortality and more likely to cause sterility (61)–(63), more likely to cause asymptomatic and chronic infections (62), and tend not to confer lasting immune responses (64). These differences are likely due to the fact that STIs are less likely to be systemic (62), and because their transmission dynamics are fundamentally unlike those of OIDs (62), (63). Typically, social contact is much more common and ephemeral than sexual contact, and so transmission opportunities for OIDs tend to be much higher and involve a more diverse and changing pool of individuals. An exception is in solitary species where interactions between conspecifics tend to only occur during mating. For example, brushtail possums (*Trichosurus vulpecula*) mostly make contact during their mating period and the bacterial infection tuberculosis (*Mycobacterium bovis*), despite being an OID, primarily behaves as an STI in this host species as its only opportunity for transmission occurs during mating (65). Social contact rates are also expected to vary more strongly with population size than sexual contact rates, and so in epidemiological models OID transmission is usually assumed to be density-dependent (per-capita social contact rates increase with population size), whereas STI transmission is often assumed to be frequency-dependent (per-capita sexual contact rates do not increase with population size). An important epidemiological consequence of frequency-dependent transmission is that STIs may persist at low host population densities, whereas OIDs will tend to be eliminated below a critical population threshold (66).

There may also be a more intimate relationship between STIs and sexual selection since STIs are inextricably linked with host mating and variation in sex ratio (67). STIs that cause host sterility may increase re-mating or lower mate fidelity, providing more opportunities for transmission (68), (69). For example, sterility could encourage an individual to find a new partner since their current mate is not producing offspring (4), (69). STIs may also cause females to increase reproductive output temporarily following infection but prior to sterilization, also known as terminal investment (63), (70). As a consequence, female terminal investment is predicted to select for lower STI resistance in males because transmitting an STI during mating will increase a female's short-term reproductive output (62).

While STIs are clearly central to the TAH, the relative importance of OIDs and STIs in the GGH and the PCH is not clear. The GGH requires heritable variation in resistance or tolerance which correspond to coevolutionary (coadaptive) cycling of genes in the parasite, but we are not aware of any comparative studies of heritable variation in resistance or tolerance between OIDs and STIs. In the context of the PCH, vertically infected offspring generally have higher fitness when the pathogen is an OID (suggesting a greater role for STIs), but OIDs are more likely to cause mortality virulence (suggesting a greater role for OIDs) (62).

## 9.5 Theoretical predictions for parasite-mediated sexual selection

Mathematical modeling plays an important role in the development of evolutionary theory by generating testable predictions and determining the mechanisms which drive selection. To date, theoretical studies have explored a wide range of scenarios involving PMSS, including the evolution of mating rates or mating system (71)–(74), mate choice (4), (30), (70)–(72), (75), and sexual conflict (5). Most of these studies consider one-sided adaptation by the host, although a number also consider counter-adaptations by the parasite (see section 9.5.4). Surprisingly, there has been very little theoretical work on OIDs and the parasite centric versions of the GGH and PCH with most studies instead focusing on models of transmission avoidance with STIs. We therefore focus on theoretical models of STIs and PMSS.

### 9.5.1 Mating rates and mating system

In general, the risk of contracting an STI depends on disease prevalence in the population, which in turn depends on mating rates and mating system structure, and so there is a feedback loop between the epidemiological dynamics of an STI and the evolution of host mating dynamics (see Box 9.2). STIs are a cost associated with mating and are therefore expected to influence the evolution of mating rates and mating system structure. Naively, one might assume that STIs will always select for lower mating rates and more cautious mating strategies, yet theoretical models suggest this is not always the case. For example, Boots and Knell (76) showed that a population can evolve contrasting levels of promiscuity in the presence of an STI, with risky (more promiscuous) and safe (less promiscuous) mating strategies coexisting when the benefits of producing more offspring by mating more frequently are offset by the increased risk of contracting an STI.

Similarly, the mating system itself may be heavily influenced by STIs, due to either ecological factors (e.g., reductions in population size) or selection acting on mating strategies. Theoretical models have shown that STIs do not necessarily select for monogamy, and the optimal strategy is not always one which minimizes disease spread

**Box 9.2  Feedbacks between STIs and host mating dynamics.**

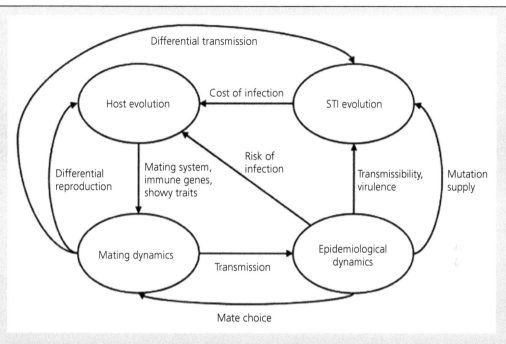

**Box 9.2, Figure 1**  Possible ecological and evolutionary feedbacks between STIs and host mating dynamics.

The epidemiological dynamics of an STI are influenced by the characteristics of both the parasite and host at an individual and population level. Resistance, transmissibility, virulence, sex ratio, the mating system, and the ways in which the hosts choose a partner can greatly affect epidemiological dynamics. For example, an STI is likely to spread further in a polygamous mating system than in monogamous populations as there are more opportunities for transmission (71).

The feedbacks between disease characteristics, disease prevalence, host characteristics, and host mating dynamics are plentiful and complex (Figure 1). For instance, if hosts choose mates based on infection status, then this will affect the spread of an STI, and in turn, selection for mate choice (30), (75). Epidemiological dynamics affect the respective fitness landscapes for both the host and parasite. As a result, epidemiology generates feedbacks in the evolution of hosts, parasites, and mating systems. For example, high disease prevalence may not select for monogamy because there is little chance of avoiding infection by mating with fewer individuals (74).

Theoretical approaches lend themselves well to investigating the intricacies of these feedbacks, as the precise mechanisms can often be hard to untangle in real systems. Mathematical modeling allows one to isolate feedbacks and determine the implications for epidemiology and evolution.

(71), (74). For example, McLeod and Day (74) show that monogamy is more likely to evolve than polygamy when an STI is cryptic (inconspicuous) and causes mortality rather than sterility.

### 9.5.2  Mate choice

Theory predicts that mate choice should evolve when the benefits of being choosy (e.g., avoiding infection, passing on resistance genes to offspring) offset the costs (e.g., wasted energy, inability to find a mate). When the aim of mate choice is to obtain good genes for offspring (GGH), females should always choose the male who appears to be in the best condition. However, when mate choice is driven by transmission avoidance or parental care, females may benefit from choosing less popular males who represent a lower risk (34), (72), (73).

Graves and Duvall (72) conjectured that heritable (additive) genetic variation in showy male traits and female choice could be maintained by STIs, as more attractive males would be more likely to contract STIs. The advantages of being showy may therefore be curtailed by an increased risk of infection. Indeed, STIs are predicted to reduce mating skews, but only if disease prevalence is not too high (73). When an STI is highly prevalent most males are infected and so there is once again an advantage of mating with the most attractive males.

If individuals are able to detect signs of STIs through visual, behavioral, or olfactory cues, then choosiness is predicted to evolve as a means of transmission avoidance provided STIs can be detected with reasonable accuracy (30), (75). If healthy mates are inadvertently avoided due to overcautious behavior, or if STIs are too difficult to detect, then mate choice may not evolve or may be weak. STIs may also affect mate choice based on other indirect cues of infection. For example, failure to produce offspring may indicate that a partner is infertile due to infection and therefore it may be advantageous to engage in extra-pair copulations or search for a different mate, potentially increasing disease spread (4).

### 9.5.3 Sexual conflict

Sexual conflict, which can lead to the evolution of coercive traits among males and defensive traits among females, occurs when males and females have different optimal mating strategies. For example, male seed beetles (*Callosobruchus maculatus*) can cause physical harm to females with the weaponized, male intromittent organ (77). In response, females have evolved various resistance traits to reduce the mating rate or cost of mating to mitigate against this potential harm. STIs are likely to influence sexual conflict either through ecological and evolutionary effects on mating rate and mate choice, and the consequences of infection may differ between males and females. Few models of parasite-mediated sexual conflict currently exist, although Thrall *et al.* (71) have shown that the optimal mating strategy may differ in the presence of an STI. Wardlaw and Agrawal (5) have also shown that an STI can potentially reduce or increase sexual

conflict depending on whether the STI increases mortality or sterility virulence. Specifically, sexual conflict (in the form of male persistence and female resistance) escalates when an STI causes mortality but de-escalates when it causes sterility. At the genetic level, inter-locus or intra-locus sexual conflict for STI resistance is predicted to occur if females engage in terminal investment (an increase in reproductive output prior to sterilization or death) following infection (70). Males may experience selection for weaker STI resistance as transmitting an infection to females may increase their short-term reproductive output.

### 9.5.4 Host–parasite coevolution

Relatively few theoretical studies have explored PMSS in a coevolutionary context. This is surprising given the extensive general literature on host–parasite coevolution (78), (84). It is important to consider how parasites may evolve in response to sexual selection in the host, as coevolution can fundamentally change modeling predictions, for example leading to fluctuating selection where the optimal strategy changes over time (e.g., the coadaptive cycles which are a prerequisite for the GGH). Early theoretical studies focused on whether the GGH as envisaged by Hamilton and Zuk is able to select for mate choice and secondary sex characteristics (79)–(81). This work revealed that although parasites may select for female preference and showy male traits in the context of cycling resistance genes, selection for female preference does not occur when it is sufficiently rare, which suggests there may be difficulties in initiating good genes mechanisms (see also (10)).

More recently, several theoretical studies have considered coevolution between STI virulence and transmission avoidance mating strategies (5), (30), (75). Ashby and Boots (30) showed that coevolution can lead to stable or fluctuating levels of host mate choosiness and STI sterility virulence (Box 9.3, Figure 1). Ashby (75) later generalized the model to consider the effects of mortality virulence and a wider range of mating dynamics, showing that high and low levels of choosiness can coexist with stable intermediate levels of STI virulence (Box 9.3, Figure 1). These findings contrast with previous

## Box 9.3  Case study of coevolution between mate choice and STI virulence

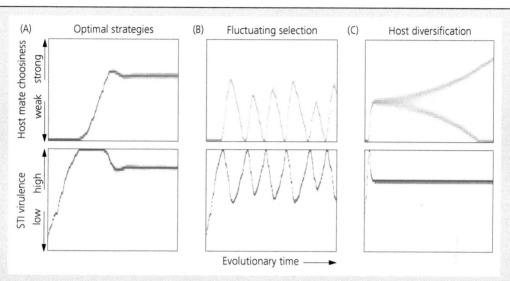

**Box 9.3, Figure 1** Coevolutionary dynamics of host mate choosiness (top row) and STI virulence (bottom row). Three qualitatively different outcomes are possible: (A) the host and STI coevolve to stable (optimal) levels of mate choice and virulence; (B) mate choice and virulence fluctuate through time; or (C) the host diversifies into high and low levels of mate choice, with stable STI virulence. Model as described in Box 9.3, adapted from (75).

The following model is adapted from (75). Consider a well-mixed host population in which there is no sex-specific variation in disease characteristics (a single hermaphroditic sex is modeled for simplicity), sexual partnerships are ephemeral (short-term), and a chronic STI causes mortality or sterility virulence. Suppose that hosts inspect prospective mates for signs of infection, but do not always assess them correctly. Choosy hosts are less likely to mate with infected individuals, and the choosier the host, the less likely a prospective mate is to be accepted, especially if they are infected. We can model the population dynamics of this system using the following ordinary differential equations

$$\frac{dS}{dt} = b(S, I, fg) - \beta(\alpha, f)[SI] - dS$$

$$\frac{dI}{dt} = \beta(\alpha, f)[SI] - dI - \alpha I$$

where $S$ and $I$ give the population densities of susceptible and infected individuals; $b(S,I,f,g)$ is the birth rate, with $f$ the strength of disease-associated sterility and $g$ the strength of mate choice; $d$ and $\alpha$ are the natural and disease-associated mortality rates; and $\beta(\alpha, f)$ is the probability that the STI is transmitted during mating, which is assumed to depend on the virulence of the STI (more transmissible STIs

are assumed to be more virulent, such that $\partial\beta/\partial\alpha > 0$ and/or $\partial\beta/\partial f > 0$). The sexual contact rate between individuals in the susceptible and infected classes is

$$[SI] = \frac{pm_s(g)m_I(\alpha, f, g)SI}{N}$$

where $p$ is the baseline sexual contact rate; $N = S + I$ is the total population size; and $m_s(g)$ and $m_I(\alpha, f, g)$ are the probabilities that a susceptible and an infected individual are accepted as mates, respectively, with the latter depending on the virulence of the infection.

We can use a process known as "Adaptive Dynamics" (83) to examine the coevolutionary dynamics of host mate choosiness ($g$) and STI virulence ($\alpha, f$). In general, three outcomes are possible: the host and STI coevolve to stable (optimal) levels of mate choice and virulence; mate choice and virulence fluctuate through time; or the host diversifies into high and low levels of mate choice, with stable STI virulence (Figure 1). By varying host and STI life-history traits, we can predict when each outcome is likely to occur. For example, fluctuations in mate choice and virulence are more likely to occur when hosts have intermediate lifespans (neither too short, nor too long), accidental avoidance of healthy individuals is rare, and the STI causes sterility virulence (75).

work based on one-sided adaptation, where STIs were predicted to evolve lower virulence to avoid detection, thus rendering mate choice ineffective (59). This added to the general perception that STIs should always evolve to become cryptic. However, coevolutionary feedbacks mean that selection may favor greater virulence as mate choice weakens, which suggests that STIs do not necessarily evolve to be cryptic in response to mate choice (30).

Wardlaw and Agrawal (5) also developed a coevolutionary model of PMSS, but in the context of sexual conflict, encompassing male persistence traits, female resistance traits, and STI virulence. The model revealed that if an STI causes mortality then sexual conflict will increase, in turn selecting for higher virulence. However, if the STI causes sterility, then virulence does not coevolve with sexual conflict. These findings demonstrate that mode of virulence can be critical to host–STI coevolution.

## 9.6 Conclusions and future directions

PMSS has been the subject of intense theoretical and empirical research since Hamilton and Zuk's seminal paper on "good genes" (7). There are now multiple hypotheses for PMSS encompassing both direct and indirect fitness effects, including theories based on transmission avoidance and parental care. These specific hypotheses, as well as PMSS more generally, have been explored both theoretically and experimentally, yet there are a number of important avenues for future research. For example, we know that STIs can select for monogamy or polygamy, but can STIs select for polygyny or polyandry? Similarly, we know that STIs can select for mate choice, but do they also select for showy traits? Are both direct and indirect fitness effects involved in PMSS? Crucially, there seems to be a general disconnect in the type of parasites that theoretical and empirical studies investigate in the context of PMSS. Theoretical studies tend to focus on the effect that STIs may have on PMSS (see section 9.5), perhaps stemming from the intuitive link between STIs and host mating dynamics, whereas empirical studies tend to focus on OIDs (see section 9.3), which may be because there are likely fewer STIs than OIDs and because determining transmission routes in wild systems can be challenging. Future theoretical and

empirical research should therefore aim to bridge this gap, with greater theoretical focus on OIDs and greater empirical focus on STIs.

Theoretical research to date has largely focused on one-sided adaptation in the host, with few studies taking a coevolutionary perspective that accounts for counter-adaptations by the parasite (5), (30), (75). This is important as coevolutionary feedbacks can lead to qualitatively and quantitatively different predictions (see Box 9.3). Most theoretical models also focus on pre-copulatory PMSS, but we lack predictions for post-copulatory PMSS. We also have few predictions for parasite-mediated sexual conflict. Finally, Vos et al. (50) and others (82) have proposed that beneficial sexually transmitted microbes are likely to influence PMSS, but this has yet to be explored theoretically.

The GGH depends on the coevolutionary cycling of host and parasite genes, but this has yet to be demonstrated empirically. Recent advances in genome sequencing may allow researchers to identify candidate genes and test for evidence of cycles (8), although observing cycles may require long-term experiments. Empirical studies to date have been heavily biased towards birds and the GGH; future research should seek to broaden the taxa of study systems as well as the hypotheses tested, including predictions for the coevolution of host mating rate or choosiness and STI virulence.

Given the difficulties of empirically testing PMSS in animals, future research might instead look to the microbial world to test certain predictions. Expelling and uptaking DNA in bacteria has been suggested as being analogous to sex in hermaphrodites (50). Following this analogy, Vos et al. (50) suggest that the CRISPR-Cas immune mechanism is akin to female mate choice, which could potentially allow for testing of coevolutionary cycling between phage virulence and bacteria choosiness. Bacteria may also manipulate the DNA uptake of other cells, potentially leading to conflict analogous to "male persistence" and "female resistance." Testing PMSS hypotheses using experimental evolution in microbial populations would be an exciting development which may provide new insights into the fundamental mechanisms underlying the relationship between parasitism and sexual selection.

# References

1. Allen BJ and Levinton JS. Costs of bearing a sexually selected ornamental weapon in a fiddler crab. *Funct Ecol.* 2007;21(1):154–61.

2. Darwin C. *On the Origin of Species by Means of Natural Selection, or the Preservation of Favoured Races in the Struggle for Life*, 1st edn. London: John Murray, 1859.

3. Martínez-Ruiz C and Knell RJ. Sexual selection can both increase and decrease extinction probability: Reconciling demographic and evolutionary factors. *J Anim Ecol.* 2017;86(1):117–27.

4. Ashby B and Gupta S. Sexually transmitted infections in polygamous mating systems. *Philos Trans R Soc Lond B: Biol Sci.* 2013;368(1613).

5. Wardlaw AM and Agrawal AF. Sexual conflict and sexually transmitted infections (STIs): Coevolution of sexually antagonistic host traits with an STI. *Am Nat.* 2019;193(1):E1–14.

6. Freeland WJ. Pathogens and the evolution of primate sociality. *Biotropica.* 1976;8(1):12–24.

7. Hamilton WD and Zuk M. Heritable true fitness and bright birds: A role for parasites? *Science* 1982;218(4570):384–7.

8. Balenger SL and Zuk M. Testing the Hamilton–Zuk hypothesis: Past, present, and future. *Integr Comp Biol.* 2014;54(4):601–13.

9. Borgia G. Satin bowerbird parasites: A test of the bright male hypothesis. *Behav Ecol Sociobiol.* 1986;19(5): 355–8.

10. Loehle C. The pathogen transmission avoidance theory of sexual selection. *Ecol Model.* 1997;103(2–3): 231–50.

11. Trivers RL. Parental investment and sexual selection in Houck LD and Drickamer LC (eds), *Foundations of Animal Behavior: Classic Papers with Commentaries*, reprinted from Campbell B (ed), Sexual Selection and the Descent of Man, 1871–1971, Chicago, IL: Aldine, 1972, 136–79.

12. Hoelzer GA. The good parent process of sexual selection. *Anim Behav.* 1989;38(6):1067–78.

13. Kirkpatrick M and Ryan MJ. The evolution of mating preferences and the paradox of the lek. *Nature.* 1991;350(6313):33–8.

14. Howard RD and Minchella DJ. Parasitism and mate competition. *Oikos.* 1990;120–2.

15. Demuth JP, Naidu A, and Mydlarz LD. Sex, war, and disease: The role of parasite infection on weapon development and mating success in a horned beetle (*Gnatocerus cornutus*). *PLoS One.* 2012;7(1):e28690.

16. Abbot P and Dill LM. Sexually transmitted parasites and sexual selection in the milkweed leaf beetle, *Labidomera clivicollis*. *Oikos.* 2001;92(1):91–100.

17. Thomas F, Renaud F, and Cézilly F. Assortative pairing by parasitic prevalence in *Gammarus insensibilis* (Amphipoda): Patterns and processes. *Anim Behav.* 1996;52(4):683–90.

18. Borgia G. Sexual Selection and the evolution of mating systems in Blum MS and Blume NA (eds), *Sexual Selection and Reproductive Competition in Insects.* Cambridge, MA: Academic Press, Inc.; 1979, 19–80.

19. Borgia G and Collis K. Female choice for parasite-free male satin bowerbirds and the evolution of bright male plumage. *Behav Ecol Sociobiol.* 1989; 25:445–53.

20. Borgia G and Collis K. Parasites and bright male plumage in the satin bowerbird (*Ptilonorhynchus violaceus*). *Am Zool.* 1990;30(2):279–86.

21. Clayton DH. The influence of parasites on host sexual selection. *Parasitol Today.* 1991;7(12):329–34.

22. Hamilton WJ and Poulin R. The Hamilton and Zuk hypothesis revisited: A meta-analytical approach. *Behaviour.* 1997;134(3–4):299–320.

23. Read AF. Sexual selection and the role of parasites. *Trends Ecol Evol.* 1988;3(5):97–102.

24. Koch RE and Hill GE. Shared biochemical pathways for ornamentation and immune function: Rethinking the mechanisms underlying honest signaling of parasite resistance. in Ezenwa VO, Altizer S, and Hall RJ (eds), *Animal Behavior and Parasitism.* Oxford: Oxford University Press, 2022. DOI: 10.1093/oso/9780192895561.003.0010.

25. Poulin R and Vickery WL. Parasite distribution and virulence: Implications for parasite-mediated sexual selection. *Behav Ecol Sociobiol.* 1993;33(6):429–36.

26. Poulin R. Mate choice decisions by parasitized female upland bullies, *Gobiomorphus breviceps. Proc R Soc B: Biol Sci.* 1994;256(1346):183–7.

27. López S. Parasitized female guppies do not prefer showy males. *Anim Behav.* 1999;57(5):1129–34.

28. Rowe L and Houle D. The lek paradox and the capture of genetic variance by condition dependent traits. *Proc R Soc B: Biol Sci.* 1996;263(1375):1415–21.

29. Able DJ. The contagion indicator hypothesis for parasite-mediated sexual selection. *PNAS.* 1996;93(5):2229–33.

30. Ashby B and Boots M. Coevolution of parasite virulence and host mating strategies. *PNAS.* 2015;112(43): 13290–5.

31. Norris K and Evans MR. Ecological immunology: Life history trade-offs and immune defense in birds. *Behav Ecol.* 2000;11(1):19–26.

32. Kavaliers M, Colwell DD, and Choleris E. Parasites and behavior: An ethopharmacological analysis and biomedical implications. *Neurosci Biobehav Rev.* 1999;23(7):1037–45.

33. Loehle C. Social barriers to pathogen transmission in wild animal populations. *Ecology*. 1995;76(2):326–35.

34. Thrall PH, Antonovics J, and Dobson AP. Sexually transmitted diseases in polygynous mating systems: Prevalence and impact on reproductive success. *Proc R Soc B: Biol Sci*. 2000;267(1452):1555–63.

35. Winterniz JC and Abbate JL. The genes of attraction: Mating behavior, immunogenetic variation, and parasite resistance. In Ezenwa VO, Altizer S, and Hall RJ (eds), *Animal Behavior and Parasitism*. Oxford: Oxford University Press, 2022. DOI: 10.1093/oso/9780192895561.003.0011.

36. Folstad I and Karter AJ. Parasites, bright males, and the immunocompetence handicap. *Am Nat*. 1992;139(3):603–22.

37. Saks L, Ots I, Hõrak P. Carotenoid-based plumage coloration of male greenfinches reflects health and immunocompetence. *Oecologia*. 2003;134(3):301–7.

38. Hõrak P, Ots I, Vellau H, Spottiswoode C, and Møller AP. Carotenoid-based plumage coloration reflects hemoparasite infection and local survival in breeding great tits. *Oecologia*. 2001;126(2):166–73.

39. del Cerro S, Merino S, Martínez-de la Puente J, Lobato E, Ruiz-de-Castañeda R, Rivero-de Aguilar J, *et al*. Carotenoid-based plumage colouration is associated with blood parasite richness and stress protein levels in blue tits (*Cyanistes caeruleus*). *Oecologia*. 2010;162(4):825–35.

40. Kolluru GR, Grether GF, South SH, Dunlop E, Cardinali A, Liu L, *et al*. The effects of carotenoid and food availability on resistance to a naturally occurring parasite (*Gyrodactylus turnbulli*) in guppies (*Poecilia reticulata*). *Biol J Linn Soc*. 2006;89(2):301–9.

41. Salvador A, Veiga JP, Martin J, Lopez P, Abelenda M, and Puertac M. The cost of producing a sexual signal: Testosterone increases the susceptibility of male lizards to ectoparasitic infestation. *Behav Ecol*. 1996;7(2): 145–50.

42. Reusch TBH, Häberli MA, Aeschlimann PB, and Milinski M. Female sticklebacks count alleles in a strategy of sexual selection explaining MHC polymorphism. *Nature*. 2001;414(6861):300–2.

43. Milinski M. Arms races, ornaments and fragrant genes: The dilemma of mate choice in fishes. *Neurosci Biobehav Rev*. 2014;46:567–72.

44. Potts WK, Manning CJ, and Wakeland EK. Mating patterns in seminatural populations of mice influenced by MHC genotype. *Nature*. 1991;352(6336):619–21.

45. Koskella B and Lively CM. Evidence for negative frequency-dependent selection during experimental coevolution of a freshwater snail and a sterilizing trematode. *Evol Int J Org Evol*. 2009;63(9):2213–21.

46. Barber I, Arnott SA, Braithwaite VA, Andrew J, and Huntingford FA. Indirect fitness consequences of mate choice in sticklebacks: Offspring of brighter males grow slowly but resist parasitic infections. *Proc R Soc B: Biol Sci*. 2001;268(1462):71–6.

47. Buczek M, Okarma H, Demiaszkiewicz AW, and Radwan J. MHC, parasites and antler development in red deer: No support for the Hamilton & Zuk hypothesis. *J Evol Biol*. 2016;29(3):617–32.

48. Getty T. Signaling health versus parasites. *Am Nat*. 2002;159(4):363–71.

49. Garamszegi LZ and Møller AP. The interspecific relationship between prevalence of blood parasites and sexual traits in birds when considering recent methodological advancements. *Behav Ecol Sociobiol*. 2012;66(1): 107–19.

50. Vos M, Buckling A, and Kuijper B. Sexual Selection in Bacteria? *Trends Microbiol*. 2019;27(12):972–81.

51. Zuk M, Thornhill R, Ligon JD, and Johnson K. Parasites and mate choice in red jungle fowl. *Am Zool*. 1990;30(2):235–44.

52. Pruett-Jones SG, Pruett-Jones MA, and Jones HI. Parasites and sexual selection in birds of paradise. *Am Zool*. 1990;30(2):287–98.

53. Kennedy CEJ, Endler JA, Poynton SL, and McMinn H. Parasite load predicts mate choice in guppies. *Behav Ecol Sociobiol*. 1987;21(5):291–5.

54. Milinski M and Bakker TCM. Female sticklebacks use male coloration in mate choice and hence avoid parasitized males. *Nature*. 1990;344(6264):330–3.

55. Worden BD, Parker PG, and Pappas PW. Parasites reduce attractiveness and reproductive success in male grain beetles. *Anim Behav*. 2000;59(3):543–50.

56. Nahrung HF and Allen GR. Sexual selection under scramble competition: Mate location and mate choice in the eucalypt leaf beetle *Chrysophtharta agricola* (Chapuis) in the field. *J Insect Behav*. 2004;17(3):353–66.

57. Webberley KM, Hurst GDD, Buszko J, and Majerus MEN. Lack of parasite-mediated sexual selection in a ladybird/sexually transmitted disease system. *Anim Behav*. 2002;63(1):131–41.

58. Jones SL, Pastok D, and Hurst GDD. No evidence that presence of sexually transmitted infection selects for reduced mating rate in the two spot ladybird, *Adalia bipunctata*. *PeerJ*. 2015;3(e1148).

59. Knell RJ. Sexually transmitted disease and parasite-mediated sexual selection. *Evolution (N Y)*. 1999;53(3):957–61.

60. Møller AP, Barbosa A, and Cuervo JJ, Lope F de, Merino S, Saino N. Sexual selection and tail streamers in the barn swallow. *Proc R Soc B: Biol Sci*. 1998;265(1394):409–14.

61. Lombardo MP. On the evolution of sexually transmitted diseases in birds. *J Avian Biol*. 1998;29(3):314.

62. Lockhart AB, Thrall PH, and Antonovics J. Sexually transmitted diseases in animals: Ecological and evolutionary implications. *Biol Rev*. 1996;71(3):415–71.

63. Knell RJ and Webberley KM. Sexually transmitted diseases of insects: distribution, Evolution, ecology and host behaviour. *Biol Rev*. 2004;79(3):557–81.

64. Russell MW, Whittum-Hudson J, Fidel PL, Hook EW, Mestecky J. Immunity to sexually transmitted infections in Mestecky J, Strober W, Russell MW, Kelsall BL, Cheroutre H, and Lambrecht BN (eds), *Mucosal Immunology*, 4th edn. Boston, MA: Elsevier; 2015, 2183–214.

65. Ramsey D, Spencer N, Caley P, Efford M, Hansen K, Lam M, *et al.* The effects of reducing population density on contact rates between brushtail possums: Implications for transmission of bovine tuberculosis. *J Appl Ecol*. 2002;39(5):806–18.

66. Anderson RM and May RM. *Infectious Diseases of Humans: Dynamics and Control*. Oxford: Oxford University Press, 1991.

67. Halimubieke N, Pirrie A, Székely T, and Ashby B. How do biases in sex ratio and disease characteristics affect the spread of sexually transmitted infections? *J Theor Biol*. 2021;527:110832.

68. Choudhury S. Divorce in birds: A review of the hypotheses. *Anim Behav*. 1995;50(2):413–29.

69. Dubois F and Cézilly F. Breeding success and mate retention in birds: A meta-analysis. *Behav Ecol Sociobiol*. 2002;52(5):357–64.

70. Johns S, Henshaw JM, Jennions MD, and Head ML. Males can evolve lower resistance to sexually transmitted infections to infect their mates and thereby increase their own fitness. *Evol Ecol*. 2019;33(2):149–72.

71. Thrall PH, Antonovics J, and Bever JD. Sexual transmission of disease and host mating systems: Within-season reproductive success. *Am Nat*. 1997;149(3):485–506.

72. Graves BM and Duvall D. Effects of sexually transmitted diseases on heritable variation in sexually selected systems. *Anim Behav*. 1995;50(4):1129–31.

73. Kokko H, Ranta E, Ruxton G, and Lundberg P. Sexually transmitted disease and the evolution of mating systems. *Evolution (N Y)*. 2002;56(6):1091–100.

74. McLeod D and Day T. Sexually transmitted infection and the evolution of serial monogamy. *Proc R Soc B: Biol Sci*. 2014;281(1796).

75. Ashby B. Antagonistic coevolution between hosts and sexually transmitted infections. *Evolution (N Y)*. 2020;74(1):43–56.

76. Boots M and Knell RJ. The evolution of risky behaviour in the presence of a sexually transmitted disease. *Proc R Soc B: Biol Sci*. 2002;269(1491):585–9.

77. Dougherty LR, van Lieshout E, McNamara KB, Moschilla JA, Arnqvist G, and Simmons LW. Sexual conflict and correlated evolution between male persistence and female resistance traits in the seed beetle *Callosobruchus maculatus*. *Proc R Soc B: Biol Sci*. 2017;284(1855):20170132.

78. Ashby B, Iritani R, Best A, White A, and Boots M. Understanding the role of eco-evolutionary feedbacks in host-parasite coevolution. *J Theor Biol*. 2019;464:115–25.

79. Kirkpatrick M. The handicap mechanism of sexual selection does not work. *Am Nat*. 1986;127(2):222–40.

80. Pomiankowski A. Sexual selection: The handicap principle does work—sometimes. *Proc R Soc B: Biol Sci*. 1987;231(1262):123–45.

81. Iwasa Y, Pomiankowski A, and Nee S. The evolution of costly mate preferences II. The "handicap" principle. *Evol*. 1991;45(6):1431–42.

82. Smith CC and Mueller UG. Sexual transmission of beneficial microbes. *Trends Ecol Evol*. 2015;30(8):438–40.

83. Geritz SAH, Kisdi É, Meszéna G, and Metz JAJ. Evolutionarily singular strategies and the adaptive growth and branching of the evolutionary tree. *Evol Ecol*. 1998;12(1):35–57.

84. Buckingham LJ and Ashby B. Coevolutionary theory of hosts and parasites. *J Evol Biol*. 2022;35:205–24.

# CHAPTER 10

# Shared biochemical pathways for ornamentation and immune function: Rethinking the mechanisms underlying honest signaling of parasite resistance

Rebecca E. Koch and Geoffrey E. Hill

## 10.1 Introduction

The complex and varied forms of display traits used in mate choice have long intrigued biologists. A primary explanation for why displays evolve is that they convey consequential and honest information about individual quality (1), (2). Under this concept of traits as indicators, the expression of various visual, vocal, and chemical displays that are assessed in mate choice correlate with aspects of performance such that a higher-quality display indicates a higher-quality individual (3), (4). Such traits hold the potential to communicate information from signaler to receiver about the direct or indirect benefits of selecting a given mate (5). Among the most important classes of information about a prospective mate are the current health state and the capacity to resist or avoid future parasite infection (6), (7).

Beginning in the last decades of the twentieth century, seemingly ubiquitous associations between display expression and aspects of individual quality have been documented in hundreds of publications in mainstream behavior, evolution, and physiology journals (reviewed in (7), (8)). The question that has bedeviled this line of investigation since its inception, however, is what prevents individuals from cheating and maximally expressing a display trait regardless of individual quality? Indeed, many hypotheses over the past several

decades of sexual selection research have focused on articulating a framework of physiological costs or constraints that impose honesty on mating display traits (1), (9).

In this chapter, we first briefly review several of the major hypotheses from the past 50 years that have proposed explanations for how display traits may reflect physiological performance and resist cheating strategies, focusing specifically on parameters related to immune defenses against parasites. We then contrast signaling theory founded on the costs of display production with emerging theory founded on vital cellular processes from which arise both displays and physiological performance. We discuss how shared physiological pathways—pathways that often involve mitochondrial processes—offer intriguing new insight into how seemingly esoteric traits like coloration or a courtship dance might convey information about core physiological condition. Lastly, we focus on the implications of the shared pathway hypothesis for the Hamilton–Zuk hypothesis, which proposes that mating display traits are associated with genes for parasite resistance (6).

## 10.2 Cost-based signaling hypotheses

The hypothesis that signal honesty is maintained by the cost of producing display traits has become entrenched in the literature (9). It is a seductive

Rebecca E. Koch and Geoffrey E. Hill, *Shared biochemical pathways for ornamentation and immune function*. In: *Animal Behavior and Parasitism*. Edited by Vanessa O. Ezenwa, Sonia Altizer, and Richard J. Hall, Oxford University Press. © Oxford University Press (2022). DOI: 10.1093/oso/9780192895561.003.0010 idea because the production of many display traits exacts a resource cost and it seems logical that such resources could alternatively be used in processes that are important to body maintenance. Hence, the hypothesis that display production necessitates costly resource trade-offs is widely cited for how expression of display traits remains an uncheatable signal of quality (e.g., (10)). The assumption that mating display traits must entail a trade-off cost to remain honest is commonly applied in assessments of the Hamilton–Zuk hypothesis (11). Despite decades of tests of this hypothesis, however, evidence in support of trade-offs between body maintenance and display production remains mixed, with some experimental observations indicating that such trade-offs may not exist (e.g., (12)).

The theory that signal honesty is maintained by costs related to resource trade-offs has roots in the concept of signals as handicaps (13). Under Zahavi's handicap hypothesis, display traits evolved wasteful exaggeration in order to maintain honesty (Table 10.1). An expansion on the original hypothesis considered the concept of differential costs: that individuals of poor physiological quality would pay higher costs for investing in a high quality signal, so individuals adjust signal expression according to capacity (14). While the original hypothesis that displays evolve as direct handicaps to survival has been widely dismissed, cost-based signaling hypotheses carry forth a form of the handicap hypothesis.

Among the most prominent of the cost-based hypotheses is the Immunocompetence Handicap Hypothesis (ICHH), which proposes that display traits can indicate parasite burden and resistance if a "biochemical substance"—such as a hormone—serves as a "double-edged sword" that simultaneously boosts display expression and compromises the immune system (15). Folstad and Karter (15) specifically outlined this hypothesis as it pertains to the androgen testosterone, which, at least in many vertebrates, can have suppressive effects on the immune system as well as stimulatory effects on display trait expression (Figure 10.1). Through this framework, display traits may be honest indicators of parasite burden because only individuals with low burden and high resistance can increase testosterone levels sufficiently to generate high-quality displays, despite concomitant immunosuppression (15). This hypothesis has ties to the concepts of

both display traits as handicaps and individuals optimizing their display trait expression according to their quality.

The ICHH has stimulated a rich body of study in the decades since its publication, but the relationships between steroid hormones, ornamentation, and immunocompetence often do not clearly conform to predictions (Figure 10.1). The effects of testosterone on either display trait or immune system performance are rarely straightforward, and can be confounded by other hormones (e.g., glucocorticoids) and by testosterone-induced changes in behavior that also alter parasite exposure (16). A full review of tests of the ICHH is beyond the scope of this chapter, though it is notable that the hypothesis remains relevant and a focus of research interest nearly 30 years after its initial definition (17).

The fields of eco-immunology and oxidative stress ecology have provided greater understanding of how other physiological states and processes may affect display trait expression. In 2007, for example, Alonso-Alvarez et al. (18) proposed a follow-up to the ICHH that considers whether risk of oxidative stress, in addition to risk of increased parasitism, may be the primary cost associated with increased testosterone in displaying males. Termed the "Oxidative Handicap Hypothesis," (OHH) (18) this framework united the ICHH with earlier considerations that oxidative stress may be a key mediator of trade-offs between testosterone and displays (19). The OHH integrates information from literatures that had largely not been considered in concert, exploring connections between testosterone, oxidative stress, parasite resistance and burden, and display expression; however, each of these relationships is itself complex, potentially bidirectional, and likely context-specific, and support for the OHH to date has been mixed.

Determining and quantifying the currency that is traded between display or self-maintenance can be challenging. The costs of mating displays may be considered from multiple perspectives—for example, we might consider the physical or energetic resources directly involved in display production or maintenance, or indirect physiological consequences of producing a display (e.g., if producing the display requires particular hormones or oxidative states). Despite the widely stated view that mating displays must be costly to be honest,

**Table 10.1** Several major hypotheses to explain display behaviors that act as honest indicators of individual quality, including parasite resistance.

| Hypothesis for display honesty | Mechanism of honesty? | Brief description | Example of immune- or parasite-specific prediction | Key reference |
|---|---|---|---|---|
| Handicap Hypothesis a) | Costs | Displays are wasteful handicaps that only high-quality individuals can withstand | Individuals with poor health are less likely to survive the burden of their display | (13) |
| Handicap Hypothesis b) | Costs | Individuals express varying quality of display depending on their ability to bear its cost | Individuals with poor health will express a lower-quality display in order to preserve resources to put toward survival | (14) |
| Immunocompetence Handicap Hypothesis (ICHH) | Costs | Only high-quality individuals can withstand the "double-edged sword" of a chemical required for display (e.g., testosterone) | The increased testosterone required to produce a high-quality display comes at the cost of immunosuppression | (15) |
| Oxidative Handicap Hypothesis (OHH) | Costs | Only high-quality individuals can withstand the oxidative challenge posed by the testosterone required for a display | The increased testosterone required to produce a high-quality display imposes both oxidative challenge and immuno-suppression such that only individuals with effective antioxidant and immune defenses (and/or low parasite burden) can produce it | (18) |
| Index Hypothesis | Index | Display quality depends directly on the same trait that is being signaled | A colorful ornament may indicate external parasite burden by being visually altered by their presence | (21) |
| Shared Pathway Hypothesis | Index | Display quality is directly and inescapably linked to physiological quality through dependence on the same mechanistic pathways | The performance of a core physiological process will affect both parasite resistance and display quality such that individuals with better-expressed displays will tend to have lower current and future parasite burdens | (22) |
| Mitochondrial Function Hypothesis | Index | Display quality is directly and inescapably linked to physiological quality through dependence on mitochondrial performance | High-performing mitochondria are necessary for both effective immune response and high-quality display production | (23) |

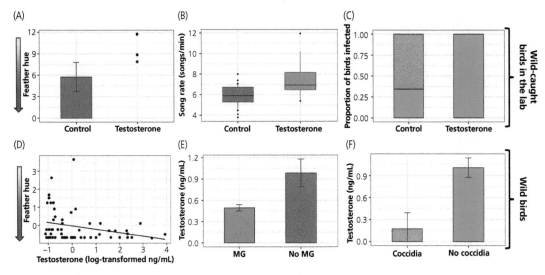

**Figure 10.1**  A series of studies on both captive (top) and wild (bottom) male house finches collectively provide a test of the ICHH, and specifically the hypothesis that the display-enhancing but immuno-suppressing properties of testosterone ensure honest signaling of immune defenses. Lab experiments using artificial testosterone (T) supplementation via implants tended to find somewhat contradictory patterns between T, mating display traits, and infection status (A–C), compared to studies measuring the naturally circulating T levels of wild birds (D–F). Experimentally increasing T had mixed effects on house finch mating display quality, increasing song rate (B) but tending to decrease ornamental feather redness (A). Among wild birds, however, individuals naturally circulating higher levels of T instead had redder (i.e., higher quality) plumage coloration (D)—the opposite pattern. Further, among captive birds, all of those with experimentally increased T were found to be infected with coccidian gut parasites (*Isospora* sp.), while fewer than a third of control birds were infected (C). Yet again, the opposite pattern was discovered in wild birds: those with coccidian gut parasites (F) or *Mycoplasma gallisepticum* (MG) bacterial infection (E) had lower average circulating T. These results are consistent with the predictions of a shared pathway hypothesis, which expects individuals with higher functionality to have both higher quality displays and superior parasite resistance (and vice versa); however, these results do not conform with the predictions of the ICHH, which would anticipate birds with higher testosterone to have better-expressed displays at the cost of immune defenses. The results depicted here, while just a narrow subset of the vast literature exploring the ICHH and related hypotheses, demonstrate complexity of testing a cost-based hypothesis and the context-dependence of the results, even within a single species. The three points for the testosterone group in (A) represent raw data, due to low sample size; the point and error bars for the control group are mean ± standard deviation. Points and error bars in (F) and (G) represent means ± standard error. Data from (74).

there is no consensus on the currency of recompense for the production of mating displays nor on the downstream consequences of expenditures.

## 10.3  Explaining indicator traits with shared pathways rather than shared costs

While the resource trade-off hypothesis is founded on the costs of shared resources and handicap-based hypotheses like the ICHH and OHH stem from the negative effects of the physiological substances necessary to produce the signal, there is an alternative to the hypotheses that signals must be costly to remain honest: index signal hypothesis. An index signal is a trait whose production is

causally linked to the property of the individual that is being signaled (20), (21). For instance, the height of scratch marks on a tree is an index signal of size; signal honesty is maintained because physical limits are inescapable, in contrast to costs that can be differentially paid. The shared pathway hypothesis proposes that many display traits are index signals because their production requires high functionality in the same vital cellular pathways that are required to sustain processes that underlie individual quality, such as immunocompetence (9), (22). This hypothesis proposes that full display expression is possible only if core life-sustaining processes are performing well; thus, display traits are causally linked to system function, and dysfunctions of vital cellular processes are revealed

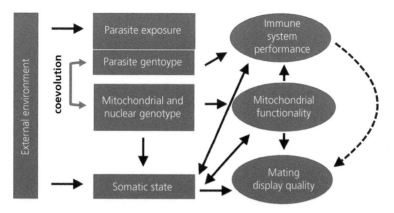

**Figure 10.2** Mitochondrial function is determined by genotype (both mitochondrial and nuclear) and current somatic state (e.g., as subject to health, external environmental effects). Because mitochondrial processes can affect both immune system performance and behavior expression, the mitochondrial function hypothesis proposes that shared pathways through the mitochondria establish a link between display trait quality and immune parameters (dashed arrow). Moreover, a mitonuclear extension of the Hamilton–Zuk hypothesis proposes that G × G × E interactions between parasites, host mitonuclear genotype, and the immune system will maintain genetic variation for parasite resistance and the traits that signal it. Adapted from (23).

in the quality of the display (22), (23). Numerous cellular pathways might be linked to production of various types of displays. For instance, the honesty of melanin-based ornaments may arise from dependency on the melanocortin system in vertebrates, which has pleiotropic effects across multiple physiological pathways, including immune responses (24). Recently, however, there has been a particular focus on mitochondrial function as a key to both whole-organism performance and display traits (23), (25)–(27).

While energy flow and whole-animal metabolism have long been key topics in ecology and physiology, the importance of cellular metabolism and mitochondrial performance has only recently been raised in the conversation. An increased interest in subcellular performance stems largely from burgeoning biomedical research that has revealed intersections between mitochondria and physiological variables of interest to ecologists and evolutionary biologists, such as immune system performance (28) and oxidative systems (19), (29). To date, our understanding of the biochemical pathways of mitochondria and their interactions with major physiological processes arises largely from biomedical studies on humans or laboratory-based model systems. Briefly, high performing mitochondria are central to organismal function due to their

role not only in harnessing energy as adenosine triphosphate (ATP), but also in transducing signals and performing biosynthesis (see (23), (26), (27) for reviews within an ecological context). Importantly, mitochondria have critical involvement in the same processes that mating displays are proposed to signal, including immune defense, oxidative state, and sex steroid hormone synthesis and function (23), (26), (27). The mechanistic role of mitochondria in processes like immunocompetence places them at the center of individual quality, and the mitochondrial function hypothesis proposes that mitochondria are the "shared pathway" between display quality and individual performance, forming the basis for honest signaling (Figure 10.2; Table 10.1). According to this hypothesis, variation among individuals in their mating displays arises not from resource trade-offs but through individual variation in the performance of cellular respiration in their mitochondria (variation arising due to environmental effects and/or genetic interactions; see below).

Perhaps the most direct demonstrations of the role of mitochondria as immune signaling hubs pertain to the innate immune response (30). Two specific signaling proteins—the mitochondrial antiviral signaling (MAVS) protein and the evolutionarily conserved signaling intermediate in Toll pathways (ECSIT)—have been found to localize to

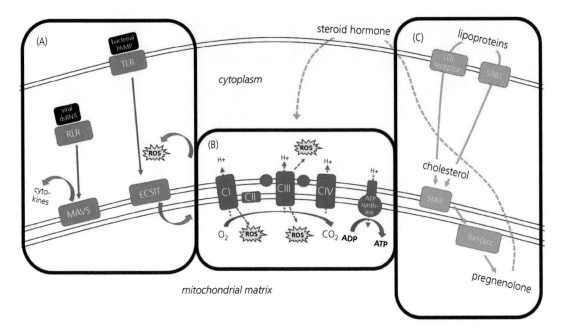

**Figure 10.3** A diagram depicting simplified pathways involving proteins that reside in or localize to mitochondria, here using terminology associated with mammalian systems. Two innate immune signaling proteins, MAVS and ECSIT, localize to the outer mitochondrial membrane in response to the stimulation of particular RLR or TLRs, respectively (A) MAVS stimulates upregulation of pro-inflammatory cytokines in response to viral double-stranded RNA detection; ECSIT interacts with the electron transport chain (B) to increase ROS production around phagocytes containing intracellular bacteria (after detection of a bacterial pathogen-associated molecular pattern, or PAMP). The electron transport chain itself comprises four main complexes (CI–CIV) that harness the energy from electrons to pump protons (H+) across the inner mitochondrial membrane and into the cytoplasm, which can then drive ATP synthesis from ADP by ATP synthase through OXPHOS (B) Electrons can "escape" from complexes I and III, producing reactive oxygen species (ROS). The initial steps of steroid hormone synthesis take place within the mitochondria of steroidogenic cells (C) Low-density (LDL) and high-density (e.g., scavenger receptor B1, or SRB1) lipoprotein receptors can import cholesterol and precursors, which can be processed and transported to the outer mitochondrial membrane; from there, the steroidogenic acute regulatory protein (StAR) can transport cholesterol to the inner mitochondrial membrane, where P450scc converts it to pregnenolone, a steroid hormone precursor. Pregnenolone can then leave the mitochondria passively to undergo further processing. In turn, steroid hormones can mediate mitochondrial OXPHOS via hormone receptor elements present on mitochondrial-associated nuclear DNA and even mitochondrial DNA itself.

mitochondrial membranes and facilitate antiviral or antibacterial innate immune defenses, respectively (Figure 10.3) (30). MAVS mediates signals between receptors that recognize viral double-stranded RNA and the factors that stimulate pro-inflammatory cytokine and type I interferon production (31). Similarly, ECSIT is a signaling intermediate between receptors that detect signs of bacteria and mitochondria, recruiting mitochondria around sites of bacterial ingestion and stimulating increased mitochondrial ROS production to damage the captured bacteria (32). The mammalian toll-like receptor-4 (TLR4) is one innate immune receptors known to signal through ECSIT (32), which is particularly relevant given that TLR4 recognizes bacterial

lipopolysaccharide (LPS), a substance often used as an experimental immune activator.

While researchers who are focused on biomedicine have established key links between mitochondrial function and immunocompetence, evolutionary physiologists are investigating links between mitochondrial function and display traits (23), (33). Such links can be as basic as a reduction in rate or quality of performing a behavioral display when mitochondria produce less ATP (to date demonstrated more broadly in activities such as exploration (34), or in the upregulation of mitochondrial proteins in processes like developing neural pathways related to vocal display production in songbirds (35), but not mating behaviors *per se*). Further, more complex links between production of pigments used in color displays and specific aspects of electron transport system function might provide very tight links between mitochondrial function and ornamental displays (Box 10.1).

If mitochondrial function affects both immune responsiveness and display behaviors, it follows that display traits might signal immune defenses or current infection status via shared mitochondrial pathways. The mitochondrial function hypothesis proposes that assessment of display traits becomes an assessment of mitochondrial function. Choosing an individual with highly functional mitochondria then becomes a strategy to avoid potential mates that are likely to perform poorly in the face of environmental challenges, including parasite exposure. Moreover, the mitochondrial function hypothesis requires no trade-offs or even direct costs to maintain signaling honesty. When mitochondria are functioning well, then both display production and immune system function proceed without difficulty; inversely, when mitochondria are performing poorly, then both display production and immune performance are compromised, and there is no strategy that can cheat such a system.

That the mitochondrial function hypothesis proposes cheating-resistant links between behavioral display and internal physiology highlights key differences in the framework of shared pathway hypotheses compared to cost-based hypotheses. The former proposes a cellular mechanism underpinning functionality of both physiology and display, while the latter focuses on diversion of finite resources from one function to another and/or mitigation of the costs associated with the display. For example, if we consider the ICHH, we might imagine an individual with a higher-quality mating display to have suppressed immune performance that must be compensated for with reduced parasite exposure or a sufficient baseline immune performance to withstand suppression safely. In contrast, under the mitochondrial function hypothesis, we instead consider quality of display to relate positively to quality of immune system performance, as both depend on pathways that require excellent mitochondrial function. Both hypotheses may make similar predictions: individuals with the best

displays may have had low exposure to parasites and/or have high enough immune defense capability to withstand such exposure with fewer negative consequences. However, the mitochondrial function hypothesis provides a specific intermediate mechanism to test (i.e., mitochondrial performance) while the ICHH centers around mechanisms of physiological trade-offs that have yet to be empirically demonstrated. The mitochondrial function hypothesis also posits specific information (mitochondrial function) as the object of female choice for mating displays (48), (49). Moreover, while testosterone-centered hypotheses like the ICHH and OHH are limited to vertebrates, the mitochondrial function hypothesis might apply to all animals.

Interestingly, mitochondria also have a primary role in the biosynthesis and function of the primary intermediate proposed by the ICHH (and the OHH) to drive trade-offs between immune and display performance: testosterone (Figure 10.3) (26), (50). Such relationships are thought to be important because many of the effects stimulated by sex steroid hormones—growth, increased activity, developmental differentiation—may pose particularly high metabolic demands. This in turn forms the basis for what has been called the aerobic activity hypothesis, which predicts that the increased aerobic respiration required for many mating displays poses an oxidative cost through metabolic reactive oxygen species (ROS) production (51). At the cellular level, aerobic respiration in the mitochondria does produce ROS as a by-product, but the dynamics of the process are such that mitochondrial ROS production does not increase predictably with rate of respiration (33), (52). Increased ATP output can in fact decrease ROS generation. ROS is produced as a side reaction as the energy in electrons oxidizes oxygen instead of pumping protons (Figure 10.3). Allowing electrons to flow more freely along the electron transport chain can reduce the probability of side reactions creating ROS, even as it yields more ATP. Such a system, with minimal ROS production, is called a more tightly coupled system (53). Moreover, even when ROS is produced, it does not necessarily mean increased oxidative damage and lasting negative effects; ROS serve as important signaling molecules, and can even induce

## Box 10.1  Case study: Carotenoid coloration in bird feathers

The carotenoid coloration of bird feathers has been extensively studied in the context of female mate choice assessing condition dependent ornamentation (36). An extensive literature indicates that females of many bird species show a mating preference for males with more saturated or more red-shifted color displays (reviewed in (37)). In addition, more saturated or more red-shifted carotenoid-based plumage coloration is linked to diverse measures of individual performance, including flight performance (38), parasite burden (39), and oxidative stress (40). This extensive line of research has made carotenoid-based feather coloration among the most frequently stated examples of condition dependent ornamentation—and specifically signals of immunocompetence and disease resistance—used in mate choice (8), (9), (36).

Despite the extensive literature on condition-dependent sexual signaling via carotenoid feather coloration, the mechanisms that might link color to performance have remained much debated and poorly understood (12). One recent advance in understanding how feather coloration can serve as an honest signal is a consideration of whether the color display is produced by pigments that are unaltered after being ingested, or if the color display is produced by metabolizing ingested carotenoids into ketolated carotenoids (25). In a meta-analysis of 50 published studies, Weaver et al. (41) found that the link between feather coloration and individual condition was strongest when birds metabolized dietary carotenoids to produce ornamental coloration. This link between carotenoid metabolism and condition dependency became more intriguing with

the discovery of CYP2J19 as the enzyme that enables the metabolic conversion of yellow dietary carotenoids to red pigments used in ornamentation (42), (43). Red carotenoids concentrate in the mitochondria of songbirds that are growing red feathers, implicating mitochondria in production of red pigments (44) and supporting hypotheses that mitochondrial function is linked to red feather coloration (25).

To look for evidence for a direct role of mitochondrial function in production of red carotenoid pigments, Hill et al. (45) compared measures of mitochondrial performance to the redness of growing feathers in molting house finches (Haemorhous mexicanus). As predicted, they found that males growing redder feathers, and therefore with the more efficient ketolation pathways, also had higher inner mitochondrial membrane potentials and higher respiratory control ratios, two key measures of mitochondrial function (Figure 1).

In experimental studies, Cantarero and Alonso-Alvarez (46) and Cantarero et al. (47) used drug treatments to manipulate mitochondrial function. In red crossbills (Loxia curvirostra), a species in the same family as house finches and with similar red feather coloration, treating birds with a chemical that affects the inner mitochondrial membrane caused naturally red individuals to get redder, while drab birds did not increase in coloration (47). In zebra finches, a species that deposits red ketolated carotenoids in its bill, experimentally altering the inner mitochondria membrane potential had significant effects on the redness of bill coloration (46).

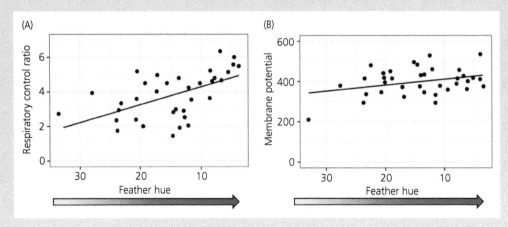

**Box 10.1, Figure 1** The mitochondria of molting wild male house finches exhibited higher energy capacity (as measured by "respiratory control ratio," an estimate of mitochondrial aerobic respiration efficiency; (A) and membrane potential (a measurement of potential energy stored across the inner mitochondrial membrane; (B) in individuals with redder ornamental feather coloration. Data from (45).

---

**Box 10.1** *Continued*

Many questions remain regarding the specific pathways that might link the metabolism of red carotenoid pigments and oxidative phosphorylation (OXPHOS) in the electron transport system. Given the known links between mitochondrial function and immune responses and red carotenoid coloration and disease resistance, the hypothesis that red coloration is inherently linked to mitochondrial function is especially intriguing.

---

a long-term benefit in a hormetic response (54), (55). These points underscore the challenge inherent to making predictions about how mating behavior may affect or in turn be affected by oxidative status, in addition to any further complexity added by hormones or immune activation. Considering shared pathways through mitochondria again offers an alternative approach: the inherent functionality of mitochondria influences mitochondrial ROS generation and how it may change with increasing metabolic demand (and with changing hormonal effects). For example, while increasing metabolic rate may pose little to no oxidative challenge with high-functioning mitochondria, the same metabolic demand may exacerbate inefficiencies and ROS production with mitochondria with mild dysfunction. Such relationships place mitochondria at the center of individual variation that is important to mating display honesty, and to individual quality itself.

In sum, a shared pathway rather than a cost-based framework provides a testable foundation for studying why many display traits reliably signal individual quality. If variation in the functionality of mitochondrial aerobic respiration drives variation in both mating display behavior and aspects of condition—such as immune system performance—then understanding that core process becomes key to understanding individual quality and honest signaling. Importantly, a shift away from a focus on costs does not mean that internal resource availability and environmental factors never play a role in driving individual variation in display behavior. For example, we might expect animals mounting an immune response or undergoing severe food restriction to have poorer expression of display behaviors relative to healthy counterparts because, regardless of potential for high performance, core

processes cannot function without substrate. However, we predict that the quality of that core process performance will affect the severity with which an individual is compromised by such a challenge, such as the likelihood that the animal becomes sick upon exposure, and effects on its functionality during and after infection. For example, *Drosophila* fruit flies with compromised mitochondrial aerobic respiration were found to decrease fecundity after a live bacterial immune challenge, whereas controls with normal mitochondria showed no such effect (56) (Figure 10.4). Though this study measured reproductive output rather than investment in a behavioral display, we might expect that the capacity of an individual's mitochondria may similarly shape its ability to mount an effective response during parasite challenge and still carry out an effective display. The functionality of core shared pathways, such as may involve the mitochondria, thereby becomes key to the long-term evolution of display traits that honestly signal underlying physiology.

## 10.4 The Hamilton–Zuk hypothesis and signals of genes for parasite resistance

Thus far, we have focused on the physiological mechanisms underlying variation in display trait expression, and the honesty of that display trait as a signal of individual quality. Considering the genetic underpinnings of such relationships and how the form and function of traits may be shaped by selection raises new questions. To date, among the most influential hypotheses for the evolution of ornamental traits is the Hamilton–Zuk hypothesis, which proposes that ornamental mating displays will be shaped by whether or not an individual

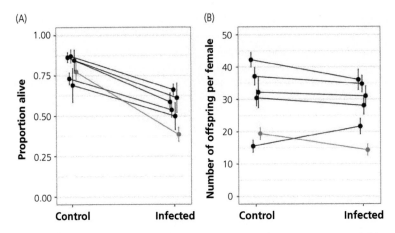

**Figure 10.4** In a 2018 study, Buchanan *et al.* (56) examined *Drosophila* fruit flies of six different distinct mitonuclear genetic combinations—one of which is known to have compromised mitochondrial oxidative phosphorylation due to genetic incompatibilities. Experimental groups of flies were injected with a natural fruit fly pathogen, the gram-negative bacterium *Providencia rettgeri*, while control groups received a sham injection containing no bacteria. Interestingly, female flies of the compromised mitonuclear genetic type (depicted in orange) had both significantly increased mortality over 10 days (A) and decreased fecundity over three days (B) when infected, compared to the uncompromised genetic types (black). These results demonstrate how a dysfunction at the level of mitochondrial respiration can affect not only immune defenses, but also ability to reproduce during immune activation—a demonstration of a mitochondrial by nuclear by environment (G × G × E) interaction. Adapted from (56).

carries specific alleles for resistance to the dominant genotype of a common pathogen (6). This hypothesis focuses on solving the conundrum that good genes for pathogen resistance are predicted to go to fixation rapidly in a stable environment, thus undercutting associations between good genes and ornamentation. This predicted depletion of good genes that will affect male quality is known as the Lek Paradox. The Hamilton–Zuk hypothesis proposed a solution to the Lek Paradox by invoking coevolutionary cycles between the resistance alleles in the host and virulence alleles in the pathogen (6). As alleles for host resistance to a most common virulence allele in the pathogen increase in a population, selection favors alternate virulence alleles in the pathogen. Selection then shifts to resistance to the rising virulence alleles, thereby increasing the frequency of alleles that protect from that pathogen genotype but leading to a decline in alleles for resistance to the original pathogen genotype. Such host–parasite coevolutionary cycles are proposed to generate perpetual genetic-based variation in pathogen resistance and honest signaling of good genes via ornamental mating displays.

Despite the importance of the Hamilton–Zuk hypothesis for stimulating research into signaling via display traits, this specific hypothesis has not been well supported by observations. Moreover, the Hamilton–Zuk hypothesis proposes an explanation for standing variation for alleles for resistance to pathogens within a population, but it does not explicitly explain why more parasite-resistant individuals would be able to produce better display traits than less resistant individuals. Nevertheless, if we take the Hamilton–Zuk hypothesis to be a statement of the need to consider the genetic bases of parasite resistance in hypotheses for mating display signaling, then this foundational hypothesis remains central to studies of display trait evolution (57). For further review of theory and empirical observations of the Hamilton–Zuk hypothesis as well as the coevolutionary processes it references, see Pirrie *et al.* (58) in this volume.

In the previous section, we describe the mitochondrial function hypothesis for honest signaling in mating display traits. By extension, the genes underlying mitochondrial phenotype become key to understanding the genetic bases of both display traits and the aspects of physiological quality that may be signaled, such as parasite resistance. This line of explanation, however, seems to lead to the same Lek Paradox that originally prompted the Hamilton–Zuk hypothesis: if mitochondria-associated genes are "good genes" for parasite resistance, then such genes should be driven to fixation, leaving little variation upon which sexual selection may act. However, mitochondrial function does not arise exclusively from nuclear genes where allelic variation might underlie fitness. Rather, the mitochondrial phenotype is shaped by the cofunction of products of both the mitochondrial genome and the nuclear genome (Box 10.2). While the Hamilton–Zuk hypothesis considers a form of gene by environment (G × E) interaction (i.e., parasite resistance genes interacting with the parasite environment), any pathways mediated by mitochondria will be shaped by gene by gene by environment (G × G × E) interactions due to epistasis between mitochondrials and nuclear genes ("mitonuclear" interactions; (59), (60)). Adding a mitonuclear perspective to the Hamilton–Zuk hypothesis leads to new a prediction: if resistance to a parasite is affected by the G × G × E interactions shaping mitochondrial phenotype, then the additive genetic variation (the good genes) underlying parasite resistance will not be exhausted because of the myriad possible G × G × E combinations; further, if expression of a mating display trait is also sensitive to mitochondrial phenotype, then the display trait may serve as a signal of parasite resistance through the shared mitochondrial pathway (48) (Figure 10.2). While these specific predictions have yet to be explored empirically, there is evidence that selection for functionality across different environments can create and maintain variation in genes associated with mitochondrial function; for example, *D. melanogaster* fruit flies collected in a latitudinal cline down Australia's eastern coast have been found to have two main mitochondrial genetic types that appear to confer higher functionality in either the subtropical northern climate or the temperate southern climate (61). Given that such different climates are likely also to harbor distinct parasites, an important next step is to explore how such variation in mitochondrial (and thereby mitonuclear) genotype influences parasite resistance.

> **Box 10.2  Mitonuclear co-evolution and coadaptation**
>
> Mitochondria are constructed from the products of genes encoded in two genomes, which presents challenges to conventional approaches to core evolutionary concepts such as how a system will respond to directional selection and the nature of epistatic interactions among genes (62). In bilaterian animals, mitochondria are encoded by 37 mitochondrial (mt) genes and more than 1,000 nuclear genes whose products function in the mitochondria (N-mt genes; (60)). mt and N-mt genes are inherited independently, and mt genomes of most bilaterian animal taxa do not engage in recombination, so mt genomes accumulate mutations rapidly (62). Despite their independent transmission across generations, mt and N-mt genes must be able to cofunction for core energy production (63). Thus, there must be perpetual coevolution to maintain coadaptation between mt and N-mt genes (62). This two-genome system complicates the manner in which mitochondrial function can respond to selection because it is not as simple as driving a nuclear allele to fixation—function is achieved via a co-adapted set of mt and N-mt genes (60). The end result is that in each generation there will be individuals with poorly matched mt and N-mt genes and hence with poor mitochondrial function (62), (63). Mitochondrial function can also change within the lifetime of an individual as mutations accrue in mitochondrial lineages. Consequently, there should arise means for choosing individuals to assess mitonuclear compatibility via mitochondrial function, and mating display traits seem to be such signals of mitochondrial function (48).

The importance of mitonuclear interactions to shaping mitochondrial (and consequently organismal) phenotype and driving broader evolutionary patterns has only relatively recently begun to be explored from an evolutionary and ecological perspective, but growing theoretical and empirical research has led to a new field of study called mitonuclear ecology (33), (62). Studies in field and laboratory systems alike have found that the same mitochondrial genes can have different effects on mitochondrial and/or organismal phenotype depending on the nuclear context in which they are expressed (63), (64). The phenotypes found to have been affected by mitonuclear genotypes are varied, ranging from fertility (65) to personality (66). Importantly, while no study has yet tested the mitonuclear extension of the Hamilton–Zuk hypothesis, mitonuclear effects have often been found to vary depending on environment (60), supporting the importance of G × G × E interactions. For example, the effects of mitonuclear genotype on longevity in *Drosophila* fruit flies vary by diet (59), and mitonuclear genetic variation in hybrid *Urosaurus* lizards altered how mitochondrial membrane potential reacted to high temperatures (67). Further, the effects of compromised mitochondrial respiration in a specific mitonuclear genotype of *Drosophila* fruit flies are exacerbated during bacterial challenge or altered nutrient availability (Figure 10.4) (56). Finally, mitonuclear genetic variation in hybrid populations of leaf beetles (*Chrysomela aenuicollis*) has been found to be linked to depressed fecundity, running speed, development rate, and male mating frequency, and these effects are more pronounced at more extreme environments (heat treatment or higher elevation) (68). This latter study is one of the few to date that has empirically linked mitonuclear effects to mating rate, if not mating display behavior itself.

Collectively, such studies suggest the potential for mitonuclear G × G × E interactions to shape individual performance along multiple axes, including response to parasites and mating success. Reframing the core concept of the Hamilton–Zuk hypothesis under a mitonuclear perspective offers a new solution to an old problem: G × G × E interactions can maintain genetic variation independently of the kinds of host–parasite cycles proposed by Hamilton and Zuk (6). Further, the connections between mitochondrial and immune performance as well as display trait expression (as described in the mitochondrial function hypothesis, earlier) provide an explanation for why such genetic parasite resistance may be signaled through mating displays.

## 10.5 Conclusions and future research

A consideration of honest signaling from the perspective of shared pathways rather than costs and resources shifts the focus toward discrete mechanisms and away from difficult-to-isolate trade-offs. Mitochondria provide one possible shared pathway to link aspects of physiological quality like immune system performance and oxidative stress to

the expression of mating display traits, and the role of mitochondria as a subcellular hub of signaling and function across many physiological pathways places them at the interface of parasite–behavior feedback itself (Box 10.3).

Technologies are increasingly becoming available to facilitate the quantification and interpretation of mitochondrial performance parameters (e.g., (33)) to begin to test the mitochondrial function hypothesis. Whereas tests of cost-based hypotheses often experimentally manipulate resource availability, shared pathway hypotheses may be better tested through assessing standing variation in the functionality of the shared pathway and the downstream processes proposed to rely on that pathway (e.g., mating display and immune system quality), or by comparing how individuals with different underlying pathway functionality may respond to a challenge. Targeted experimental manipulations that are designed to alter the shared pathway itself can also be used to simulate conditions of varying pathway functionality, but it can be difficult to interpret the results of such manipulations because such pathways, by definition, connect to a wide variety of processes. For example, an experimental "uncoupler" that decreases relative amounts of ATP generated during mitochondrial aerobic respiration can be dosed to animals (72); however, ATP availability can be maintained in mitochondria with increased uncoupling by increasing respiration rate, potentially leaving little detectable effect of treatment on traits of interest, though increasing respiration rate may in turn affect other parameters, like metabolic heat production (33). Similarly, mitochondrial-targeted antioxidants or pro-oxidants are available that can localize specifically to mitochondria to increase or decrease oxidative stress. Such manipulations may be useful means to test pathways; for example, dosing zebra finches (*Taeniopygia guttata*) with a mitochondrial-targeted antioxidant increased the expression of a display trait—red bill coloration—supporting the hypothesis that mitochondrial processes may be related to ornamental red pigmentation (46).

More generally, an important step to investigating a shared pathway hypothesis like the mitochondrial function hypothesis is to evaluate sources and

---

**Box 10.3  Parasites, behavior, and the mitochondrial shared pathway**

Mitochondrial function has been proposed to be a "shared pathway" not only between mating display traits and immune system performance but also among broader aspects of behavior, as may be expected given the role of mitochondrial processes in sensorimotor and neural development as well as function, and in synthesis of and response to hormones (26), (27). Given the bidirectional interactions between parasite exposure/infection and behavior, the roles of mitochondria in fueling and shaping behaviors may place them at the interface of parasite-behavior feedback (69).

Perhaps one of the most intriguing connections is the suggestion that mitonuclear genotype—and, therefore, mitochondrial phenotype—can influence animal activity levels, and even personality. For example, male fruit flies (*D. melanogaster*) were found to exhibit different levels of activity—a trait related to mate searching and success in this species—depending on mitochondrial genotype (70); and seed beetles (*Callosobruchus maculatus*) with different mitonuclear genotypes were consistently different in activity levels and patterns of behavior, collectively termed personality (66). When considered along with the role of mitochondria in the synthesis and function of testosterone—a hormone generally associated with increased activity, be it territory defense or mate searching—then mitochondria become key players in supporting and potentially even driving patterns of activity in animals.

These connections are particularly interesting because animal activity patterns, such as increased locomotory behavior, exploration, and mate searching, can influence parasite exposure. Yet, if high functioning mitochondria are necessary to support the energetic needs of an especially active individual and high-functioning mitochondria are also key to mounting strong immune defenses, then it is plausible that individuals with active behavioral phenotypes may also be more resistant to parasite infection. However, the seed beetle lines with more active personalities in the study described above were also found to have shorter lifespans and lower body mass, which suggests that their personality types were more associated with a "live fast, die young" strategy than an overall boost to performance (66). These relationships between mitochondria, behavior, and parasite exposure remain to be tested within an ecological context.

While mitochondria may mediate some interesting relationships between behavior and parasite exposure and/or resistance, any potential effects of parasites on behavior via the mitochondria are less clear. Interestingly, many viruses have been found to alter mitochondrial signaling during the innate immune response to prevent apoptosis of viral host cells (71); however, it remains to be tested whether such effects may in turn alter behavior.

---

levels of variation in that pathway's performance, particularly within ecologically relevant settings. Much of our understanding of the role of mitochondria in parasite defense, for instance, comes from biomedical studies that focus on high levels of mitochondrial dysfunction from genetic knockouts or invasive experimental manipulations. Presumably, individuals with severely compromised mitochondria would not survive long enough to be relevant to questions of mating display signaling. Significant natural variation in mitochondrial performance is expected, however, due to G × G E effects (described earlier) and damage that may accumulate over the lifespan of an individual (e.g., due to replication error) (73).

Whether or not the results of future research support the mitochondrial function hypothesis, shared pathway hypotheses based on physiological constraints offer an important alternative perspective

to cost-based hypotheses like the Zahavi handicap hypothesis, the ICHH, and the OHH by shifting focus from resources to core functionality. Such a perspective has the potential to bring together the wide and varied literature investigating the costs and trade-offs underlying variation in mating display behaviors by presenting a common mechanism with links to different research foci (e.g., oxidative stress ecology, eco-endocrinology, and eco-immunology).

## References

1. Higham JP. How does honest costly signaling work? *Behav Ecol.* 2014;25:8–11.
2. Andersson M. Sexual selection, natural selection and quality advertisement. *Biol J Linn Soc.* 1982;17:375–93.
3. Kodric-Brown A and Brown JH. Truth in advertising: The kinds of traits favored by sexual selection. *Am Nat.* 1984;124:309–23.

4. Johnstone RA. Sexual selection, honest advertisement and the handicap principle: Reviewing the evidence. *Biol.* 1995;70:1–65.

5. Hill GE. Sexiness, individual condition, and species identity: The information signaled by ornaments and assessed by choosing females. *Evol Biol.* 2015;42:251–9.

6. Hamilton W and Zuk M. Heritable true fitness and bright birds: A role for parasites? *Science.* 1982;218:384–7.

7. Andersson M. *Sexual Selection.* Princeton, NJ: Princeton University Press, 1994.

8. Cotton S, Small J, and Pomiankowski A. Sexual selection and condition-dependent mate preferences. *Curr Biol* 2006;16:R755–65.

9. Weaver RJ, Koch RE, and Hill GE. What maintains signal honesty in animal colour displays used in mate choice? *Philos Trans R Soc Lond B: Biol Sci.* 2017;372:20160343.

10. Koch RE and Hill GE. Do carotenoid-based ornaments entail resource trade-offs? An evaluation of theory and data. *Funct. Ecol.* 2018;32:1908–20.

11. Megía-Palma R, Barrientos R, Gallardo M, Martínez J, and Merino S. Brighter is darker: The Hamilton–Zuk hypothesis revisited in lizards. *Biol J Linn Soc.* 2021;134(2):461–73.

12. Koch RE, Staley M, Kavazis AN, Hasselquist D, Toomey MB, and Hill GE. Testing the resource trade-off hypothesis for carotenoid-based signal honesty using genetic variants of the domestic canary. *J Exp Biol.* 2019;222(6):jeb188102.

13. Zahavi A. Mate selection—A selection for a handicap. *J Theor Biol.* 1975;53:205–14.

14. Zahavi A. The cost of honesty. *J Theor Biol.* 1977;67:603–5.

15. Folstad I and Karter AJ. Parasites, bright males, and the immunocompetence handicap. *Am Nat.* 1992;139:603–22.

16. Ezenwa VO, Stefan Ekernas L, and Creel S. Unravelling complex associations between testosterone and parasite infection in the wild. *Funct Ecol.* 2012;26:123–33.

17. Newhouse DJ and Vernasco BJ. Developing a transcriptomic framework for testing testosterone-mediated handicap hypotheses. *Gen Comp Endocrinol.* 2020;298:113577.

18. Alonso-Alvarez C, Bertrand S, Faivre B, Chastel O, and Sorci G. Testosterone and oxidative stress: The oxidation handicap hypothesis. *Proc R Soc B: Biol Sci.* 2007;274:819–25.

19. von Schantz T, Bensch S, Grahn M, Hasselquist D, and Wittzell H. Good genes, oxidative stress and condition–dependent sexual signals. *Proc R Soc B: Biol Sci.* 1999;266:1–12.

20. Biernaskie JM, Grafen A, and Perry JC. The evolution of index signals to avoid the cost of dishonesty. *Proc R Soc B: Biol Sci.* 2014;281:20140876.

21. Maynard-Smith J and Harper DGC. Animal signals: Models and terminology. *J Theor Biol.* 1995;177:305–11.

22. Hill GE. Condition-dependent traits as signals of the functionality of vital cellular processes. *Ecol Lett,* 2011;14:625–34.

23. Hill GE. Cellular respiration: The nexus of stress, condition, and ornamentation. *Integr Comp Biol.* 2014;54:645–57.

24. Ducrest A, Keller L, and Roulin A. Pleiotropy in the melanocortin system, coloration and behavioural syndromes. *Trends Ecol Evol.* 2008;23:502–10.

25. Hill GE and Johnson JD. The Vitamin A–Redox Hypothesis: A biochemical basis for honest signaling via carotenoid pigmentation. *Am Nat.* 2012;180:E127–50.

26. Koch RE, Josefson CC, and Hill GE. Mitochondrial function, ornamentation, and immunocompetence. *Biol.* 2017;92:1459–74.

27. Koch RE and Hill GE. Behavioural mating displays depend on mitochondrial function: A potential mechanism for linking behaviour to individual condition. *Biol.* 2018;93:1387–98.

28. Breda CN de S, Davanzo GG, Basso PJ, Câmara NOS, and Moraes-Vieira PMM. Mitochondria as central hub of the immune system. *Redox Biology* 2019;26:101255.

29. Costantini D, Rowe M, Butler MW, and McGraw KJ. From molecules to living systems: Historical and contemporary issues in oxidative stress and antioxidant ecology. *Funct Ecol.* 2010;24:950–9.

30. West AP, Shadel GS, Ghosh S. Mitochondria in innate immune responses. *Nature Reviews Immunology* 2011;11(6):389–402.

31. Koshiba T. Mitochondrial-mediated antiviral immunity. *Biochim Biophys Acta Mol Cell Res.* 2013;1833:225–32.

32. West AP, Brodsky IE, Rahner C, Woo DK, Erdjument-Bromage H, Tempst P, *et al.* TLR signalling augments macrophage bactericidal activity through mitochondrial ROS. *Nature.* 2011;472(7344):476–80.

33. Koch RE, Buchanan K, Casagrande S, Crino O, Dowling DK, Hill GE, *et al.* Integrating mitochondrial aerobic metabolism into ecology and evolution. *Trends Ecol Evol.* 2021;11(14):9791–803.

34. Le Roy A, Mazué GPF, Metcalfe NB, and Seebacher F. Diet and temperature modify the relationship between energy use and ATP production to influence behavior in zebrafish (*Danio rerio*). *Ecol Evol.* 2021. DOI: 10.1002/ece3.7806.

35. Adret P, Margoliash D. Metabolic and neural activity in the song system nucleus robustus archistriatalis: Effect of age and gender. *J Comp Neurol.* 2002;454: 409–23.

36. Svensson PA, Wong BM. Carotenoid-based signals in behavioural ecology: A review. *Behaviour* 2011;148:131–89.

37. Hill GE. Female mate choice for ornamental coloration in Hill GE, McGraw KJ (eds). *Bird Coloration Volume II. Function and Evolution.* Cambridge, MA: Harvard University Press, 2006, 137–200.

38. Mateos-Gonzalez F, Hill G, and Hood W. Carotenoid coloration predicts escape performance in the House Finch (*Haemorhous mexicanus*). *Auk* 2014;131(3):275–81.

39. Thompson CW, Hillgarth N, Leu M, and McClure HE. High parasite load in House Finches (*Carpodacus mexicanus*) is correlated with reduced expression of a sexually selected trait. *Am Nat.* 1997;149:170–94.

40. Simons MJP, Cohen AA, and Verhulst S. What does carotenoid-dependent coloration tell? Plasma carotenoid level signals immunocompetence and oxidative stress state in birds-a meta-analysis. *PLoS One* 2012;7:e43088.

41. Weaver RJ, Santos ESA, Tucker AM, Wilson AE, and Hill GE. Carotenoid metabolism strengthens the link between feather coloration and individual quality. *Nature Commun.* 2018;9:73.

42. Lopes RJ, Johnson JD, Toomey MB, Ferreira MS, Araujo PM, Melo-Ferreira J, *et al.* Genetic basis for red coloration in birds. *Current Biol.* 2016;26:1427–34.

43. Mundy NI, Stapley J, Bennison C, Tucker R, Twyman H, Kim KW, *et al.* Red carotenoid coloration in the Zebra Finch is controlled by a cytochrome P450 gene cluster. *Current Biol.* 2016;26:1435–40.

44. Ge Z, Johnson JD, Cobine PA, McGraw KJ, Garcia R, and Hill GE. High concentrations of ketocarotenoids in hepatic mitochondria of *Haemorhous mexicanus. Physiol Biochem Zool.* 2015;88:444–50.

45. Hill GE, Hood WR, Ge Z, Grinter R, Greening C, Johnson JD, *et al.* Plumage redness signals mitochondrial function in the house finch. *Proc R Soc B: Biol Sci.* 2019;286(1911):20191354.

46. Cantarero A and Alonso-Alvarez C. Mitochondria-targeted molecules determine the redness of the zebra finch bill. *Biol Lett.* 2017;13:20170455.

47. Cantarero A, Mateo R, Camarero PR, Alonso D, Fernandez-Eslava B, and Alonso-Alvarez C. Testing the shared-pathway hypothesis in the carotenoid-based coloration of red crossbills. *Evol.* 2020;74: 2348–64.

48. Hill GE and Johnson JD. The mitonuclear compatibility hypothesis of sexual selection. *Proc R Soc B: Biol Sci.* 2013;280:20131314.

49. Hill GE. Mitonuclear mate choice: A missing component of sexual selection theory? *BioEssays* 2018;40:1700191.

50. Miller WL. Steroid hormone synthesis in mitochondria. *Mol Cell Endocrinol.* 2013;379:62–73.

51. Baldo S, Mennill DJ, Guindre-Parker S, Gilchrist HG, and Love OP.The oxidative cost of acoustic signals: Examining steroid versus aerobic activity hypotheses in a wild bird. *Ethology* 2015;121: 1081–90.

52. Speakman JR and Garratt M. Oxidative stress as a cost of reproduction: Beyond the simplistic trade-off model. *BioEssays* 2014;36:93–106.

53. Salin K, Auer SK, Rudolf AM, Anderson GJ, Cairns AG, Mullen W, *et al.* Individuals with higher metabolic rates have lower levels of reactive oxygen species *in vivo. Biol Lett.* 2015;11:2015–18.

54. Hood WR, Zhang Y, Mowry AV, Hyatt HW, and Kavazis AN. Life history trade-offs within the context of mitochondrial hormesis. *Integr Comp Biol.* 2018;58:567–77.

55. Zhang Y and Hood WR. Current versus future reproduction and longevity: A re-evaluation of predictions and mechanisms. *J Exp Biol.* 2016;219:3177–89.

56. Buchanan JL, Meiklejohn CD, and Montooth KL. Mitochondrial dysfunction and infection generate immunity–fecundity tradeoffs in *Drosophila. Integr Comp Biol.* 2018;58:591–603.

57. Balenger SL and Zuk M. Testing the Hamilton–Zuk hypothesis: Past, present, and future. *Integr Comp Biol.* 2014;54:601–13.

58. Pirrie A, Chapman H, and Ashby B. Parasite-mediated sexual selection: To mate or not to mate? in Ezenwa VO, Altizer S, and Hall RJ (eds), *Animal Behavior and Parasitism.* Oxford: Oxford University Press, 2022. DOI: 10.1093/oso/9780192895561.003.0008.

59. Zhu C-T, Ingelmo P, and Rand DM. G×G×E for lifespan in *Drosophila*: Mitochondrial, nuclear, and dietary Interactions that modify longevity. *PLoS Genetics* 2014;10:e1004354.

60. Rand DM and Mossman JA. Mitonuclear conflict and cooperation govern the integration of genotypes, phenotypes and environments. *Philos Trans R Soc Lond B: Biol Sci.* 2020;375(1790):20190188.

61. Camus MF, Wolff JN, Sgrò CM, and Dowling DK. Experimental support that natural selection has shaped the latitudinal distribution of mitochondrial haplotypes in Australian *Drosophila melanogaster. Mol Biol Evol.* 2017;34:2600–12.

62. Hill GE. *Mitonuclear Ecology.* Oxford: Oxford University Press, 2019.

63. Sunnucks P, Morales HE, Lamb AM, Pavlova A, and Greening C. Integrative approaches for studying

mitochondrial and nuclear genome co-evolution in oxidative phosphorylation. *Front Genet*. 2017;8: 1–12.

64. Barreto FS, Watson ET, Lima TG, Willett CS, Edmands S, Li W. *et al*. Genomic signatures of mitonuclear coevolution across populations of *Tigriopus californicus*. *Nature Ecol Evol*. 2018;2: 1250–7.

65. Wolff JN, Tompkins DM, Gemmell NJ, and Dowling DK. Mitonuclear interactions, mtDNA-mediated thermal plasticity, and implications for the Trojan Female Technique for pest control. *Sci Rep*. 2016;6(1):1–7.

66. Løvlie H, Immonen E, Gustavsson E, Kazancioğlu E, and Arnqvist G. The influence of mitonuclear genetic variation on personality in seed beetles. *Proc R Soc Biol B: Biol Sci*. 2014;281:20141039.

67. Haenel GJ and Del Gaizo Moore V. Functional divergence of mitochondria and coevolution of genomes: Cool mitochondria in hot lizards. *Physiol Biochem Zool*. 2018;91:1068–81.

68. Rank NE, Mardulyn P, Heidl SJ, Roberts KT, Zavala NA, Smiley JT, *et al*. Mitonuclear mismatch alters performance and reproductive success in naturally introgressed populations of a montane leaf beetle. *Evol*. 2020;74:1724–40.

69. Ezenwa VO, Archie EA, Craft ME, Hawley DM, Martin LB, Moore J, *et al*. Host behaviour-parasite feedback: An essential link between animal behaviour and disease ecology. *Proc R Soc B: Biol Sci*. 2016;283(1828):20153078.

70. Dean R, Lemos B, and Dowling DK. Context-dependent effects of Y chromosome and mitochondrial haplotype on male locomotive activity in *Drosophila melanogaster*. *J Evol Biol*. 2015;28:1861–71.

71. Neumann S, El Maadidi S, Faletti L, Haun F, Labib S, Schejtman A, *et al*. How do viruses control mitochondria-mediated apoptosis? *Virus Res*. 2015;209:45–55.

72. Caldeira Da Silva CC, Cerqueira FM, Barbosa LF, Medeiros MH, and Kowaltowski AJ. Mild mitochondrial uncoupling in mice affects energy metabolism, redox balance and longevity. *Aging Cell* 2008;7: 552–60.

73. Hood WR, Williams AS, and Hill GE. An ecologist's guide to mitochondrial DNA mutations and senescence. *Integr Comp Biol*. 2019;59:970–82.

74. Hill GE. *A Red Bird in a Brown Bag: The Function and Evolution of Colorful Plumage in the House Finch*. Oxford: Oxford University Press, 2002.

# CHAPTER 11

# The genes of attraction: Mating behavior, immunogenetic variation, and parasite resistance

Jamie C. Winternitz and Jessica L. Abbate

## 11.1 Introduction

Mating preferences are ubiquitous in the animal kingdom and were remarked on by Charles Darwin over 150 years ago (1). Darwin observed that females appear to favor certain males with elaborate showy traits and competitive ability. If these male traits signal high quality and the ability to provide material benefits to the female, such as protection, resources, or paternal care, then it is fairly obvious why females are choosy: they receive direct fitness benefits from their choices. However, there are many species where certain males are preferred by females but provide only sperm (indirect "genetic" benefits). Some of the most obvious cases of mate choice for genetic benefits are those of lekking species, spanning birds, mammals, fish, frogs, and insects, where males congregate in leks to display to discriminating females (2), (3). In these cases, female preferences are harder to explain. One challenge posed by indirect genetic benefits models of female choice is explaining how genetic variation for the male traits females prefer is maintained. This problem, known as the "Lek Paradox," considers why females remain choosy for male traits when their choice erodes genetic variation in those traits, dissolving the incentive to be choosy. To break this down, if all choosy females select the same high-quality males, and quality is genetically determined, then eventually higher-quality males with "good genes" will reproduce more and lower-quality genes will be purged from the population. Over time, there will be no genetic variation among males, preference genes in females will no longer be beneficial, and random mating will eventually dominate in the population. Thus, how can genetic variation in mate quality persist in a population with sexual selection for "good genes"? An intriguing solution proposed by Hamilton and Zuk's "good genes hypothesis" (4) is that genetic variation in male condition is maintained by continuously evolving parasites that change the relative fitness advantage conferred by pathogen-specific resistance alleles. In effect, it is the arms race between hosts and parasites that drives the continuous evolution of parasites to evade immune detection. In turn, the arms race selects for the evolution of novel host immune genes that parasites have not yet encountered, maintaining variation at immune loci and thus male quality.

Why parasites? Parasites, broadly defined to include macroparasites (i.e., helminths, arthropods, protozoa) and microbial pathogens (i.e., viruses, bacteria, and fungi), are ubiquitous in natural populations (5), (6), including humans (7). While some parasites can exert strong selection on survival—exemplified by prominent deadly infectious disease epidemics in humans and wildlife (8–11)—many natural populations are more commonly infected by parasites that exert selection on fecundity (12), which is also essential to an individual's fitness. Parasites are of course not the only selective force

Jamie C. Winternitz and Jessica L. Abbate, *The genes of attraction*. In: *Animal Behavior and Parasitism*. Edited by Vanessa O. Ezenwa, Sonia Altizer, and Richard J. Hall, Oxford University Press. © Oxford University Press (2022). DOI: 10.1093/oso/9780192895561.003.0011 affecting survival or fecundity, but other models of mate choice for genetic benefits besides the model of Hamilton and Zuk (4) have also required fluctuating environments or external perturbations in gene frequencies to maintain genetic variation at condition-determining genes (13)–(15). While many factors in a host's environment change over time, little does so more rapidly, widely, and consistently than the genetic identity of its parasites (16).

Immune genes, which code for a variety of host defenses against pathogens, are excellent candidates for genes underpinning sexual selection—both affecting condition and frequently changing—for several reasons. First, they include some of the most polymorphic genes in the genomes of vertebrates and invertebrates (17), suggesting their role in mate choice could be general across animals. For vertebrates, the major histocompatibility complex (MHC), responsible for recognizing foreign molecules and presenting them to stimulate important arms of the adaptive immune system, stands out for its unparalleled diversity. For example, 28,938 alleles at the human leukocyte antigen (HLA) system, the MHC in humans, have been described (as of February 9, 2021 <https://www.ebi.ac.uk/ipd/imgt/hla/stats.html>, (18)). Second, diversity at many immune genes is thought to be maintained by pathogen-mediated balancing selection (17), (19), (20), and other forms of pathogen-mediated selection (PMS), such as positive (diversifying) selection, have also been observed in immune genes conserved across birds, mammals, and other vertebrates (21), (22). Third, condition often depends on infection status, providing an honest signal for heritable parasite resistance alleles that increase offspring fitness (23).

The three attributes of immune genes mentioned above—high polymorphism, maintenance by pathogen-mediated selection, and determining host infection status and therefore condition—allow immune genes to meet the assumptions of "good genes" genetic benefit models (24). What are these models? In the sexual selection literature, "good genes" models are genetic benefit models that assume preferred genes are paternally heritable (conferring additive effects on offspring genotype) and increase offspring viability. In this case, alleles that protect against parasitism are passed from father to offspring. In contrast, genes that do not increase offspring viability but increase their attractiveness are consistent with Fisherian models (24). In addition, "compatible genes" genetic benefit models refer to non-additive genetic variation where the fitness of offspring depends on both maternal and paternal genotype. One example is where two different alleles from each the mother and father give rise to heterozygous offspring with greater parasite resistance. Neff and Pitcher (25) recommend that the term "good gene" be used exclusively to refer to additive genetic variation in fitness, "compatible gene" be used to refer to nonadditive genetic variation in fitness, and "genetic quality" be defined as the sum of the two effects. Immune genes are thus considered important candidate loci for genetic benefit models of mate choice because their variation is maintained by antagonistic coevolution with parasites, and pathogen-driven mechanisms to diversify immune genes should favor the evolution of mating preferences for condition-dependent traits.

In this chapter, we review the hypotheses for immune gene-based mating preferences and the theoretical prerequisites for models of female preference, including good genes, heterozygous males, compatible/dissimilar males, and similar males. We then discuss the mechanisms by which mate choice may act on immunogenetic variation, focusing on pre-mating and post-mating selection. Next, we perform a systematic literature search to identify and summarize examples of sexual selection acting on immune genes from across both vertebrate and invertebrate systems. The purpose of our systematic review is to improve understanding of the taxonomic diversity and types of immune genes for which relationships between mate choice and immunogenetic diversity have been investigated. Finally, we identify some of the important contextual factors that can complicate the discovery of sexually selected immune genes and lay out the challenges to testing the generality of this phenomenon. Identifying if and how immune genes are involved in mate choice is important conceptually because it addresses the problem of maintaining genetic variation in sexually selected traits. Practically, it is important for conservation because sexual selection is believed to increase population viability by promoting "good genes" and filtering out undesirable phenotypes from the population. By studying sexual selection on immune genes, we can test how different models of female preference meet this prediction, and thus provide more informed strategies for conservation management.

## 11.2 Hypotheses for immune gene-based mating preferences

There are three main adaptive hypotheses for immune gene-based mate choice (Figure 11.1): Mate choice to (i) produce offspring with specific resistant alleles (Figure 11.1A), (ii) produce heterozygous offspring that have a greater chance of responding to a wide range of parasites (Figure 11.1B), and (iii) avoid inbreeding (Figure 11.1C), would all serve to reduce levels of infectious or genetic disease in offspring. These three hypotheses fit evolutionary models of female mate choice with different assumptions and consequences for genetic diversity (Table 11.1) and therefore merit elaboration.

(1) **To produce offspring with specific resistance alleles**, females must prefer males that can pass down those alleles (Figure 11.1A). This hypothesis fits under the good genes models, including a

focused preference for males that are homozygous for the resistance allele (ensuring its presence in offspring) and a broad preference for males that may be homozygous or heterozygous for the resistance allele (assuming dominance). In these cases, fitness is additive (the effect of the allele is independent of the genomic context), and female preferences are congruent, so selection is directional (25). Directional selection will reduce genetic diversity at these loci. However, assuming mutation preserves some variation and heterozygotes do not have superior condition, female choice can evolve. When female preference is broad (AA or Aa), female choice can evolve regardless of heterozygote relative condition, but only when choice is cost-free (15).

Another suitable model for producing offspring with specific resistance alleles is female preference for similar alleles (assortative mating model). In this case, female choice for similar immune genotypes can evolve when homozygotes have superior pathogen resistance relative to heterozygotes (e.g., in matching alleles (MA) infection models where parasites must be capable of attaching to a host protein in order to infect, and homozygous

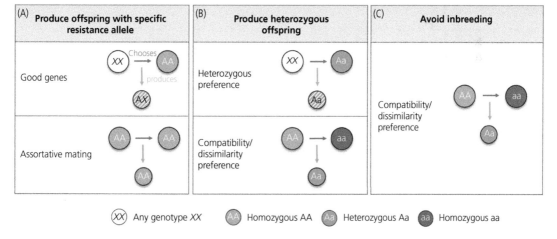

**Figure 11.1** Immune gene-based mate choice hypotheses and potential indirect benefits models that support them. Blue boxes represent the main hypotheses put forth to explain potential genetic benefits of mate choice that would also apply to immune gene-based mate choice. Within the hypotheses are preference models that would produce the desired output. (A) Under the hypothesis that the adaptive genetic benefit comes from producing offspring with specific resistance alleles, compatible preference models include good genes and assortative mating models. (B) Under the hypothesis that the adaptive genetic benefit comes from producing heterozygous offspring, compatible models include heterozygous preference and compatible/dissimilar preference models. (C) Under the hypothesis that the adaptive genetic benefit comes from avoiding inbreeding, compatible models include compatible/dissimilar preference models. Larger circle pairs represent parental genotypes and smaller circles beneath them represent desired offspring genotype. Colors indicate individual genotypes. Striped circles indicate a mix of genotypes possible.

**Table 11.1** Summary of genetic quality models and main predictions

| Female preference | Female preference hypothesis | Heritable (fitness type) | Female preference; sexual selection type | Diversity outcome at immune genes | Model prediction for evolutionary potential | Model references |
|---|---|---|---|---|---|---|
| Good genes: focused preference (AA) | Offspring with specific resistance alleles | Yes (additive) | Congruent; directional | Reduced | Female choice can invade and persist in a randomly mating population when the heterozygote condition is inferior or equal to the homozygote's | (15) |
| Good genes: broad preference (AA or Aa) | Offspring with specific resistance alleles | Yes (additive) | Congruent; directional | Maintained | Female choice can invade and persist in a randomly mating population regardless of heterozygote relative condition but only when choice is cost-free | (15) |
| Heterozygous males | Heterozygous offspring | It depends* (additive when rare alleles are dominant) | Congruent; balancing | Maintained | Female choice can invade and persist in a randomly mating population if heterozygote condition is superior to either homozygote, and the relative frequencies of the homozygotes change (e.g., due to rare-allele advantage against parasites or to biased mutation rates) | (14), (15) *(31) |
| Compatible males (disassortative mating) | Heterozygous offspring; avoid inbreeding | No (non-additive) | Incongruent; balancing | Maintained | Female choice can invade and reach equilibrium with random mating (determined by the balance between sexual and natural selection) when the heterozygote condition is superior to either homozygote (overdominance). Populations with higher number of immune gene alleles should exhibit stronger levels of disassortative mating. | (15), (26), (27), (32), (33) |
| Similar males (assortative mating) | Offspring with specific resistance alleles | No (non-additive) | Incongruent; directional | Reduced | Female choice can invade and persist in a randomly mating population when homozygotes are favored by parasite-mediated selection (e.g., in MA models where parasites must mimic host "self" proteins to infect them). | (15), (26), (27) |

hosts with less diverse proteins present a smaller range of potential targets (26), (27)). Under the assortative mating model, female mate preferences are incongruent, and fitness is non-additive because it depends on the identities of the mother and father. However, selection on immune genes will be directional and polymorphism reduced because the frequency of homozygotes with superior resistance alleles will increase in the population. This could lead to faster adaptation against virulent and prevalent pathogens but could also lead to lower immune genotype diversity and increased susceptibility to novel pathogens (28). Thus, assortative mating preferences in populations are expected to be limited by negative feedback between lower immune gene diversity resulting from preferences based on current conditions and selection imposed by the emergence of new pathogens that can escape widespread immunity (Box 11.1).

(2) **To produce heterozygous offspring**, females can prefer either heterozygous males or compatible males (mate disassortatively) (Figure 11.1B). Heterozygosity can increase the spectra of parasites to which the immune system is capable of responding if each allele confers resistance only to a discreet subset of parasite genotypes (e.g., via antigen recognition). Under negative frequency-dependent parasite selection, where parasites adapt resistance to common alleles, heterozygotes may have higher condition as well if they are more likely to have rare alleles that confer higher resistance to a coevolving parasite (29). For both heterozygous and disassortative female preference models, female preferences will promote balancing selection and thus preserve genetic diversity at immune genes.

Under the preference for heterozygote males model, females have congruent preferences because males that are heterozygous at immune loci will have the highest condition (13). While heterozygosity is not heritable *per se*, individual genetic diversity is correlated between parents and their offspring (30). This generally occurs because individuals with rare alleles are more likely to be heterozygous, and because they pass rare alleles to their offspring, they also tend to have heterozygous offspring (31). If rare alleles are dominant, then heterozygous parents and offspring will resemble each other at immune phenotypes, leading to additive genetic variance. For female choice to evolve under

this model, the heterozygote condition must be superior to either homozygote (overdominance) and the relative frequencies of the homozygotes must fluctuate (e.g., due to rare-allele advantage against parasites or to biased mutation rates) (14), (15).

Under the preference for compatible males model, females prefer dissimilar male immune genotypes and thus have mate preferences that are specific to individual females. This results in non-additive fitness benefits because offspring condition depends on the interaction between male and female genotypes. Disassortative mating will lead to balancing selection and maintain immune gene diversity in the population. Similar to the model for heterozygote preference above, female choice can evolve only when heterozygote condition is superior to either homozygote (32), (33). Interestingly, under these assumptions disassortative mate preference cannot become fixed in the population but will reach a stable polymorphism with random mating determined by the balance between parasite-mediated selection and sexual selection (15), (26), (27), (33). This is because of negative feedback between parasite-mediated selection and mate preferences for dissimilar mates. Essentially, parasite-mediated selection favors the immune variation of heterozygotes, and therefore also the disassortative mating preferences that produced them. However, as the frequency of heterozygotes rises in the population, the frequency of preferred dissimilar genotypes declines, leading to a limit on the advantage to females with adaptive mating preferences. Disassortative mating can evolve to higher frequency in populations with higher allelic diversity because this increases the chance that heterozygote males are dissimilar from choosy heterozygote females (27).

Specific disassortative mating rules are important to consider as they can lead to strikingly different outcomes in terms of the evolution of number of alleles per individual (via loci copy number). For example, computer simulations of MHC-based mate choice found that if preferences for avoiding identical alleles are strong, selection favors a smaller number of MHC alleles per individual than would be expected under random mating, even if this number decreases resistance to pathogens. This is because the more alleles an individual carries, the

---

**Box 11.1 The role of evolutionary feedbacks between parasites and mating behavior in immune gene-based mate choice models**

In hypotheses for the evolution of immune gene-based mate choice (Figure 11.1; to produce offspring with specific resistant alleles and to produce heterozygous offspring), feedbacks with parasites are expected, but rarely investigated. This is likely to be because most behavioral researchers focus their attention on host behavior and genetics, while identifying parasites mediating sexual selection over many generations would require enormous effort. Mathematical modeling can be useful to anticipate the reciprocal evolutionary effects of host behavior on pathogen populations, and vice versa, even when empirical evidence is lacking. Theoretical studies have shown that evolutionary feedbacks may function as follows for mating preferences that have evolved to:

**(1) Produce offspring with specific resistance alleles:** If resistant females prefer males with the same resistance alleles (assortative mating model) this could lead to faster adaptation against virulent and prevalent pathogens, but also to lower immune gene diversity, increasing population susceptibility to novel pathogens (28). Thus, negative

feedback between sexual selection that increases the commonness of specific resistance alleles and natural selection imposed by novel parasites able to escape widespread immunity is expected to limit the evolution of assortative mate choice.

**(2) Produce heterozygous offspring:** Disassortative mating preferences to produce heterozygous offspring are also expected to be limited by negative feedback between parasite-mediated selection and sexual selection. Essentially, parasite-mediated selection favors more resistant heterozygotes and the disassortative mating preferences that produced them. As these choosy heterozygotes increase in frequency in the population, the frequency of dissimilar genotypes with less diverse immunity declines. Thus, a very high frequency of heterozygotes in the population limits the advantage conferred to females with dissimilar mating preferences (27).

To our knowledge, parasite–behavior feedbacks have not yet been considered in relation to the third hypothesis, the evolution of mate choice for inbreeding avoidance.

---

higher the chance that one of those alleles will be shared by a potential mate. In contrast, if the mating rule simply minimizes the proportion of shared alleles in partners, selection favors higher MHC allele number per individual than under random mating (34). This is because a greater number of alleles per individual increases the likelihood that fewer alleles are shared between potential mates, selecting for ever higher numbers of loci.

(3) **To avoid inbreeding**, females can prefer compatible/dissimilar males (Figure 11.1C). The assumptions and predictions for this model have been discussed earlier in this chapter. The ultimate function of immune genes is to discriminate between self and non-self. If immune genotype could encode genetic identity in mating cues, as demonstrated by studies on odors influenced by the MHC (reviewed in (35)) then highly polymorphic immune genes could signal genetic relatedness and facilitate kin discrimination (reviewed in (36)).

In general, with all three hypotheses, it is important to note that models of female choice evolution that provide additive fitness benefits are also

consistent with providing direct fitness benefits. In this case, congruent female mate preferences for high-condition "good genes" and "good-genes-as-heterozygosity" (13) males may secure direct benefits from these males in the form of additional resources, protection, parental care, or avoidance of infectious diseases.

**Take-away messages from models (Table 11.1):**

(1) Non-additive (compatibility) models require some form of over-dominant selection, where heterozygote fitness is superior to that of either homozygote. This could occur if different immune alleles confer resistance to different pathogens, resulting in higher fitness for heterozygotes in environments where multiple infections are common and virulent (37).

(2) Mating preferences for the same male (congruent preferences) do not necessarily result in directional selection and reduced immune genetic diversity. A broad preference for "good immune genes" (AA or Aa) and a preference

for heterozygotes will lead to maintenance of diversity at immune genes.

Outcomes for the evolution of mate choice depend on the underlying genetic model of infection and resistance (38), in other words, *how immune genes recognize parasites matters* and can predict whether mating will be assortative, disassortative, or random (26), (27). Identifying molecular hotspots of host–parasite coevolution, for instance using pathogen-challenge transcriptome data (21), will help to improve parameterization of coevolutionary models of pathogen-driven sexual selection. The mate preference models described earlier are not mutually exclusive, particularly given the context of multiple selective pressures (see section 11.5), and evidence of concurrent mechanisms in a single system has been found (e.g., (39), (40)). In section 11.3, we explore such mechanisms underlying female mate choice linked to immune gene variation.

## 11.3 Mechanisms of mate choice on immunogenetic variation

The mechanisms driving female mate choice can act both before and after mating. Establishing the genetic link between these mechanisms and variation in immunity is key to finding systems in which the models listed earlier (good genes, assortative mating, heterozygous preference, and compatibility/dissimilarity preference) can be tested. Here, we give some examples of known mechanisms by which mate choice can select for immunogenetic variation.

### 11.3.1 Pre-mating selection

The most well-studied pre-mating cues for which immunogenetic determination has been identified are those with visual and olfactory links to the MHC. While both specific MHC alleles and levels of MHC diversity have been associated with preferred traits that signal parasite resistance (reviewed in (41)), signals of MHC composition also exist that are (or have become) relatively independent of condition, such as human body odor. Box 11.2 describes the results of a systematic review and meta-analysis on some of these cues in humans and

non-human primates (42). MHC-linked pre-mating cues are also abundant in a range of other animals (e.g., spur length in pheasants (43) and nest quality in sticklebacks (44)). Mate preferences associated with immune functions beyond the MHC have also been identified (such as interleukin 4 (*IL4*) involved in humoral immunity and reproductive success in lemurs (45)), though the genes coding for those functions are not always explored (but see section 11.5 and Table 11.2). While establishing a link between immune function and sexual selection is a necessary step in identifying genes under parasite-mediated sexual selection, the influence of sexual selection on immunogenetic variation cannot be understood without studying the evolution of the underlying genes. This is generally true for any trait, but it is particularly true for traits under multiple selective pressures (here, parasite-mediated selection (PMS) and parasite-mediated sexual selection (PMSS)) and, for example, for innate immune functions where the genotype cannot typically be inferred by measuring expression of the trait. Furthermore, evidence supporting selection for immunogenetic similarity or dissimilarity is obscured when the female's genotype is unknown.

### 11.3.2 Post-mating selection

Post-mating mechanisms of female choice linked to immune genes have most often been identified as acting through three cryptic processes: sperm selection, oocyte fertilization, and selective abortion (46). As with pre-mating selection for mates, the involvement of the immune system in these processes is related to self/non-self recognition functions, but in this case female immunity targets the removal of gametes or zygotes of disadvantageous males. Wong *et al.* (47) found signals of selection among genes involved in the breakdown of proteins, some of which contribute to defense against pathogens, expressed in the male and female reproductive organs of *Drosophila* spp. These signals fluctuated over time in asynchronous patterns that are consistent with the hypothesis that they are under shifting PMSS pressures. In passerine birds, positive selection was also detected for genes with known immune function that are expressed in seminal fluids, suggesting a potential role in creating

**Table 11.2** Summary of studies investigating candidate immune genes that do not include the MHC

| Taxa | Nb. immune genes | Animal species | Mate choice/sexual selection question | Study |
|------|------------------|----------------|----------------------------------------|-------|
| Insect | 5 | budworm moth (*Heliothis virescens*) | Do sexual attractiveness and immunity compete for the same resource pool and thus negatively affect each other in terms of gene expression and sexual signaling? | (84) |
| | 12 | fruit fly (*Drosophila melanogaster*) | Does intensity of postmating up-regulation of immune transcripts depend on an interaction between female and male genotype? | (59) |
| | 12 | red flour beetle (*Tribolium castaneum*) | Is paternal immune priming information transferred to offspring via sperm or seminal fluid? | (76) |
| | 10 | fruit fly (*Drosophila melanogaster*) | Are immune genes differentially expressed in females stimulated with song typical of either conspecific or heterospecific males? | (60) |
| | 14 | fruit fly (*Drosophila melanogaster*) | Does sexual selection on female seminal storage receptacle genes lead to rapid evolution of defense/immune genes? | (62) |
| | 4 | tramp ant (*Cardiocondyla obscurior*) | Does sympatric and allopatric mating differently affect ant queen fitness, and if so, are physiological mechanisms affected by immune gene expression? | (61) |
| | 5 | fruit fly (*Drosophila melanogaster*) | Are seminal fluid proteases targets of selection due to their roles in between-sex interactions and immune processes? | (63) |
| | 7 | fruit fly (*Drosophila melanogaster*) | What are the patterns of evolution for reproductive proteins with roles in immune defense in Drosophila? | (47) |
| Fish | 18 | Seven cichlid fishes | Does higher maternal investment increase sexual dimorphism in immune parameters? | (74) |
| | 14 | broad-nosed pipefish (*Syngnathus typhle*) | Does the strength of offspring immune defense depend only on paternal or maternal effects, or their interaction, in sex-role-reversed species? | (77) |
| Bird | 7 | Stewart Island robin (*Petroica australis*) | Can selection operating on toll-like receptor (TLR) diversity overcome genetic drift that occurs during population establishment and is this due to nonrandom mating? | (67) |
| | 3 | great snipe (*Gallinago media*) | Can comparing gene expression profiles and nucleotide variation in relation to mating success/status identify candidate genes that are related to sexual selection? | (65) |
| | 9 | Two wild sparrow species | Is selection effecting seminal fluid protein diversification between sympatric populations of house and Spanish sparrow, and are positively selected genes enriched in immune-related pathways? | (48) |

| | 202 | red grouse (*Lagopus scoticus*) | What is the effect of experimentally increased testosterone levels and infection status on transcription levels of genes involved in immune response? | (66) |
|---|---|---|---|---|
| Mammal | 1 | red-fronted lemur (*Eulemur fulvus rufus*) | Is there a functional role of *IL4* polymorphism on male reproductive success? | (45) |
| | 5 | 12 primate species | Does rate of immune gene evolution correlate with mating system? | (72) |
| | 312 | California mouse (*Peromyscus californicus*) | Are reproductive partners more dissimilar than randomly generated male-female pairs at MHC, innate or nonimmune loci? | (68) |
| | 3 | Three wild mice species | Is expression of immune genes greater in polygynandrous species compared to monogamous species? | (71) |
| | 15 | 16 primate species per gene (avg.) | Has mating system (i.e., promiscuity) influenced the evolution of immune genes in primates? | (73) |
| | 5 | domestic pig (*Sus scrofa domesticus*) | Does strong artificial sexual selection effect inbreeding in the genome and positive selection at immune genes? | (85) |

or maintaining a post-mating reproductive barrier between intermating species (48). In Chinook salmon, male fertilization success was found to be positively associated with MHC class II dissimilarity between mates (49), while in Atlantic salmon the evidence for cryptic female choice is mixed (50), (51). In humans, the role of immunity in post-fertilization success when the zygote interacts with the mother's immune system, particularly MHC compatibility, is well-known (reviewed in (52)). These examples illustrate the range of mechanisms by which selection on immune genes is involved in the behaviors determining reproductive success in animals.

## 11.4 Empirical examples of mate choice linked to immune genes from a systematic review of the literature

It has been speculated that MHC genes are extraordinarily diverse relative to even other immune genes because MHC diversity is maintained by mating preferences that produce heterozygotes, while other immune genes experience only parasite-mediated selection, which is directional, and thus reduces genetic diversity (57). However, recent advances in genome sequencing have identified high diversity in other immune genes as well (reviewed by (17)). We speculate that until recently, prohibitively high costs in terms of both money and effort dissuaded researchers from investigating other candidate immune loci that may be involved in mate choice. Thus, we conducted a systematic literature review to understand the diversity of study systems and types of immune genes where relationships between mate choice and immunogenetic diversity have been investigated. We did not aim for an exhaustive search, nor did we aim to review the results of those studies; instead we present a systematic and unbiased sample of the empirical literature on this topic to identify if certain taxa or gene systems are disproportionately represented.

### 11.4.1 Methods

We had three criteria for including a study in our review: the study (i) uses or discusses animals (i.e., not plants, bacteria, or fungi); (ii) characterizes immune genotypes of study subjects (e.g., sequencing or expression analysis of specific immune genes); and (iii) tests for mating behavior or sexual selection related to immune genotype (e.g., differential expression of genes in females that depends on male genotype is relevant, but

## Box 11.2  Meta-analysis of primate MHC-linked mate choice

Winternitz *et al.* (42) conducted a review and meta-analysis of MHC-based mating behavior studies in humans and non-human primates. MHC-linked pre-mating cues that have so far been identified in animals are primarily linked to visual and olfactory cues. In primates, visual cues often involve sexually dimorphic traits and behavioral displays, such as canine size (53) or vocal frequency (54). In humans, studies on visual cues for mate preference have specifically focused on facial attractiveness, a trait which is unlikely to be greatly influenced by environmental stressors (55), (56). Olfactory cues in animals play an important role in both attraction and kin recognition (i.e., to avoid inbreeding), but particular

MHC peptides have also been found to play a direct role in odor recognition and production.

Results from 30 studies on mate choice for MHC-diversity are in line with predictions that humans—and likely also non-human primates—prefer mates with generally higher MHC diversity (Figure 1). In contrast, no consistent preference was found for MHC-dissimilarity. However, several factors appear to modulate preferences based on similarity or dissimilarity. The two most prominent are female hormones (menstrual cycle and contraceptive pill use) and social-cultural factors related to the identity of the chooser and diversity of the mating pool.

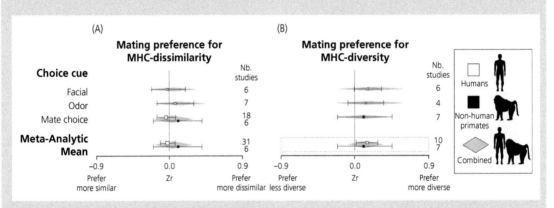

**Box 11.2, Figure 1** Effect of choice mechanism (facial and odor preference, and mate choice) on strength of MHC-associated mating outcome. This figure is a forest plot of effect size for categorical moderators for (A) preference for MHC dissimilarity and (B) preference for MHC diversity. Positive estimates of Fisher's Z transform of the correlation coefficient (Zr) indicate positive associations, whereas negative estimates indicate negative associations between MHC target and mating outcome. Colored boxes show the mean estimate from the model for humans (white box), non-human primates (black box), and both combined (gray diamond), and error bars represent the 95% highest posterior density (HPD) interval. The meta-analytic mean is from the intercept-only model run with study ID and phylogeny as random effects (and study ID as the random effect for the human data model). Significant results are emphasized with a dashed border. Figure modified from (42).

differential expression of genes in virgin versus mated females is not relevant). We also considered studies exploring associations between condition or condition-related sexual signals and immune genotype. We concentrated on empirical studies because we were interested in the study species and immune gene focus of individual investigations. Thus, we excluded reviews, pure modeling studies, and meta-analyses that used previously published articles.

Searches were conducted in Web of Science on November 3, 2020 (304 unique records refined to CATEGORIES: Ecology, Evolutionary Biology,

Genetics Heredity, Zoology and Behavioral Sciences). Our search terms for TOPIC were as follows: ((((((("mate choice" OR "mate preference*" OR "mating" OR "sexual selection") AND (("parasit*" NEAR/5 "resistance*") OR ("pathogen*" NEAR/5 "resistance*") OR "immune response" OR "immunocompetence*" OR "immunit*" OR "immunogen*") AND (("gene*" AND "diversity") OR ("immun*" NEAR/3 "gen*") OR "genotype" OR "sequenc*" OR "allel*") NOT ("plant*" OR "vegetable*" OR "crop*")))))). Records were retained if the LANGUAGE was English and the DOCUMENT TYPE was article or review, for

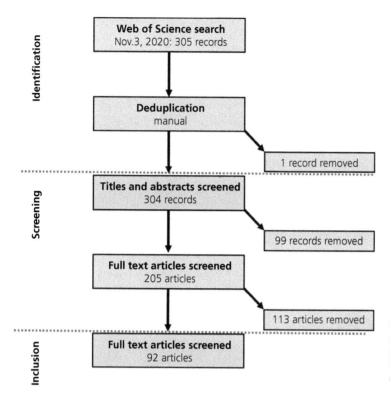

**Figure 11.2** PRISMA diagram for systematic review showing the steps of the literature search, including identification, screening, and inclusion. Reasons for inclusion and exclusion of records are also provided.

TIMESPAN of all years (1900–2020). For the 304 unique records, we screened titles and abstracts and retained 205 records. Of these records, full text screening identified 92 articles for inclusion in our review (steps of the literature search are shown in Figure 11.2).

For included studies, we extracted the following information: the category of gene investigated (MHC/non-MHC); the number of genes investigated per category; species name; taxa (i.e., insect, fish, amphibian, reptile, bird, mammal); year; and citation. Study data, following PRISMA reporting guidelines (58), and including all full text articles screened and reasons for exclusions, are available in Electronic Supplementary Material 1 (available on Figshare at <https://doi.org/10.6084/m9.figshare.16616524>).

### 11.4.2 Results

In total, approximately 78% of studies (72 of 92) investigated MHC genes and only 22% (20 studies) looked at non-MHC immune genes (Table 11.2). Our review uncovered three key results: (i) the MHC was the first to be investigated for its role in PMSS (in 1997 based on our search terms), and studies beyond this gene family were minimal over the following decade. From 2011 onwards, the search for immune genes linked to mate choice beyond the MHC expanded, and the number of those genes ($n = 1072$) is now more than those targeted within the MHC ($n = 413$; Figure 11.3); (ii) MHC studies typically focus on a small number of genes (mean = 2.4, median = 1), whereas those beyond the MHC typically involve wider screening techniques resulting in larger numbers of genes (mean = 33.5, median = 8; Figure 11.3), presumably due to the more exploratory nature of studies hunting for less well-described immune genes as targets of sexual selection; (iii) the number of immune genes targeted beyond the MHC are comparatively enriched in mammals and birds and completely lacking in amphibians and reptiles (Figure 11.3).

Focusing on the 20 studies that investigated non-MHC genes in relation to sexual selection/mate choice (Table 11.2), some common themes are

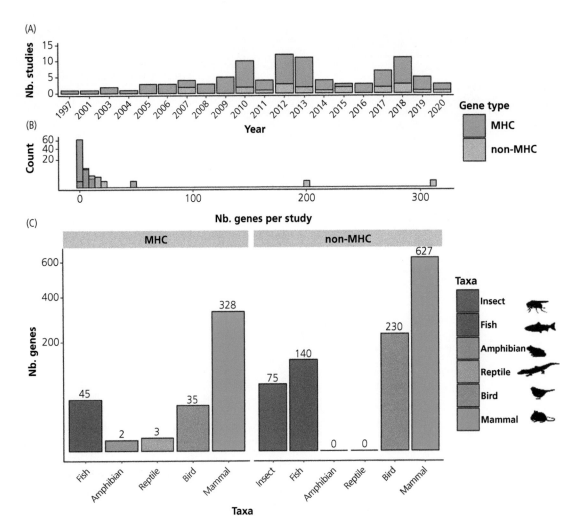

**Figure 11.3** Distribution of empirical research on studies of immune genes involved in mate choice/sexual selection. (A) Number of studies that investigated each gene type by year. (B) Histogram of the number of genes included in each study by gene type, where count is number of studies. (C) The number of MHC and non-MHC genes investigated by taxon.

apparent. Almost half of these studies involved insects, mostly fruit flies (*Drosophila melanogaster*) but also ants and beetles. These studies investigated between 5 and 14 immune genes, typically looking at differential expression related to male and/or female genotype (59)–(61). Many looked for signatures of positive selection on reproductive protein genes that have dual functions in immunity (47), (62), (63), speculating that female immune responses may assess sperm quality or compatibility (64). In contrast to the lab-based studies on

insects, studies on birds and mammals were mostly field-based and looked for a wider array of immune genes (1 to 312) and their associations with reproductive success (45), (65), testosterone and parasitism (66), and patterns of non-random mating (67), (68).

Regarding the outcomes of the 20 studies, it is worth keeping in mind that studies with significant results tend to be overrepresented in the scientific literature because they are published more frequently or sooner (i.e., publication bias, (69), (70)).

Thus, the majority of studies revealed at least *some* evidence for the role of immune genes other than the MHC in PMSS, such as linkage of mate choice, secondary sexual characteristics, and/or offspring fitness with immune genotypes. Some of these examples include genes coding for antimicrobial peptides (AMPs), proteases, and other types of foreign protein recognition receptors, most of which feature as part of primitive or innate immune systems (e.g., (59), (61), (65), (71)).

A number of studies used multi-species comparisons to test evolutionary predictions. Those on primates and wild mice tested if the rate of immune gene evolution correlated with mating system under the expectation that higher promiscuity would increase exposure to parasites, and thus select for faster immune gene evolution (71)–(73). A multi-species comparison with birds investigated genes underlying the process of speciation. In this study, Rowe *et al.* (48) tested for positive selection on seminal fluid protein diversification between sympatric populations of house and Spanish sparrows and found that positively selected genes are enriched in immune-related pathways. These results raise the possibility that immunological mechanisms may contribute to reproductive incompatibilities in sympatric species. Finally, a study on seven cichlid fish species tested if parental care (maternal or biparental) affected sexual dimorphism in immune gene expression (74). The authors predicted that due to trade-off allocation theory, females of species with maternal care would show lower expression of immune-related genes due to greater investment of resources into offspring care. Supporting their predictions, the authors found species with maternal care are more sexually dimorphic in immune parameters, and females generally have lower expression of immune-related genes. In contrast, species with biparental care, which share costs of parental care, showed upregulation of immune-related genes. Interestingly, only adaptive immune system genes were related to parental investment and sexual dimorphism, implying that the adaptive immune system is more costly than the innate immune system (75) and thus more affected by resource allocation trade-offs.

A final theme among the studies was related to transgenerational paternal/maternal immune priming of offspring, or the transfer of immune stimulation from the parent to the offspring generation. Two studies investigated the contribution of paternal effects to the strength of offspring immune defense in red flour beetles (76) and pipefish (77). These studies found immune gene expression of offspring was related to paternal identity and infection status, respectively. While not explicitly considering immune gene-based mate choice, immune priming, which depends on paternal or maternal immune stimulation, could still be linked to mate choice if the underlying parental genotype provides greater protection to offspring in environments with high pathogen prevalence.

To summarize, the majority of studies on immune genes involved in sexual selection investigated the MHC, though since 2011 exploratory analyses have expanded investigation to immune genes beyond the MHC. These non-MHC genes have diverse functions and various levels of involvement in aspects of sexual selection, including reproductive success, mate choice, secondary sexual traits, mating system, speciation, parental care, and offspring fitness and immunity. The study systems investigated strongly favor insects, mammals, and birds, while amphibians and reptiles had no studies identified by our systematic review. Thus, the impression of the vertebrate MHC as the best candidate of mate choice for genetic benefits may be incomplete, since other immune genes have been relatively understudied and only in a few mostly model taxa.

## 11.5 Contextual factors complicating the identification of immune genes involved in mate choice

Despite the rise in interest for discovering immune genes involved in mate choice across a wider range of animals and immune functions noted in our systematic review (see section 11.4), significant challenges remain that hamper this progress.

**Figure 11.4** Contextual factors complicating the identification of immune genes involved in mate choice. (A) Selection pressures from parasite-mediated selection (PMS) and from parasite-mediated sexual selection (PMSS) can be conflicting and oppose each other. This could result in shifted distributions of the frequency of allelic diversity and/or selection coefficients (e.g., measured yearly) in the population, represented as dotted to solid shifts in distribution. The overall effect of two opposing selection pressures may also increase the variance, which would reduce researchers' ability to detect statistical significance in genes of interest. In contrast, synergistic selection where PMS and PMSS act in the same direction would increase allelic diversity/selection coefficients and could reduce the variance as well, increasing the ability to detect significant genes. (B) Pleiotropy and epistasis may also affect the ability to identify relevant immune genes involved in mate choice. Pleiotropy occurs when a gene affects multiple traits and epistasis occurs when traits Affected by a gene (Trait A) are modified (Trait A*) by the presence of one or more other genes. (C) Mechanisms of infection shaping host–pathogen coevolutionary dynamics. Under matching infection models like the matching alleles (MA) model, infection occurs when genotypes match, while a mismatch is observed as resistance. This model fits the strategy of pathogens that mimic host molecules to gain attachment and entry. Under targeted recognition models like the inverse matching alleles (IMA) model, hosts have a suite of recognition receptors that can bind particular pathogen molecules. Under IMA, infection fails when genotypes match, while a mismatch leads to infection. This model is appropriate for the vertebrate MHC system and the pattern recognition receptors of the innate immunity system that can recognize pathogen-associated molecular patterns (PAMPs). Created with BioRender.com.

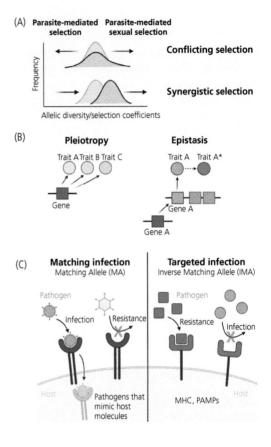

Identifying the genes under PMSS and interpreting the evidence expected to support the genetic benefit models detailed above (Table 11.1) can be complicated by a number of important contextual factors. Identification may be particularly challenging for immune genes which are necessarily expected to be under dual pathogen-mediated sexual *and* natural selection (Figure 11.4). Here, we examine four broad (but non-exhaustive) categories of such contextual factors.

(1) **Conflicting selection pressures.** Natural selection from PMS and sexual selection from PMSS may act in opposing directions, which can appear in the data as a lack of significant effect (Figure 11.4A). Mate choice and parasitism may be in conflict if higher immune allelic diversity is promoted by

mate choice, for example via mating rules favoring high dissimilarity between partners or because high allelic diversity allows greater discrimination of kin, while immune allelic diversity is constrained by T-cell depletion or other outcomes that reduce parasite resistance (78). It could also be possible that in smaller populations, PMS results in strong directional selection for specific resistance alleles that reduces genetic diversity while PMSS in the form of disassortative mating results in stabilized immune allele frequencies and maintains genetic diversity (79). The overall effect of two opposing selection pressures would obscure signals of selection estimated from allele frequencies, either by opposing shifts in allele frequencies or by increasing the variance in allele frequencies in populations exposed to different degrees of PMS and PMSS. Either case would reduce the ability to detect statistical significance in genes of interest.

(2) **Synergistic selection pressures.** PMS and PMSS forces may act in the same direction, increasing allelic diversity and strengthening signals of selection, thereby increasing the ability to detect significant genes (Figure 11.4A). This could occur, for example, if both parasites and mating preferences select for higher immune gene allelic diversity, or if increased attractiveness helps to overcome the costs of investment in immune defense. However, very careful experimental studies are needed to tease out whether both forces are occurring on the same set of immune genes. For instance, observational studies test for sexual selection by using differential frequency of specific alleles in mated versus unmated males but they do not control for current infection status, and thus may not be able to distinguish whether patterns are due to mate choice or to reduced competitive ability of infected individuals. Mate choice experiments should be conducted (pre- and post-copulatory) to tease apart what genotypes lead to mating success because they confer preferred traits independent of infection status, and what genotypes lead to mating success because they confer parasite resistance. However, this can be tricky to evaluate in artificial lab settings with unrealistically low levels of immune gene diversity and using only proxies of natural infection. On the other hand, wild systems have their own challenges in pinpointing mechanisms of selection. In particular, when selection is strong, it reduces the variation available for identifying— and maintaining—choice, highlighting a negative feedback between mate choice, parasite-mediated selection, and immune gene diversity. For example, in common yellowthroats, MHC diversity is so high that females may not be able to use signals of diversity as a marker for mate quality because all individuals are equally diverse (80). Both PMS and PMSS can drive this diversity, though only PMS is expected to be able to maintain diversity on its own.

(3) **Pleiotropy and epistasis.** Theoretical studies of condition-based mating preference models have mainly focused their analyses using a single 2-allele locus for simplicity (Figure 11.1). In the real world, genes do not act in isolation and a single gene can influence multiple phenotypic traits (pleiotropy), as well as interact with multiple other genes (epistasis) (Figure 11.4B). While this complexity may actually increase the feasibility of genetic benefit models that require perpetual genetic variation, interactions among genes and traits can make it difficult to identify which immune genes are functionally involved in mate choice. Take for instance a result from sticklebacks in which Clough *et al.* (45) found higher than expected male reproductive success for individuals with a genetic mutation in *IL4* associated with lower resistance to nematode infection. Since sexual selection should not favor lower parasite resistance, the authors speculate that a trade-off exists between Th1/Th2 immune responses, where infection by the nematode protects the male from harmful Th1-mediated immunopathology in response to other stimuli. Thus, the involvement of *IL4* in sexual selection could be hidden in the absence of the nematode.

(4) **Mechanism of infection.** Each member of a parasite community interacts with the host immune system when it attempts to first enter and then multiply within the host. These two stages are each characterized by an *infection matrix*, a set of rules based on a general mechanistic model that defines the underlying genetic basis for how infection occurs and how resistance is achieved (38), (81). The two most commonly referenced types of infection matrices, explained in detail shortly, lead to opposite expectations for the coevolution of diversity in host and pathogen. Although this is beyond the scope of our chapter, we want to highlight that the infection matrix has a dominant role in shaping host–pathogen coevolutionary dynamics and so expectations should be carefully defined using the mechanistic model that is most appropriate given the putative immune gene under investigation.

Briefly, the two most commonly used general mechanistic models for infection are *matching infection* and *targeted recognition* (reviewed in (38), Figure 11.4C). Under matching infection models, like the MA model, infection occurs when genotypes match, while a mismatch is observed as resistance. This model fits the strategy of pathogens that mimic host molecules to gain attachment and entry. All the host must do is lose the matching molecule. Under targeted recognition models, like the inverse matching alleles (IMA) model, hosts have a suite of recognition receptors that can bind particular pathogen molecules. This model is appropriate for the vertebrate MHC system of the adaptive immune system and the pattern recognition receptors of the innate immune system that can recognize pathogen-associated molecular patterns (PAMPs) once the pathogen has gained entry and attempts to multiply. Under IMA, infection fails when genotypes match, while a mismatch leads to infection. Here, the host must maintain a diverse repertoire to recognize invaders. As noted earlier in section 11.2, how immune genes recognize parasites matters for evolutionary outcomes of PMSS and can predict whether mating will be assortative, disassortative, or random.

These four broad categories of contextual factors are neither exhaustive nor are they mutually exclusive. Additional factors specific to the biology or behavior of the organism can lead to unique needs for consideration, such as the effects of birth control pill use in humans, or paternal investment in offspring care in cichlids. The details of these caveats are important in designing investigations meant to discover genes involved in PMSS because they define the null expectations necessary for accurate interpretation of results. Whether it is the direction and magnitude of acting selective forces, genomic architecture, or mechanisms of infection, consideration of these contextual factors is the only way to avoid spurious or inconclusive results.

## 11.6 Conclusions and future directions

In this chapter, we have shown how pathogen-mediated natural (PMS) and sexual (PMSS) selection are expected to directly involve genes of the immune system. While the most well-studied of these genes by far are those of the vertebrate MHC, the generality of immune gene involvement extends to genes underlying other facets of vertebrate immunity and to those acting in invertebrates.

Sequencing technologies have advanced to the point where study species do not necessarily need to be model organisms like fruit flies. The "omics" age has facilitated investigation beyond candidate gene sequences, including *de novo* candidate gene approaches interrogating whole genomes, comparing differential transcription profiles, and differential protein abundance for putatively involved genes. A wealth of interesting studies published on animal preferences for immuno-competent mates via visual, olfactory, and even auditory pre-mating cues provide many likely candidates for discovering more of these immune genes on which PMSS acts (e.g., those underlying melanin production in insects, reviewed in (64)). Once candidates are identified, studies should take care to test assumptions of genetic benefit models fully by measuring mating preferences, measuring offspring viability, distinguishing between additive and non-additive fitness benefits, and controlling for manipulation of differential female investment in offspring sired by attractive males (24). The greater the diversity of animals and populations in which these processes can be tested, the easier it will be to find prevailing patterns that can guide ongoing research. One glaring result from our systematic review was the complete lack of non-MHC genes investigated within amphibians and reptiles. Unquestionably, there can be serious consequences for our understanding of mate choice when whole taxa are absent from studies, including biases in interpretation and inflated generality of findings.

Practical justification given for research on PMSS is that sexual selection is believed to increase population viability by promoting "good genes" and filtering out undesirable phenotypes from the population. However, this justification is not always accurate for immune genes. For example, a meta-analysis of 315 effect sizes from 90 studies on 55 species tested the key assumption of "good genes," that male attractiveness was heritable and genetically correlated with fitness (82). Results showed positive correlations between sire attractiveness and offspring physiological traits such as immuno-competence and condition, providing some support for the "good genes" hypothesis (82). These results were typically present for a single generation or cohort, not over multiple generations, and thus, evolutionary trajectory was not considered. Interestingly, another meta-analysis of 459 effect sizes from 65 studies on 7 species (5 invertebrates plus mouse and guppy) used an experimental evolution approach to manipulate sexual selection over multiple generations and test if sexual selection improves population fitness over time (83). They found that indeed, sexual selection tends to increase population fitness, particularly when populations are exposed to novel environmental conditions. However, they also found that increasing the strength of sexual selection typically resulted in comparatively weaker immunity in the population (83). These findings might be explained by a trade-off between sexually selected phenotypes and immunity, and they highlight that evidence for PMSS does not mean that populations will be generally more resistant to infectious disease. Going forward, the implications of PMSS for conservation, wildlife management, and long-term viability of populations, particularly as climate change threatens to impose accelerating challenges, mean it is important to consider the models of mate choice for genetic benefits and contextual factors to guide interpretation.

# References

1. Darwin C. *The Descent of Man, and Selection in Relation to Sex.* London: John Murray, 1871.
2. Borgia G. Sexual selection and the evolution of mating systems in Blum MS and Blum NA (eds), *Sexual Selection and Reproductive Competition in Insects.* New York, NY: Academic Press, 1979, 19–80.
3. Kirkpatrick M and Ryan MJ. The evolution of mating preferences and the paradox of the lek. *Nature* 1991;350:33–8.
4. Hamilton WD and Zuk M. Heritable true fitness and bright birds: A role for parasites? *Science* 1982;218:384–7.
5. Arneberg P, Skorping A, Grenfell B, and Read AF. Host densities as determinants of abundance in parasite communities. *Proc R Soc B: Biol Sci.* 1998;265:1283–89.
6. Shaw D, Grenfell B, and Dobson A. Patterns of macroparasite aggregation in wildlife host populations. *Parasitology* 1998;117:597–610.
7. Taylor LH, Latham SM, and Mark E. Risk factors for human disease emergence. *Philos Trans R Soc Lond B Biol Sci.* 2001;356:983–9.
8. Daszak P, Cunningham AA, and Hyatt AD. Emerging infectious diseases of wildlife—threats to biodiversity and human health. *Science* 2000;287:443–9.
9. Fisher MC, Henk DA, Briggs CJ, Brownstein JS, Madoff LC, McCraw SL, *et al.* Emerging fungal threats to animal, plant and ecosystem health. *Nature* 2012;484:186–94.
10. Cunningham AA, Daszak P, and Wood JLN. One health, emerging infectious diseases and wildlife: Two decades of progress? *Philos Trans R Soc Lond B: Biol Sci.* 2017;372:20160167.
11. Morens DM and Fauci AS. Emerging pandemic diseases: How we got to COVID-19. *Cell* 2020;182:1077–92.
12. Tompkins D and Begon M. Parasites can regulate wildlife populations. *Parasitol. Today* 1999;15:311–13.
13. Brown JL. A theory of mate choice based on heterozygosity. *Behav Ecol.* 1997;8:60–5.
14. Irwin AJ and Taylor PD. Heterozygous advantage and the evolution of female choice. *Evol Ecol Res.* 2000;2:107–18.
15. Lehmann L, Keller LF, and Kokko H. Mate choice evolution, dominance effects, and the maintenance of genetic variation. *J Theo. Biol.* 2007; 244:282–95.
16. Preston D and Johnson P. Ecological consequences of parasitism. *Nature Education Knowledge* 2010;3:47–52.
17. Ebert D and Fields PD. Host–parasite co-evolution and its genomic signature. *Nat Rev Genet.* 2020;21:754–68.
18. Robinson J, Barker DJ, Georgiou X, Cooper MA, Flicek P, and Marsh SG. Ipd-imgt/hla database. *Nucleic Acids Res.* 2020;48:D948–D955.
19. Piertney SB and Oliver MK. The evolutionary ecology of the major histocompatibility complex. *Heredity* 2006;96:7–21.
20. Bitarello BD, de Filippo C, Teixeira JC, Schmidt JM, Kleinert P, Meyer D, *et al.* Signatures of long-term balancing selection in human genomes. *Genome Biol Evol.* 2018;10:939–55.
21. Shultz AJ and Sackton TB. Immune genes are hotspots of shared positive selection across birds and mammals. *eLife* 2019;8:e41815.
22. Furlong R and Yang Z. Diversifying and purifying selection in the peptide binding region of DRB in mammals. *J Mol Evol.* 2008;66:384–94.
23. Sánchez CA, Becker DJ, Teitelbaum CS, Barriga P, Brown LM, Majewska AA, *et al.* On the relationship between body condition and parasite infection

in wildlife: A review and meta-analysis. *Ecol Lett.* 2018;21:1869–84.

24. Achorn AM and Rosenthal GG. It's not about him: Mismeasuring "good genes" in sexual selection. *Trends Ecol Evol.* 2020;35:206–19.

25. Neff BD and Pitcher TE. Genetic quality and sexual selection: An integrated framework for good genes and compatible genes. *Mol Ecol.* 2005;14:19–38.

26. Nuismer SL, Otto SP, and Blanquart F. When do host–parasite interactions drive the evolution of non-random mating? *Ecol Lett.* 2008;11:937–46.

27. Greenspoon PB and M'Gonigle LK. Host–parasite interactions and the evolution of nonrandom mating. *Evol.* 2014;68:3570–80.

28. Campbell L, Head M, Wilfert L, and Griffiths A. An ecological role for assortative mating under infection? *Conserv Genet.* 2017;18:983–94.

29. Phillips KP, Cable J, Mohammed RS, Herdegen-Radwan M, Raubic J, Przesmycka KJ, et al. Immuno-genetic novelty confers a selective advantage in host–pathogen coevolution. *PNAS.* 2018;115:1552–7.

30. Mitton JB, Schuster WSF, Cothran EG, and De Fries JC. Correlation between the individual heterozygosity of parents and their offspring. *Heredity* 1993;71: 59–63.

31. Nietlisbach P and Hadfield JD. Heritability of heterozygosity offers a new way of understanding why dominant gene action contributes to additive genetic variance. *Evol.* 2015;69:1948–52.

32. Hedrick PW. Female choice and variation in the major histocompatibility complex. *Genetics* 1992;132: 575–81.

33. Howard RS and Lively CM. Good vs complementary genes for parasite resistance and the evolution of mate choice. *BMC Evol Biol.* 2004;4:48.

34. Bentkowski P and Radwan J. Evolution of major histocompatibility complex gene copy number. *PLoS Comp. Biol.* 2019;15:e1007015.

35. Schubert N, Nichols HJ, and Winternitz JC. How can the MHC mediate social odor via the microbiota community? A deep dive into mechanisms. *Behav Ecol.* 2021;32:359–73.

36. Ruff JS, Nelson AC, Kubinak JL, and Potts WK. Mhc signaling during social communication in López-Larrea C (ed.), *Self and Nonself.* New York, NY: Springer, 2012, 290–313.

37. Radwan J, Babik W, Kaufman J, Lenz TL, and Winternitz J. Advances in the evolutionary understanding of *mhc* polymorphism. *Trends Genet.* 2020;36:298–311.

38. Dybdahl MF, Jenkins CE, Nuismer SL. Identifying the molecular basis of host-parasite coevolution: Merging models and mechanisms. *Am Nat.* 2014;184:1–13.

39. Bonneaud C, Chastel O, Federici P, Westerdahl H, and Sorci G. Complex mhc-based mate choice in a wild passerine. *Proc R Soc B: Biol Sci.* 2006;273:1111–16.

40. Huang W, Pilkington JG, and Pemberton JM. Patterns of mhc-dependent sexual selection in a free-living population of sheep. *Mol Ecol.* 2021. DOI: 10.1111/mec.15938.

41. Milinski M. The major histocompatibility complex, sexual selection, and mate choice. *Annu Rev Ecol Evol Syst.* 2006;37:159–86.

42. Winternitz J, Abbate J, Huchard E, Havlíček J, and Garamszegi L. Patterns of mhc-dependent mate selection in humans and nonhuman primates: A meta-analysis. *Mol Ecol.* 2017;26:668–88.

43. Von Schantz T, Wittzell H, Goransson G, and Grahn M. Mate choice, male condition-dependent ornamentation and MHC in the pheasant. *Hereditas* 1997;127: 133–40.

44. Jäger I, Eizaguirre C, Griffiths SW, Kalbe M, Krobbach C, Reusch T, et al. Individual MHC class I and MHC class 2b diversities are associated with male and female reproductive traits in the three-spined stickleback. *J Evo. Biol.* 2007;20:2005–15.

45. Clough D, Kappeler PM, and Walter L. Genetic regulation of parasite infection: Empirical evidence of the functional significance of an *IL4* gene SNP on nematode infections in wild primates. *Front Zool.* 2011;8: 1–9.

46. Ziegler A, Kentenich H, and Uchanska-Ziegler B. Female choice and the mhc. *Trends Immunol.* 2005;26:496–502.

47. Wong A, Turchin MC, Wolfner MF, and Aquadro CF. Temporally variable selection on proteolysis-related reproductive tract proteins in *Drosophila. Mol Biol Evol.* 2012;29:229–38.

48. Rowe M, Whittington E, Borziak K, Ravinet M, Eroukhmanoff F, Sætre G-P, et al. Molecular diversification of the seminal fluid proteome in a recently diverged passerine species pair. *Mol Biol Evol.* 2020;37:488–506.

49. Gessner C, Nakagawa S, Zavodna M, and Gemmell N. Sexual selection for genetic compatibility: The role of the major histocompatibility complex on cryptic female choice in chinook salmon (*Oncorhynchus tshawytscha*). *Heredity* 2017;118:442–52.

50. Promerová M, Alavioon G, Tusso S, Burri R, and Immler S. No evidence for mhc class ii-based non-random mating at the gametic haplotype in Atlantic salmon. *Heredity* 2017;118:563–67.

51. Yeates SE, Einum S, Fleming IA, Megens H-J, Stet RJ, Hindar K, et al. Atlantic salmon eggs favour sperm in competition that have similar major histocompatibility

alleles. *Proc R Soc B: Biol Sci.* 2009;276: 559–66.

52. Deshmukh H and Way SS. Immunological basis for recurrent fetal loss and pregnancy complications. *Annu Rev Pathol.* 2019;14:185–210.

53. Plavcan JM. Sexual dimorphism in primate evolution. *Am J Phys Anthropol.* 2001;116:25–53.

54. Puts DA, Hill AK, Bailey DH, Walker RS, Rendall D, Wheatley JR, *et al.* Sexual selection on male vocal fundamental frequency in humans and other anthropoids. *Proc R Soc B: Biol Sci.* 2016;283:20152830.

55. Kleisner K, Tureček P, Roberts SC, Havlíček J, Valentova JV, Akoko RM, *et al.* How and why patterns of sexual dimorphism in human faces vary across the world. *Sci Rep.* 2021;11:5978.

56. Petersen RM and Higham JP. The role of sexual selection in the evolution of facial displays in male nonhuman primates and men. *Adapt Hum Behav Physiol.* 2020;6:249–76.

57. Penn DJ and Potts WK. The evolution of mating preferences and major histocompatibility complex genes. *Am Nat.* 1999;153:145–64.

58. Moher D, Liberati A, Tetzlaff J, Altman DG, and Group P. Preferred reporting items for systematic reviews and meta-analyses: The PRISMA statement. *PLoS Medicine* 2009;6:e1000097.

59. Delbare SY, Chow CY, Wolfner MF, and Clark AG. (2017) Roles of female and male genotype in postmating responses in drosophila melanogaster. *J Hered.* 2017;108:740–53.

60. Immonen E and Ritchie MG. The genomic response to courtship song stimulation in female *Drosophila melanogaster. Proc R Soc B: Biol Sci.* 2012;279: 1359–65.

61. Schrempf A, Von Wyschetzki K, Klein A, Schrader L, Oettler J, and Heinze J. Mating with an allopatric male triggers immune response and decreases longevity of ant queens. *Mol Ecol.* 2015;24:3618–27.

62. Prokupek A, Eyun SI, Ko L, Moriyama EN, and Harshman LG. Molecular evolutionary analysis of seminal receptacle sperm storage organ genes of *Drosophila melanogaster. J Evol Biol.* 2010;23:1386–98.

63. Wong A, Turchin MC, Wolfner MF, and Aquadro CF. Evidence for positive selection on *Drosophila melanogaster* seminal fluid protease homologs. *Mol Biol Evol.* 2007;25:497–506.

64. Lawniczak MK, Barnes AI, Linklater JR, Boone JM, Wigby S, and Chapman T. Mating and immunity in invertebrates. *Trends Ecol Evol.* 2007;22:48–55.

65. Höglund J, Wang B, Sæther SA, Blom MPK, Fiske P, Halvarsson P, *et al.* Blood transcriptomes and de novo identification of candidate loci for mating success

in lekking great snipe (*Gallinago media*). *Mol Ecol.* 2017;26:3458–71.

66. Wenzel MA, Webster LM, Paterson S, Mougeot F, Martínez-Padilla J, and Piertney SB. A transcriptomic investigation of handicap models in sexual selection. *Behav. Ecol Sociobiol.* 2013;67:221–34.

67. Grueber CE, Wallis GP, and Jamieson IG. Genetic drift outweighs natural selection at toll-like receptor (TLR) immunity loci in a re-introduced population of a threatened species. *Mol Ecol.* 2013;22:4470–82.

68. Meléndez-Rosa J, Bi K, and Lacey EA. Genomic analysis of mhc-based mate choice in the monogamous California mouse. *Behav Ecol.* 2018;29:1167–80.

69. Møller AP and Jennions MD. Testing and adjusting for publication bias. *Trends Ecol Evol.* 2001;16:580–6.

70. Nakagawa S, Lagisz M, Jennions MD, Koricheva J, Noble DWA, Parker TH, *et al.* Methods for testing publication bias in ecological and evolutionary meta-analyses. *Methods Ecol Evol.* 2021. Available from: <https://doi.org/10.1111/2041-210X.13724>.

71. Meléndez-Rosa J, Bi K, and Lacey EA. Differential gene expression in relation to mating system in peromyscine rodents. *Ecol Evol.* 2019;9:5975–90.

72. Hurle B, Swanson W, Green ED, and Program NCS. Comparative sequence analyses reveal rapid and divergent evolutionary changes of the WFDC locus in the primate lineage. *Genome Res.* 2007;17:276–86.

73. Wlasiuk G and Nachman MW. Promiscuity and the rate of molecular evolution at primate immunity genes. *Evol.* 2010;64:2204–20.

74. Keller IS, Bayer T, Salzburger W, and Roth O. Effects of parental care on resource allocation into immune defense and buccal microbiota in mouthbrooding cichlid fishes. *Evol.* 2018;72:1109–23.

75. Boots M and Bowers RG. The evolution of resistance through costly acquired immunity. *Proc R Soc B: Biol Sci.* 2004;271:715–23.

76. Eggert H, Kurtz J, and Diddens-de Buhr MF. Different effects of paternal trans-generational immune priming on survival and immunity in step and genetic offspring. *Proc R Soc B: Biol Sci.* 2014;281:20142089.

77. Roth O, Klein V, Beemelmanns A, Scharsack JP, and Reusch TB. Male pregnancy and biparental immune priming. *Am Nat.* 2012;180:802–14.

78. Kubinak JL, Nelson AC, Ruff JS, Potts WK. Trade-offs limiting mhc heterozygosity in Demas G, Nelson R, and Nelson RJ (eds), *Ecoimmunology.* Oxford: Oxford University Press, 225–58.

79. Ejsmond MJ, Radwan J, and Wilson AB. Sexual selection and the evolutionary dynamics of the major histocompatibility complex. *Proc R Soc B: Biol Sci.* 2014;281:20141662.

80. Bollmer JL, Dunn PO, Freeman-Gallant CR, and Whittingham LA. Social and extra-pair mating in relation to major histocompatibility complex variation in common yellowthroats. *Proc R Soc B: Biol Sci.* 2012;279:4778–85.

81. Ashby B and Boots M. Multi-mode fluctuating selection in host–parasite coevolution. *Ecol Lett.* 2017;20:357–65.

82. Prokop ZM, Michalczyk Ł, Drobniak SM, Herdegen M, and Radwan J. (2012) Meta-analysis suggests choosy females get sexy sons more than "good genes". *Evol.* 2012;66:2665–73.

83. Cally JG, Stuart-Fox D, and Holman L. Meta-analytic evidence that sexual selection improves population fitness. *Nature Commun.* 2019;10:1–10.

84. Barthel A, Staudacher H, Schmaltz A, Heckel DG, and Groot AT. Sex-specific consequences of an induced immune response on reproduction in a moth. *BMC Evol Biol.* 2015;15:1–12.

85. Zhang Z, Zhang Q, Xiao Q, Sun H, Gao H, Yang Y, *et al.* Distribution of runs of homozygosity in Chinese and Western pig breeds evaluated by reduced-representation sequencing data. *Anim Genet.* 2018;49:579–91.

# Parasite Modification of Host Behavior

# CHAPTER 12

# Host manipulation by parasites: From individual to collective behavior

Stephanie S. Godfrey and Robert Poulin

## 12.1 Introduction

Animal behavior is a key determinant of infection risk (1). Behavioral traits evolved for other functions may coincidentally increase the risk of parasite acquisition, just as behaviors evolved specifically to avoid parasites will lower that risk. At the same time, however, infection by parasites often leads to changes in an animal's behavior. In particular, many parasite species actively manipulate the behavior of their host in ways that benefit the parasite. We define host manipulation as any alteration in host phenotype, induced by parasite infection, that has fitness benefits for the parasite (i.e., increased reproductive success, transmission rate, or dispersal) but not for the host. This implies that the altered host traits are either directly or indirectly modulated by genes in the parasite genome, and comprise the parasite's extended phenotype (2). Host manipulation by parasites has been documented in hundreds of different host–parasite systems across the tree of life, with manipulated hosts found among all major metazoan phyla, and manipulative parasites including viruses, prokaryotes, and both uni- and multicellular eukaryotes (3)–(5). Even by the most conservative estimates, host manipulation must have evolved independently at least 20 to 30 separate times among different parasite lineages (4).

Natural selection has favored the evolution of host manipulation in parasites with various transmission modes where opportunities for transmission can be maximized, or losses of infective stages can be minimized, by small changes in the way hosts behave (Figure 12.1). Manipulative parasites occur widely among species of parasites with trophic transmission (6) and vector-borne transmission (see Chapter 13 of this book (7)). They also occur in taxa where the parasite must emerge from its host in a particular location in order to continue its development and/or release its dispersal stages in the external environment (8), with manipulated behaviors being expressed post-emergence in some cases (9). Manipulative behavior can also be found in parasitoids, where the behavior of the infected host enhances the survival of the parasitoids' progeny (10). Host manipulation is also reported among parasites with contact and/or sexual transmission (see Chapters 9 and 14 of this book (11), (12)), but only rarely and with less convincing evidence, possibly because selection for manipulation is weaker in these parasites since normal host behavior may suffice to guarantee opportunities for contacts with new hosts. Overall, the extent of behavioral changes observed in infected animals can range from subtle (e.g., minor changes in activity levels) to dramatic (e.g., a fatal attraction to water), regardless of their ultimate fitness consequences for the parasite. Similarly, at a proximate level, the underlying mechanisms causing those changes can range from simple to very complex (Box 12.1).

In this chapter, we explore the consequences of host manipulation by parasites for both individual and group behavioral patterns. After a summary of the types of behavioral traits commonly modified by manipulative parasites, we discuss the known

Stephanie S. Godfrey and Robert Poulin, *Host manipulation by parasites*. In: *Animal Behavior and Parasitism*. Edited by Vanessa O. Ezenwa, Sonia Altizer, and Richard J. Hall, Oxford University Press. © Oxford University Press (2022). DOI: 10.1093/oso/9780192895561.003.0012

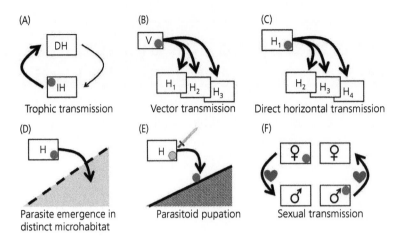

**Figure 12.1** Transmission modes in which a parasite (circle) can manipulate the behavior of its host (rectangle); the thick arrow indicates the transmission step where manipulation occurs. (A) Trophic transmission: altered behavior or appearance of the intermediate host, IH, increases the probability of transmission by predation to the definitive host, DH. (B) Vector transmission: altered feeding behavior of the vector, V, leads to transmission to a greater number of hosts. (C) Direct horizontal transmission: altered behavior of infected individuals increases contact with and transmission to conspecific hosts. (D) Parasite emergence in distinct microhabitat: movement of the host into an environment other than the one where it normally lives allows parasite exit and survival. (E) Parasitoid pupation: the host protects the parasitoid after the latter egresses from the host and pupates attached to an external surface. (F) Sexual transmission: altered behavior of infected males or females increases mating encounters and therefore opportunities for parasite transmission. Figure is adapted from Poulin 2010 (4).

and possible implications of manipulative parasites for two frameworks widely used to study behavior: animal personality and social networks. We also propose ways to test the effects of manipulative parasites on personality traits, behavioral syndromes, and social network structure. All animals have parasites, and many of the latter are manipulators: we advocate integrating the potential phenotypic effects of infection in all future studies of animal behavior.

## 12.2 Behavioral targets of manipulation

### 12.2.1 Changes in activity

Parasite manipulation of host behavior can target a wide range of behaviors in a variety of different ways. One of the most common ways in which host behavior is altered is through changes in host activity, where the level of activity (higher or lower), or the timing (day vs night) of host activity may be altered (24). Such changes have mainly evolved in parasites with complex lifecycles, where changes to activity patterns lead to an increased likelihood of the host encountering the next host in its lifecycle (usually via predation).

Amphipods have become a common system to examine how parasites manipulate host behavior, including changes in activity. For example, the amphipod *Echinogammarus stammeri* infected with an acanthocephalan (*Pomphorhynchus laevis*) were more active (made more laps of the mesocosm) than uninfected amphipods (25). This response also depended on the presence of predator cues. The presence of fish scent in the mesocosm led to a decrease in activity in uninfected *E. stammeri* amphipods while infected amphipods showed no change in activity level (25). In a similar way, changes in the timing of activity may increase contact between an infected host and the predatory definitive host. For instance, in the amphipod *Corophium volutator*, daytime surface activity was significantly higher in amphipods infected by the nematode *Skrjabinoclava morrisoni*, while nighttime surface activity was not significantly different between infected and uninfected amphipods (26). The effect of infection on daytime behavior is thought to correspond with the main activity period of the predator and final host for this parasite, the semipalmated sandpiper (*Calidris pusilla*). While it is assumed that these parasite-induced changes in activity increase the likelihood of the parasite

---

**Box 12.1  Mechanistic basis of host manipulation**

While parasite manipulation is a well-known phenomenon, the mechanisms that underpin these manipulations are poorly understood. In some cases, parasites can induce marked changes in the behavior of their hosts simply by lodging themselves within and damaging muscles or sensory organs (13), thereby impairing host movements or responses to external stimuli. Their mere size relative to the host can also cause physiological stress and associated behavioral changes in the host (14). However, in most cases, host manipulation appears to involve interference by the parasite at a molecular level (15). This can take many forms. First, parasites can disrupt the communication between the immune and central nervous systems (16). For example, infection by *Toxoplasma gondii*, before leading to well-documented changes in host behavior (see Box 12.2), triggers a local immune response in the central nervous system, and the release of a cascade of cytokines which then influence neuromodulator levels and neuronal activity (17). Amphipods infected by the juvenile stages of various species of acanthocephalans display aberrant responses to light, which can facilitate their transmission to definitive hosts. The abnormal phototaxis of infected amphipods appears linked to a surge in serotonin levels induced by the parasite (18), which is itself facilitated by a partial deactivation

of the phenoloxidase cascade, the main immune mechanism in crustaceans used to combat parasites like acanthocephalans (19). Secondly, parasites can modify gene expression in the host, via epigenetic processes or otherwise, leading to under- or overexpression of certain host genes (20). In particular, proteomic studies have revealed different protein profiles in the brains of uninfected arthropods compared to those of conspecifics harboring manipulative hairworms and nematodes (21), indicating that these parasites modulate the transcriptome of their hosts. Precisely how parasites exert molecular changes in their hosts remains unclear. In some cases, they may act as neuropharmacologists, by secreting and releasing hormones, neurotransmitters or other active substances in their hosts (16), (22). In other cases, they appear to have co-opted the services of symbiotic microorganisms. For example, parasitoid wasps that induce their beetle host to protect them post-emergence as they pupate on vegetation have harnessed the paralytic effects of a virus to achieve this manipulation: adult female wasps inject the virus within the host at the same time as their eggs, leaving the future of their progeny in the hands of their viral partner (23). Research on the mechanisms used by parasites to usurp the behavior of their hosts remains the least understood aspect of host manipulation (5, 15).

---

encountering their next host via predation of the infected host, this largely remains untested.

## 12.2.2  Responses to environmental cues

### 12.2.2.1  Light and color preferences

Another form of behavioral manipulation involves changing the way the host responds to environmental cues. For instance, a commonly affected trait is how the host responds to light, with most hosts losing their photophobic response, and even being attracted to light (27). This response to light probably acts to attract (or at least not repel) hosts from locations where they are easily visible, and more detectable to predators. In a similar way, infected hosts may change their preference for substrate color, again making them more conspicuous. For example, isopods (*Armadillidium vulgare*) infected with an acanthocephalan parasite (*Plagiorhynchus cylindraceus*) showed a stronger

preference for lighter-colored substrates than uninfected isopods, and infected isopods were predated on more frequently by the definitive host (starlings, *Sturnus vulgaris*) (28). Alternatively, a parasite may alter the appearance of its host, either directly, by altering its color (29) or indirectly, by impairing the color change response of the host to camouflage with its background (30). While changes in appearance may not strictly be considered behavioral changes in the same way discussed earlier, they have the same net effect of making the host more conspicuous to predators.

### 12.2.2.2  Elevation and moisture cues

Parasites may also manipulate host preference for other environmental variables. For instance, parasites can manipulate a host's elevation preferences, most often causing the host to seek higher elevation locations. Snails (*Stagnicola elodes*) infected with the trematode *Plagiorchis elegans* spent significantly

**Figure 12.2** Hairworm *Gordius* sp. (Nematomorpha) emerging from the posterior end of its host, the wētā *Heimidena* sp. (Orthoptera) in a New Zealand stream. Photo: Jean-François Doherty.

more time at the water's surface when the parasite's infective stages (cercariae) were emerging compared to uninfected snails (31). Parasites are also known to induce an attraction to water. Some of the most well-known examples of this include the effect of horsehair worms (Nematomorpha) on orthopteran hosts (Figure 12.2); a host that normally shows no attraction to water is suddenly compelled to seek water where the parasite emerges from its host to complete its lifecycle (8). Similar responses are seen in humans infected by guinea worms (*Dracunculus medicinis*), where a burning/itching lesion on the leg is only relieved when it is placed in water. The parasite then emerges from the leg and releases its larvae into the water body (32). Thus, parasites can induce changes in how a host responds to environmental variables and their environmental preferences.

### 12.2.3 Changes to social or reproductive behavior

The third main behavioral target of parasite host manipulation is social or reproductive behavior, which can be affected in a variety of ways. In most cases, these changes in behavior lead to reductions in aggregation or social isolation (29). For instance, reductions in aggregative behavior have been found in the amphipod *Gammarus pulex* infected with *Pomphorhynchus* laevis when a predator scent is present (33). It is hypothesized that the inhibition of

aggregation in response to predator stimuli makes infected hosts more vulnerable to predation. Reproductive behaviors may also be altered by parasites, where sexual selection may be altered, or in extreme cases, reversed. For instance, in the bush cricket *Requena verticalis* courtship roles become reversed when female crickets are infected with a protozoan gut parasite (34). The adaptive significance of changes in reproductive behaviors induced by parasites is less well-understood compared to other more obvious forms of manipulation. Nevertheless, changes to these behaviors may have consequences for the host's social organization and population fitness.

### 12.2.4 Manipulation of more than one trait

Parasites can target a wide range of behaviors as a part of host manipulation, and in many cases, a single parasite may target more than one behavior, culminating in infected hosts having altered behavior or appearance along multiple phenotypic dimensions (35). For example, a parasite may manipulate both activity (increasing activity) and microhabitat selection (choosing more open habitats) to increase the likelihood of predation by the definitive host. As a result, these simple targets of behavioral manipulation may have consequences for other behavioral phenomena in host populations, such as personality and behavioral syndromes.

## 12.3 Manipulation of host personality and behavioral syndromes

Animal personalities and behavioral syndromes provide a powerful framework to investigate individual behavior. They involve consistent among-individual differences in behavior, temporal, and contextual repeatability of behavior within individuals, and correlations between various behaviors, in particular activity, boldness, exploration, aggressiveness, and sociability (48, 49). Animal personality and behavioral syndromes also represent an ideal platform to explore the interactions and feedback between parasitism and behavior (50, 51). On the one hand, certain personality types may incur greater susceptibility to infection. For example, correlational and experimental evidence suggests that more active and bolder individuals encounter more infective stages of parasites transmitted by either contact or ingestion, in animals ranging from fish to mammals (52, 53). On the other hand, infection by parasites may alter the expression of personality traits or the correlation among behaviors, either through mere pathology, or through adaptive manipulation of host personality with fitness benefits for the parasite. In general, feedbacks may develop between personality and parasitism, where infection by a parasite alters personality in a way that changes the host's infection risk (see Chapter 3 of this book, (54)).

### 12.3.1 Parasite manipulation of host personality traits

Highly virulent parasites do not necessarily have the greatest impact on the personality of their hosts (55). In contrast, the recognition that manipulative parasites do not target single host traits, but instead a suite of more-or-less interrelated traits (35), makes it almost inevitable that they should influence host personality. First, in the simplest scenario, a manipulative parasite can change the expression of a host behavioral trait (Figure 12.3A), including any of the usual behaviors used to characterize personality such as boldness, aggressiveness, or sociability. This can manifest as an increase or decrease in the average value of the trait across multiple observations, or in its repeatability in a given context (51). There are multiple examples

of manipulative parasites causing increases or decreases in the magnitude of personality traits. For instance, microsporidian infection renders sticklebacks more social, which could facilitate the horizontal transmission of the parasite to conspecific fish (56) (see also Box 12.3). Altered boldness has also been observed in individuals infected by other types of parasites, but whether or not this results from manipulation or pathology is unclear, as transmission benefits are not always apparent (57). Some of the most intriguing examples of altered personality traits induced by a manipulative parasite are those associated with *Toxoplasma gondii* infections in humans (see Box 12.2), even if humans are not the normal intermediate host of this trophically transmitted parasite.

### 12.3.2 Parasite effects on the repeatability of behaviors

Beyond acting on the average value of behavioral traits, natural selection should also restrict their actual expression to only a narrow range of values that maximize fitness under particular environmental conditions (58). Thus, faced with a predator, only a limited range of responses along the shy–bold continuum will allow escape. A trophically transmitted parasite using a prey organism as an intermediate host that is able to decrease the repeatability (increase the variance) in its host's response may increase its own probability of transmission via predation (50). Decreased repeatability in host behavior could create mismatches between stimulus and response, with inappropriate anti-predator responses being more likely to occur (51). However, few studies have tested the effects of manipulative parasites on the repeatability of behaviors. For example, Coats *et al.* (59) have shown that crustaceans infected with trophically transmitted stages of trematodes show lower behavioral repeatability than their uninfected conspecifics (see also Box 12.3). This is certainly an area at the intersection of parasitism and animal personality that requires further study.

#### 12.3.2.1 Parasite manipulation of reaction norms

Manipulative parasites may also modify the consistency of personality traits across different environmental contexts (51). Seen as a reaction

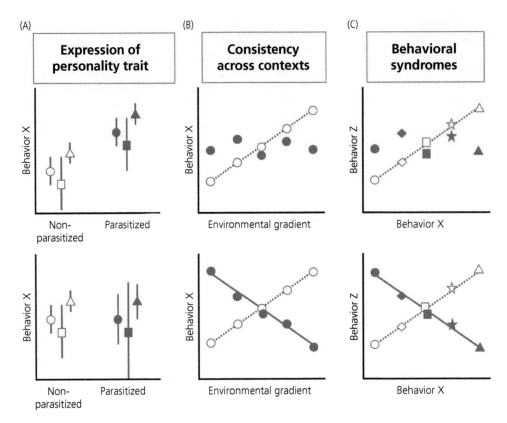

**Figure 12.3** Effects of manipulative parasites on host personality. Different symbols represent different individuals that are either infected (filled symbols) or uninfected (open symbols). (A) Infection may alter the average value of the trait across multiple observations over time in different host individuals (top), change its within-individual variance, that is lower its repeatability (bottom), or both. (B) Behavioral measurements taken on the same individual across a gradient of environmental or ecological contexts can be changed by infection, from a well-adjusted responsiveness to immediate conditions, to the complete loss of context-dependence (top), or the expression of maladaptive responses mismatched to the current context (bottom). (C) Infection can modify behavioral syndromes, such that a correlation between two behavioral traits across different individuals is broken and the two behaviors become uncoupled and independent of each other (top), or the correlation is drastically different from that seen pre-infection (bottom).

norm, that is a two-dimensional phenotype-by-environment space, this would mean that the phenotypic expression of a given genotype across a spectrum of environmental conditions is altered by parasite infection (Figure 12.3B). The responsiveness of particular behaviors to changing external conditions is itself a component of personality. The average behavioral score, and possibly its variance, shown by an individual post-infection may be the same as pre-infection; however, the slope of the reaction norm may be changed. Manipulative parasites may improve their transmission success by making the responses of the host completely independent from the environmental context, or

by reversing the relationship between context and response. Although empirical examples are lacking, it is easy to imagine, for example, a trophically transmitted parasite benefiting from a change in the reaction norm characterizing the expression of boldness or activity as a function of hunger or perceived risk.

### 12.3.3 Parasite manipulation of behavioral syndromes

Manipulative parasites may affect behavioral syndromes, especially if they modify certain behaviors but not others, resulting in the decoupling

---

**Box 12.2  Behavioral modifications of hosts by a widespread generalist. *Toxoplasma gondii* as a case study**

---

*Toxoplasma gondii* (Apicomplexa) provides one of the most famous examples of host manipulation, whereby rodents infected with *T. gondii* show an attraction to cat urine (36). Because felines are the definitive host for this parasite (they need to consume *T. gondii* encysted tissue to become infected), an attraction to cat urine may increase encounter rates of the intermediate host with the final, predatory host. But the raft of behavioral modifications goes beyond a mere attraction to cat urine, with motor coordination, memory and learning, and anxiety among some of the traits that are reportedly different in infected rodents (summarized by Worth (37)). One of the major gaps in our knowledge on this system however relates to whether infection with *T. gondii* actually leads to a greater risk of being predated on by cats, and consequently, aids the transmission of *T. gondii*.(38) Despite this gap in knowledge, it remains a well-known example of host manipulation.

While much of the focus of this host–parasite association has centered on rodents, associations between *T. gondii* infection and behavior have been reported in non-rodent host species as well. For instance, *T. gondii* seropositive spotted hyena (*Crocuta crocuta*) approached lions (*Panthera leo*) more closely, and tended to die more frequently from lion attacks than seronegative hyena (39). Similarly, *T. gondii* seropositive chimpanzees (*Pan troglodytes*) spent more time inspecting leopard urine (*Panthera pardus*) than seronegative chimpanzees (40). A spate of correlational

studies have also been carried out in humans, linking entrepreneurship (41), traffic accidents (42), and even culture (43) with seropositivity with *T. gondii*. Of course, the problem of studying associations between infection and behavior in these non-model systems is that it is difficult to tease apart correlation from causation.

Since *T. gondii* is associated with changes in key behavioral traits such as activity, exploration, and a loss of fear, these *T. gondii*-induced behavioral changes may also have consequences for host personality. While associations between *T. gondii* infection and human personality traits such as impulsiveness and rule-breaking have been found (44), surprisingly the effects of *T. gondii* on animal personality have not received much attention. Webster *et al.* (45) found that *T. gondii* seropositive rats were less neophobic than seronegative rats. Similarly, Berdoy *et al.* (46) found that *T. gondii* seropositive rats were more active, exploratory, and more easily trapped than seronegative rats. However, none of these studies have examined whether these behaviors are repeatable, whether different behaviors are correlated to form behavioral syndromes, nor how the personality of the host changes after infection. Thus, there is significant opportunity to explore this area of research further. Given the high prevalence of this parasite in some wildlife populations (47), *T. gondii* could well have significant effects on the behavioral composition of animal populations.

---

of previously linked traits or a change in how traits correlate with each other (Figure 12.3C). Again, consider a trophically transmitted parasite having to interfere with its host's anti-predator tactics to reach its definitive host. To achieve protection against predators, prey rely on a suite of behaviors (e.g., altered microhabitat selection, decreased activity, decreased boldness); however, these are most likely to be effective only if working in conjunction. Uncoupling these behaviors and disrupting behavioral syndromes may therefore increase the probability of transmission. Potentially, the effects on behavioral syndromes could be quite nuanced; decoupling behavioral syndromes enough to increase the likelihood of transmission,

while maintaining other behavioral traits at their optimum to enable the survival of the host until an opportunity for transmission occurs. Only a few studies to date have tested for differences in behavioral syndromes between parasitized and non-parasitized conspecific hosts (50, 51) (see also Box 12.3). Experimental studies will allow stronger insights into whether parasites can decouple these behavioral syndromes, and the consequences for transmission. Despite the limited empirical research currently available, it is clear that animal personality research provides both a conceptual basis and a simple analytical toolkit to explore the impact of manipulative parasites on individual host behavior.

> **Box 12.3 Manipulating fish personality—*Diplostomum* trematodes**

**Box 12.3, Figure 1** Metacercaria of the diplostomid trematode *Tylodelphys darbyi* in the eye of its intermediate host, the freshwater fish *Gobiomorphus cotidianus*, from a New Zealand lake. Photo: Brandon Ruehle.

Diplostomid trematodes infect freshwater fish worldwide (Figure 1) and have been well studied in the context of their effects on fish behavior. These parasites have a complex three-host lifecycle. Adult worms live in the gastrointestinal tract of various piscivorous birds, from which they release their eggs in host feces. Larvae hatched from these eggs proceed to infect snails, in which they multiply asexually to produce cercariae, the free-swimming infective stages that emerge from snails to seek suitable fish hosts. After penetrating the skin of a fish, cercariae migrate to the brain or the eyes of the host (depending on the species), where they encyst as metacercariae, and await ingestion by a bird to complete the lifecycle. Therefore, individual parasites capable of manipulating fish personality in ways that enhance the risk of avian predation should reap transmission benefits. These host parasite systems also provide great opportunities to seek parasite–behavior feedbacks: the same behaviors that are manipulated by the parasite may in turn determine the exposure of the fish to further infection by the parasite's free-swimming cercariae.

When experimentally infected by the brain-encysting *Diplostomum phoxini*, the Eurasian minnow *Phoxinus phoxinus* showed no change in its average boldness, activity, and exploration. However, the repeatability of these personality traits was changed: infection caused an increase in the temporal repeatability of boldness and activity, but a reduction in that of exploration (60). In addition, a behavioral syndrome

was broken. Indeed, the correlation between exploration and the other two traits disappeared among fish with a high intensity of infection (60). These results highlight the power of personality analysis to reveal subtle effects of parasites on animal behavior: if only mean trait expression had been compared between infected and non-infected fish, the influence of the parasites would have remained completely undetected. It is worth noting that in wild populations, the opposite pattern was detected, where a population experiencing both high infection and predation risk had higher boldness but lower activity (61). This result highlights the importance of combining field and laboratory studies in fully understanding how parasite manipulation works in natural systems.

In contrast, *Diplostomum pseudospathaceum*, which encysts in the eye lens of salmonid fishes, causing cataracts above a certain intensity of infection, appears to have no impact on the standard metrics of animal personality. In an experimental study with rainbow trout (*Oncorhynchus mykiss*), infection did not affect the repeatability of boldness, exploration, or activity, or the correlations among these traits (62). Infected fish tended to be a little less bold than uninfected controls, but only in one of three rounds of observation (62). In this host–parasite system, it seems that personality can determine infection risk (63), whereas infection cannot in turn modify personality. Interestingly,

---

**Box 12.3** *Continued*

even if it does not influence personality, *D. pseudospathaceum* can modify some specific aspects of the fish's anti-predator response (13). Despite their shared phylogenetic history and similar position within the host's neurosensory apparatus, *D. phoxini* and *D. pseudospathaceum* do not mediate similar changes in host personality, suggesting that the evolution of the ability to manipulate host personality may be highly species-specific.

---

## 12.4 Manipulation of host social networks and collective behavior

### 12.4.1 Manipulation of aggregative behavior

#### 12.4.1.1 Parasites that reduce aggregative behavior

As outlined earlier in this chapter, parasites are capable of manipulating host social behavior, which may have implications for host social organization or collective behaviors. Usually, aggregation tendencies are affected, with infected individuals less likely to aggregate, particularly in the presence of predator signals (33). This reduction in the propensity to aggregate seems to be a commonly observed response in amphipods infected with acanthocephalans (64) or trematodes (65). A reduction in aggregation behavior is thought to increase the vulnerability of amphipods to predation (66), presumably by the definitive host for the parasite. This parasite-induced disruption to aggregative behavior could have implications for the social behavior of amphipods. Amphipod social behavior is not well understood, although some studies have hinted at relatively complex social organization, involving maternal care and kin recognition (summarized by Beermann *et al.* (67)).

#### 12.4.1.2 Parasites that increase aggregative behavior

In other cases, aggregative behavior in infected hosts increases. For example, three-spined sticklebacks infected with an ectoparasite (*Glugea anomala*) were more sociable (spending more time in proximity of a mirror) than uninfected fish (56). This may favor the transmission of this microsporidian, which releases infective spores into the water when tumors containing microsporidia rupture.

An interesting case of parasite-induced aggregation occurs in the paper wasp *Polistes dominula*, where the strepsipteran parasite *Xenos vesparum* causes infected wasps to abandon their colony and form extranidal aggregations (68). The authors suggest this manipulation acts to increase mating opportunities for the emerged parasite, which would otherwise get attacked within the nest. Thus, while this parasite induces an aggregation response in the host, the impacts of this parasite on *P. dominula* social organization may be more noticeable at the colony level, where colony workers are lost through nest abandonment.

#### 12.4.1.3 Effects of manipulative parasites on social organization

In the above examples, while aggregation behavior is affected by manipulative parasites, whether this results in changes in social organization is unknown. For instance, changes to aggregative behavior may just interrupt aggregation tendencies, but does it affect social preferences or social bonds as well? There are few examples of this kind of effect of parasites on social preferences. For instance, uninfected three-spined sticklebacks preferred to join shoals composed of both uninfected fish and fish infected by the eye fluke *Diplostomum pseudospathaceum*, compared to shoals comprising only uninfected fish (69). While the adaptive significance of this behavior is unknown, it demonstrates the potential for manipulative parasites to alter social preferences of infected hosts. This may have consequences for animal social networks (see Box 12.4), and the corresponding transmission pathways of parasites, both those directly affected by the manipulation, as well as other parasites or diseases (summarized by Poulin (70)).

## Box 12.4  Parasite manipulation of social networks—an example using three-spined stickleback societies

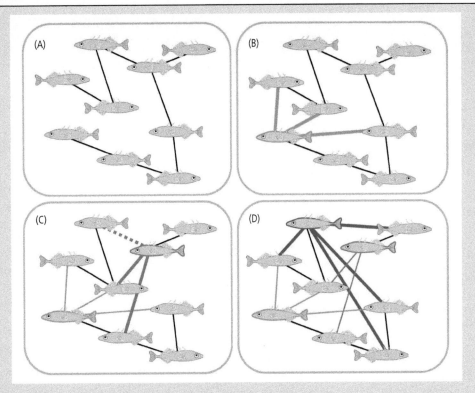

**Box 12.4, Figure 1**  A theoretical social network of three-spined sticklebacks (*Gasterosteus aculeatus*) showing how manipulative parasites may impact social networks in different ways. (A) Shows the original (completely fictional) social network; (B) shows the effect microsporidia (*Glugea anomala*) have on conspecific attraction, leading to the infected fish (in green) generating additional network connections; (C) shows the effect eye flukes (*Diplostomum pseudospathaceum*) have on altering social preferences of the infected fish (in orange), creating new connections, and omitting existing associations (dotted line); (D) shows the effect cestodes (*Schistocephalus solidus*) have on associations; cestodes make fish bolder, and other fish are attracted to bold fish, again, generating new connections to the infected fish (in blue).

While no single study has yet examined how parasite manipulation may alter social networks, several studies on the three-spined stickleback (*Gasterosteus aculeatus*) provide insights into how parasites alter social networks (and the processes that occur on them). Three-spined sticklebacks display stable partner associations within shoals (74), and by looking at who shoals with who, researchers have demonstrated that these shoaling preferences form non-random associations in highly connected social networks (75) (Figure 1A). Three-spined stickleback are also host to parasites that manipulate host behavior. One of these manipulates the social behavior of the host. The microsporidian *Glugea anomala* affects social attraction, with infected fish being faster to approach a mirror and spending more time in proximity of the mirror than uninfected fish (56). Pre-

sumably, this would lead to stronger associations between fish (since infected fish spend more time in proximity of other fish), and potentially a higher number of connections (if infected fish are faster to approach other fish) within stickleback social networks (Figure 1B). Thus, infection with *G. anomala* may lead to more closely connected social networks that provide increased opportunities for the transmission of this and other parasites.

Another manipulative parasite infecting sticklebacks has been found to alter shoaling preferences. The eye fluke *Diplostomum pseudospathaceum* has a complex lifecycle where the fish acts as an intermediate host for the trematode, which needs to be eaten by a bird (the definitive host) to complete the lifecycle. Rahn *et al.* (69) recently found this parasite can alter shoaling preferences of fish.

| **Box 12.4** *Continued* |
| --- |

In experimental tanks, fish were given the option of shoaling with either a shoal composed of all uninfected fish, or a shoal with both infected and uninfected fish. Uninfected fish showed a significant preference for mixed shoals over uninfected shoals, while infected fish showed no preference. While the adaptive significance of this behavioral difference between infected and uninfected fish is unclear, it demonstrates that these flukes which live in the eye lens of the fish have the potential to alter shoaling preferences of three-spined sticklebacks (Figure 12.5C).

Another parasite is known to manipulate the personality of sticklebacks. The cestode *Schistocephalus solidus* has a complex lifecycle, for which three-spined sticklebacks are a second intermediate host, with the final host being a predatory bird. Notably, *S. solidus* has been found to increase the risk-taking behavior of three-spined sticklebacks, with infected fish swimming closer to the surface and emerging sooner after a simulated predatory attempt (76). More recent work has revealed that infected sticklebacks are bolder; taking less time to return to a food bait after a simulated predator attack than uninfected sticklebacks (77). Similarly, infected fish moved more slowly and predictably, and when paired with another infected fish they moved less cohesively than uninfected pairs (78).

While these behavioral manipulations presumably act to make sticklebacks more vulnerable to predation by the definitive host of *S. solidus*, they also have implications for host personality. Stickleback personality has been found to have a strong influence on social network structure, with bolder fish having significantly fewer, but more even interactions in their social networks than shy fish, who had more, heterogeneous interactions (79). In addition, both bold and shy fish have been found to prefer shoaling with bold fish (80). Thus, effects of *S. solidus* on boldness traits could have significant consequences for shoaling preferences and social network structure in sticklebacks, leading to more connections with bolder fish (Figure 1D).

While the link between parasite infection and social network structure has not been explicitly investigated in this host–parasite system, parasites have effects on the types of behaviors that are important determinants of social network structure. These parasite-induced changes to network structure conceivably may impact other processes that occur in networks, including communication and the transmission of other infections. The persistence of these effects will depend on how long the host survives while being infected. Given that most manipulative parasites are transmitted via predation of the definitive host, ultimately the impacts of manipulative parasites within host populations could be relatively fleeting.

## 12.4.2 Manipulation of collective behavior

For species that show synchronized, collective behaviors, infection with manipulative parasites may disrupt behavioral synchrony (see Chapter 5 of this book, (71)). For example, schooling behavior in European minnows, *Phoxinus phoxinus*, may become disrupted when fish are infected with the cestode *Ligula intestinalis*, with infected fish not schooling as tightly as uninfected fish (72). Since schooling behavior has anti-predator benefits, deviation from normal schooling behavior is likely to lead to increased predation rates of infected fish. On the other hand, parasites may increase swarming behaviors. For example, swarming behavior increased in the crustacean *Artemia parthenogenetica* in response to infection with the cestode *Flamingolepis liguloides* (73). This change in behavior is also coupled with an increase in red color intensity, making these hosts more conspicuous to the definitive host (the greater flamingo, *Phoenicopterus roseus*) for this parasite (73). *Artemia parthenogenetica* and *A. franciscana* are also infected by two microsporidian species (*Anostracospora rigaudi* and *Enterocytospora artemiae*), and infection by these microsporidia also leads to more swarming and surfacing behaviors, which allows increased spore transmission (73). Thus, in this system, manipulation by two different parasites with different transmission methods both resulted in increased swarming of infected hosts. The effects of parasite manipulation on collective behaviors such as swarming and schooling are likely to be dependent on the mode of transmission; transmission via contact should be favored by increased aggregation/swarming behavior, while parasites with complex lifecycles may be favored by reduced aggregation/swarming. Testing these predictions would require more data on a range of comparable systems.

### 12.4.3 Implications of manipulative parasites for social behavior and feedbacks

Overall, changes to individual behavioral tendencies by manipulative parasites, particularly those linked to aggregation or social behavior, may have significant consequences for social organization and collective behaviors. Social networks are increasingly being used as a framework to model animal social structure, including how behavioral changes as a result of parasitic infection may lead to changes in network structure (see Chapter 5 of this book, (71); also see Box 12.4). However, social networks are currently underutilized in understanding the effects of manipulative parasites on social structure. Incorporating a network approach to characterize aggregative behavior and social preferences may help reveal the population-level effects of parasite manipulation on social network structure.

## 12.5 Conclusions and future research

Animal personality and social networks have recently become the two most widely used conceptual frameworks for the study of animal behavior. In this chapter, we have summarized the many ways through which manipulative parasites can alter host personality and social interactions. If parasites are ignored by researchers, they act as an unknown source of behavioral variation. If they are included in a study, they represent an additional explanatory variable, sometimes playing a more important role than the usual drivers of behavior such as hunger or predation risk. The most glaring shortcoming of past research in this area has been its reliance on correlative evidence. Indeed, most studies to date have used naturally infected animals, rather than experimentally infected ones (5). The ideal study design would involve experimental infections, with before-and-after comparisons of infected individuals and uninfected controls, for investigations of both individual-level and group-level behaviors.

A full understanding of the interactions between any aspect of animal behavior and parasitism requires a comprehensive and explicit consideration

of the reciprocal feedbacks between the two (81). The causal arrow can point in both directions: just as the behavior of individual animals can influence their probability of acquiring parasitic infections, these infections can in turn affect behavior. On evolutionary timescales, positive feedbacks between host behavior and parasitism can have self-reinforcing effects, with parasite-mediated selection shaping the evolution of host behavior, and vice versa, in other words host behaviors shaping parasite evolution. In addition, other trade-offs might emerge as a result of host–parasite manipulation (e.g., costs of producing manipulation cues versus parasite reproductive fitness). These trade-offs could lead to feedbacks between hosts and parasites that might ultimately alter the fitness of manipulative parasites in hosts, and the selective advantage of parasite manipulation of host behavior. As we highlighted in this chapter, manipulative parasites are particularly likely to act as selection agents, given their strong proximate impacts on host behavior. Although not widely acknowledged, they exert subtle influences on the expression of host behavior, which may filter through to affect personality traits and the structure of host social networks. While a growing number of studies are starting to look at this, our understanding of the topic remains relatively superficial.

The conceptual frameworks of both animal personality and social networks provide ideal platforms to begin to advance our understanding of the wider consequences of parasite manipulation on host behavior, but remain underutilized. A limitation around their use lies in suitable model systems. Fish are a particularly promising group for incorporating manipulative parasites into these frameworks, since they can be individually marked and their associations and behavior can be tracked (see Box 12.4). It should become standard to consider the potential phenotypic effects of infection by manipulative parasites in all future studies of animal behavior in the wild. Animals most likely to be affected by manipulative parasites include those often used as intermediate hosts by trophically transmitted parasites (invertebrates and small-bodied vertebrates), and insects and spiders known to be victims of parasitoids. We suggest these might be useful starting points for exploring the impacts of parasite manipulation on behavioral systems.

# References

1. Altizer S, Nunn CL, Thrall PH, Gittleman JL, Antonovics J, Cunningham AA, *et al*. Social organization and parasite risk in mammals: Integrating theory and empirical studies. *Annu Rev Ecol Evol Syst.* 2003;34:517–47.

2. Hughes DP, Brodeur J, and Thomas F. *Host Manipulation by Parasites*. Oxford: Oxford University Press, 2012.

3. Moore J. *Parasites and the Behavior of Animals*. Oxford: Oxford University Press, 2002.

4. Poulin R. Parasite manipulation of host behavior: An update and frequently asked questions. *Adv Stud Behav.* 2010;41:151–86.

5. Poulin R and Maure F. Host manipulation by parasites: A look back before moving forward. *Trends Parasitol.* 2015;31:563–70.

6. Lafferty KD. The evolution of trophic transmission. *Parasitol Today* 1999; 15:111–15.

7. Cator LJ. Altered feeding behaviors in disease vectors. In Ezenwa VO, Altizer S, and Hall RJ (eds), *Animal Behavior and Parasitism*. Oxford: Oxford University Press, 2022. DOI: 10.1093/oso/9780192895561.003.0013.

8. Thomas F, Schmidt-Rhaesa A, Martin G, Manu C, Durand P, and Renaud F. Do hairworms (Nematomorpha) manipulate the water seeking behaviour of their terrestrial hosts? *J Evol Biol.* 2002;5:356–61.

9. Maure F, Daoust SP, Brodeur J, Mitta G, and Thomas F. Diversity and evolution of bodyguard manipulation. *J Exp Biol.* 2013;216:6–42.

10. Grosman AH, Janssen A, de Brito EF, Cordeiro EG, Colares F, Fonseca JO, *et al*. Parasitoid increases survival of its pupae by inducing hosts to fight predators. *PLoS One* 2008;3:e2276.

11. Pirrie A, Chapman H, and Ashby B. Parasite-mediated sexual selection: To mate or not to mate? in Ezenwa VO, Altizer S, and Hall RJ (eds), *Animal Behavior and Parasitism*. Oxford: Oxford University Press, 2022. DOI: 10.1093/oso/9780192895561.003.0009.

12. Lopes PC, French SS, Woodhams DC, and Binning SA. Infection avoidance behaviors across vertebrate taxa: Patterns, processes and future directions. In Ezenwa VO, Altizer S, and Hall RJ (eds), *Animal Behavior and Parasitism*. Oxford: Oxford University Press, 2022. DOI: 10.1093/oso/9780192895561.003.0014.

13. Seppälä O, Karvonen A, and Valtonen ET. Eye fluke-induced cataracts in natural fish populations: Is there potential for host manipulation? *Parasitol.* 2011;138:209–14.

14. Grécias L, Valentin J, and Aubin-Horth N. Testing the parasite mass burden effect on alteration of host behaviour in the *Schistocephalus*–stickleback system. *J Exper Biol.* 2018;221:jeb174748.

15. Herbison REH. Lessons in mind control: Trends in research on the molecular mechanisms behind parasite-host behavioral manipulation. *Front Ecol Evol.* 2017;5:102.

16. Klein SL. Parasite manipulation of the proximate mechanisms that mediate social behavior in vertebrates. *Physiol Behav.*2003;79:441–9.

17. Henriquez SA, Brett R, Alexander J, Pratt J, and Roberts CW. Neuropsychiatric disease and *Toxoplasma gondii* infection. *Neuroimmunomodulation* 2009; 6:22–33.

18. Tain L, Perrot-Minnot MJ, and Cézilly F. Altered host behaviour and brain serotonergic activity caused by acanthocephalans: Evidence for specificity. *Proc R Soc B: Biol Sci.* 2006;273:3039–45.

19. Helluy S. Parasite-induced alterations of sensorimotor pathways in gammarids: Collateral damage of neuroinflammation? *J Exper Biol.* 2013;216:67–77.

20. Hari Dass SA and Vyas A. *Toxoplasma gondii* infection reduces predator aversion in rats through epigenetic modulation in the host medial amygdala. *Mol Ecol.*2014;23:6114–22.

21. Herbison R, Evans S, Doherty J-F, Algie M, Kleffmann T, and Poulin R. A molecular war: Convergent and ontogenetic evidence for adaptive host manipulation in related parasites infecting divergent hosts. *Proc R Soc B: Biol Sci.* 2019; 86:20191827.

22. Berger CS and Aubin-Horth N. The secretome of a parasite alters its host's behaviour but does not recapitulate the behavioural response to infection. *Proc R Soc B: Biol Sci.* 2020;287:20200412.

23. Dheilly NM, Maure F, Ravallec M, Galinier R, Doyon J, Duval D, *et al*. Who is the puppet master? Replication of a parasitic wasp-associated virus correlates with host behaviour manipulation. *Proc R Soc B: Biol Sci.* 2015;282:20142773.

24. Lafferty KD and Shaw JC. Comparing mechanisms of host manipulation across host and parasite taxa. *J Exp Biol.* 2013;216:56–66.

25. Dezfuli BS, Maynard BJ, and Wellnitz TA. Activity levels and predator detection by amphipods infected with an acanthocephalan parasite, *Pomphorhynchus laevis*. *Folia Parasitologica* 2003;50:129–34.

26. McCurdy DG, Forbes MR, and Boates JS. Evidence that the parasitic nematode *Skrjabinoclava* manipulates host *Corophium* behavior to increase transmission to the sandpiper, *Calidris pusilla*. *Behav Ecol.* 1999;10: 351–7.

27. Benesh DP, Duclos LM, and Nickol BB. The behavioral response of amphipods harboring *Corynosoma constrictum* (Acanthocephala) to various components of light. *J Parasitol.* 2005;91:731–6.

28. Moore J. Responses of an avian predator and its isopod prey to an acanthocephalan parasite. *Ecol.* 1983;64:1000–15.

29. Krause J and Godin JGJ. Influence of parasitism on the shoaling behaviour of banded killifish, *Fundulus diaphanus. Can J Zool.* 1994;72:1775–9.

30. Seppälä O, Karvonen A, and Valtonen ET. Impaired crypsis of fish infected with a trophically transmitted parasite. *Anim Behav.* 2005;70:895–900.

31. Lowenberger CA and Rau ME. *Plagiorchis elegans*: Emergence, longevity and infectivity of cercariae, and host behavioural modifications during cercarial emergence. *Parasitol.* 1994;109:65–72.

32. Fenwick A. Waterborne infectious diseases—could they be consigned to history? *Science.* 2006;313: 1077–81.

33. Durieux R, Rigaud T, and Médoc V. Parasite-induced suppression of aggregation under predation risk in a freshwater amphipod: Sociality of infected amphipods. *Behav Processes* 2012;91:207–13.

34. Simmons LW. Courtship role reversal in bush crickets: Another role for parasites? *Behav Ecology* 1994;5: 259–66.

35. Thomas F, Poulin R, and Brodeur J. Host manipulation by parasites: A multidimensional phenomenon. *Oikos.* 2010;119:1217–23.

36. Berdoy M, Webster JP, and Macdonald DW. Fatal attraction in rats infected with *Toxoplasma gondii. Proc R Soc B: Biol Sci.* 2000;267: 591–4.

37. Worth AR, Thompson RA, and Lymbery AJ. Reevaluating the evidence for *Toxoplasma gondii*-induced behavioural changes in rodents. *Advances in Parasitol.* 2014;85:109–42.

38. Worth AZ, Lymbery AJ, and Thompson RA. Adaptive host manipulation by *Toxoplasma gondii*: Fact or fiction? *Trends Parasitol.* 2013;29:150–5.

39. Gering E, Laubach Z, Weber P, Hussey G, Turner J, Lehmann K, *et al.* Time makes you older, parasites make you bolder—*Toxoplasma Gondii* infections predict hyena boldness toward definitive lion hosts in Banzhaf W, Cheng BHC, Deb K, Holekamp KE, Lenski RE, Ofria, C, *et al.* (eds), *Evolution in Action: Past, Present and Future.* Cham; Springer, 2020, 205–24.

40. Poirotte C, Kappeler PM, Ngoubangoye B, Bourgeois S, Moussodji M, and Charpentier MJ. Morbid attraction to leopard urine in *Toxoplasma*-infected chimpanzees. *Curr Biol.* 2016; 6:R98–R99.

41. Johnson SK, Fitza MA, Lerner DA, Calhoun DM, Beldon MA, Chan ET, *et al.* Risky business: Linking *Toxoplasma gondii* infection and entrepreneurship behaviours across individuals and countries. *Proc R Soc B: Biol Sci.* 2018;285:20180822.

42. Flegr J, Havlícek J, Kodym P, Malý M, and Smahel Z. Increased risk of traffic accidents in subjects with latent toxoplasmosis: A retrospective case-control study. *BMC Infect Dis.* 2002;2:11.

43. Lafferty KD. Can the common brain parasite, *Toxoplasma gondii*, influence human culture? *Proc R Soc B: Biol Sci.* 2006;273:2749–55.

44. Flegr J, Zitkova S, Kodym P, and Frynta D. Induction of changes in human behaviour by the parasitic protozoan *Toxoplasma gondii. Parasitol.* 1996;113:49–54.

45. Webster JP, Brunton CFA, and Macdonald DW. Effect of *Toxoplasma gondii* upon neophobic behaviour in wild brown rats, *Rattus norvegicus. Parasitol.*1994;109:37–43.

46. Berdoy M, Webster JP, and MacDonald DW. The manipulation of rat behaviour by *Toxoplasma gondii. Mammalia* 1995;59:605–14.

47. Lindsay DS and Dubey JP. Toxoplasmosis in wild and domestic animals in Weiss L and Kim K (eds), *Toxoplasma gondii.* New York, NY: Academic Press, 2020, 293–320.

48. Wolf, M. and Weissing FJ. Animal personalities: Consequences for ecology and evolution. *Trends Ecol Evol.* 2012;27:452–61.

49. Sih A, Mathot KJ, Moiron M, Montiglio PO, Wolf M, *et al.* Animal personality and state-behaviour feedbacks: A review and guide for empiricists. *Trends Ecol Evol.* 2015;30:50–60.

50. Barber I and Dingemanse NJ. Parasitism and the evolutionary ecology of animal personality. *Philos Trans R Soc Lond B Biol Sci.* 2010;365:4077–88.

51. Poulin R. Parasite manipulation of host personality and behavioural syndromes. *J Exp Biol.* 2013;216: 18–26.

52. Wengström N, Wahlqvist F, Näslund J, Aldvén D, Závorka L, Österling ME, *et al.* Do individual activity patterns of brown trout (*Salmo trutta*) alter the exposure to parasitic freshwater pearl mussel (*Margaritifera margaritifera*) larvae? *Ethology* 2016;122:769–78.

53. Santicchia F, Romeo C, Ferrari N, Matthysen E, Vanlauwe L, Wauters LA, *et al.* The price of being bold? Relationship between personality and endoparasitic infection in a tree squirrel. *Mammalian Biol.* 2019;97: 1–8.

54. Sadoughi B, Anzà S, Defolie C, Manin V, Müller-Klein N, Murillo T, *et al.* Parasites in a social world: Lessons from primates. In Ezenwa VO, Altizer S, and Hall RJ (eds), *Animal Behavior and Parasitism.* Oxford: Oxford University Press, 2022. DOI: 10.1093/oso/9780192895561.003.0003.

55. Turner J and Hughes WOH. The effect of parasitism on personality in a social insect. *Behav Processes* 2018;157:532–9.

56. Petkova I, Abbey-Lee RN, and Løvlie H. Parasite infection and host personality: *Glugea*-infected three-spined sticklebacks are more social. *Behav Ecol Sociobiol.* 2018;72:173.

57. Seaman B and Briffa M. Parasites and personality in periwinkles (*Littorina littorea*): Infection status is associated with mean-level boldness but not repeatability. *Behav Processes* 2015;115:132–4.

58. Dingemanse NJ and Réale D. Natural selection and animal personality. *Behav.* 2005;142:1159–84.

59. Coats J, Poulin R, and Nakagawa S. The consequences of parasitic infections for host behavioural correlations and repeatability. *Behav.* 2010;147:367–82.

60. Kekäläinen J, Lai Y-T, Vainikka A, Sirkka I, and Kortet R. Do brain parasites alter host personality? Experimental study in minnows. *Behav Ecol Sociobiol.* 2014;68:197–204.

61. Kortet R, Sirkka I, Lai Y-T, Vainikka A, and Kekäläinen J. Personality differences in two minnow populations that differ in their parasitism and predation risk. *Front Ecol Evol.* 2015;3:9.

62. Klemme I, Kortet R, and Karvonen A. Parasite infection in a central sensory organ of fish does not affect host personality. *Behav Ecol.* 2016;27:1533–8.

63. Mikheev VN, Pasternak AF, Morozov AY, Morozov AY, and Taskinen J. Innate antipredator behavior can promote infection risk in fish even in the absence of predators. *Behav Ecol.* 2020;31:267–76.

64. Lewis SE, Hodel A, Sturdy T, Todd R, and Weigl C. Impact of acanthocephalan parasites on aggregation behaviour of amphipods (*Gammarus pseudolimnaeus*). *Behav Processes* 2012;91:159–63.

65. Arnal A, Droit A, Elguero E, Ducasse H, Sanchez MI, Lefevre T, *et al.* Activity level and aggregation behavior in the crustacean gammarid *Gammarus insensibilis* parasitized by the manipulative trematode *Microphallus papillorobustus*. *Front Ecol Evol.* 2015;3:109.

66. Thünken T, Baldauf SA, Bersau N, Bakker T, Kullmann H, and Frommen JG. Impact of olfactory non-host predator cues on aggregation behaviour and activity in *Polymorphus minutus* infected *Gammarus pulex*. *Hydrobiologia* 2010;654:137–45.

67. Beermann J, Dick JT, and Thiel M. Social recognition in amphipods: An overview in Aquiloni L and Tricarico E (eds), *Social Recognition in Invertebrates*. Cham: Springer, 2015, 85–100.

68. Hughes DP, Kathirithamby J, Turillazzi S, and Beani, L. Social wasps desert the colony and aggregate outside if parasitized: Parasite manipulation? *Behav Ecol.* 2004;15:1037–43.

69. Rahn AK, Vitt S, Drolshagen L, Scharsack JP, Rick IP, and Bakker TC. Parasitic infection of the eye lens affects shoaling preferences in three-spined stickleback. *Biol J Linn Soc.* 2018;123: 377–87.

70. Poulin R. Modification of host social networks by manipulative parasites. *Behav.* 2018;155:671–88.

71. Keiser CN. Collective behavior and parasite transmission. In Ezenwa VO, Altizer S, and Hall RJ (eds), *Animal Behavior and Parasitism*. Oxford: Oxford University Press, 2022. DOI: 10.1093/oso/ 9780192895561.003.0005.

72. Barber I and Huntingford FA. Parasite infection alters schooling behaviour: Deviant positioning of helminth-infected minnows in conspecific groups. *Proc R Soc B: Biol Sci.* 1996;263:1095–102.

73. Rode NO, Lievens EJ, Flaven E, Segard A, Jabbour-Zahab R, Sanchez MI, *et al.* Why join groups? Lessons from parasite-manipulated *Artemia*. *Ecol Lett.* 2013;16:493–501.

74. Ward AJ, Botham MS, Hoare DJ, James R, Broom M, Godin JGJ, *et al.* Association patterns and shoal fidelity in the three–spined stickleback. *Proc R Soc B: Biol Sci.* 2002;269:2451–5.

75. Croft DP, James R, Ward AJW, Botham MS, Mawdsley D, and Krause J. Assortative interactions and social networks in fish. *Oecologia* 2005;143:211–19.

76. Giles N. Behavioural effects of the parasite *Schistocephalus solidus* (Cestoda) on an intermediate host, the three-spined stickleback, *Gasterosteus aculeatus* L. *Animal Behav.* 1983;31:1192–4.

77. Talarico M, Seifert F, Lange J, Sachser N, Kurtz J, and Scharsack JP. Specific manipulation or systemic impairment? Behavioral changes of three-spined sticklebacks (*Gasterosteus aculeatus*) infected with the tapeworm *Schistocephalus solidus*. *Behav Ecol Sociobiol.* 2017; 1:1–10.

78. Jolles JW, Mazué GP, Davidson J, Behrmann-Godel J, and Couzin ID. *Schistocephalus* parasite infection alters sticklebacks' movement ability and thereby shapes social interactions. *Sci Rep.* 2020;10:1–11.

79. Pike TW, Samanta M, Lindström J, and Royle NJ. Behavioural phenotype affects social interactions in an animal network. *Proc R Soc B: Biol Sci.* 2008;275: 2515–20.

80. Harcourt JL, Sweetman G, Johnstone RA, and Manica A. Personality counts: The effect of boldness on shoal choice in three-spined sticklebacks. *Animal Behav.* 2009;77:1501–05.

81. Ezenwa VO, Archie EA, Craft ME, Hawley DM, Martin LB, Moore J, *et al.* Host behaviour–parasite feedback: An essential link between animal behaviour and disease ecology. *Proc R Soc B: Biol Sci.* 2016;283: 20153078.

CHAPTER 13

# Altered feeding behaviors in disease vectors

Lauren J. Cator

## 13.1 Introduction

It has long been noted that infection can impact animal foraging, feeding patterns, and food choice (1), (2). Early work documented anorexic responses in animals such as infected mice (3) and hedgehogs (4) and that forcing these animals to feed during this period resulted in higher mortality. Further work, particularly in insects, has revealed that immune activation, even in the absence of infection, can impact feeding phenotypes (5)–(7) and that these changes in behavior appear to be correlated with the ability to resist and tolerate infection (6)–(11). There is an ongoing discussion about the evolutionary processes that have led to these changes in feeding behavior and the ultimate explanations for connections between immune performance, host behavior, and infection outcome (12).

In parallel, but noticeably distinct from this body of work, there are many studies showing that infected vectors exhibit altered feeding behaviors. Until very recently, the role of the vector in these behaviors was relatively ignored, with most explanations invoking parasite manipulation (13). This chapter explores the evidence for changes in vector feeding behavior associated with infection and highlights what is known about the physiological mechanisms underlying these changes. It concludes with a discussion of the potential evolutionary explanations for these behavioral changes and identification of priorities for future research.

## 13.2 Vector-borne disease transmission

The transmission cycles of many human, animal, and plant pathogens involve a vector. In the majority of cases, a small ectothermic organism, such as an insect, tick, or mite, is required to transmit the pathogen between plant or animal hosts. These transmission cycles are as diverse as the pathogens involved in them but share some common features. First, the vector is most commonly infected or picks up the pathogen when it feeds on an infectious host. Once associated with the vector, many pathogens must undergo a period of development in the vector known as the extrinsic incubation period (EIP) (Figure 13.1). This period can vary from almost non-existent, as for non-circulative plant viruses (14), to a period of several days, as for many protozoan parasites (15). At the end of the EIP, the vector is able to pass on the pathogen to a new host. In most cases, infection is passed both to and from the vector through the act of feeding or at least in association with the feeding event (e.g., *Trypanosoma cruzi* transmitted by triatomine bugs). During the EIP, the vector continues to undergo its own lifecycle, including foraging which in many vectors is tied to discrete, sometimes complex, reproductive cycles. The interplay between the timing of vector feeding events, vector host choice, vector mortality, and pathogen development leads to the complex and captivating transmission dynamics of these diseases (Figure 13.1).

Lauren J. Cator, *Altered feeding behaviors in disease vectors*. In: *Animal Behavior and Parasitism*. Edited by Vanessa O. Ezenwa, Sonia Altizer, and Richard J. Hall, Oxford University Press. © Oxford University Press (2022). DOI: 10.1093/oso/9780192895561.003.0013

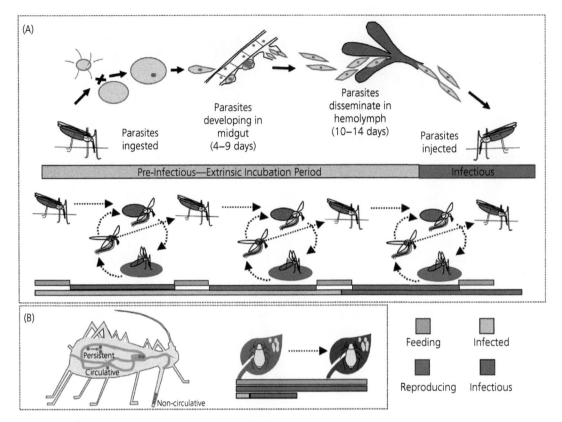

**Figure 13.1** Variation among vector types in feeding and reproduction and associated pathogen development. Vector behaviors during pathogen development may influence vector survival until infectious, while behaviors during the infectious period may influence the probability of onward transmission. Many animal pathogens have long extrinsic incubation periods in which they undergo obligate replication and life stages inside of the vector. Upper panel shows the obligate development of malaria parasites which spend more than a week developing inside of the mosquito vector (yellow bars) prior to reaching their infectious stage (red bars) and migrating to the mosquito salivary glands. Meanwhile, like many blood-feeding vectors, mosquitoes undergo their own lifecycle which involves discrete feeding (green bars) and reproductive (blue bars) cycles shown in the lower panel. These feeding events are often significant sources of mortality, and thus feeding, and associated reproduction, can greatly affect transmission by reducing the number of vectors surviving to the infectious stage (red bars). In contrast, many plant pathogens have short or non-existent EIPs and, depending on their biology, may or may not be infectious for the lifetime of the vector. The panel on the left highlights the differences between non-circulative, circulative, and persistent plant pathogens in an aphid vector (adapted from (14)). Non-circulative pathogens interact only with the vector's mouthparts, with transmission duration determined by the amount of pathogen on these exterior mouthparts. Circulative pathogens are ingested by the vector and can be found in the hemocoel, with transmission duration depending on vector lifespan and the amount of pathogen ingested. Persistent pathogens not only circulate within the vector but also replicate, and thus pathogen load can increase over time; transmission duration is then primarily dependent on vector lifespan. Unlike blood-feeding vectors, many plant-feeding vectors feed and reproduce more or less continuously, meaning that the proportion of infected vectors surviving until infectiousness is relatively high (right panel). Interestingly, there have been more reports of infection altering dispersal and aggregation of feeding vectors in plant systems (84,85) than in animal systems where the timing and frequency of feeding events are more often altered by infection.

## 13.3 Overview of altered feeding behaviors in disease vectors

Before exploring the evidence for infection-associated changes in vector behavior there are two important points to clarify. First, this chapter is focused on the effect of infection on vector physiology, behavior, and life history and its implications for transmission. This does not include changes in the attractiveness of hosts regardless of vector infection status. While this phenomenon also impacts the contact rate between vectors and

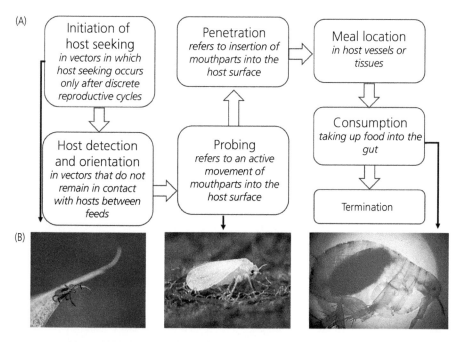

**Figure 13.2** (A) Sequence of feeding behaviors. "Feeding" actually encompasses several behaviors, many of which have been reported to be affected by infection. Some events such as host-seeking, detection, and orientation may be missing in vectors, such as aphids, that remain in contact with hosts between meals. Infection may determine individual events in this sequence or modify all parts of it. (B) There is evidence that infection alters many of these behaviors in a wide range of vectors. For example, questing behavior in ticks (21) (photo source: Skyler Ewing/Pexels), probing behavior in whiteflies (36) (photo source: Public Domain/US Agriculture Research Service via Wikimedia Commons), and consumption in fleas (48) (photo source: Public Domain/Centers for Disease Control via Wikimedia Commons) have all been suggested to be modulated by pathogen infection.

hosts, the underlying causes of changes in host attractiveness are differences in host physiology and behavior, not vector traits, and are reviewed elsewhere (16)–(18).

Second, "feeding" actually encompasses several behaviors (19). This review is organized into effects of infection on host–vector contact, including the initiation of host-seeking, host detection and orientation, and feeding behaviors given a successful contact which includes "at host" behaviors such as probing, penetration, consumption, and termination (Figure 13.2). It is important to consider that feeding behaviors occur in this sequential pattern (host-seeking initiation through to termination) and isolating subsets of behaviors out of sequence can lead to inaccurate characterization. For example, if infection alters host detection such that infected vectors do not ever contact the host, changes in probing behavior may be both irrelevant but also not representative. Care should be taken when interpreting

studies in which one of these behaviors has been measured in isolation.

### 13.3.1 Effects on vector–host contact rates

In order to make contact with the appropriate host, a vector must first initiate host-seeking behavior. There are several examples of infection altering the propensity of vectors to initiate host-seeking behavior from a wide range of systems. For example, whiteflies (*Bemisia tabaci*) infected with Tomato Yellow Leaf Curl Virus (TYLCV) have been found to feed more readily than uninfected individuals (20). Ticks (*Ixodes scapularis*) infected with *Borrelia burgdorferi* exhibit attraction to vertical surfaces which may indicate increased questing behavior (21). *Aedes aegypti* mosquitoes infected with dengue virus were reported to initiate feeding more quickly when offered a host than uninfected mosquitoes of the same age (22).

In many instances, changes in the likelihood of host-seeking correspond to the developmental stage of the pathogen inside of the vector. In these studies, increases in host-seeking responses were specifically associated with the transmissible stage of the pathogen and in some cases decreased host-seeking responses were associated with pre-infectious stages (23)–(25). In a particularly compelling example, Gleave et al. (2016) reported that *Ae. aegypti* mosquitoes infected with filarial parasites were less likely to approach hosts than uninfected controls while parasite larvae were developing inside them. Once larvae had matured to the infectious stage, however, infected females were more likely to approach hosts (25). Interestingly, changes in host-seeking responses were also qualitatively dependent on infection load. Nonresponsive mosquitoes harbored higher loads of developing parasites and responsive mosquitoes harbored higher loads of transmissible stage larvae. Similar patterns of host-seeking have been reported for malaria infected mosquitoes in some studies (23), (26).

In addition to feeding at the appropriate time, successful transmission requires that the vector feed on the appropriate hosts. When vectors are generalists and host species vary in their ability to support and transmit infection, changes in preferences between host species can have large impacts on transmission dynamics (27). For example, a recent study measuring host choice in aphids infected with Barley Yellow Dwarf Virus (BYDV) documented an infection-associated shift in preference for perennial host plants compared to annual hosts with infection (28). An associated mathematical model predicted that this infection-induced shift towards specialist feeding behavior would increase transmission intensity within perennial hosts, decreasing net transmission by concentrating infection in a specific host group. Work in animal pathogens has not found strong evidence for this kind of infection-induced shift in host preference (29). Given the potential importance of changes in preferences between host species, particularly for the spillover of pathogens, more studies addressing the effect of infection on host choice are needed.

More than altering preferences for different host species, vector infection status has also been reported to alter preferences between uninfected and infected hosts of the same species (recently reviewed in (18)). Again, the majority of the examples here come from plant disease systems with the effect of infection on preference varying with pathogen. For example, aphids infected with Potato Leaf Roll Virus (PLRV) prefer uninfected plants, while those infected with Barley and Cereal Yellow Virus (BCYV) exhibit preferences for BCYV-infected plants (30). In at least two cases, there is an apparent interaction between vector and plant infection status. Aphids infected with BYDV prefer uninfected plants, while uninfected aphids prefer BYDV-infected plants (31). Similarly, whiteflies infected with TYLCV also showed a preference for uninfected plants and uninfected whiteflies showed a preference for infected plants (32), (33). This pattern was predicted to enhance transmission probability greatly by maximizing contacts between infectious vectors and uninfected hosts and uninfected vectors and infectious hosts (34).

Again, evidence that infection affects vector preferences is largely lacking for animal and human disease vectors. While there is growing evidence that infection alters the odor profile of vertebrate hosts to make them more attractive to vectors (18), there is no evidence that pathogen inside the vector affects host preference. One study testing for changes in host preferences in *Culex* mosquitoes infected with bird malaria (*Plasmodium relictum*) found no difference in the preferences of infected and uninfected vectors (35).

## 13.3.2 Effects on feeding behaviors given an appropriate contact

There are a great deal more data on how infection can alter vector feeding behaviors once a vector has responded to and located a host. Infection has been shown to affect how long it takes for an individual to start feeding once it has reached the host. For example, infection leads to shorter latency to initiate feeding once reaching a host in dengue virus infected mosquitoes (22), while latency was longer in whiteflies infected with TYLCV (36).

In several plant and animal systems, infection increases the duration of the feeding event and the number of probes during the feeding period. Examples of these types of changes have been documented in whiteflies infected with TYLC (20), (36), thrips (*Frankliniella occidentalis*) infected with Tomato Spotted Wilt Virus (TSWV) (37), and potato psyllids infected with the bacteria *Candidatus Liberibacter solanacearum* (38). In animal systems, increased feeding duration and number of probes have been reported in tsetse flies (*Glossina* spp.) infected with *Trypanosoma* parasites (39), (40), sandflies (*Phlebotomus* spp.) infected with Leishmania parasites (41), and mosquitoes infected with La Crosse (42) and dengue (43), (44) viruses. In addition, for some mosquito–parasite combinations (45), mosquitoes infected with malaria sporozoites have been reported to have increased probing duration (46) and frequency (47).

Across the board, this increase in probing duration and frequency does not appear to correspond to increased food uptake. There is evidence from several systems that infection interferes with feeding success and efficiency (40)–(42), (48)–(56), but see (22). In some cases, this decreased feeding efficiency seems to lead to vectors feeding more persistently and in at least two examples infected vectors were even less deterred by chemical repellents (57), (58). Persistent feeders are less likely to abort feeding attempts after being unsuccessful and infection has resulted in reports of increased rates of vectors feeding from multiple hosts during a single feeding bout (22), (39), (56), (59), (60).

## 13.4 Mechanisms underlying changes in feeding behavior

Across all types of transmission systems (i.e., trophic, directly transmitted, sexually transmitted) the mechanisms that underlie infection-induced changes in host behavior remain poorly characterized (61). The mechanisms that underlie changes in vector behavior in vector-borne transmission systems are no different. However, there is some convincing evidence that some pathogens directly and physically block the feeding apparatus or digestive system of vectors to affect probing, refeeding, and feeding efficiency associated with behaviors described in section 13.3.2. One of the most iconic examples of a manipulated vector are the "blocked fleas" that transmit the plague bacteria (48). Early

work described the blockage of the fore and midgut with a solid culture of plague bacteria. This blockage leads to fleas that are able to suck, but unable to ingest blood. During feeding attempts the esophagus distends and blood escapes back through the mouth parts (48).

Similar kinds of blockages have also been observed in the final stage of development of Leishmania parasites in sandflies. The infectious form of the parasite has been reported to damage (62) and secrete a gel which physically blocks (52), (63) the stomodeal valve between the crop and midgut that regulates blood intake. This is hypothesized to increase transmission by increasing the probability that pathogens are spit back into the wound while feeding occurs (64). In addition to interfering with mechanoreceptors important for feeding (39), a recent study found that *G. morsitans* infected with *T. brucei* exhibit significant decreases in expression of genes encoding for salivary proteins when infected with *T. brucei*. The decreased expression is associated with a 70% reduction in protein content in saliva and results in significant impairment of anti-hemostatic (wound healing) activity (40). Decreases in activity of another important protein, apyrase, has been implicated in infection associated changes in mosquito probing, but it remains unclear if this varies with mosquito-parasite combinations (65). There appears to be a lack of explanation for how viral and bacterial infections affect the probing behavior and duration of plant disease vectors. While these pathogens appear to trigger very similar changes in feeding behavior, they do not appear to physically interfere with feeding mechanisms in the same way.

Newer work has provided potential mechanisms by which infection can alter host-seeking and longer-range attraction. Most of this work is in mosquito vectors. Mosquitoes infected with *Plasmodium* exhibit stage specific changes in electrophysiological sensitivity to host odor (23). Females become less sensitive to host odors than uninfected females while parasites are developing and more sensitive to host odors once parasites reach the infectious stage (23), (26). Analysis of gene expression in Zika infected *Aedes aegypti* suggests that similar suppression and enhancement of host-seeking may occur in this system. Several genes responsible for odorant binding receptor proteins are sequentially suppressed and enhanced depending on whether the virus is transmissible (66). Lefèvre *et al.* (67) found altered levels of proteins reported to regulate energy metabolism in the heads of infectious *Anopheles* mosquitoes and proposed that these proteins may alter host-seeking directly as neuromodulators or indirectly through detrimental effects on neuronal cells. Further work into the effects of infection on vector olfaction is needed. Not only would such studies provide a better understanding of the proximate causes of observed changes in host-seeking behavior but they would also offer a tantalizing possibility of developing control tools that specifically target the small proportion of the vector population that are actually responsible for transmission prior to contacting hosts.

## 13.5 Are changes in feeding behavior with infection exclusively manipulation?

Altered feeding behaviors have been explained as host adaptations, parasite manipulations, and infection by-products (12). Parasite manipulation refers to instances in which the pathogen enhances its own transmission by altering host or vector behavior (68). Infection-associated changes in vector behavior have been almost exclusively framed as parasite manipulations. While it is true that many of these behavioral changes in vectors should increase transmission (either directly by increasing contact between relevant hosts or indirectly by increasing the proportion of vectors that survive parasite development (18), (69), (70)), the effect of these behaviors on vector fitness is usually not addressed. Non-vector hosts in other disease systems have also been reported to adjust feeding behavior in similar ways as part of resistance or tolerance to infection (12). A life-history framework is regularly applied to changes in feeding in other insects to understand these behaviors, but this is rarely applied to vector species. This may be because of a belief that pathogens should not affect vector fitness; however, when pathogens alter feeding behavior, host fitness may be impacted directly or via trade-offs with other life-history processes (Box 13.1). Viewing these behaviors as infection phenotypes that are

a result of both the pathogen and vector optimizing responses to each other and their environment could lead to informed predictions of how these behaviors will evolve and vary with context.

In reality, animals have needs for multiple nutrients and some traits, for example components of the immune response, have differing nutritional requirements. More recent work has suggested that trade-offs between traits can be made at ingestion rather than allocation because the blend of nutrients available determine relative performance of the traits (5). If immune trait "peaks" differ in infected and naïve insects, then infected individuals can alter diet to optimize immune response function, modify the allocation of consumed nutrients for a given diet, or both. Infection can alter food preferences (6), (8), (73) and overall intake (reviewed by (12)). Studies testing how infection may affect qualitative changes in diet in vectors are noticeably missing. Many of the changes in behavior that have been described in vector species result in changes in nutritional profiles. For example, in several plant disease systems, where infection alters preference between infected and uninfected hosts, it has been shown that the nutritional profiles of infected and uninfected plants differ and affect vector reproduction (16). In mosquitoes, it has been shown that infection can incite shifts in metabolism (74). Further, many blood-feeding vectors have distinct dietary sources of carbohydrates (plant sugars) and protein (blood) macronutrients. For example, in the malaria system, there is evidence that infection can alter mosquito responses to sugar sources (75) and carbohydrate metabolism (76). The balance between these different nutritional resources can have large knock-on effects for transmission.

While pathogens in general exert ubiquitous pressure on animal populations, in most cases only a small proportion of the vector population will ever encounter a pathogen which they are able to transmit. In the absence of a clear and specific mechanism for the changes observed in host-seeking and host choice, and given the evidence that infection can alter these exact behaviors in non-vector species, we must consider that these shifts in behavior are the product of evolution of the vector immune response separate from their role as vectors (13).

## Box 13.1  Vector–pathogen feedbacks

Vector-feeding behavior is critical to both pathogen and vector fitness. While timing and frequency of feeding dictates transmission success, these events are also critical for determining vector reproduction and life-history strategies. When processes such as reproduction, growth, or immunity rely on a common pool of resources, increasing investment in one process comes at the cost of the others. Thus, altered feeding can alter transmission either directly by affecting vector traits, or indirectly by altering how resources are allocated amongst life-history processes (Figure 1). For example, changes in feeding behavior could alter transmission by increasing contact rates between vectors and hosts, resulting in direct feedbacks between infection and behavior (blue arrows). Changes in feeding behavior could also impact transmission by altering survival of infected vectors. For example, infection can lead to changes in feeding behavior which decreases vector survival directly (red arrows) as plague infection does in some fleas (71). Alternatively, behavioral changes could increase survival by altering life-history trade-offs. For example, infection could decrease the costs associated with reproduction (gold arrows), as the case may be for malaria infected mosquitoes that skip pre-infectious bloodmeals (72). In both examples, the effects of behavioral changes will feedback on pathogen fitness via host survival.

Very few studies have attempted to tackle predicting these dynamics. Koella (70) looked at expected evolutionary relationships that might explain manipulations of biting rate in malaria by modeling the impact of biting rate on parasite transmission and host reproductive success. The analysis

suggests that parasite fitness is constrained by conflict between increasing transmission (through increased biting rate) and increasing mosquito survival (through decreasing biting rate). Recently, Gandon (18) used models to explore the evolution of host choice manipulation. Not only was the analysis able to recover a range of host choice manipulations documented across plant and animal vectors, Gandon notably found that the evolution of changes in vector preference is primarily governed by the risk of dying before reproducing and the risk of producing fewer eggs. Evolution in vector preference can feedback to pathogen fitness because the relative prevalence of infection depends on preference as explored by Shoemaker and colleagues (28). Theoretical approaches linking parasite and host fitness with epidemiological consequences could be adapted to vector transmitted infections to provide insight into how these phenotypes might evolve and respond to environmental conditions (12).

In addition to further theoretical work, more experiments are also required to further improve our understanding of these feedbacks. In particular, we need to improve our understanding of:

- The degree to which changes in vector behavior regulate resource allocations between life-history processes in the absence of infection.
- Variation in the feedbacks between vector behavior and pathogens with vector age and how this might drive fluctuation in individual behavior over vector lifespan.
- The factors (body condition, environment) that mediate infection-induced shifts in behavior.

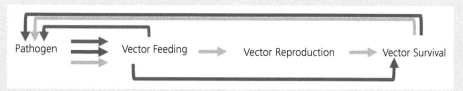

**Box 13.1, Figure 1** Feedbacks between pathogen infection and vector feeding behaviors can influence pathogen transmission and host fitness. Infection effects on feeding behavior can influence transmission directly through changes to contacts with hosts (blue arrows), or indirectly by modifying vector survival (and thus the duration of transmission) directly (red arrows), or via effects on vector reproduction (yellow arrows).

## 13.6 Conclusions and future research

There are two key gaps in our understanding of infection-induced changes in vector behavior that limit our ability to understand this phenomenon.

First, in the majority of examples, the effect of behavioral changes of vector and pathogen fitness have not been quantified or estimated. For example, the most consistent effects are seen in probing behavior and it remains unclear how probing relates to parasite transmission (77)–(79). Even in the "blocked flea" example, the effect of plague biofilm on transmission is controversial. While it increases biting rate and infectivity it also increases flea mortality (71). In addition to not fully understanding the implications of behavioral changes for pathogen fitness, very few studies have measured the impact of altered behaviors on vector fitness (either positive or negative). Given the mounting evidence from other insects that altered foraging is a key component of the host response to immune challenge, this is an important knowledge gap. The effect of behavioral changes on pathogen and host fitness are critical for distinguishing between host adaptation and parasite manipulation. Without understanding the impact of altered behaviors on fitness, the possibility remains that these are simply non-adaptive "by-products" of infection that benefit neither party. Finally, for both pathogen and vector, there has been very little attention paid to the costs of manipulation. Without the benefits and costs of these changes in behavior to both pathogen and vector it is difficult to have an informed discussion about vector–pathogen feedbacks (see Box 13.1).

Second, while changes in vector physiology and gene expression are an excellent start towards uncovering underlying mechanisms of infection-related behavioral change, these data don't inform the degree to which the pathogen alters behavior. In most cases, we still have only a correlation between pathogen presence and an observed change. This makes it exceedingly difficult to distinguish between purely adaptive responses for the pathogen or host and other consequences of infection (sometimes referred to as by-products) (80). While there are some good examples of pathogen-specific mechanisms for changes in probing behavior and feeding efficiency (40), (48), (52), other behaviors, such as host-seeking, lack this kind of explanation. This lack of mechanistic explanation for changes leaves us stuck in a semantic argument. Vectors may alter feeding behavior to reallocate resources away from life-history processes such as reproduction toward immunity. Equally, pathogens may manipulate host "compensatory" responses by increasing demands of immune response (81). It is not clear how to differentiate between these without detailed understanding of the mechanisms at play.

If altered feeding phenotypes are mediated or largely driven by vector life-history allocation, as early evidence suggests some may be (12), then we would expect selection on these trade-offs to be important for understanding and predicting variation in manipulation phenotypes. Better understanding of the relationships between behavioral changes, infection, and vector life history will allow for better predictions for how genotype and environment will alter these manipulative phenotypes. This may explain why, for example, we find evidence for altered host-seeking in some parasite-vector combinations and conditions (23), (26) but not in others (82). There is great potential for experimental evolution approaches which have been used in other manipulation systems to identify the key genes involved in these processes and their roles (83).

These research priorities are shared with understanding infection associated behaviors in other non-vector species. Unique to vectors, however, is that these behavioral changes and their impacts on transmission dynamics have critical public health, agriculture, and economic consequences. These changes in behavior offer the opportunity to target the small proportion of the population that is responsible for transmission. Understanding the evolutionary forces shaping these phenotypes also has important implications for the potential evolutionary response of these systems to disease control strategies (12), (18). Taking a broader view of these phenotypes as a product of the selective pressures on both pathogens and vectors will be critical in the future.

## Acknowledgements

I would like to thank the editors, in particular R Hall, and two thoughtful reviewers for excellent feedback. I would also like to thank AM Milner for useful discussion of the ideas in this chapter.

# References

1. Hart BL. Biological basis of the behavior of sick animals. *Neurosci Biobehav Rev*. 1988;12(2):123–37.
2. Kyriazakis I, Tolkamp BJ, and Hutchings MR. Towards a functional explanation for the occurrence of anorexia during parasitic infections. *Anim Behav*. 1998 Aug;56(2):265–74.
3. Murray MJ and Murray AB. Anorexia of infection as a mechanism of host defense. *Am J Clin Nutr*. 1979 Mar;32(3):593–6.
4. Edwards JT. Discussion in nutrition and its effects on infectious disease. *J R Soc Med*. 1937;30:1039–52.
5. Cotter SC, Simpson SJ, Raubenheimer D, and Wilson K. Macronutrient balance mediates trade-offs between immune function and life history traits. *Funct Ecol*. 2011;25(1):186–98.
6. Povey S, Cotter SC, Simpson SJ, and Wilson K. Dynamics of macronutrient self-medication and illness-induced anorexia in virally infected insects. *J Anim Ecol*. 2014;83(1):245–55.
7. Dinh H, Mendez V, Tabrizi ST, and Ponton F. Macronutrients and infection in fruit flies. *Insect Biochem Mol Biol*. 2019;110:98–104.
8. Lee KP, Cory JS, Wilson K, Raubenheimer D, and Simpson SJ. Flexible diet choice offsets protein costs of pathogen resistance in a caterpillar. *Proc R Soc B: Biol Sci*.2006;273(1588):823–9.
9. Ayres JS and Schneider DS. The role of anorexia in resistance and tolerance to infections in *Drosophila*. *PLoS Biol*. 2009 Jul;7(7).
10. Adamo SA, Bartlett A, Le J, Spencer N, and Sullivan K. Illness-induced anorexia may reduce trade-offs between digestion and immune function. *Anim Behav*. 2010;79(1):3–10.
11. Ponton F, Morimoto J, Robinson K, Kumar SS, Cotter SC, Wilson K, *et al*. Macronutrients modulate survival to infection and immunity in *Drosophila*. *J Anim Ecol*. 2020;89(2):460–70.
12. Hite JL, Pfenning AC, and Cressler CE. Starving the enemy? feeding behavior shapes host–parasite interactions. *Trends Ecol Evol*. 2020;35(1):68–80.
13. Murdock CC, Luckhart S, and Cator LJ. Immunity, host physiology, and behaviour in infected vectors. *Curr Opin Insect Sci*. 2017;20:28–33.
14. Froissart R, Doumayrou J, Vuillaume F, Alizon S, and Michalakis Y. The virulence–transmission trade-off in vector-borne plant viruses: A review of (non-) existing studies. *Philos Trans R Soc Lond B: Biol Sci*. 2010;365(1548):1907–18.

15. Bellan S. The importance of age dependent mortality and the extrinsic incubation period in models of mosquito-borne disease transmission and control. *PLoS One*. 2010;5(4):e10165.
16. Eigenbrode SD, Bosque-Pérez NA, and Davis TS. Insect-borne plant pathogens and their vectors: Ecology, evolution, and complex interactions. *Annu Rev Entomol*. 2018 07;63:169–91.
17. Mauck KE and Chesnais Q. A synthesis of virus-vector associations reveals important deficiencies in studies on host and vector manipulation by plant viruses. *Virus Res*. 2020 Aug 1;285:197957.
18. Gandon S. Evolution and manipulation of vector host choice. *Am Nat*. 2018;192(1):23–34.
19. Friend WG and Smith JJB. Factors affecting feeding by bloodsucking insects. *Annu Rev Entomol*. 1977;22(1):309–31.
20. Liu B, Preisser EL, Chu D, Pan H, Xie W, Wang S, *et al*. Multiple forms of vector manipulation by a plant-infecting virus: *Bemisia tabaci* and Tomato Yellow Leaf Curl Virus. *J Virol*. 2013 May 1;87(9): 4929–37.
21. Lefcort H and Durden LA. The effect of infection with Lyme disease spirochetes (*Borrelia burgdorferi*) on the phototaxis, activity, and questing height of the tick vector *Ixodes scapularis*. *Parasitol*. 1996;113(2):97–103.
22. Maciel-de-Freitas R, Sylvestre G, Gandini M, and Koella JC. The influence of dengue virus serotype-2 infection on *Aedes aegypti* (Diptera: Culicidae) motivation and avidity to blood feed. *PloS One*. 2013;8(6):e65252.
23. Cator LJ, George J, Blanford S, Murdock CC, Baker TC, Read AF, *et al*. "Manipulation" without the parasite: Altered feeding behaviour of mosquitoes is not dependent on infection with malaria parasites. *Proc R Soc B: Biol Sci*. 2013; 280(1763).
24. Ferguson LV, Hillier NK, and Smith TG. Influence of Hepatozoon parasites on host-seeking and host-choice behaviour of the mosquitoes *Culex territans* and *Culex pipiens*. *Int J Parasitol Parasites Wildl*. 2013;2:69–76.
25. Gleave K, Cook D, Taylor MJ, and Reimer LJ. Filarial infection influences mosquito behaviour and fecundity. *Sci Rep*. 2016 Oct 31;6:36319.
26. Smallegange RC, van Gemert G-J, van de Vegte-Bolmer M, Gezan S, Takken W, Sauerwein RW, *et al*. Malaria infected mosquitoes express enhanced attraction to human odor. *PloS One*. 2013;8(5):e63602.
27. Ostfeld RS and Keesing F. Effects of host diversity on infectious disease. *Annu Rev Ecol Evol Syst*. 2012;43.
28. Shoemaker LG, Hayhurst E, Weiss-Lehman CP, Strauss AT, Porath-Krause A, Borer ET, *et al*. Pathogens manipulate the preference of vectors, slowing disease spread in a multi-host system. *Ecol Lett*. 2019;22(7):1115–25.

29. Berret J and Voordouw M. Lyme disease bacterium does not affect attraction to rodent odour in the tick vector. *Parasit Vectors*. 2015;8:249.

30. Rajabaskar D, Bosque-Pérez NA, and Eigenbrode SD. Preference by a virus vector for infected plants is reversed after virus acquisition. *Virus Res*. 2014 Jun 24;186:32–7.

31. Ingwell LL, Eigenbrode SD, and Bosque-Pérez NA. Plant viruses alter insect behavior to enhance their spread. *Sci Rep*. 2012;2(578).

32. Legarrea S, Barman A, Marchant W, Diffie S, and Srinivasan R. Temporal effects of a Begomovirus infection and host plant resistance on the preference and development of an insect vector, *Bemisia tabaci*, and implications for epidemics. *PLoS One*. 2015;10(11): e0142114.

33. Fereres A, Peñaflor MFGV, Favaro CF, Azevedo KEX, Landi CH, Maluta NKP, *et al*. Tomato infection by whitefly-transmitted circulative and non-circulative viruses induce contrasting changes in plant volatiles and vector behaviour. *Viruses*. 2016;8(8):225.

34. McElhany P, Real LA, and Power AG. Vector preference and disease dynamics: A study of barley yellow dwarf virus. *Ecology*. 1995;444–57.

35. Robinson A, Busula AO, Voets MA, Beshir KB, Caulfield JC, Powers SJ, *et al*. *Plasmodium*-associated changes in human odor attract mosquitoes. *PNAS*. 2018;115(18):E4209–18.

36. Cornet S, Nicot A, and Gandon S. Malaria infection increases bird attractiveness to uninfected mosquitoes. *Ecol Lett*. 2013;16(3):323–9.

37. Moreno-Delafuente A, Garzo E, Moreno A, and Fereres A. A plant virus manipulates the behavior of its whitefly vector to enhance its transmission efficiency and spread. *PLoS One*. 2013;8(4).

38. Stafford CA, Walker GP, and Ullman DE. Infection with a plant virus modifies vector feeding behavior. *PNAS*. 2011;108(23):9350–5.

39. Nalam VJ, Han J, Nachappa P, and Szczepaniec A. Drought stress and pathogen infection alter feeding behavior of a phytopathogen vector. *Entomol Exp Appl*. 2020;168(8).

40. Jenni L, Molyneux DH, Livesey JL, and Galun R. Feeding behaviour of tsetse flies infected with salivarian trypanosomes. *Nature*. 1980;283:383–5.

41. Van Den Abbeele J, Caljon G, De Ridder K, De Baetselier P, and Coosemans M. *Trypanosoma brucei* modifies the tsetse salivary composition, altering the fly feeding behavior that favors parasite transmission. *PLoS Pathog*. 2010;6(6):e1000926.

42. Chung H, Feng L, and Feng S. Observations concerning the successful transmission of Kala-Azar in North China by the bites of naturally infected *Phlebotomus*

*chinensis*. *Peking Nat Hist Bull*. 1950;19(Pts. 2/3):302–26.

43. Grimstad PR, Ross QE, and Craig GB. *Aedes triseriatus* (Diptera: Culicidae) and La Crosse virus. II. Modification of mosquito feeding behavior by virus infection. *J Med Entomol*. 1980;17(1):1–7.

44. Platt KB, Linthicum KJ, Myint KS, Innis BL, Lerdthusnee K, and Vaughn DW. Impact of dengue virus infection on feeding behavior of *Aedes aegypti*. *Am J Trop Med Hyg*. 1997;57(2):119.

45. Sylvestre G, Gandini M, and Maciel-de-Freitas R. Age-dependent effects of oral infection with dengue virus on *Aedes aegypti* (Diptera: Culicidae) feeding behavior, survival, oviposition success and fecundity. *PLoS One*. 2013;8(3):e59933.

46. Li X, Sina B, and Rossignol PA. Probing behaviour and sporozoite delivery by *Anopheles stephensi* infected with *Plasmodium berghei*. *Med Vet Entomol*. 1992;6(1): 57–61.

47. Rossignol PA, Ribeiro JM, and Spielman A. Increased intradermal probing time in sporozoite-infected mosquitoes. *Am J Trop Med Hyg*. 1984;33(1):17.

48. Wekesa JW, Copeland RS, Mwangi RW, *et al*. Effect of *Plasmodium falciparum* on blood feeding behavior of naturally infected *Anopheles* mosquitoes in western Kenya. *Am J Trop Med Hyg*. 1992;47(4):484.

49. Bacot AW and Martin CJ. LXVII. Observations on the mechanism of the transmission of plague by fleas. *J Hyg (Lond)*. 1914;13(Suppl):423–39.

50. Killick-Kendrick R, Leaney AJ, Ready PD, and Molyneux DH. Leishmania in phlebotomid sandflies. IV. The transmission of Leishmania mexicana amazonensis to hamsters by the bite of experimentally infected *Lutzomyia longipalpis*. *Proc R Soc B: Biol Sci*. 1977;196(1122):105–15.

51. Beach R, Kiilu G, and Leeuwenburg J. Modification of sand fly biting behavior by Leishmania leads to increased parasite transmission. *Am J Trop Med Hyg*. 1985;34(2):278–82.

52. Koella JC, Packer MJ, *et al*. Malaria parasites enhance blood-feeding of their naturally infected vector *Anopheles punctulatus*. *Parasitol*. 1996;113:105–10.

53. Rogers ME, Ilg T, Nikolaev AV, Ferguson MAJ, and Bates PA. Transmission of cutaneous leishmaniasis by sand flies is enhanced by regurgitation of fPPG. *Nature*. 2004;430(6998):463–7.

54. Bockarie MJ and Dagoro H. Are insecticide-treated bednets more protective against *Plasmodium falciparum* than *Plasmodium vivax*-infected mosquitoes? *Malar J*. 2006;5(1):15.

55. Rogers ME and Bates PA. Leishmania manipulation of sand fly feeding behavior results in enhanced transmission. *PLoS Pathog*. 2007;3(6):e91.

56. Bennett KE, Hopper JE, Stuart MA, West M, and Drolet BS. Blood-feeding behavior of vesicular stomatitis virus infected *Culicoides sonorensis* (Diptera: Ceratopogonidae). *J Med Entomol*. 2008;45(5):921–6.

57. Jackson, BT, Brewster CC, and Paulson SL. La Crosse virus infection alters blood feeding behavior in *Aedes triseriatus* and *Aedes albopictus* (Diptera: Culicidae). *J Med Entomol*. 2012;49(6):1424–9.

58. Thiévent K, Hofer L, Rapp E, Tambwe MM, Moore S, and Koella JC. Malaria infection in mosquitoes decreases the personal protection offered by permethrin-treated bednets. *Parasit Vectors*. 2018;11(1):1–10.

59. Belova OA, Burenkova LA, and Karganova GG. Different tick-borne encephalitis virus (TBEV) prevalences in unfed versus partially engorged ixodid ticks—evidence of virus replication and changes in tick behavior. *Ticks Tick-Borne Dis*. 2012;3(4): 240–6.

60. Anderson RA, Koella JC, and Hurd H. The effect of *Plasmodium yoelii nigeriensis* infection on the feeding persistence of *Anopheles stephensi* Liston throughout the sporogonic cycle. *Proc R Soc B: Biol Sci*. 1999;266(1430):1729–33.

61. Koella JC, Sorensen FL, and Anderson RA. The malaria parasite, *Plasmodium falciparum*, increases the frequency of multiple feeding of its mosquito vector, *Anopheles gambiae*. *Proc R Soc B: Biol Sci*. 1998;265(1398):763–8.

62. Houte S van, Ros VID, and van Oers MM. Walking with insects: Molecular mechanisms behind parasitic manipulation of host behaviour. *Mol Ecol*. 2013;22(13):3458–75.

63. Schlein Y, Jacobson RL, and Messer G. Leishmania infections damage the feeding mechanism of the sandfly vector and implement parasite transmission by bite. *PNAS*. 1992 Oct 15;89(20):9944–8.

64. Titus RG and Ribeiro JM. Salivary gland lysates from the sand fly *Lutzomyia longipalpis* enhance Leishmania infectivity. *Science*. 1988 Mar 11;239(4845):1306–8.

65. Jefferies D, Livesey JL, and Molyneux DH. Fluid mechanics of bloodmeal uptake by Leishmania-infected sandflies. *Acta Trop*. 1986 Mar;43(1):43–53.

66. Hurd H. Manipulation of medically important insect vectors by their parasites. *Annu Rev Entomol*. 2003;48:141–61.

67. Etebari K, Hegde S, Saldaña MA, Widen SG, Wood TG, Asgari S, *et al.* Global transcriptome analysis of *Aedes aegypti* mosquitoes in response to Zika virus infection. *mSphere*. 2017;2(6):e00456–17.

68. Lefevre T, Thomas F, Schwartz A, Levashina E, Blandin S, Brizard J-P, *et al.* Malaria *Plasmodium* agent induces alteration in the head proteome of their *Anopheles* mosquito host. Proteomics. 2007;7(11): 1908–15.

69. Thomas F, Adamo S, and Moore J. Parasitic manipulation: Where are we and where should we go? *Behav Processes*. 2005 Mar 31;68(3):185–99.

70. Koella JC. An evolutionary view of the interactions between anopheline mosquitoes and malaria parasites. *Microbes Infect Inst Pasteur*. 1999 Apr;1(4):303–8.

71. Cator LJ, Lynch PA, Thomas MB, and Read AF. Alterations in mosquito behaviour by malaria parasites: Potential impact on force of infection. *Malar J*. 2014;13(1):164.

72. Hinnebusch BJ, Jarrett CO, and Bland DM. "Fleaing" the plague: Adaptations of *Yersinia pestis* to its insect vector that lead to transmission. *Annu Rev Microbiol*. 2017;71:215–32.

73. Ohm JR, Teeple J, Nelson WA, Thomas MB, Read AF, and Cator LJ. Fitness consequences of altered feeding behavior in immune-challenged mosquitoes. *Parasit Vectors*. 2016;9:113.

74. Povey S, Cotter SC, Simpson SJ, Lee KP, and Wilson K. Can the protein costs of bacterial resistance be offset by altered feeding behaviour? *J Anim Ecol*. 2009;78(2): 437–46.

75. Pietri JE, Pakpour N, Napoli E, Song G, Pietri E, Potts R, *et al.* Two insulin-like peptides differentially regulate malaria parasite infection in the mosquito through effects on intermediary metabolism. *Biochem J*. 2016 Oct 15;473(20):3487–503.

76. Nyasembe VO, Teal PEA, Sawa P, Tumlinson JH, Borgemeister C, and Torto B. *Plasmodium falciparum* infection increases *Anopheles gambiae* attraction to nectar sources and sugar uptake. *Curr Biol*. 2014 Jan 20;24(2):217–21.

77. Zhao YO, Kurscheid S, Zhang Y, Liu L, Zhang L, Loeliger K, *et al.* Enhanced survival of *Plasmodium*-infected mosquitoes during starvation. *PLoS One*. 2012;7(7):e40556.

78. Ponnudurai T, Lensen AHW, Van Gemert GJA, Bolmer MG, and Meuwissen JHE. Feeding behaviour and sporozoite ejection by infected *Anopheles stephensi*. *Trans R Soc Trop Med Hyg*. 1991;85(2):175–80.

79. Jiang YX, de Blas C, Barrios L, and Fereres A. Correlation between whitefly (Homoptera: Aleyrodidae) feeding behavior and transmission of tomato yellow leaf curl virus. *Ann Entomol Soc Am*. 2000;93(3):573–9.

80. Aleshnick M, Ganusov VV, Nasir G, Yenokyan G, and Sinnis P. Experimental determination of the force of malaria infection reveals a non-linear relationship to mosquito sporozoite loads. *PLoS Pathog*. 2020 May 26;16(5):e1008181.

81. Herbison R, Lagrue C, and Poulin R. The missing link in parasite manipulation of host behaviour. *Parasit Vectors*. 2018;11(1):222.

82. Lefèvre T, Roche B, Poulin R, Hurd H, Renaud F, and Thomas F. Exploiting host compensatory responses: The "must" of manipulation? *Trends Parasitol.* 2008 Oct;24(10):435–9.

83. Vantaux A, Hien de S, François D, Yaméogo B, Dabiré KR, Thomas F, *et al.* Host-seeking behaviors of mosquitoes experimentally infected with sympatric field isolates of the human malaria parasite *Plasmodium falciparum*: No evidence for host manipulation. *Front Ecol Evol.* 2015;3.

84. Hafer-Hahmann N. Experimental evolution of parasitic host manipulation. *Proc R Soc B: Biol Sci.* 2019;286(1895):20182413.

85. Strauss AT, Henning JA, Porath-Krause A, Asmus AL, Shaw AK, Borer ET, *et al.* Vector demography, dispersal and the spread of disease: Experimental epidemics under elevated resource supply. *Funct Ecol.* 2020;34(12):2560–70.

86. Zhang X-S, Holt J, and Colvin J. A general model of plant-virus disease infection incorporating vector aggregation. *Plant Pathol.* 2000;49(4):435–44.

# PART VI

# Behavioral Defenses

# Infection avoidance behaviors across vertebrate taxa: Patterns, processes, and future directions

Patricia C. Lopes, Susannah S. French, Douglas C. Woodhams, and Sandra A. Binning

## 14.1 Introduction

Sick animals may act, look, sound, taste, and smell different than healthy animals. Similarly, habitats harboring infectious agents, or infectious agents themselves, may leave conspicuous cues that potential hosts can detect. Recognizing signals of diseases and displaying aversion towards conspecifics, habitats, or other organisms exhibiting those signals should, therefore, be evolutionarily advantageous by reducing an individual's likelihood of becoming infected, so long as the repercussions of prophylactic or aversion behaviors are not a net cost to host fitness.

An organism's first line of defense against infectious agents is avoiding infection by reducing exposure to those agents. Recent research proposes that many organisms actively try to avoid infection by navigating a "landscape of disgust," whereby animals change their behavior and/or location to avoid potentially infected conspecifics, or pathogen-contaminated food and water sources or habitats (1). Immune defenses are, therefore, often a second line of defense against many infectious agents. Initiating a robust immune defense to control or eliminate infection comes at a cost to the host (e.g., resources needed to support immune function, such as energy and nutrients (2)), as does engaging in infection avoidance behaviors (e.g., nutritional resources lost due to avoidance of contaminated foods (3)).

Buck *et al.* (3) proposed a conceptual "cost-benefit" framework for understanding when avoidance behaviors should occur, whereby the benefits of engaging in avoidance strategies must be balanced against the potential costs of avoidance, and against the relative costs of infection and costs of immunological responses (Figure 14.1). For example, Raffel *et al.* (4) showed that at low densities, tadpoles of *Bufo americanus* that reduced activities that attracted the attention of predators were more likely to become parasitized with a trematode, *Echinostoma trivolvis*. In this example, the avoidance behavior (increased activity) would lead to increased predation, creating a trade-off between infection or predation. In the túngara frog, *Engystomops pustulosus*, infections with *Trypanosoma tungarae* are vectored by frog-biting midges, *Corethrella* spp., that are attracted to the frog mating calls. These infections are most common in males (5) and lead to a trade-off between attracting mates and parasitism. This type of trade-off is felt not only by the calling male, but extends to non-signaling individuals, such as satellites and females, and other potential hosts in mixed species aggregations (6), highlighting that social distancing can curtail the risk of parasitism even for non-communicable diseases or parasites (7).

Patricia C. Lopes et al., *Infection avoidance behaviors across vertebrate taxa*. In: *Animal Behavior and Parasitism*. Edited by Vanessa O. Ezenwa, Sonia Altizer, and Richard J. Hall, Oxford University Press. © Oxford University Press (2022). DOI: 10.1093/oso/9780192895561.003.0014

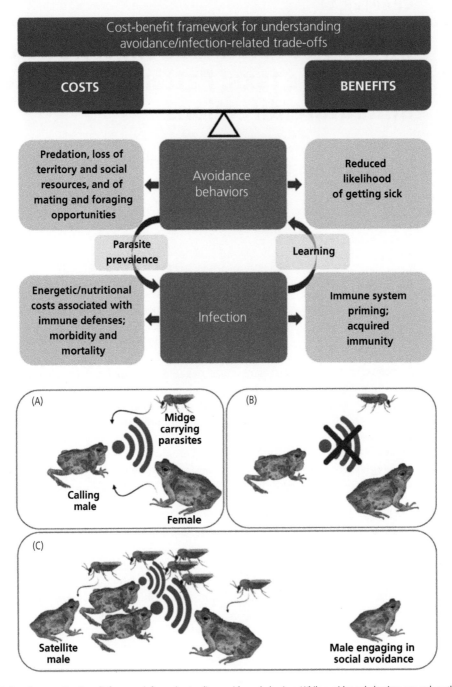

**Figure 14.1** Top diagram: Cost-benefit framework for understanding avoidance behaviors. While avoidance behaviors can reduce the likelihood of becoming infected, these behaviors can also affect mating and foraging opportunities and lead to loss of social status or resources (such as territories). In some cases, these behaviors can even expose hosts to increased predation. When avoidance behaviors are not engaged in, the animal may become infected/parasitized. The costs associated with the particular infection need to be weighed when considering avoidance behaviors. Infections will always have some detrimental effects in terms of energy or nutritional resources associated with activation of immune defenses. However, certain infections can lead to more extensive tissue damage, and even be fatal. Recovering from an infection or developing tolerance to a pathogen can have benefits, such as preparing the immune system for future infections. Avoidance behaviors can feedback onto parasite prevalence (e.g., if avoidance of diseased animals is strong in a social group, parasite prevalence in that group should go down). Conversely, the experience of infection can feedback onto avoidance behaviors through learning (within the lifetime of the individual) or through selection for avoidance behaviors (evolutionary feedback).

*(Continued)*

In this chapter, we examine how individuals alter their behavior to avoid becoming infected across non-mammalian vertebrate groups. We bring focus to non-mammalian vertebrates (specifically, avian, reptile, amphibian, and fish taxa), which tend to be under-represented in the literature, but note that there are many examples of avoidance behaviors in invertebrates (8), (9). We begin by summarizing how infection leads to disease cues, the physiology underlying disgust, and the triggers of avoidance behaviors (10). Next, we synthesize the different kinds of infection avoidance behaviors observed in vertebrate groups and use the cost-benefit framework to explain when avoidance behaviors are likely to occur in these groups. We conclude by highlighting some knowledge gaps in understanding cross-species patterns of behavioral avoidance, and how ongoing environmental change might shape infection avoidance behaviors in the Anthropocene.

## 14.2 How infection leads to disease cues

The presence of parasites and pathogens can introduce changes in several aspects of hosts (Figure 14.2) and their habitats, which can be used as disease cues by other animals. Behaviorally, a sick animal can bear little resemblance to a healthy version of itself. Infected animals frequently eat and drink less, sleep more, and alter their sexual and social interactions (11). Disease-associated behavioral changes can come in two general forms: they can be host-induced or parasite-induced. Host-induced behavioral changes post-infection are collectively called "sickness behaviors" and these changes are considered to be a consequence of the inflammatory response to the invader (12). Sickness behaviors are thought to have evolved because they increase host survival (11). Alternatively, parasite-induced changes in host behavior tend to be beneficial to the parasite by increasing the probability

of transmission and/or by facilitating completion of the parasite's lifecycle (13). These latter behavioral changes are known to involve parasite-secreted molecules and disruptions of normal nervous system communication/anatomy (13), but more studies exploring these mechanisms are needed (14). Note that some behavioral changes may also be due to disease-associated pathology with no clear benefit to either host or parasite (13). The distinctions between parasite versus host-induced changes in behavior are not always well defined, especially concerning costs and benefits. For example, prolonged sickness behaviors will end up being detrimental to the host (15). Similarly, some forms of parasite manipulation can increase the host's inclusive fitness by leading the infected host away from its kin (16).

Infections can alter host vocalizations, appearance, and odor. For example, infected animals tend to vocalize less (e.g., (17), (18), but see (19)), and can have less consistent songs (20). Behavioral responses to parasitism and infections can also be associated with changes in the appearance or other physical attributes of the host. One example of this is the formation of blisters or wounds associated with excessive scratching. Parasitism and infections can also cause swelling of glands (e.g., mumps), and lead to the formation of rashes (e.g., chickenpox), warts (e.g., herpes), and tumors, changes in skin or eye coloration (jaundice), skin thickening, and loss or altered appearance of hair, fur, feathers, or scales. Changes in microbiota and pathological processes can also create imbalances in the ratios of volatile organic compounds (VOCs) normally produced or lead to production of new VOCs, thereby changing the odors of sick animals (21).

Similarly, contaminated habitats and food and water sources may have visual, odor, or other characteristics that allow animals to detect contamination. For example, feces constitute a good

**Figure 14.1** (*Continued*)
Bottom illustration: Illustrative example of cost-benefit framework. (A) Male túngara frog calls attract females, as well as parasite-carrying midges. (B) Males could engage in parasite avoidance behaviors by reducing calls, but that would come at the cost of losing mating opportunities. (C) Calling in a chorus may increase the number of midges attracted to the location, creating a cost of grouping behavior. Calling behavior can lead to parasitism not only of the calling male, but also of satellite males (grayed out males) and even females that are attracted by the calls. A social distance approach (frog to the right) avoids parasitism but once again carries social and mating costs. Image created with BioRender.com, using artwork (frogs) by Nina McDonnell.

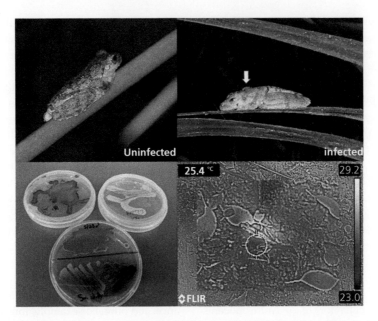

**Figure 14.2** Indications of vertebrate infection status include behavioral changes discussed in the text, as well as visual cues such as posture, color, or activity; auditory cues such as vocalization frequency or effort; chemical cues including microbiota and immune defense compounds; and tactile cues including sloughing skin or temperature. An uninfected green-eyed treefrog, *Litoria serrata* (upper left) is shown in contrast to a lethargic, diseased frog (upper right) with hunched posture and half-open eyes in an exposed position. This frog has a high burden of *Batrachochytrium dendrobatidis* (microscopic skin fungus) and *Batrachomyia* (parasitic fly, lump in dorsal lymph sac indicated by arrow). Cultures of symbiotic skin bacteria (lower left) produce volatile organic compounds that can function in immune defense. Body temperature is mediated by behavioral responses to infection (lower right: thermal map of a froglet (located in the middle of the image) displaying a core temperature (approx. 28°C) higher than the surrounding environment (25.4°C). Photo credits: DC Woodhams.

risk indicator for pathogens (22). Humans are universally disgusted by feces (22) and other primates have been found to be more selective when consuming fecally contaminated food or water (e.g., they consume more high quality relative to low quality fecally contaminated food (23), or they prefer uncontaminated waterholes to contaminated ones, as long as they are equidistant (24)).

## 14.3 What is disgust and how are disease avoidance behaviors triggered?

Growing evidence suggests that animals have systems that facilitate disease detection and promote behavioral avoidance of pathogens. Even though we are here focusing on avoidance of disease-causing organisms, avoidance of non-biological disease-causing agents, such as radiation and mutagenic chemicals (which can lead to cancer) has also

been described (25). In humans and non-human primates, research on pathogen avoidance has been conducted under a framework referred to as the "behavioral immune system" (26); see also Chapter 15 (27). This term describes a system of disease detection that activates behavioral responses aimed at diminishing pathogen exposure. Another overlapping term for disease avoidance behavior is the "disgust adaptive system" and this term is sometimes used interchangeably with "behavioral immune system" (28).

According to the idea that animals have a "disgust adaptive system," disgust is a response to signals that are reliably associated with diseases or pathogen/parasite abundance (29). Others have suggested that cutaneous responses may occur in addition to disgust as part of an ectoparasite defense system (30). The detection of these signals can trigger behavioral mechanisms to avoid exposure. A study in humans, for instance, showed that odor samples and photographs of faces of sick donors (injected with an antigen) were less desirable than samples from healthy donors (injected with saline) and that the sickness status of the samples led to heightened activation of odor- and face-perception neural networks (31). The mechanisms for disease detection are likely much less conserved across vertebrates than the mechanisms regulating avoidance behaviors. This is because detection mechanisms should depend on the environmental conditions in which the species lives, be related to trophic level, social structure, and species-specific social recognition systems, and might be biased relative to the sensory organs most used for detection. For example, rodents rely heavily on odors to recognize and avoid infected conspecifics (32), while birds seem able to detect disease symptoms visually (e.g., (33)). Primates likely use a combination of both visual and olfactory cues (e.g., (22)–(24), (30)). These detection mechanisms might also be subject to learning, as demonstrated in rats that developed taste aversion for a flavor associated with an experimental parasitic infection (34).

## 14.4 Behavioral avoidance of infection throughout vertebrate taxa

For a behavior to be described as "infection avoidance," typically two requirements must be met (35).

First, infectious agents must reduce host fitness. Second, the behavior should allow the host to reduce infection intensity or escape infection altogether and/or avoid the negative costs of infection. Traditionally four types of behavioral avoidance of infection have been described and have empirical support (Table 14.1; Figure 14.3): avoidance of infected conspecifics, avoidance of high-risk habitats, avoidance of infected or contaminated food sources, and avoidance of parasites and their vectors (29), (35)–(37), although other types of disease avoidance behaviors (e.g., migratory escape and niche modification) exist. For avoidance behaviors to occur, the potential host must first be able to detect an infected conspecific, the infectious agent, or a high-risk area (35).

### 14.4.1 Avoidance of infected conspecifics

Studies in mammals indicate that, in many instances, parasitized conspecifics are avoided (32). Allogrooming, an important social behavior in primate (38), is reduced towards conspecifics parasitized with orofacially transmitted parasites in mandrills (*Mandrillus sphinx*) (39). Allogrooming under these circumstances would increase the probability of acquiring the parasites from parasitized individuals. Contrastingly, vampire bats do not reduce allogrooming towards groupmates

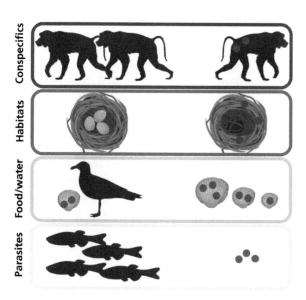

**Figure 14.3** Schematic examples of main avoidance behaviors. Animals may avoid parasitized conspecifics, contaminated habitats and food/water sources, and avoid parasites or vectors themselves. Infectious agents are represented as red dots. Examples illustrated come from the text and include mandrills reducing allogrooming towards parasitized mates; numerous bird species that avoid parasitized nests; oystercatchers showing preference for consumption of intermediate sizes of parasitized cockles, avoiding consumption of larger but more intensely parasitized cockles; and fish shoaling more tightly with conspecifics in order to reduce parasite exposure. Image created with BioRender.com.

**Table 14.1** Avoidance behaviors synthesized across vertebrate taxa. We indicate whether avoidance (avoid) or attraction (attract) are found, except for niche modification and migratory escape (Box 14.1), where we indicate whether it takes place (+) or does not (−) in each taxa. ? denotes no known examples. Examples are given as superscript numbers. Details and additional examples are found in the text.

| Vertebrate group | Infected conspecifics | Contaminated habitats | Contaminated food/water | Parasites and Vectors | Niche modification | Migratory escape |
|---|---|---|---|---|---|---|
| Mammal | Avoid[39-42] | Avoid[63, 64] | Avoid[29-95] | Avoid[106] | +[122] | +[79-133] |
| Bird | Avoid[43], Attract[44] | Avoid[65-68] | Avoid[96, 135] Attract[97] | ? | +[117-120] | +[45] |
| Reptile | Avoid[47, 136] | ? | ? | ? | ? | ? |
| Amphibian | Avoid[49] | Avoid[69] | ? | Avoid[101] | ? | +[82] |
| Fish | Avoid[53-57] | Avoid[21-74] | Attract[102-104] | Avoid[110-115] | ? | +[81, 134] |

that received an experimental immune challenge (40). Because vampire bats use allogrooming to maintain long-term cooperative relationships that involve food donations from other bats, the cost of avoidance of an infected conspecific here might be larger than the costs for mandrills. The social costs to infection avoidance are therefore important to predict its occurrence.

Sexual contact is an important way in which parasites can be acquired and parasitism should therefore influence sexual behavior. In rodents, mate choice is affected by parasitism and tends to be driven by females, who prefer males that are not infected with parasites with direct transmission modes (41). Female olive baboons (*Papio anubis*) avoid mating with males carrying a symptomatic sexually transmitted disease caused by the bacterium *Treponema pallidum* (42).

Similar responses are seen in avian taxa. Female satin bowerbirds (*Ptilonorhynchus violaceus*) prefer to mate with males with fewer lice (*Myrsidea ptilonorhynchid*) (43) and female sage grouse (*Centrocercus urophasianus*) avoid males presenting with visual symptoms (experimentally applied artificial hematomas) of lice infestation (33). Infection can alter physical attributes of birds, potentially providing visual cues of disease. One example in which this has been studied is conjunctivitis caused by the bacterial pathogen *Mycoplasma gallisepticum*. This is a directly transmissible pathogen, and it causes visible symptoms around the eye in house finches

(*Carpodacus mexicanus*), as well as lethargy. However, rather than avoiding infected conspecifics, male house finches seem to prefer to feed near diseased conspecifics, which the authors linked to a reduction in aggression by the diseased animals (44). It is therefore possible that disease symptoms are not always interpreted as a threat but also as an opportunity. Aggregation behavior may depend on scale (Box 14.1), and in house finches there is evidence for conspecific attraction at local scales, but at larger scales there is support for the idea that house finches migrate away from aggregation sites, potentially reducing exposure (45). There is also evidence that parasitic infestation can alter odors of birds. For example, the odors of great tits infected with malarial parasites (*Plasmodium* spp.) were less attractive to the mosquito vector (*Culex pipiens*) of those parasites relative to odors of uninfected birds (46). While mosquitos therefore use this cue to avoid an infected food source, it is not known whether conspecifics (i.e., other great tits) can themselves pick up on those cues to avoid infected group mates behaviorally.

Avoidance of conspecifics has also been described in lizards, but only during breeding. For example, female lizards of the species *Psammodromus algirus* are more attracted to uninfected conspecifics (47), which may be related to detectable chemical cues (i.e., lipid secretions from the femoral pores) that are tied to parasite load. In contrast, one study in common lizards, *Zootoca vivipara*, found no evidence for avoidance of lizards parasitized with mites (48). The contrast between these two studies may derive from the fact that the study in the common lizards used yearlings, which are not yet reproductive, and that, when individuals congregate for mating, avoidance of infection becomes more critical.

Similar to reptilian species, avoidance of infected conspecifics in amphibians seems to be linked to life-history stages when individuals are more likely to congregate. American bullfrog (*Rana catesbeiana*) tadpoles avoid conspecifics infected with a pathogenic yeast (*Candida humicola*) or chemical cues emitted by these infected conspecifics (49). For amphibians and other organisms, mate choice based on body size and advertisement calls may be supplemented by chemical signals that are produced by the host directly and through skin microbiota. These chemical signals can differ with infection status and thus impact behavioral attraction or avoidance. Brunetti *et al.* (50) demonstrated that VOCs were produced from amphibian skin bacteria and differed between males and females. In addition, amphibian skin microbes produce VOCs functioning in pathogen defense (51). Thus, microbiota offer information about host reproductive status and immunity. In amphibians, infection is associated with changes in skin microbiota compared to uninfected amphibians (52). Thus, individuals may select mates or avoid congregating based on direct detection of odor cues coming from immune molecules or symbiotic microbes.

Due to widespread shoaling behaviors across fish species, which create ample opportunity for transmission of contagious pathogens and parasites, much research regarding infection avoidance in fishes has focused on shoaling behaviors. Both infected and uninfected banded killifish (*Fundulus diaphanous*) prefer to shoal with uninfected conspecifics even though the trematode parasites causing black spots on infected individuals are not transmitted directly between hosts (53). Juvenile sticklebacks (*Gasterosterus aculeatus*) avoid schools with individuals infected by an ectoparasite (*Argulus canadensis*) (54). Adult sticklebacks

also prefer shoals of uninfected conspecifics over shoals infected with monogenean flatworm ectoparasites *Gyrodactylus* spp. (55) and shoals infected with the microsporidian parasite, *Glugea anomala* (56). Trinidadian guppies (*Poecilia reticulata*) will also actively avoid shoaling with conspecifics infected with gyrodactylid ectoparasites (57) and can detect and avoid both chemical and visual cues of late-stage infected conspecifics (58).

Avoidance of infected conspecifics seems to occur throughout vertebrates (Table 14.1). Furthermore, whether species are social or solitary, sexual reproduction brings conspecifics into close contact and is, therefore, a particularly sensitive period for detection and behavioral avoidance of infected conspecifics. In addition to costs of infection to the host, mating with infected conspecifics may also have intergenerational costs, both in terms of infectious agents that can be transmitted from parent of offspring (i.e., vertical transmission; (59)), but also in terms of a less effective immune response being inherited by the offspring (e.g., (60)). Avoidance of conspecifics in other contexts is, however, not always observed (e.g., vampire bats; (61)) and, in some cases, attraction towards diseased conspecifics has even been described (e.g., house finches; (44)). It is likely that the costs of social avoidance in non-reproductive contexts are not as significant (or are only high in more social species) because when animals are not reproducing, avoiding conspecifics does not result in missed mating opportunities and fitness consequences. However, in some highly social species, avoidance in non-reproductive contexts could conceivably accrue high costs, such as when social networks are necessary for survival (e.g., reduced protection from predators). Therefore, costs and benefits of avoidance may vary across life-history contexts and lead to differences in the presence of avoidance behaviors among species. The social costs associated with avoidance (Figure 14.1) and the benefits accrued by outcompeting diseased conspecifics deserve additional research in order to predict when avoidance of infected conspecifics should occur.

## 14.4.2 Avoidance of contaminated habitats

Habitats containing a high density of parasites seem to be generally avoided across taxa. For example, mammal hosts eat less in areas with high tick abundance (62). Colonial female Bechstein's bats (*Myotis bechsteinii*) also prefer day-roosting sites with lower relative prevalence of the bat fly, *Basilia nana* (63). Raccoons (*Procyon lotor*) use communal defecation sites, or latrines, that concentrate both eggs from racoon parasites (the roundworm *Baylisascaris procyonis*) and potential food (such as seeds) for other species. One study used wildlife cameras at latrines to investigate whether animals avoid use of these contaminated sites (64). The study found that avoidance of latrines was strong in species such as rabbits (*Sylvilagus* spp.) and other small native mammals and birds for which the parasite causes severe pathological effects. Conversely, species more tolerant to the parasite, such as raccoons and rats, were attracted to latrines. This example illustrates how important the cost of infection is in determining behavioral avoidance.

Numerous studies have demonstrated that birds avoid nest sites containing parasites. Species exhibiting this behavior include great tits (*Parus major*) (65), blue tits (*Parus caeruleus*), and cliff swallows (*Hirundo pyrrhonota*) (66). In great tits, parasite abundance also affects the choice of a roosting site (67). Nest desertion in response to parasitism has been found in a variety of species, such as cliff swallows (66), great tits (65), Guanay cormorants (*Phalacrocorax bougainvillii*), the Peruvian booby (*Sula variegata*), and the Peruvian brown pelican (*Pelecanus occidentalis thagus*) (68). Nest parasitism can have strong negative effects on nestling survival and growth (e.g., (66)), which is a likely reason for this widespread avoidance behavior in birds.

No studies regarding avoidance of infected habitats in reptiles exist, constituting a major gap in knowledge. In amphibians, female grey tree frogs (*Hyla versicolor*) avoided ovipositing in pools with uninfected snails that are intermediate hosts of trematode parasites, and increased avoidance of pools containing cercariae-releasing snails (69). Daversa *et al.* (70) showed that alpine newts

(*Ichthyosaura alpestris*) heavily infected with *Batrachochytrium dendrobatidis* (*Bd*) spent less time in water leading to reduced infection burdens, a beneficial sickness behavior, but when given a choice, did not avoid *Bd* in the aquatic habitat.

In fish, more examples of habitat avoidance are available. For example, in laboratory experiments, rainbow trout (*Oncorhynchus mykiss*) will avoid areas with high densities of trematode cercaria (*Diplostomum spathaceum*), which decreases their chance of being parasitized (71). Sea trout (*Salmo trutta trutta*) can learn to avoid visual cues (artificial color signs experimentally manipulated by researchers) associated with the presence of trematode cercaria (*Diplostomum pseudospathaceum*) (72). Juvenile blackspotted stickleback (*Gasterosteus wheatlandi*) change their microhabitat preference to reduce the chance of infection by the Branchiurid ectoparasite *Argulus canadensis* (73). Avoidance of infected microhabitats may be enhanced by group living. Rainbow trout swimming in groups were better able to avoid parasitized habitats, and as such, acquired fewer trematode parasites (*D. pseudospathaceum*) than solitary individuals (74).

Overall, with the exception of reptiles, evidence exists that shows that contaminated habitats are avoided by vertebrates (Table 14.1). While the examples known for mammals are more related to reducing time spent feeding near sites with high parasite abundance, examples in birds and one in amphibians highlight the avoidance of parasitized sites used for reproduction (i.e., ovipositing sites). More generally, movement between habitats, via migration, a common behavior across taxa, can impart numerous infection-related benefits to migrants, and may be considered an infection avoidance behavior in some cases ((75), (76); see also Chapter 7 in this book (77)). Escape from parasites is likely an important driver of animal migration in a range of taxa (36) (Box 14.1, Figure 1). Regardless of avoidance mechanism, the consistent avoidance of contaminated habitats across vertebrate species suggests that the benefits of avoiding contaminated habitats are likely to outweigh the costs, as long as food resources and reproductive opportunities are evenly distributed in space.

### 14.4.3 Avoidance of contaminated food and drinking sources

Mammalian herbivores generally avoid feeding near areas contaminated by feces and many mammals have dedicated defecation sites or defecate away from sleeping or nesting sites (88), (89). Humans (29), Japanese macaques (*Macaca fuscata*; (23)), chimpanzees (*Pan troglodytes*; (90)), and bonobos (*Pan paniscus*; (91)), avoid food items contaminated with feces. Red-fronted lemurs (*Eulemur rufifrons*) show preference for clean relative to fecal-contaminated water (24). Interestingly, grey mouse lemurs (*Microcebus murinus*) show a sex-biased parasite avoidance strategy, where females, but not males, avoid fecally contaminated food, water, and nests.(92) This observation in mouse lemurs could be related the social system of this species (females are more gregarious than males in

---

### Box 14.1  Migration as a parasite avoidance behavior

Migration, the movement of individuals between different habitats, is a common animal behavior that has important implications for disease transmission and spread (78). Although increased exposure to new parasites is commonly noted as a cost of migration (e.g., migratory exposure (78)), growing empirical and theoretical research suggests that movement between habitats can also impart numerous infection-related benefits to migrants, and may be considered an infection avoidance behavior in some cases (75), (76). For example, Loehle (79) suggested that migratory species benefit from moving away from infected habitats for a part of the year as infested areas will be clean again when the migrants return ("environmental migratory escape;" Figure 1; (75)): reindeer (*Rangifer tarandus*) experience reduced infection from warble fly larvae (*Hypoderma tarandi*) as they migrate further away from their calving grounds (80). When migration involves the spatial separation of individuals in a population, migration allows hosts to escape infection by reducing direct transmission of parasites from infected conspecifics, a phenomenon known as "social migratory escape." Theoretical work suggests that migrating may benefit house finches (*Carpodacus mexicanus*) by allowing them to seasonally escape from infected individuals congregating around bird feeders in the winter (45). Relatedly, "migratory allopatry" suggests that susceptible juveniles benefit from being spatially separated from infected adults during this vulnerable life stage through reduced parasite transmission among cohorts (81), (82). Juvenile pink salmon (*Oncorhynchus gorbuscha*) experience low infection prevalence from harmful salmon lice (*Lepeophtheirus salmonis*) during months when they occupy different marine environments from adults. Prevalence of sea lice infection increases dramatically in juveniles when adult salmon migrate back to coastal habitats from the open sea

and are no longer allopatric with younger cohorts (81). Similarly in amphibians that migrate away from aquatic larval habitats, infection risk may decrease. For example, Eastern newts (*Notophthalmus viridescens*) have a terrestrial eft phase with lower infection prevalence of the chytrid fungus (*Batrachoshytrium dendrobatidis*) which is very common in the skin of aquatic adults (83). Reduced host density and increased habitat complexity during the terrestrial phase may also limit pathogen transmission (84).

Parasite presence and infection risk can affect migratory decisions unidirectionally (i.e., without feedback loops). For example, facultative parasites could lead hosts to migrate, but the parasites could remain prevalent in the environment. However, in some cases, feedbacks between infection and migration may result in changes in parasite prevalence and/or infection risk, which in turn, may affect migratory decisions (76), (85), (86). For example, if environmental migratory escape is effective at cleaning indirectly-transmitted parasites from an area in a given time step, reduced parasite infection risk may shift the costs and benefits of migrating in subsequent time steps. This could lead to changes in migration timing or distance or lead to partial migration, where only a fraction of the population migrates in a given year. Very little empirical work has explored parasite–behavior feedbacks in the context of migration. Interestingly, Butler and Roper (87) found that experimental ectoparasite removal in European badger (*Meles meles*) led to a decrease in the frequency of nest-switching behavior, suggesting that parasite infestation can drive movement behavior decisions in some mammals. Migration of one of the species exploited by a parasite could also have feedbacks on increased parasitism on other species that remain resident. More manipulative experiments on larger-scale movement behaviors are needed to further explore these ideas.

**Box 14.1** *Continued*

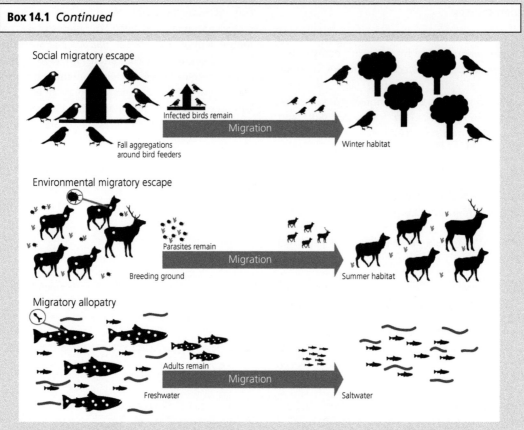

**Box 14.1, Figure 1** Diagram illustrating three putative mechanisms through which migration can act as an infection-avoidance behavior in a range of systems (adapted from (76), by permission of John Wiley and Sons, © 2020 British Ecological Society). Social migratory escape occurs when uninfected individuals migrate away from infected conspecifics, thus reducing direct transmission of parasites. Environmental migratory escape occurs when migrating individuals leave infected habitats, which reduces the risk of infection from indirectly transmitted parasites. Migratory allopatry occurs when adults and juveniles separate spatially in order to reduce parasite transmission between cohorts. Illustrated examples (in top-down order) include house finches migrating away from conspecifics infected with conjunctivitis (45), red deer migrating away from habitats infested with ticks (75,133), and larval Galaxid fish migrating away from natal grounds where adults are infested with trematodes (134).

terms of sleeping associations) but could also be related to sex differences in behavioral avoidance strategies linked to underlying immunological sex differences (93).

There is little evidence of avoidance of parasitized prey in omnivorous or carnivorous species (in mammals or other vertebrates). Researchers have suggested that this is due to reduced pathogenicity to predatory final hosts, which would not impose a high enough cost to avoid parasitized prey (94).

However, mammalian carnivore scavengers have been found generally to avoid consuming carcasses of other carnivores, particularly conspecifics (cannibalism), and the authors of the study suggest that this behavior may have evolved for parasite avoidance (95).

There is some suggestion that birds may be able to avoid consuming parasitized prey. Oystercatchers (*Haematopus ostralegus*) preferentially consume intermediate sizes of a parasitized prey (cockles containing helminth parasites) when larger sizes would be predicted to maximize energy intake but are more intensely parasitized (96). This represents a trade-off between nutrition and parasitism mediated by behavior. By contrast, final host birds preyed more on parasitized killifish (*Fundulus parvipinnis*) than they did on unparasitized ones (97). In this latter example, the authors suggest that the effects are driven by parasite-induced behavioral modifications of the killifish, which make them more conspicuous, and thus easier to catch, by avian predators.

Evidence of avoidance of contaminated food or water is limited in amphibians. Han *et al.* (98) showed that predatory salamanders, *Taricha granulosa* and *Ambystoma macrodactylum*, consumed tadpole stages of three anuran species at similar rates for *Bd* infected and uninfected tadpoles, even though contact with infected tadpoles may increase infection risk to the salamanders. However, tadpoles of the leopard frog (*Rana pipiens*) reduced foraging in the presence of freeze-killed trematode cercariae (99).

There is also currently no evidence to suggest that reptiles and fishes avoid contaminated food such as first intermediate hosts for trophically transmitted parasites (35), (100), (101). Indeed, the few studies that have explored this question in fishes suggest that potential hosts often select infected prey, possibly because infected prey are more easily detectable and/or catchable than uninfected prey. For instance, sticklebacks are more likely to feed on amphipods infected with larval stages of acanthocephalan worms over uninfected ones, probably because infected amphipods are more conspicuous and less photophobic than uninfected ones (102). Stickleback also prefer to eat copepods infected with *Schistocephalus solidus* procercoids, which increase copepod activity and decrease their swimming ability making them easier for fish to catch (103). Butterfly fish (*Chaetodon multicinctus*) have been shown to preferentially feed on trematode-infected coral polyps, which are larger and more noticeable than uninfected ones (104). Selection could favor the preferential consumption of infected prey if the benefits of such behaviors outweigh the costs (105). For instance, Aeby (104) also showed that feeding on infected polyps provided *C. multicinctus* with more

coral tissue per bite and that the rate of successful trematode establishment was low.

As underscored by these examples, avoidance of parasitized prey is rare (Table 14.1) and is likely to be due to parasitized prey being easier to catch and to detect than unparasitized prey, but also potentially due to low risk of pathogenicity to the predator (low cost of infection, Figure 14.1). Therefore, unlike habitat avoidance, there is little overall evidence of avoidance of contaminated food and drinking water sources across vertebrates. The consistency in this pattern across species suggests that the costs of avoiding key food and water resources outweigh the benefits of avoiding potential infection. One exception is that fecal contamination of foods appears to be a more important deterrent of consumption in mammals, suggesting that there may indeed be greater benefits than costs of avoidance in this particular instance. Overall, more research is needed to determine whether avoidance of contaminated water sources for consumption is prevalent across species.

### 14.4.4 Avoidance of parasites and their vectors

A recent study in European badgers (*Meles meles*) suggests that badgers minimize parasite transmission by congregating in areas of parasite prevalence (106), lending to the idea that parasite presence in the environment leads to "landscapes of disgust" for potential hosts (1).

It is likely that the examples presented here for avoidance of parasitized nests in avian species are also examples of direct avoidance of parasites and their vectors. It is hard to disentangle whether birds make nest avoidance decisions because they directly detect the parasites or because they detect correlates of infestation.

No evidence of avoidance of parasites or their vectors has been reported in reptiles. Amphibians can avoid aquatic trematode cercariae by increasing activity after detecting tactile or chemical cues (reviewed in (101)). Increased activity as a parasite avoidance behavior can lead to trade-offs. For example, Raffel *et al.* (4) showed that American toads (*Bufo americanus*) avoiding trematode infection via increased activity can lead to a trade-off when activity attracts predators. There are other amphibian behaviors that lead to trade-offs with parasitism. For instance, Túngara frogs (*Engystomops pustulosus*) and other amphibian species need to trade-off vigorous mating calls and sociality with reducing vector-borne disease and infection risk from biting midges (6), (107). Subsequent to exposure, oak toads (*Anaxyrus quercicus*) learn to avoid areas with to *Bd* (108). Because infection can lead to host immune responses and altered microbiomes (52), (109), indirect detection of pathogens based on these downstream effects on hosts is an untested but likely cue resulting in avoidance behaviors.

Similar to amphibians, changes in activity patterns may also help potential fish hosts avoid infection (101). Previously parasitized fathead minnows (*Pimephales promelas*) reduce activity in the presence of visual and chemical cues of *Ornithodiplostomum ptychocheilus* cercaria (110), a behavior that is absent from parasite-naïve minnows. Since encounter rate with cercaria increases with minnow activity in this system, reduced activity in the presence of parasites likely reduces infection risk. This experiment is interesting because it indicates that avoidance behaviors can be learned after experience of infection.

Across vertebrate taxa, there are examples of how aggregation behaviors can have dilution effects on parasite intensity and prevalence by reducing per-individual exposure to infection. For example, clustering behavior in reindeer reduces the number of fly attacks that each animal receives (111). In Galapagos marine iguanas (*Amblyrhynchus cristatus*), night-time grouping congregations reduced individual risk of parasitic infection by mobile ticks (Figure 14.4) (112). In fathead minnows, individuals located within the center of shoals have

fewer parasitic worms than those restricted to the shoal periphery suggesting that changes in shoaling behavior and positioning confers infection-related benefits to fish hosts (113). Parasites and their vectors may even promote polyspecific associations that enhance encounter dilution, such as the ones observed in mangabeys (*Cercocehus alhigena*) and other primate species, which aggregate during times of increased biting fly activity (114). These studies did not, however, examine whether the presence of parasites or their vectors was the cause of the aggregation behaviors or the changes in group positioning. A study in fathead minnows showed that these fishes can recognize trematode cercaria (*Ornithodiplostomum ptychocheilus*) and respond by shoaling more tightly with conspecifics in order to reduce parasite exposure (113). Similarly, Poulin and Fitzgerald (115) found that shoal size increases in three-spined stickleback when *Argulus* ectoparasites are present. Aggregation can therefore occur as an avoidance behavior.

Combining the examples within this section, it appears that either more studies of parasite avoidance have been done in aquatic environments or that this behavior is more widespread or pronounced in aquatic environments. Therefore, in the cost-benefit framework of avoidance behavior we may posit that the benefits of avoidance may be greater in aquatic environments and outweigh the costs, whereas in terrestrial settings perhaps the opposite is true. One possibility is that parasite mobility and detectability is greater in water than in other settings. To address these questions it will be important to expand these studies into different habitats and also to test more extensively whether avoidance or aggregation behaviors occur as a direct

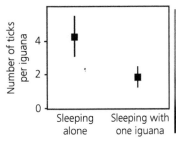

**Figure 14.4** An experimental manipulation of nighttime group size in marine iguanas illustrates how, for a highly mobile parasite (*Ornithodoros* ticks), aggregation behaviors can have dilution effects, where per-individual risk of parasitism is reduced in larger groups (adapted from (112), by permission of Oxford University Press). Iguanas sleeping alone had higher parasite burden than iguanas sleeping in groups of two. Photograph of iguana with tick (red arrow) by SS French.

response to parasite presence. Similar to rats that learned to associate parasite infection with a particular taste (34), amphibians and fish demonstrate learned avoidance behaviors after initial parasite exposure. These behaviors are not typically studied but may be widespread among vertebrates.

### 14.4.5 Other disease avoidance behaviors

Even though we have focused thus far on disease avoidance behaviors and on how susceptible hosts can remove themselves or avoid pathogenrich environments, other forms of avoidance behaviors are possible. Parasite avoidance can also be achieved by different ways in which animals build or modify their environments (116) or niche modification (Table 14.1). For instance, animals use a variety of materials when building their nests that reduce exposure to parasites. Examples include the use of specific aromatic plants that reduce parasite load in nests by certain bird species (117)–(119). Birds also engage in nest sanitation. One example of this behavior occurs in house wrens (*Troglodytes aedon*) where prior to nest-building, males will remove old nests from their nest-boxes leading to a decrease in mite numbers in the nest (120). In mammals, very few examples of use of anti-parasite products are known. Two species of slow lorises (*Nycticebus javanicus* and *Nycticebus coucang*) produce a venom that when applied to their own fur, reduces ectoparasite load (121). Dusky-footed woodrats (*Neotoma fuscipes*) place fresh bay leaves (*Umbellularia californica*) into or near their nests, a behavior that is suggested to reduce the number of flea larvae (122). It will be exciting to expand infection-driven niche modification research into additional vertebrate taxa, where nothing is known (Table 14.1).

## 14.5 Conclusions and future research

The research reviewed in this chapter makes it clear that, under many circumstances, hosts display infection avoidance behaviors. Here, we discuss future research that may allow us to illuminate common principles that predict avoidance behaviors, starting with a proximate perspective, followed by a cost-benefit framework. We conclude with

predictions for how global change may affect avoidance behaviors.

While the cues and systems that allow for disease detection may vary drastically across species, it is possible that the modulators of disease avoidance behaviors overlap, but this is not currently known. In vertebrates, the bulk of the research on this topic has been carried out in rodents, where the neuropeptide oxytocin appears to be a key mediator of the avoidance (and recognition) of infected individuals (10). As oxytocin or oxytocin-like nonapeptides occur in almost all vertebrates and are involved in regulation of socially relevant physiology and behavior in a wide range of taxa (123), (124), future research should examine whether nonapeptides are also important in modulating disease avoidance behaviors in non-rodent taxa. Inspiration from findings in invertebrate model organisms, such as the nematode *Caenorhabditis elegans*, may also prove fruitful in advancing the understanding of the proximate mechanisms underlying behavioral avoidance of parasites and pathogens (125)–(127).

Based on the currently available studies, the most pervasive modes of avoidance across taxa seem to be avoidance of infected conspecifics and of contaminated habitats, as well as migratory escape (Table 14.1). A critical consideration for developing a unifying framework for the occurrence of avoidance behaviors is that of potential costs and benefits associated with avoidance, and how those trade-off with costs and benefits of infections (Figure 14.1).(128) In general, the most important costs associated with avoidance behaviors can be thought of as loss of resources, such as loss of reproductive opportunities, or loss of suitable habitats or of territories, or loss of life in the case of increased predation.

Examples throughout the chapter indicate that costs and benefits associated with mating opportunities, with mate choice and with investment in offspring should impact avoidance behaviors. Because for most sexually reproducing species social interactions (even if brief) are needed for fertilization, additional research focused on avoidance behaviors during reproductive activity are likely to be an important step in developing cross-taxonomic frameworks for infection avoidance behaviors. This type of research, combined with research on sex differences in immune function (93), (129)–(131), may also help clarify the existence of sex biases in avoidance behaviors in certain systems.

There are also costs associated with resource competition. When resources, such as food or suitable habitats, are limited, resource competition should limit avoidance behaviors more strongly. It would be interesting to expand research that manipulates resource availability to examine its impact on avoidance behaviors and to understand how plastic these behaviors are. It may be that the avoidance behaviors have evolved under resource abundance and cannot be changed if resources become quickly limited.

One way in which animals will be exposed to drastic changes in resource availability is through global change. The consequences of global change on animal behavior broadly are increasingly gaining attention. Through an increase in the costs associated with avoidance, we predict that global change will impact landscapes of disgust at both small and large spatial scales and through time.

At a local scale, decreases in availability of food patches and watering holes may put individuals in more frequent contact during times when avoiding conspecifics may have infection-related benefits. We can consider, for example, the túngara frog aggregations illustrated in Figure 14.1B. Small ponds will force individuals to aggregate in even greater densities or for more extended periods of time, which may increase the number of parasite-carrying midges attracted to that location. Similarly, as food and water become limited during periods of drought, severe floods, or scarcity, the benefits of avoiding contaminated food or water may be outweighed by the costs of forgoing sources of food or hydration. This has already been indirectly illustrated in red-fronted lemurs' (*Eulemur rufifrons*), for which shorter travel distances to waterholes are prioritized over avoidance of waterholes that are fecally contaminated, despite a strong preference for clean water (24).

Global change will also affect landscapes of disgust at broader spatial and temporal scales. Since different environmental cues, such as temperature, food availability, and photoperiod guide the onset of breeding differently in different species, global change could decouple the timing of breeding (or replication) of parasites, their vectors, and their various hosts. This could, in turn, lead to inefficient migratory escape strategies, whereby hosts arrive at their destinations at times of higher parasite or vector abundance. Additionally, longer growing seasons and warmer winter temperatures are shifting the distributions, dispersal patterns, and migratory routes of many species (132). For instance, blackcaps (*Sylvia atricapilla*) are increasingly overwintering closer to their breeding grounds as warmer winters reduce the need to migrate long distances to track favorable climates. These types of reductions in migratory behavior may decrease the effectiveness of parasite avoidance strategies such as social and environmental migratory escape (Box 14.1; see also Chapter 7 (77)). How these altered movement patterns interact with parasite avoidance strategies is an area in need of more investigation. In sum, global change has the potential to alter current parasite avoidance strategies by disrupting the costs and benefits of engaging in certain behaviors in certain contexts. These trade-offs remain poorly understood and understudied, presenting rich areas of future research.

## References

1. Weinstein SB, Buck JC, and Young HS. A landscape of disgust. *Science*. 2018 Mar 16;359(6381):1213–14.
2. Zuk M and Stoehr AM. Immune defense and host life history. *Am Nat*. 2002 Oct 1;160(S4):S9–22.
3. Buck JC, Weinstein SB, and Young HS. Ecological and evolutionary consequences of parasite avoidance. *Trends Ecol Evol*. 2018 Aug 1;33(8):619–32.
4. Raffel TR, Hoverman JT, Halstead NT, Michel PJ, and Rohr JR. Parasitism in a community context: Trait-mediated interactions with competition and predation. *Ecol*. 2010;91(7):1900–7.
5. Bernal XE and Pinto CM. Sexual differences in prevalence of a new species of trypanosome infecting túngara frogs. *Int J Parasitol. Parasites Wildl*. 2016;5(1):40–7.
6. Trillo PA, Benson CS, Caldwell MS, Lam TL, Pickering OH, and Logue DM. The influence of signaling conspecific and heterospecific neighbors on eavesdropper pressure. *Front Ecol Evol*. 2019;7:292.
7. Lopes PC. We are not alone in trying to be alone. *Front Ecol Evol*. 2020;8:172.

8. Cremer S, Armitage SAO, and Schmid-Hempel P. Social immunity. *Current Biol.* 2007 Aug 21;17(16):R693–702.

9. Grüter C, Jongepier E, and Foitzik S. Insect societies fight back: The evolution of defensive traits against social parasites. *Phil Trans R Soc B.* 2018 Jul 19;373(1751):20170200.

10. Kavaliers M, Ossenkopp K-P, and Choleris E. Social neuroscience of disgust. *Genes, Brain Behav.* 2019;18(1):e12508.

11. Hart BL. Biological basis of the behavior of sick animals. *Neurosci Biobehav Rev.* 1988 Jun;12(2):123–37.

12. Dantzer R. Cytokine-induced sickness behaviour: A neuroimmune response to activation of innate immunity. *Eur J Pharmacol.* 2004 Oct;500(1–3):399–411.

13. Klein SL. Parasite manipulation of the proximate mechanisms that mediate social behavior in vertebrates. *Physiol Behav.* 2003 Aug 1;79(3):441–9.

14. Herbison R, Lagrue C, and Poulin R. The missing link in parasite manipulation of host behaviour. *Parasites & Vectors.* 2018 Apr 3;11(1):222.

15. Lopes PC. When is it socially acceptable to feel sick? *Proc R Soc B: Biol Sci.* 2014 Aug 7;281(1788):20140218.

16. McAllister MK and Roitberg BD. Adaptive suicidal behaviour in pea aphids. *Nature.* 1987 Aug;328(6133):797–9.

17. Lopes PC and König B. Choosing a healthy mate: Sexually attractive traits as reliable indicators of current disease status in house mice. *Anim Behav.* 2016 Jan 1;111:119–26.

18. Møller AP. Parasite load reduces song output in a passerine bird. *Anim Behav.* 1991 Apr 1;41(4):723–30.

19. Roznik EA, Sapsford SJ, Pike DA, Schwarzkopf L, and Alford RA. Condition-dependent reproductive effort in frogs infected by a widespread pathogen. *Proc R Soc B: Biol Sci* 2015 Jul 7;282(1810):20150694.

20. Gilman S, Blumstein DT, and Foufopoulos J. The effect of hemosporidian infections on White-Crowned sparrow singing behavior. *Ethology.* 2007;113(5):437–45.

21. Shirasu M and Touhara K. The scent of disease: Volatile organic compounds of the human body related to disease and disorder. *J Biochem.* 2011 Sep 1;150(3):257–66.

22. Curtis V and Biran A. Dirt, disgust, and disease: Is hygiene in our genes? *Perspect Biol Med.* 2001;44(1):17–31.

23. Sarabian C and MacIntosh AJJ. Hygienic tendencies correlate with low geohelminth infection in free-ranging macaques. *Biol Lett.* 2015 Nov 30;11(11):20150757.

24. Amoroso CR, Kappeler PM, Fichtel C, and Nunn CL. Fecal contamination, parasite risk, and waterhole use by wild animals in a dry deciduous forest. *Behav Ecol Sociobiol.* 2019 Nov 21;73(11):153.

25. Vittecoq M, Ducasse H, Arnal A, Møller AP, Ujvari B, Jacqueline CB, *et al.* Animal behaviour and cancer. *Anim Behav.* 2015 Mar 1;101:19–26.

26. Ackerman JM, Hill SE, and Murray DR. The behavioral immune system: Current concerns and future directions. *Soc Personal Psychol Compass* 2018;12(2):e12371.

27. Poirotte C and Charpentier MJE. Inter-individual variation in parasite avoidance behaviors and its epidemiological, ecological, and evolutionary consequences. In Ezenwa VO, Altizer S, and Hall RJ (eds), *Animal Behavior and Parasitism.* Oxford: Oxford University Press, 2022. DOI: 10.1093/oso/9780192895561.003.0015

28. Curtis V, de Barra M, and Aunger R. Disgust as an adaptive system for disease avoidance behaviour. *Philos Trans R Soc Lond B Biol Sci.* 2011 Feb 12;366(1563):389–401.

29. Curtis VA. Infection-avoidance behaviour in humans and other animals. *Trends in Immunology.* 2014 Oct 1;35(10):457–64.

30. Kupfer TR and Fessler DMT. Ectoparasite defence in humans: Relationships to pathogen avoidance and clinical implications. *Phil Trans R Soc B.* 2018 Jul 19;373(1751):20170207.

31. Regenbogen C, Axelsson J, Lasselin J, Porada DK, Sundelin T, Peter MG, *et al.* Behavioral and neural correlates to multisensory detection of sick humans. *PNAS.* 2017 Jun 13;114(24):6400–5.

32. Kavaliers M and Choleris E. The role of social cognition in parasite and pathogen avoidance. *Phil Trans R Soc B.* 2018 Jul 19;373(1751):20170206.

33. Spurrier MF, Boyce MS, and Manly BFJ. Effects of parasites on mate choice by captive sage grouse in Loye JE and Zuk E (eds), *Ecology, Behavior and Evolution of Bird–Parasite Interactions.* Oxford: Oxford University Press, 1991, 389–98.

34. Keymer A, Crompton DWT, and Sahakian BJ. Parasite-induced learned taste aversion involving *Nippostrongylus* in rats. *Parasitol.* 1983;86(3):455–60.

35. Wisenden BD, Goater CP, and James CT. Behavioral defenses against parasites and pathogens in Zaccone G, Perrière C, Mathis A, and Kapoor BG (eds), *Fish Defenses: Pathogens, Parasites and Predators,* 1st edn. Enfield, NH: Science Publishers, 2009, 151–68.

36. Binning SA, Shaw AK, and Roche DG. Parasites and host performance: Incorporating infection into our understanding of animal movement. *Integr Comp Biol.* 2017 Jul 8;57(2):267–80.

37. Perrot-Minnot M-J, Cézilly F. Parasites and behaviour in Thomas F, Guégan J-F, and Renaud F (eds), *Ecology*

*and Evolution of Parasitism.* Oxford: Oxford University Press, 2009, 49–68.

38. Silk JB, Alberts SC, and Altmann J. Social bonds of female baboons enhance infant survival. *Science.* 2003 Nov 14;302(5648):1231–4.

39. Poirotte C, Massol F, Herbert A, Willaume E, Bomo PM, Kappeler PM, *et al.* Mandrills use olfaction to socially avoid parasitized conspecifics. *Science Advances.* 2017 Apr 1;3(4):e1601721.

40. Stockmaier S, Bolnick DI, Page RA, and Carter GG. An immune challenge reduces social grooming in vampire bats. *Anim Behav.* 2018 Jun 1;140:141–9.

41. Beltran-Bech S and Richard F-J. Impact of infection on mate choice. *Anim Behav.* 2014 Apr 1;90:159–70.

42. Paciência FMD, Rushmore J, Chuma IS, Lipende IF, Caillaud D, Knauf S, *et al.* Mating avoidance in female olive baboons (*Papio anubis*) infected by *Treponema pallidum. Sci Adv.* 2019 Dec 1;5(12):eaaw9724.

43. Borgia G and Collis K. Female choice for parasite-free male satin bowerbirds and the evolution of bright male plumage. *Behav Ecol Sociobiol.* 1989;25(6): 445–53.

44. Bouwman KM and Hawley DM. Sickness behaviour acting as an evolutionary trap? Male house finches preferentially feed near diseased conspecifics. *Biol Lett.* 2010 Aug 23;6(4):462–5.

45. Hurtado P. The potential impact of disease on the migratory structure of a partially migratory passerine population. *Bull Math Biol.* 2008 Nov;70(8):2264–82.

46. Lalubin F, Bize P, van Rooyen J, Christe P, and Glaizot O. Potential evidence of parasite avoidance in an avian malarial vector. *Anim Behav.* 2012 Sep 1;84(3):539–45.

47. Martín J, Civantos E, Amo L, and López P. Chemical ornaments of male lizards *Psammodromus algirus* may reveal their parasite load and health state to females. *Behav Ecol Sociobiol.* 2007 Aug 15;62(2):173–9.

48. Sorci G, de Fraipont M, and Clobert J. Host density and ectoparasite avoidance in the common lizard (*Lacerta vivipara*). *Oecologia.* 1997 Jul 1;111(2):183–8.

49. Kiesecker JM, Skelly DK, Beard KH, and Preisser E. Behavioral reduction of infection risk. *PNAS.* 1999 Aug 3;96(16):9165–8.

50. Brunetti AE, Lyra ML, Melo WGP, Andrade LE, Palacios-Rodríguez P, Prado BM, *et al.* Symbiotic skin bacteria as a source for sex-specific scents in frogs. *PNAS.* 2019 Feb 5;116(6):2124–9.

51. Woodhams DC, LaBumbard BC, Barnhart KL, Becker MH, Bletz MC, Escobar LA, *et al.* Prodigiosin, Violacein, and volatile organic compounds produced by widespread cutaneous bacteria of amphibians can inhibit two *Batrachochytrium* fungal pathogens. *Microb Ecol.* 2018 May;75(4):1049–62.

52. Jani AJ and Briggs CJ. The pathogen *Batrachochytrium dendrobatidis* disturbs the frog skin microbiome during a natural epidemic and experimental infection. *PNAS.* 2014 Nov 25;111(47):E5049–58.

53. Krause J and Godin J-GJ. Influence of parasitism on the shoaling behaviour of banded killifish, *Fundulus diaphanus. Can J Zool.* 1994 Oct 1;72(10):1775–9.

54. Dugatkin LA, FitzGerald GJ, and Lavoie J. Juvenile three-spined sticklebacks avoid parasitized conspecifics. *Environ Biol Fishes.* 1994 Feb;39(2): 215–18.

55. Rahn AK, Hammer DA, and Bakker TCM. Experimental infection with the directly transmitted parasite Gyrodactylus influences shoaling behaviour in sticklebacks. *Anim Behav.* 2015 Sep;107:253–61.

56. Ward AJW, Duff AJ, Krause J, and Barber I. Shoaling behaviour of sticklebacks infected with the microsporidian parasite, *Glugea anomala. Eviron Biol Fishes.* 2005 Feb;72(2):155–60.

57. Croft DP, Edenbrow M, Darden SK, Ramnarine IW, van Oosterhout C, and Cable J. Effect of gyrodactylid ectoparasites on host behaviour and social network structure in guppies *Poecilia reticulata. Behav Ecol Sociobiol.* 2011 Jul 23;65(12):2219–27.

58. Stephenson JF, Perkins SE, and Cable J. Transmission risk predicts avoidance of infected conspecifics in Trinidadian guppies. *J Anim Ecol.* 2018 Nov 1;87(6):1525–33.

59. Busenberg S and Cooke K. *Vertically Transmitted Diseases: Models and Dynamics.* Berlin: Springer-Verlag, 1993.

60. Brindley PJ, He S, Sitepu P, Pattie WA, and Dobson C. Inheritance of immunity in mice to challenge infection with *Nematospiroides dubius. Heredity.* 1986 Aug;57(1):53–8.

61. Stockmaier S, Bolnick DI, Page RA, and Carter GG. Sickness effects on social interactions depend on the type of behaviour and relationship. *J Anim Ecol.* 2020;89(6):1387–94.

62. Fritzsche A and Allan BF. The ecology of fear: Host foraging behavior varies with the spatio-temporal abundance of a dominant ectoparasite. *EcoHealth.* 2012 Mar 1;9(1):70–4.

63. Reckardt K and Kerth G. Roost selection and roost switching of female Bechstein's bats (*Myotis bechsteinii*) as a strategy of parasite avoidance. *Oecologia.* 2007 Dec;154(3):581–8.

64. Weinstein SB, Moura CW, Mendez JF, and Lafferty KD. Fear of feces? Tradeoffs between disease risk and foraging drive animal activity around raccoon latrines. *Oikos.* 2018;127(7):927–34.

65. Oppliger A, Richner H, and Christe P. Effect of an ectoparasite on lay date, nest-site choice, desertion,

and hatching success in the great tit (*Parus major*). *Behav Ecol*. 1994 Jul 1;5(2):130–4.

66. Brown CR and Brown MB. Ectoparasitism as a cost of coloniality in cliff swallows (*Hirundo Pyrrhonota*). *Ecology*. 1986;67(5):1206–18.

67. Christe P, Oppliger A, and Richner H. Ectoparasite affects choice and use of roost sites in the great tit, *Parus major. Anim Behav*. 1994 Apr 1;47(4):895–8.

68. Duffy DC. The ecology of tick parasitism on densely nesting Peruvian seabirds. *Ecology*. 1983;64(1):110–19.

69. Kiesecker JM and Skelly DK. Choice of oviposition site by Gray treefrogs: The role of potential parasitic infection. *Ecology*. 2000;81(10):2939–43.

70. Daversa DR, Manica A, Bosch J, Jolles JW, and Garner TWJ. Routine habitat switching alters the likelihood and persistence of infection with a pathogenic parasite. *Funct Ecol*. 2018 May;32(5):1262–70.

71. Karvonen A, Seppälä O, and Valtonen ET. Parasite resistance and avoidance behaviour in preventing eye fluke infections in fish. *Parasitol*. 2004 Aug;129(2):159–64.

72. Klemme I and Karvonen A. Learned parasite avoidance is driven by host personality and resistance to infection in a fish–trematode interaction. *Proc R Soc B: Biol Sci* 2016 Sep 14;283(1838):20161148.

73. Poulin R and FitzGerald GJ. Risk of parasitism and microhabitat selection in juvenile sticklebacks. *Can J Zool*. 1989 Jan 1;67(1):14–18.

74. Mikheev VN, Pasternak AF, Taskinen J, and Valtonen TE. Grouping facilitates avoidance of parasites by fish. *Parasites Vectors*. 2013 Oct 17;6(1).

75. Mysterud A, Qviller L, Meisingset EL, and Viljugrein H. Parasite load and seasonal migration in red deer. *Oecologia*. 2016 Feb 1;180(2):401–7.

76. Shaw AK and Binning SA. Recovery from infection is more likely to favour the evolution of migration than social escape from infection. *J Anim Ecol*. 2020 Jun 1;89(6):1448–57.

77. Hall RJ, Altizer S, Peacock SJ, and Shaw AK. Animal migration and infection dynamics: Recent advances and future frontiers. In Ezenwa VO, Altizer S, and Hall RJ (eds), *Animal Behavior and Parasitism*. Oxford: Oxford University Press, 2022. DOI: 10.1093/oso/9780192895561.003.0015.

78. Altizer S, Bartel R, and Han BA. Animal migration and infectious disease risk. *Science*. 2011 Jan 21;331(6015):296–302.

79. Loehle C. Social barriers to pathogen transmission in wild animal populations. *Ecol*. 1995;76(2):326–35.

80. Folstad I, Nilssen AC, Halvorsen O, and Andersen J. Parasite avoidance: The cause of post-calving migrations in Rangifer? *Can J Zool*. 1991 Sep 1;69(9):2423–9.

81. Krkošek M, Gottesfeld A, Proctor B, Rolston D, Carr-Harris C, and Lewis MA. Effects of host migration, diversity and aquaculture on sea lice threats to Pacific salmon populations. *Proc R Soc B: Biol Sci*. 2007 Dec 22;274(1629):3141–9.

82. Todd BD. Parasites lost? An overlooked hypothesis for the evolution of alternative reproductive strategies in amphibians. *Am Nat*. 2007 Nov;170(5):793–9.

83. Robinson CW, McNulty SA, and Titus VR. No safe space: Prevalence and distribution of *Batrachochytrium dendrobatidis* in amphibians in a highly-protected landscape. *Herpetol Cinserv Biol*. 2018;13(2):373–382.

84. Malagon DA, Melara LA, Prosper OF, Lenhart S, Carter ED, Fordyce JA, *et al*. Host density and habitat structure influence host contact rates and *Batrachochytrium salamandrivorans* transmission. *Scientific Reports*. 2020 Mar 27;10(1):5584.

85. Shaw AK, Craft ME, Zuk M, and Binning SA. Host migration strategy is shaped by forms of parasite transmission and infection cost. *J Anim Ecol* 2019 Oct 1;88(10):1601–12.

86. Naven Narayanan, Binning SA, and Shaw AK. Infection state can affect host migratory decisions. *Oikos*. 2020 Oct 1;129(10):1493–503.

87. Butler JM and Roper TJ. Ectoparasites and sett use in European badgers. *Anim Behav*. 1996 Sep 1;52(3):621–9.

88. Coulson G, Cripps JK, Garnick S, Bristow V, and Beveridge I. Parasite insight: Assessing fitness costs, infection risks and foraging benefits relating to gastrointestinal nematodes in wild mammalian herbivores. *Phil Trans R Soc B*. 2018 Jul 19;373(1751):20170197.

89. Hart BL. Behavioral adaptations to pathogens and parasites: Five strategies. *Neurosci Biobehav Rev*. 1990;14(3):273–94.

90. Sarabian C, Ngoubangoye B, and MacIntosh AJJ. Avoidance of biological contaminants through sight, smell and touch in chimpanzees. *R Soc Open Sci*. 2017;4(11):170968.

91. Sarabian C, Belais R, and MacIntosh AJJ. Feeding decisions under contamination risk in bonobos. *Phil Trans R Soc B*. 2018 Jul 19;373(1751):20170195.

92. Poirotte C and Kappeler PM. Hygienic personalities in wild grey mouse lemurs vary adaptively with sex. *Proc R Soc B: Biol Sci*. 2019 Aug 14;286(1908):20190863.

93. Klein SL and Flanagan KL. Sex differences in immune responses. *Nat Rev Immunol*. 2016 Oct;16(10):626–38.

94. Øverli Ø and Johansen IB. Kindness to the final host and vice versa: A trend for parasites providing easy prey? *Front Ecol Evol*. 2019;7.

95. Moleón M, Martínez-Carrasco C, Muellerklein OC, Getz WM, Muñoz-Lozano C, Sánchez-and Zapata JA. Carnivore carcasses are avoided by carnivores. *J Anim Ecol*. 2017;86(5):1179–91.

96. Norris K. A trade-off between energy intake and exposure to parasites in oystercatchers feeding on a bivalve mollusc. *Proc R Soc B: Biol Sci*. 1999 Aug 22;266(1429):1703–9.

97. Lafferty KD and Morris AK. Altered behavior of parasitized killifish increases susceptibility to predation by bird final hosts. *Ecol*. 1996;77(5):1390–7.

98. Han BA, Searle CL, and Blaustein AR. Effects of an infectious fungus, *Batrachochytrium dendrobatidis*, on amphibian predator-prey interactions. *PLoS One*. 2011 Feb 2;6(2):e16675.

99. Koprivnikar J and Penalva L. Lesser of two evils? Foraging choices in response to threats of predation and parasitism. *PLoS One*. 2015;10(1):e0116569.

100. Barber I, Hoare D, and Krause J. Effects of parasites on fish behaviour: A review and evolutionary perspective. *Rev Fish Biol Fish*. 2000 Jun 1;10(2):131–65.

101. Behringer DC, Karvonen A, and Bojko J. Parasite avoidance behaviours in aquatic environments. *Phil Trans R Soc B*. 2018 Jul 19;373(1751):20170202.

102. Bakker TCM, Mazzi D, and Zala S. Parasite-induced changes in behavior and color make *Gammarus pulex* more prone to fish predation. *Ecology*. 1997 Jun;78(4):1098.

103. Wedekind C and Milinski M. Do three-spined sticklebacks avoid consuming copepods, the first intermediate host of *Schistocephalus solidus*?—an experimental analysis of behavioural resistance. *Parasitol*. 1996 Apr;112(4):371–83.

104. Aeby G. Trade-offs for the butterflyfish, *Chaetodon multicinctus*, when feeding on coral prey infected with trematode metacercariae. *Behav Ecol Sociobiol*. 2002 Jul 1;52(2):158–65.

105. Lafferty KD. Foraging on prey that are modified by parasites. *Am Nat*. 1992 Nov 1;140(5):854–67.

106. Albery GF, Newman C, Ross JB, MacDonald DW, Bansal S, and Buesching C. Negative density-dependent parasitism in a group-living carnivore. *Proc R Soc B: Biol Sci*. 2020 Dec 23;287(1941):20202655.

107. Toledo LF, Ruggeri J, Leite Ferraz de Campos L, Martins M, Neckel-Oliveira S, and Breviglieri CPB. Midges not only suck, but may carry lethal pathogens to wild amphibians. *Biotropica*. 2021;53(3):722–5.

108. McMahon TA, Sears BF, Venesky MD, Bessler SM, Brown JM, Deutsch K, *et al*. Amphibians acquire resistance to live and dead fungus overcoming fungal immunosuppression. *Nature*. 2014 Jul;511(7508):224–7.

109. Song SJ, Woodhams DC, Martino C, Allaband C, Mu A, Javorschi-Miller-Montgomery S, *et al*. Engineering the microbiome for animal health and conservation. *Exp Biol Med (Maywood)*. 2019 Apr;244(6):494–504.

110. James CT, Noyes KJ, Stumbo AD, Wisenden BD, and Goater CP. Cost of exposure to trematode cercariae and learned recognition and avoidance of parasitism risk by fathead minnows *Pimephales promelas*. *J Fish Biol*. 2008 Dec;73(9):2238–48.

111. Helle T and Aspi J. Does herd formation reduce insect harassment among reindeer? A field experiment with animal traps. *Acta Zoologica Fennica*. 1983;175:129–31.

112. Wikelski M. Influences of parasites and thermoregulation on grouping tendencies in marine iguanas. *Behav Ecol*. 1999 Jan 1;10(1):22–9.

113. Stumbo AD, James CT, Goater CP, and Wisenden BD. Shoaling as an antiparasite defence in minnows (*Pimephales promelas*) exposed to trematode cercariae. *J Anim Ecol*. 2012 Jul 9;81(6):1319–26.

114. Freeland WJ. Blood-sucking flies and primate polyspecific associations. *Nature*. 1977 Oct;269(5631):801–2.

115. Poulin R an, FitzGerald GJ. Shoaling as an anti-ectoparasite mechanism in juvenile sticklebacks (*Gasterosteus* spp.). *Behav Ecol Sociobiol*. 1989 Apr;24(4):251–5.

116. Pinter-Wollman N, Jelić A, and Wells NM. The impact of the built environment on health behaviours and disease transmission in social systems. *Philos Trans R Soc Lond B Biol Sci*. 2018 Aug 19;373(1753):20170245.

117. Lafuma L, Lambrechts MM, and Raymond M. Aromatic plants in bird nests as a protection against blood-sucking flying insects? *Behav Processes*. 2001 Nov 1;56(2):113–20.

118. Petit C, Hossaert-McKey M, Perret P, Blondel J, and Lambrechts MM. Blue tits use selected plants and olfaction to maintain an aromatic environment for nestlings. *Ecol Lett*. 2002;5(4):585–9.

119. Yang C, Ye P, Huo J, Møller AP, Liang W, and Feeney WE. Sparrows use a medicinal herb to defend against parasites and increase offspring condition. *Curr Biol*. 2020 Dec 7;30(23):R1411–12.

120. Pacejka AJ, Santana E, Harper RG, and Thompson CF. House wrens *Troglodytes aedon* and nest-dwelling ectoparasites: Mite population growth and feeding patterns. *J Avian Biol*. 1996;27(4):273–8.

121. Grow NB, Wirdateti, and Nekaris KAI. Does toxic defence in *Nycticebus* spp. relate to ectoparasites? The lethal effects of slow loris venom on arthropods. *Toxicon*. 2015 Mar 1;95:1–5.

122. Hemmes RB, Alvarado A, Hart BL. Use of California bay foliage by wood rats for possible fumigation

of nest-borne ectoparasites. *Behav Ecol.* 2002 May 1;13(3):381–5.

123. Goodson JL and Thompson RR. Nonapeptide mechanisms of social cognition, behavior and species-specific social systems. *Curr Opin Neurobiol.* 2010 Dec 1;20(6):784–94.

124. Goodson JL. Deconstructing sociality, social evolution and relevant nonapeptide functions. *Psychoneuroendocrinol.* 2013 Apr 1;38(4): 465–78.

125. Reddy KC, Andersen EC, Kruglyak L, and Kim DH. A Polymorphism in npr-1 is a behavioral determinant of pathogen susceptibility in C. *elegans. Science.* 2009 Jan 16;323(5912):382–4.

126. Chang HC, Paek J, and Kim DH. Natural polymorphisms in C. *elegans* HECW-1 E3 ligase affect pathogen avoidance behaviour. *Nature.* 2011 Dec;480(7378):525–9.

127. Anderson A and McMullan R. Neuronal and non-neuronal signals regulate *Caernorhabditis elegans* avoidance of contaminated food. *Phil Trans R Soc B.* 2018 Jul 19;373(1751):20170255.

128. Amoroso CR and Antonovics J. Evolution of behavioural resistance in host–pathogen systems. *Biol Lett.* 2020 Sep 30;16(9):20200508.

129. Casimir GJ, Lefèvre N, Corazza F, and Duchateau J. Sex and inflammation in respiratory diseases: A clinical viewpoint. *Biol Sex Differ.* 2013 Sep 1; 4(1):16.

130. Kadel S and Kovats S. Sex hormones regulate innate immune cells and promote sex differences in respiratory virus infection. *Front Immunol.* 2018 Jul 20;9.

131. Lasselin J, Lekander M, Axelsson J, and Karshikoff B. Sex differences in how inflammation affects behavior: What we can learn from experimental inflammatory models in humans. *Front Neuroendocrinol.* 2018 Jul 1;50:91–106.

132. Pulido F and Berthold P. Current selection for lower migratory activity will drive the evolution of residency in a migratory bird population. *PNAS.* 2010 Apr 20;107(16):7341–6.

133. Qviller L, Risnes-Olsen N, Bærum KM, Meisingset EL, Loe LE, Ytrehus B, *et al.* Landscape level variation in tick abundance relative to seasonal migration in red deer. *PLOS One.* 2013 Aug 9;8(8): e71299.

134. Poulin R, Closs GP, Lill AWT, Hicks AS, Herrmann KK, and Kelly DW. Migration as an escape from parasitism in New Zealand galaxiid fishes. *Oecologia.* 2012 Jan 24;169(4):955–63.

135. Fenton A, Magoolagan L, Lill AWT, Hicks AS, Herrmann KK, and Spencer KA. Parasite-induced warning coloration: a novel form of host manipulation. *Animal behaviour.* 2011;81(2):417–22.

136. Bull CM, Burzacott DA. The influence of parasites on the retention of long-term partnerships in the Australian sleepy lizard, Tiliqua rugosa. *Oecologia.* 2006;146(4):675–80.

# Inter-individual variation in parasite avoidance behaviors and its epidemiological, ecological, and evolutionary consequences

Clémence Poirotte and Marie JE. Charpentier

## 15.1 Introduction

In response to parasite threats, hosts have evolved different types of defense mechanisms including those that reduce parasite exposure (avoidance), parasite load during or after invasion (resistance), or the negative effects of infection on fitness (tolerance) (1)–(3). This chapter focuses on the host's "avoidance behavior repertoire" which functions to decrease exposure to parasites.[1] Animals, from worms to primates, use, indeed, a wide range of avoidance behaviors that may be classified into three categories (2): (i) avoidance of contacts with contaminated or potentially contaminated substrates; (ii) direct avoidance of parasites present in the environment, if hosts are able to detect them; and (iii) avoidance of infected or potentially infected conspecifics. These strategies have been likened to a "behavioral immune system"(4), which refers to a suite of mechanisms allowing animals to detect cues associated with the presence of parasites, triggering emotional responses such as disgust which facilitate a functional behavioral reaction. Although

understudied, these mechanisms may represent cost-effective defense strategies to avoid contamination (2) because they reduce parasite encounter, alleviating hosts from costly investment in physiological defenses. As such, these defense strategies should be under strong directional selection.

Despite the obvious benefits of fending off parasites, the few recent studies on parasite avoidance behaviors all indicate a tremendous heterogeneity in the intensity and frequency at which hosts exhibit such behaviors, at various scales: across closely related species, across populations of the same species, across individuals in the same population, and within individuals, across time (Figure 15.1). For example, during choice tests, trout (*Salmo trutta trutta*), but not closely related Atlantic salmon (*Salmo salar*), avoid areas infested with a trematode eye-fluke (*Diplostomum pseudospathaceum*) (5). The low locomotory activity of salmon possibly reduces their rate of parasite encounter, making it less crucial for this species to actively avoid parasitized areas compared to trout (but see (6)). Laboratory mice (*Mus musculus*) strongly avoid contacts with odors from conspecifics infected by gastrointestinal parasites (7) and house mice (*M. musculus domesticus*) exhibit "burrow cleaning" behavior (8), in contrast to wild mice (*Peromyscus leucopus* and *P. maniculatus*) that do not exhibit fecal avoid-

---

[1] We use here a broad definition of parasite as any organism that feeds on another organism to live thus including microparasites such as viruses, bacteria, fungi, and protozoa, as well as macroparasites such as endo-parasitic worms and ectoparasites (arthropod, blood-sucking insects).

Clémence Poirotte and Marie JE. Charpentier, *Inter-individual variation in parasite avoidance behaviors and its epidemiological, ecological, and evolutionary consequences*. In: *Animal Behavior and Parasitism*. Edited by Vanessa O. Ezenwa, Sonia Altizer and Richard J. Hall, Oxford University Press.

**VARIATION OF INVESTMENT IN PARASITE AVOIDANCE BEHAVIORS AT DIFFERENT SCALES**

**BETWEEN CLOSE SPECIES**

*Diplostomum pseudopathaceum*      *Salmo trutta trutta*

Trout avoid areas infested with a trematode eye-fluke.

*Salmo salar*

Salmon do not avoid infested areas.

**BETWEEN POPULATIONS**

*Ovis aries*

Free-ranging sheep from St Kilda prefer short non-contaminated grass over tall contaminated tussock.

 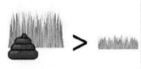

*Ovis aries*

Captive sheep prefer tall contaminated tussock over short non-contaminated grass.

**BETWEEN INDIVIDUALS**

*Mandrillus sphinx*   *Macaca fuscata fuscata*   *Microcebus murinus*

Some mandrills, Japanese macaques, and grey mouse lemurs avoid fecally contaminated substrates.

Some mandrills, Japanese macaques, and grey mouse lemurs do not avoid fecally contaminated substrates.

**WITHIN INDIVIDUALS, ACROSS TIME**

*Lagothrix lagotricha poeppigii*

Juvenile woolly monkeys avoid fecally contaminated food more than adults.

**Figure 15.1** Variation in parasite avoidance behaviors at different scales: between closely related species (5), between populations of the same species (15, 18), between individuals of the same population (10, 11, 37), and within individuals, across time (41).

ance or selective use of nesting material (9). Some Japanese macaques (*Macaca fuscata fuscata*) prefer to discard fecally contaminated food items during feeding tests while others do not. Interestingly, the "hygienic" individuals (i.e., those performing parasite avoidance behaviors) are also those that actively manipulate food covered with sand before ingestion (10). In grey mouse lemurs (*Microcebus murinus*), about half of all individuals consistently avoid fecal contamination when given the choice between contaminated and non-contaminated food items, water, and nests, and further avoid defecating in their nests, while the other half either do not or only weakly exhibit these hygienic traits (11).

Beyond describing these differences, a crucial next step is to understand the determinants of this variation in the avoidance-behavior repertoire that individuals, populations, or species exhibit. In the first section of this chapter, we present recent evidence suggesting that such variation is rooted in a cost-benefit framework shaped by ecological characteristics and individual traits. The second section of the chapter deals with the epidemiological consequences of variation in avoidance behaviors as well as the impact of these behaviors on host ecology and evolution, species interactions, and ecosystem structure.

## 15.2 The avoidance behavior repertoire: Sources of variation

### 15.2.1 A cost-benefit framework

Trade-off theory is a central concept in the study of animal life-history and behavioral traits: a trait

may be strongly selected for a given purpose, at the expense of another trait (12). This theory helped illuminate, for example, how animals make foraging decisions in complex environments, balancing the benefits of seeking good-quality resource patches with travelling costs and predation risks (12). Trade-off theory has also been used in the field of eco-immunology, improving our understanding of natural variation in physiological immune functions (13). Indeed, the costs associated to the maintenance and activation of the physiological immune system results in trade-offs between investment in immune function and other life-history traits, such as reproduction.

Similarly, trade-off theory could be applied to parasite avoidance behaviors to improve our understanding of variation in behavioral immunity. While avoidance behaviors are often considered as an economical defense option against parasite threats compared to physiological defenses, they may come with energetic and other costs (Figure 15.2). For example, avoiding ectoparasitic mites (*Macrocheles muscaedomesticae*) increases metabolic demand by 70% in fruit flies (*Drosophila hydei*; Figure 15.2B; (14)). Moreover, engaging in avoidance behaviors may cause animals to miss other opportunities to fulfill their biological needs (Figure 15.2A, Figure 15.2D). Indeed, animals face various parasite threats in almost all aspects of their lives: when foraging, seeking for a place to rest, or interacting with conspecifics. All of these vital behaviors may also favor the transmission of parasites, generating behavioral trade-offs: animals should therefore have to choose constantly between engaging

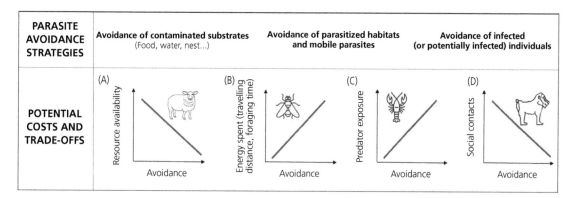

**Figure 15.2** Parasite avoidance strategies and associated trade-offs in species where such trade-offs occur. Parasite avoidance strategies can (A) deprive animals of valuable resources, such as food, water, or nests, (B) incur energy costs, (C) expose hosts to predators, and (D) deprive animals of valuable social contacts.

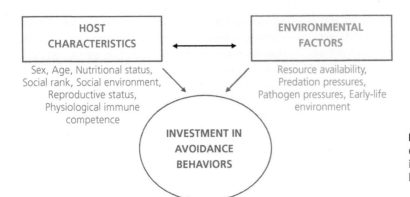

**Figure 15.3** Individual and environmental factors influencing investment in parasite avoidance behaviors.

or not engaging in potentially risky, but otherwise essential, behaviors. Here, we highlight how this behavioral trade-off framework can help us understand how animals behave in the face of parasite threats and predict when parasite avoidance is likely to occur. We provide examples of parasite avoidance behaviors following the classification proposed by Curtis (2) to illustrate that the costs and benefits associated with avoidance behavior are modulated by ecological context and individual traits (Figure 15.3). We place a particular emphasis on the role of host sex and physiological immunocompetence. As such, we show that they are important determinants of optimal investment into avoidance behavior, generating distinct repertoires across species, population, individuals, and time.

## 15.2.2 Avoiding contaminated substrates

One way to avoid parasites is to avoid contaminated substrates. Fecal avoidance in a foraging context is a common strategy observed throughout the animal kingdom, providing a first line of defense against a large array of gastrointestinal parasites (2). An extensive literature on selective foraging in herbivores has paved the way towards understanding trade-offs between nutritional demands and fecal avoidance (15). Indeed, such a general avoidance strategy may decrease the amount of palatable food resources available (Figure 15.2A). Herbivores, for example, can maximize nutrient intake through the selection of tall swards, but tall swards generally contain a greater number of parasites than short swards (15). During feeding experiments,

captive sheep (*Ovis aries*) prefer tall, contaminated swards over short, non-contaminated swards, suggesting that feeding is prioritized over parasite avoidance (15). Yet, mammalian herbivores generally exhibit strong avoidance of contaminated pastures in agricultural systems (16). In natural systems, fecal avoidance is also observed across herbivores (17), despite likely costs related to nutrient intake. Interestingly and contrary to captive sheep (15), wild Soay sheep from St Kilda prefer short, non-contaminated swards over tall, contaminated swards (Figure 15.1; (18)). In addition, foraging selectivity in natural populations varies with food availability: the tendency of male grey mouse lemurs to avoid fecally contaminated food resources decreases when food is scarce in the environment (Figure 15.1; (11)). If the energetic costs of parasitism are moderate, predators also generally prefer feeding on parasitized but more easily captured prey (19). As such, several bird species prey on fish infected by trophically transmitted parasites instead of avoiding them (20), presumably because the benefits of an easy catch outweigh the costs of infection.

Host internal condition, such as nutritional or physiological state, may also influence trade-offs associated with selective foraging behavior. For instance, grey mouse lemurs in good physical shape avoid contaminated food items more than those in poor body condition (11). In addition, these lemurs exhibit seasonal shifts in their avoidance behavior because they engage in different activities throughout the year. Males and females, however, face periods of high nutritional demands at different times (11). Female feeding motivation is highest during the early dry season, just before hibernation, explaining a reported decrease in fecal avoidance behavior at that time of the year. By contrast, male feeding motivation is highest at the end of the dry season when they need to store energy for the upcoming mating season. Accordingly, males do not show feeding selectivity at the end of the dry season in order to meet the energetic requirements associated to reproduction. Such ecologically induced behavioral flexibility allows animals to display avoidance behaviors only when they result in positive cost-benefit trade-offs. Selective foraging should thus strike a balance between the costs of infection and the benefits of nutrition. Where this balance falls is likely to depend on the context, including food availability and individual condition.

### 15.2.3 Avoiding parasitized habitats

Another way to avoid parasites is to select parasite-free habitats with the possible cost of being deprived of high-quality sites. Fruit flies (*D. nigrospiracula*) usually detect and avoid sites parasitized by mites (*Macrocheles subbadius*) in homogenous environments. However, in heterogeneous environments with variable resource quality, flies prefer to oviposit in sites with abundant cactus plants irrespective of the level of parasitism (21). Site quality with plenty of cactus thus overcomes mite avoidance in this system, despite the negative effects of mites on fly reproductive success. Avoiding visible parasites in the habitat might also expose animals to other natural and lethal enemies such as predators. Several larval amphibians avoid areas infested with free-swimming stages of trematodes and increase their activity level to prevent parasite establishment (22). However, such behaviors increase the risk of predation by larval odonates or fish. During experiments, tadpoles avoid parasites as much as predator cues, but they do not exhibit parasite avoidance behaviors if both threats are present simultaneously (23). Thus, in natural settings, both host life history and ecological factors could shape variation in anti-parasite behaviors across species and populations to balance host defenses against several types of natural enemies. Grey tree frogs

(*Hyla versicolor*) and northern leopard tree frogs (*Lithobates pipiens*), for example, do not exhibit avoidance behaviors towards trematodes despite their high susceptibility to these parasites, possibly because they are highly exposed and palatable to a wide range of predators. By contrast, American toads (*Bufo americanus*), which are unpalatable to most amphibian predators, show high levels of parasite avoidance (22). Future studies should therefore consider species with contrasting life-history traits and under various ecological conditions, including predation pressures, to elucidate the circumstances under which hosts are likely to exhibit habitat avoidance behaviors in response to parasites.

### 15.2.4 Avoiding contacts with infected conspecifics

Animals, including humans, can further limit pathogen exposure by avoiding contacts with conspecifics, especially those showing signs of infection (24), generating trade-offs between the benefits of group living and the risk of contamination with contagious diseases. The unprecedent measures taken in face of the Coronavirus 2019 disease (COVID-19) pandemic have caused depression and anxiety in a large number of individuals (25), revealing how costly "social distancing"[2] can be (24). Similarly, animals avoiding social contacts to decrease contagious risk are deprived of numerous advantages that have driven the evolution of sociality in the first place (Figure 15.2D). Healthy three-spined sticklebacks (*Gasterosteus aculeatus*) avoid shoaling with conspecifics harboring ectoparasites (*Argulus canadensis*; (26)). Such changes in shoaling behavior may, however, increase the risk of predation because grouping in fish usually offers efficient anti-predator protection (27). Similarly, by avoiding shelters occupied by conspecifics infected with a lethal virus (*Panulirus argus* virus 1—PaV1), social lobsters (*Panulirus argus*) are more preyed upon (Figure 15.2C; (28)). Animals may further disperse from their social group when parasite pressure is

---

[2] Social distancing includes the avoidance of infected individuals, the active self-exclusion of potentially infected individuals or disease-induced sickness behaviors (96).

high although dispersal generally incurs energetic costs, increases mortality risks, delays breeding, and reduces the advantage of being in a familiar social unit (29). Female gorillas (*Gorilla gorilla gorilla*) emigrate from their social group when the prevalence of a skin disease caused by a *Treponema* species is high, but only when they can immigrate into a healthy social group to avoid remaining solitary for long periods of time (30).

The costs of reduced sociality might have promoted the evolution of more fine-tuned behavioral avoidance mechanisms (24). Elaborated forms of avoidance could, indeed, efficiently protect individuals while alleviating the costs of complete social isolation. For instance, animals may avoid only some specific risky social contacts (see Box 15.1 and Figure 15.4). As such, female olive baboons (*Papio anubis*) avoid copulating only with those mating partners that exhibit visible ulcerated genitals caused by a sexually transmitted bacterium (*Treponema pallidum*; (31)). Alternatively, animals may decrease costs associated with social avoidance by exhibiting behavioral plasticity. For instance,

in mandrills (*Mandrillus sphinx*) social interactions with contagious individuals are only possible with close kin but avoided with distant or non-kin, possibly because maintaining such bonds is more important than avoiding contagious diseases (see Box 15.1 and Figure 15.4; (32)). More generally, ecological and individual factors are expected to influence the cost-benefit trade-off of such avoidance behaviors, and individuals that have much to gain from a social interaction should favor social contacts over pathogen avoidance. For instance, although empirical data are virtually non-existent, demographic parameters, such as operational sex ratio, could influence the propensity to avoid contagious partners for mating, and low-ranking individuals could be less selective than high-ranking individuals. As contagious diseases transmitted by social contacts pose a serious threat for a wide range of taxa, it is crucial that future studies focus on how habitat and host characteristics might constrain or promote the evolution of social avoidance traits and the expression of these traits at the individual level.

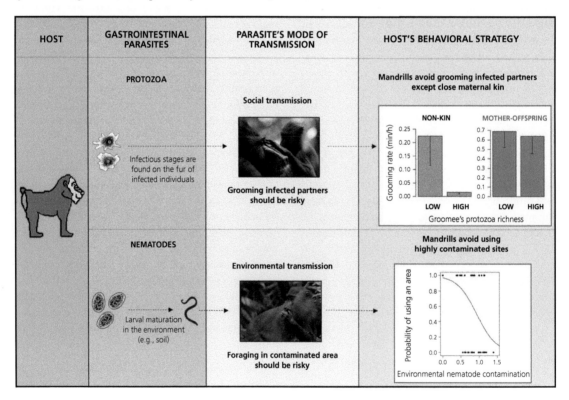

**Figure 15.4** Parasite avoidance strategies for two different types of gastrointestinal parasites in mandrills (32, 60, 67).

## Box 15.1 The mandrill's case

Mandrills are Old World primates living in large social groups in dense equatorial rainforests where parasite pressures are intense. Several recent studies revealed that these primates exhibit a rich array of fine-tuned behavioral strategies to avoid infection from a variety of gastro-intestinal parasites (Figure 15.4).

In a large natural population living in southern Gabon, individuals infected by contagious gastrointestinal protozoa harbor infective cysts on their bodies, mainly around the peri-anal region. Interestingly, mandrills do not spatially avoid contagious groupmates but do avoid grooming them, particularly around the peri-anal "risky" area. In addition, grooming rates return to normal following experimental treatment of these infected groupmates (67). Such a behavioral strategy probably efficiently decreases protozoa transmission, as infective cysts are easily ingested by groomers during grooming events. This social avoidance is, at least partly, guided by an olfactory mechanism: protozoa influence the chemical composition of volatile compounds found in feces and captive mandrills discriminate fecal samples of infected conspecifics from those collected from non-infected mandrills (67). The costs of avoiding non-contagious groupmates might be particularly high in this social species explaining such an elaborate detection mechanism. Nevertheless, this strategy may still have detrimental effects because it deprives individuals of possible valuable social interactions. Interestingly, mandrills modulate social avoidance of contagious individuals according to their focal social partner: individuals do not avoid grooming close maternal kin when contagious, although they do avoid contagious distant kin or non-kin (32). These behavioral adaptations reveal that selecting safe social partners may help these primates cope with parasite-mediated costs of sociality and that behavioral immunity likely plays a crucial role in the coevolutionary dynamics between hosts and their parasites.

In their natural habitat, mandrills are also exposed to gastrointestinal nematodes that require a maturation period in the environment to reach the infective larvae stage that infects hosts through ingestion of soiled items. To avoid contamination, mandrills should therefore avoid contact with fecal material containing infective stages and/or contaminated habitats. During feeding experiments, captive mandrills, especially females, avoid eating food items soiled with fecal material although neither the presence of nematodes nor their developmental stage influence the level of avoidance observed (37). Such a generalist strategy may limit exposure to a wide range of fecal parasites, not only nematodes. Furthermore, group movement analyses indicate that wild mandrills return less frequently and after longer time lags to sites, including sleeping sites, highly contaminated by nematodes compared to less contaminated sites (60).

Overall, these studies demonstrate that mandrills use social distancing strategies to avoid socially transmitted parasites, and also avoid contaminated substrates and habitats to limit exposure with environmentally transmitted parasites. These behaviors are intimately linked to the mode of parasite transmission, emphasizing the close relationship between parasite ecology and host responses, with consequences for mandrill social structure, foraging strategy, and ranging patterns.

### 15.2.5 Inter-sexual variation

Trade-off theory has helped explain a common observation that females invest more into physiological immune defenses than males (1). Indeed, in most species, males generally benefit from increasing investment into short-term mating success at the expense of investment into immune function. By contrast, females usually benefit from investing in reproduction over the long-term through increased longevity (33), resulting in high investment into physiological immunity because accumulating parasites might challenge their survival (33). Females should thus invest more than males into behavioral mechanisms of parasite avoidance. In humans, indeed, women show greater disgust towards situations or objects associated with the spread of infectious diseases (34). Women also express more hygienic behaviors (35) and, in response to COVID-19, engage more in social distancing than men (36). Similarly, in non-human primates, female Japanese macaques and mandrills outperform males in avoiding fecal material during feeding tests (10), (37). In grey mouse lemurs, only females avoid fecally contaminated food, water, and nests, and avoid contaminating their nests by defecating outside (11). However, there is no such sex bias for fecally contaminated water in red fronted lemurs (*Eulemur rufifrons* (38)), or fecally-contaminated food in bonobos (*Pan paniscus* (39)) and chimpanzees (*Pan troglodytes* (40)). More studies are thus required to understand why some species exhibit such a sex bias in avoidance while others do not. However, it is important to note that sex differences may be detectable only when resource availability allows females to be more selective than males. For instance, in woolly monkeys (*Lagothrix lagotricha poeppigii*), females discard more contaminated food than males only when food is abundant (41).

### 15.2.6 Multifaceted anti-parasite defenses

As shown earlier in this chapter, trade-off theory helps explain variation in investment into parasite avoidance behaviors. Host life-history traits and ecological conditions all determine how an individual responds to parasite threats, and this is embedded within a cost-benefit framework because individuals are facing different biological challenges and resources are finite. In this context, the quality of the host immune system constitutes one major determinant of individual avoidance behavior. The "pathogen defense optimization" hypothesis (42) posits that physiologically immunocompetent individuals should invest less into avoidance behaviors than more susceptible individuals, and vice versa. The physiological immune system comes with obvious costs, for example on growth or reproduction (13), while the behavioral immune system may involve missed foraging or social opportunities, as demonstrated earlier in this chapter. As such, individuals should face trade-offs in the deployment of these two defense systems.

Several examples help illustrate this potential trade-off. In trout and salmon, individuals that physiologically resist a trematode eye-fluke the most also display the lowest levels of behavioral avoidance (6). In house finches (*Carpodacus mexicanus*), individuals with low levels of circulating antibodies and immune proteins are those that avoid sick individuals, while those with high levels of circulating antibodies do not (43). In

humans, people chronically concerned with diseases and who perceive themselves as highly vulnerable to infection exhibit ethnocentric attitudes (44), a behavioral trait that might have evolved as a parasite avoidance strategy (45); they further exhibit high disgust towards pictures of potentially contaminated items and risky situations (46). Perceived self-vulnerability to disease has been also associated with social distancing behaviors since the beginning of the COVID-19 pandemic (36). Also, temporary immune deficiency during early pregnancy probably causes the increased disgust response (47) frequently observed during this period. Despite these examples, empirical studies on trade-offs between physiological and behavioral immunity are rare. Furthermore, they tend to be correlative, and mostly evaluate only a few physiological and behavioral immune parameters. Thus, one promising avenue for better understanding variation in behavioral avoidance mechanisms is to consider avoidance behavior as one component of an individual's defense repertoire, including further, more nuanced, investigations of the trade-offs between physiological versus behavioral defenses.

## 15.3 Consequences of variation in parasite avoidance behaviors

### 15.3.1 Epidemiological outcomes

Although possibly as efficient as physiological immune responses, only a few studies have quantified the efficacy of avoidance behaviors in reducing infection risk, probably because behavioral defenses are harder to manipulate experimentally than physiological immune defenses and are also challenging to study in natural populations. A few recent studies have, however, successfully compared different behavioral phenotypes and their epidemiological consequences on hosts. For example, in Japanese macaques (10) and grey mouse lemurs (11), individuals that avoid contaminated food are less infected by soil-transmitted nematodes than those that do not. Under laboratory conditions, the efficacy of avoidance behaviors has been further measured by manipulating host habitats. In small tanks with no possibility to escape trematodes (*D. spathaceum*), trout are more parasitized than individuals living in big tanks with the possibility to behaviorally avoid contacts with parasites (48).

Most studies on the epidemiological consequences of behavioral responses to parasites and their efficacy have focused on contagious diseases and the impact of social distancing at the population level. The basic reproductive number of a contagious pathogen, corresponding to the average number of secondary cases produced by an infectious individual, depends on three parameters: (i) the contact rate between infected and healthy individuals, (ii) the probability of transmission, and (iii) the duration of the infectious period (49). Social distancing behaviors have the potential to profoundly alter the first two parameters, impacting disease dynamics. In wild populations of lobsters, for example, the prevalence of PaV1, a lethal virus with a transmission rate of ~60% in the lab, reaches only 7% in the wild because infected individuals are avoided by their conspecifics (50). Simulation models on these lobsters further show that avoiding infected conspecifics influences the virus reproductive number, decreasing the intensity and duration of outbreaks and has probably prevented some populations from going extinct (28), (51). In humans, reduced mobility of people infected with the influenza (A/H1N1) decreases the virus reproductive number to about one-quarter compared to scenarios without modification of contact patterns (52). Similarly, recent analyses on the spread of COVID-19 have shown that the reproductive number and the duration of the initial exponential phase of the disease are both strongly negatively associated with social distancing efforts (53). Another recent study on ants (*Lasius niger*) showed that after experimental infection of some foragers by a fungus (*Metarhizium brunneum*), several properties of the social network are modified, inhibiting disease spread and reducing mortality rates (54). In particular, treated colonies exhibit increased segregation between foragers on the one hand, and nurses and the queen on the other hand. Simulations confirmed that such network plasticity decreases the chances that nurses and the queen

receive a high pathogen load. Interestingly, these behavioral effects decrease disease-induced mortality rates, but low pathogen loads additionally confer immunity to individuals, protecting them during future exposure with the same pathogen and potentially impacting the evolution of disease resistance.

These studies all reveal that different behavioral avoidance strategies may efficiently protect individuals and populations from contagious diseases. In particular, the recent literature on social distancing demonstrates that such behavioral practices greatly reduce disease spread by decreasing the reproductive number of pathogens. While of great value, these studies still ignore inter-individual variation in behavioral immunity, although such variation is very likely to impact disease dynamics. Considering inter-individual variation along with different ecological factors in epidemiological models will provide a better predictive power for understanding the course of epidemics. For example, considering food availability might completely change model predictions of disease outbreak size, with larger epidemics expected during periods of food scarcity (55). As the risk of emerging diseases increases, considering variation in behavioral immunity will be crucial to help predict the dynamics of infectious diseases at the population level and inform disease management policy.

### 15.3.2 Ecological and evolutionary consequences

Beyond the epidemiological consequences of parasite avoidance behaviors, host behavioral responses to parasites may also have broad impacts on host ecology and evolution, species interactions, and ecosystem structure. Here, we illustrate how the three sets of parasite avoidance behaviors considered above may drive such ecological and evolutionary processes.

Moreover, host behavioral responses have evolutionary consequences for the parasite as well, and ultimately, co-evolution between the host and parasite may be an important phenomenon driving variation in host avoidance behaviors (Box 15.2).

## Box 15.2 Implications of host parasite avoidance behaviors for parasite evolution

Host investment into resistance and tolerance to parasites has profound consequences for parasite evolution (69). Host avoidance behaviors could similarly drive the evolution of parasite life-history traits, although this field of research remains under-investigated. By negatively affecting parasite fitness, avoidance behaviors generate interspecific conflicts between hosts and parasites, selecting for parasite counter-adaptations. Such counter-adaptations may involve undetectable infective stages or asymptomatic infections in hosts. In particular, social distancing should select for less virulent parasites that inhibit pro-inflammatory cytokines in hosts (70). The avoidance of infected substrates may further drive the evolution of mobile parasite stages to move away from these substrates and colonize surrounding environments, and/or parasite stages that persist in the environment even after the disappearance of cues detected by hosts. Accordingly, non-infectious eggs of nematodes emitted in feces survive long periods in the environment and vary greatly in the maturation time required to reach the infective larvae stage, creating unpredictability in habitat contamination level. In addition, nematode larvae can survive long periods without potential hosts and migrate out of feces to travel relatively long distances (71).

Although few studies have specifically tested how avoidance behaviors drive parasite virulence, mathematical modeling on the effects of host physiological resistance helps to formulate hypotheses because, similarly to resistance, avoidance decreases transmission rates. A theoretical study showed that host resistance could either increase or decrease parasite virulence, according to the type of resistance: "qualitative" resistance that prevents infection, decreases parasite virulence, while "quantitative" resistance that limits within-host growth rate, increases parasite virulence (72). Avoidance behaviors should produce similar effects as qualitative resistance. In line with this, theoretical models show that virulence of sexually transmitted pathogens decreases if hosts exhibit mating choosiness to avoid infected partners (73). In addition, avoidance behaviors reduce the probability of infection from multiple parasites, possibly decreasing parasite competition within hosts, leading to lower host exploitation strategies and, consequently, lower virulence (72). By contrast, other behavioral defenses, such as the use of plants with anti-microbial properties in nests (2)

that decrease parasite growth rate, should produce similar effects as quantitative resistance, increasing virulence.

Parasite adaptations should in turn select for rare resistance traits in hosts and the possible evolution of new behavioral strategies to escape from parasites. Hosts and parasites are thus involved in a never-ending arms race, generating feedback (74). As predicted by the Red Queen Hypothesis (75), hosts should increase their capacity to resist their parasites, while parasites should improve their ability to infect their hosts. Considering this feedback perspective in studies of parasite avoidance behaviors is an essential step towards broadening our understanding of the evolution of host–parasite systems (74).

### 15.3.2.1 Consequences of parasite-mediated selective foraging

Selective grazing behaviors of herbivores affect herbage structure in the environment which has knock-on effects for the distribution and abundance of various invertebrates (56). When food is not a limiting resource, avoidance of fecally contaminated areas leads to structural heterogeneity of grasslands with short swards and high tussocks (57). Such heterogeneity provides different ecological niches affecting invertebrate species richness and abundance (56), with further impacts at higher trophic levels of biodiversity, including birds and mammals. Similarly, carnivores that selectively scavenge on carcasses of non-carnivorous species, rather than on other carnivores that represent a higher risk for parasite transmission, impact arthropod communities, availability of soil nutrients, vegetation growth, and herbivore distributions (58).

Selective foraging in response to parasites can further impact the evolutionary trajectory of species that are not directly involved in the host–parasite interaction. Several pollinator species, for example, avoid flowers infected by a fungus (*Ascosphaera apis*) responsible for chalkbrood disease in bees. Avoidance leads to a decrease in the reproductive success of the infected plants (59). Interestingly, plant species with hairy or spiky flowers are less avoided when contaminated possibly because parasites are less detectable. Such avoidance behaviors that depend on floral morphology could lead to changes in the floral community structure, impacting floral trait evolution (59).

### 15.3.2.2 Consequences of parasite-mediated habitat selection

Avoiding contaminated habitats influences animal movements at various geographical scales, from micro scales to large migration events, shaping patterns of species distribution, with cascading ecosystem-level consequences. At the micro scale, environmental contamination influences recursive movements (i.e., returns to previously visited areas) of several species. Mandrill ranging patterns change as a function of the level of nematode contamination in the soil: they avoid using highly contaminated areas and sleeping sites (see Box 15.1 and Figure 15.4; (60)). Avoiding habitats that are seasonally infested by environmental parasites, such as biting insects, might also influence animal movements at large scales, driving seasonal migrations (61). For example, reindeer (*Rangifer tarandus tarandus*) migrate from calving grounds in summer, when a large number of larvae from warble flies (*Hypodema tarandi*) are infesting the ground. They return to these habitats after the death of most emerging adult flies (62). Interestingly, the higher the warble fly intensity, the longer the distance reindeer migrate. These long-distance movements have far-reaching consequences for the whole ecosystem, altering energy flow, food webs, and trophic interactions, as well as the structure and dynamics of metacommunities (63). For instance, the transport of nutrients via, for example, excreta (urine and feces) during animal migration increases resource availability (64), and migrants disperse plant seeds between distant localities (63). Both phenomena affect primary productivity, plant biomass, and the structure of plant communities (64).

### 15.3.2.3 Consequences of parasite-mediated social distancing

Finally, social distancing may have important consequences for several aspects of animal social systems, including host social organization, social structure, and mating system. In ants, for example, key transmission-inhibitory properties of contact networks (i.e., diameter, modularity, clustering) were reinforced following the introduction of a contagious fungus (54). In addition to temporary modifications of social networks, parasite-mediated selection pressures may have also shaped the social organization of many species. Indeed, a meta-analysis revealed that subgrouping patterns, which efficiently reduce parasite prevalence during simulated outbreaks (65), increase with group size in seven primate species (66). Patterns of social interactions, and the resulting social structure, are also greatly influenced by the avoidance of contagious or potentially contagious groupmates. In wild mandrills, grooming patterns are strongly shaped by contagious protozoa because individuals are, on average, three times less likely to be groomed when they are infected (see Box 15.1; (67)). Ultimately, parasite pressures may pervasively decrease sociality. In line with this, since the emergence of the PaV1 virus, the natural attraction that spiny lobsters exhibit for conspecific odorants has significantly declined over the past two decades, particularly in populations living in highly parasitized areas, decreasing aggregation of lobsters within dens (68).

## 15.4 Conclusions and future directions

With this chapter, we have reviewed the empirical evidence on the causes and consequences of variation in investment into parasite avoidance behaviors at different scales. We show, in particular, how a cost-benefit framework may help explain such variation and identify individual and environmental conditions promoting the expression of avoidance behaviors. To improve our understanding of the circumstances under which behavioral avoidance can evolve, future studies need to focus on a broader range of host–parasite systems, and simultaneously measure host investment in behavioral and physiological immunity. An ongoing challenge remains integrating inter-individual variation in behavioral avoidance into epidemiological frameworks to shed new light on host–parasite dynamics and improve infectious disease control and species conservation. Achieving this goal and fully understanding the importance of behavioral immunity in shaping the dynamics and patterns of pathogen distributions will require collaborations between researchers from various fields including field ecologists, immunologists, and mathematical modelers.

# References

1. Schulenburg H, Kurtz J, Moret Y, and Siva-Jothy MT. Introduction. Ecological immunology . *Philos Trans R Soc Lond B: Biol Sci.* 2009;364(1513):3–14.

2. Curtis VA. Infection-avoidance behaviour in humans and other animals. *Trends Immunol.* 2014;35(10): 457–64.

3. Gangestad SW and Grebe NM. Pathogen avoidance within an integrated immune system: Multiple components with distinct costs and benefits. *Evol Behav Sci.* 2014;8(4):226–34.

4. Schaller M. Parasites, behavioral defenses, and the social psychological mechanisms through which cultures are evoked. *Psychol Inq.* 2006;17(2):96–137.

5. Klemme I and Karvonen A. Vertebrate defense against parasites: Interactions between avoidance, resistance, and tolerance. *Ecol Evol.* 2017;7(2):561–71.

6. Klemme I, Hyvärinen P, and Karvonen A. Negative associations between parasite avoidance, resistance and tolerance predict host health in salmonid fish populations. *Proc R Soc B: Biol Sci.* 2020;287(1925):20200388.

7. Kavaliers M. Odours of parasitized males induce aversive responses in female mice. *Anim Behav.* 1995;50(5):1161–9.

8. Schmid-Hlmes A, Lee C, Sessions AMI, and Lynn L. Burrows and burrow-cleaning behavior of house mice (*Mus musculus domesticus*). *Am Midl Nat.* 2021;146(1):53–62.

9. Walsh PT, McCreless E, and Pedersen AB. Faecal avoidance and selective foraging: Do wild mice have the luxury to avoid faeces? *Anim Behav.* 2013;86(3): 559–66.

10. Sarabian C and Macintosh AJJ. Hygienic tendencies correlate with low geohelminth infection in free-ranging macaques. *Biol Lett.* 2015;11(11):20150757.

11. Poirotte C and Kappeler PM. Hygienic personalities in wild grey mouse lemurs vary adaptively with sex. *Proc R Soc B: Biol Sci.* 2019;286(1908):20190863.

12. Dugatkin LA. *Principles of animal behavior.* Chicago, IL: University of Chicago Press, 2020.

13. Schmid-Hempel P. Variation in immune defence as a question of evolutionary ecology. *Proc R Soc B: Biol Sci.* 2003;270(1513):357–66.

14. Luong LT, Horn CJ, and Brophy T. Mitey costly: Energetic costs of parasite avoidance and infection. *Physiol Biochem Zool.* 2017;90(4):471–7.

15. Hutchings MR, Kyriazakis I, Papachristou TG, Gordon IJ, and Jackson F. The herbivores' dilemma: Trade-offs between nutrition and parasitism in foraging decisions. *Oecologia.* 2000;124(2):242–51.

16. Cooper J, Gordon IJ, and Pike AW. Strategies for the avoidance of faeces by grazing sheep. *Appl Anim Behav Sci.* 2000;69(1):15–33.

17. Ezenwa VO. Selective defecation and selective foraging: Antiparasite behavior in wild ungulates? *Ethology.* 2004;110(11):851–62.

18. Hutchings MR, Milner JM, Kyriazakis I, and Jackson F. Grazing decisions of Soay sheep, *Ovis aries*, on St Kilda: A consequence of parasite distribution? *Oikos.* 2002;96(2):235–44.

19. Lafferty KD. Foraging on prey that are modified by parasites. *Am Nat.* 1992;140(5):854–67.

20. Lafferty K and Morris K. Altered behavior of parasitized killifish increases susceptibility to predation by bird final hosts. *Ecology.* 1996;77(5):1390–7.

21. Mierzejewski MK, Horn CJ, and Luong LT. Ecology of fear: Environment-dependent parasite avoidance among ovipositing drosophila. *Parasitol.* 2019;146(12):1564–70.

22. Koprivnikar J, Redfern JC, and Mazier HL. Variation in anti-parasite behaviour and infection among larval amphibian species. *Oecologia.* 2014;174(4):1179–85.

23. Koprivnikar J, Penalva L. Lesser of two evils? Foraging choices in response to threats of predation and parasitism. *PLoS One.* 2015;10(1):1–11.

24. Townsend AK, Hawley DM, Stephenson JF, and Williams KEG. Emerging infectious disease and the challenges of social distancing in human and non-human animals. *Proceedings Biol Sci.* 2020;287(1932):20201039.

25. Holmes EA, Connor RCO, Perry VH, Tracey I, Wessely S, Arseneault L, *et al.* Multidisciplinary research priorities for the COVID-19 pandemic: A call for action for mental health science. *Lancet Psychiat.* 2020; 7(6): 547–60.

26. Dugatkin LA, FitzGerald GJ, and Lavoie J. Juvenile three-spined sticklebacks avoid parasitized conspecifics. *Environ Biol Fishes.* 1994;39(2):215–18.

27. Krause J and Godin JGJ. Predator preferences for attacking particular prey group sizes: Consequences for predator hunting success and prey predation risk. *Anim Behav.* 1995;50(2):465–73.

28. Butler MJ, Behringer DC, Dolan TW, Moss J, and Shields JD. Behavioral immunity suppresses an epizootic in Caribbean spiny lobsters. *PLoS One.* 2015;10(6):e0126374.

29. Bonte D, Van Dyck H, Bullock JM, Coulon A, Delgado M, Gibbs M, *et al.* Costs of dispersal. *Biol Rev.* 2012;87(2):290–312.

30. Baudouin A, Gatti S, Levréro F, Genton C, Cristescu RH, Billy V, *et al.* Disease avoidance, and breeding

group age and size condition the dispersal patterns of western lowland gorilla females. *Ecology.* 2019;100(9):e02786.

31. Paciência FMD, Rushmore J, Chuma IS, Lipende IF, Caillaud D, Knauf S, et al. Mating avoidance in female olive baboons (*Papio anubis*) infected by *Treponema pallidum. Sci Adv.* 2019;5(12): eaaw9724.

32. Poirotte C and Charpentier MJE. Unconditional care from close maternal kin in the face of parasites. *Biol Lett.* 2020;16(2):1–6.

33. Rolff J. Bateman's principle and immunity. *Proc R Soc B: Biol Sci.* 2002;269(1493):867–72.

34. Tybur JM, Bryan AD, Lieberman D, Caldwell AE, and Merriman LA. Sex differences and sex similarities in disgust sensitivity. *Pers Individ Dif.* 2011;51(3): 343–8.

35. Al-Shawaf L, Lewis DMG, and Buss DM. Sex differences in disgust: Why are women more easily disgusted than men? *Emot Rev.* 2018;10(2):149–60.

36. Makhanova A and Sheperd M. Behavioral immune system linked to responses to the threat of COVID-19. *Pers Individ Dif.* 2020;167:110221.

37. Poirotte C, Sarabian C, Macintosh AJJ, and Charpentier MJE. Faecal avoidance differs between the sexes but not with nematode infection risk in mandrills. *Anim Behav.* 2019;149:97–106.

38. Amoroso CR, Frink AG, and Nunn CL. Water choice as a counterstrategy to faecally transmitted disease: An experimental study in captive lemurs. *Behaviour.* 2017;154(13–15):1239–58.

39. Sarabian C, Belais R, Macintosh AJJ, and Sarabian C. Feeding decisions under contamination risk in bonobos. *Philos Trans R Soc Lond B: Biol Sci.* 2018;373(1751):20170195.

40. Sarabian C, Ngoubangoye B, and MacIntosh AJJ. Avoidance of biological contaminants through sight, smell and touch in chimpanzees. *R Soc Open Sci.* 2017;4(11):170968.

41. Philippon J, Serrano-Martínez E, and Poirotte C. Environmental and individual determinants of fecal avoidance in semi-free ranging woolly monkeys (*Lagothrix lagotricha poeppigii*). *Am J Phys Anthropol.* 2021;176(4):64–24.

42. Zylberberg M, Klasing KC, and Hahn TP. In house finches, *Haemorhous mexicanus*, risk takers invest more in innate immune function. *Anim Behav.* 2014;89: 115–22.

43. Zylberberg M, Klasing KC, and Hahn TP. House finches (*Carpodacus mexicanus*) balance investment in behavioural and immunological defences against pathogens. *Biol Lett.* 2013;9(1):20120856–20120856.

44. Navarrete CD and Fessler DMT. Disease avoidance and ethnocentrism: The effects of disease vulnerability and disgust sensitivity on intergroup attitudes. *Evol Hum Behav.* 2006;27(4):270–82.

45. Faulkner J, Schaller M, Park JH, and Duncan LA. Evolved disease-avoidance mechanisms and contemporary xenophobic attitudes. *Gr Process Intergr Behav.* 2004;7(4):333–53.

46. Mortensen CR, Becker DV, Ackerman JM, Neuberg SL, and Kenrick DT. Infection breeds reticence: The effects of disease salience on self-perceptions of personality and behavioral avoidance tendencies. *Psychol Sci.* 2010;21(3):440–7.

47. Fleischman DS and Fessler DMT. Progesterone's effects on the psychology of disease avoidance: Support for the compensatory behavioral prophylaxis hypothesis. *Horm Behav.* 2011;59(2):271–5.

48. Karvonen A, Seppälä O, and Valtonen ET. Parasite resistance and avoidance behaviour in preventing eye fluke infections in fish. *Parasitol.* 2004;129(2):159.

49. Diekmann O, Heesterbeek JAP, and Metz JAJ. On the definition and the computation of the basic reproduction ratio R0 in models for infectious diseases in heterogeneous populations. *J Math Biol.* 1990;28(4):365–82.

50. Behringer DC, Butler MJ, and Shields JD. Avoidance of disease by social lobsters. *Nature.* 2006;441(25):421.

51. Dolan TW, Butler MJ, and Shields JD. Host behavior alters spiny lobster-viral disease dynamics: A simulation study. *Ecology.* 2014;95(8):2346–61.

52. Van Kerckhove K, Hens N, Edmunds WJ, and Eames KTD. The impact of illness on social networks: Implications for transmission and control of influenza. *Am J Epidemiol.* 2013;178(11):1655–62.

53. Khataee H, Scheuring I, Czirok A, and Neufeld Z. Effects of social distancing on the spreading of COVID-19 inferred from mobile phone data. *Sci Rep.* 2021;11(1):1–9.

54. Stroeymeyt N, Grasse A V, Crespi A, Mersch DP, Cremer S, and Keller L. Social network plasticity decreases disease transmission in a eusocial insect. *Science.* 2018;362(6417):941–5.

55. Becker DJ, Streicker DG, and Altizer S. Linking anthropogenic resources to wildlife-pathogen dynamics: A review and meta-analysis. *Ecol Lett.* 2015;18(5):483–95.

56. Jerrentrup JS, Wrage-Mönnig N, Röver KU, and Isselstein J. Grazing intensity affects insect diversity via sward structure and heterogeneity in a long-term experiment. *J Appl Ecol.* 2014;51(4):968–77.

57. Hutchings MR, Gordon IJ, Kyriazakis I, and Jackson F. Sheep avoidance of faeces-contaminated patches leads to a trade-off between intake rate of forage and parasitism in subsequent foraging decisions. *Anim Behav.* 2001;62(5):955–64.

58. Moleón M, Martínez-Carrasco C, Muellerklein OC, Getz WM, Muñoz-Lozano C, and Sánchez-Zapata JA.

Carnivore carcasses are avoided by carnivores. *J Anim Ecol.* 2017;86(5):1179–91.

59. Yousefi B and Fouks B. The presence of a larval honey bee parasite, *Ascosphaera apis*, on flowers reduces pollinator visitation to several plant species. *Acta Oecologica.* 2019;96:49–55.

60. Poirotte C, Benhamou S, Mandjembe A, Willaume E, Kappeler PM, and Charpentier MJE. Gastrointestinal parasitism and recursive movements in free-ranging mandrills. *Anim Behav.* 2017;134: 87–98.

61. Altizer S, Bartel R, and Han BA. Animal migration and infectious disease risk. *Science.* 2011;331(6015): 296–302.

62. Folstad I, Nilssen AC, Halvorsen O, and Andersen J. Parasite avoidance: The cause of post-calving migrations in Rangifer? *Can J Zool.* 1991;69(9):2423–9.

63. Nathan R, Schurr FM, Spiegel O, Steinitz O, Trakhtenbrot A, and Tsoar A. Mechanisms of long-distance seed dispersal. *Trends Ecol Evol.* 2008;23(11):638–47.

64. Bauer S and Hoye BJ. Migratory animals couple biodiversity and ecosystem functioning worldwide. *Science.* 2014;344(6179).

65. Nunn CL, Jordan F, McCabe CM, Verdolin JL, and Fewell JH. Infectious disease and group size: More than just a numbers game. *Philos Trans R Soc Lond B: Biol Sci.* 2015;370(1669):20140111–20140111.

66. Griffin RH and Nunn CL. Community structure and the spread of infectious disease in primate social networks. *Evol Ecol.* 2012;26(4):779–800.

67. Poirotte C, Massol F, Herbert A, Willaume E, Bomo PM, Kappeler PM, *et al.* Mandrills use olfaction to socially avoid parasitized conspecifics. *Sci Adv.* 2017;3(4):e1601721.

68. Childress MJ, Heldt KA, and Miller SD. Are juvenile Caribbean spiny lobsters (*Panulirus argus*) becoming less social? *J Mar Sci.* 2015;72(1):i170–6.

69. Miller MR, White A, and Boots M. The evolution of parasites in response to tolerance in their hosts: The good, the bad, and apparent commensalism. *Evolution.* 2006;60(5):945–56.

70. Stockmaier S, Stroeymeyt N, Shattuck EC, Hawley DM, Meyers LA, and Bolnick DI. Infectious diseases and social distancing in nature. *Science.* 2021;371(6533).

71. Stromberg BE. Environmental factors influencing transmission. *Vet Parasitol.* 1997;72(3–4):247–64.

72. Gandon S and Michalakis Y. Evolution of parasite virulence against qualitative or quantitative host resistance. *Proc R Soc B: Biol Sci.* 2000;267(1447):985–90.

73. Ashby B and Boots M. Coevolution of parasite virulence and host mating strategies. *Proc Natl Acad Sci.* 2015;112(43):13290–5.

74. Ezenwa VO, Archie E a, Craft ME, Hawley DM, Martin LB, Moore J, *et al.* Host behaviour—parasite feedback: An essential link between animal behaviour and disease ecology . *Proc R Soc B: Biol Sci.* 2016;283(1828):20153078.

75. Van Valen L. A new evolutionary law. *Evol Theory.* 1973;1:1–30.

# CHAPTER 16

# Behavioral defenses against parasitoids: Genetic and neuronal mechanisms

Shaun Davis and Todd Schlenke

## 16.1 Introduction

Host organisms are confronted by a diversity of parasites, including microbes like viruses and bacteria, and macroparasites like helminths and parasitoids. Hosts have evolved to recognize parasite presence and respond to the threats they pose by altering their internal physiology and external behaviors. With respect to behavioral defenses, the sensory systems hosts use to detect parasites and the neuronal circuits that cause hosts to induce the behaviors are of particular interest. Are there conserved genetic and neuronal mechanisms that underlie anti-parasite defense behaviors across host lineages? Do limits on sensory systems and neuronal circuitry constrain the types of defense behaviors that hosts can deploy? How do the host mechanisms of parasite recognition and response influence the ecology of host–parasite interactions? In this chapter, we focus on a particular class of insect parasites called parasitoids. We survey the different types of anti-parasitoid behaviors that hosts have evolved, and what is known of the neurogenetic mechanisms underlying those behaviors in *Drosophila*. Although there is still much left to learn, we are excited about the prospect of developing a holistic understanding—from molecular interactions to community structure—of host–parasitoid interactions.

## 16.2 Overview of parasitoids

Parasitoids are organisms that lay their eggs on the surface or inside the body of hosts resulting in their offspring consuming host resources until the host dies. This lifestyle is similar to that of a typical parasite in that the parasitoid lives in association with the host for a significant period of time, but it is also similar to that of a typical predator in that the parasitoid kills its host during the course of its lifecycle. Parasitoid size and generation times are often of the same order of magnitude as the hosts, whereas typical parasites are much smaller and complete their lifecycles much quicker than their hosts. The intimate developmental relationship between parasitoids and their hosts has led parasitoids to evolve intricate strategies for manipulating host physiology and behavior to their benefit. For example, parasitoids can alter the developmental timing of their hosts to match their own, induce hosts to consume excess food, and manipulate hosts in numerous other more complicated ways (1), (2).

Parasitoids have been described from a variety of insect taxa, including Hymenoptera, Diptera, and Coleoptera, but parasitoid wasps are the most numerous and diverse with more than 60,000 species described. Most insect species, and many non-insect arthropods, are infected by parasitoids, including generalist parasitoids with broad host

Shaun Davis and Todd Schlenke, *Behavioral defenses against parasitoids*. In: *Animal Behavior and Parasitism*.
Edited by Vanessa O. Ezenwa, Sonia Altizer, and Richard J. Hall, Oxford University Press. © Oxford University Press (2022).
DOI: 10.1093/oso/9780192895561.003.0016

ranges and more specialist parasitoids that only infect a small number of host species. A number of parasitoids even infect other parasitoids, a process known as hyperparasitism. Furthermore, large proportions of host populations can become infected by parasitoids. By controlling host population numbers, parasitoids often act as keystone species in natural ecosystems, and can be extremely useful as biological control agents for managing pest species (3).

Because insect/arthropod hosts are so often infected by parasitoids, and because infection usually leads to death, host species have experienced strong selection pressures to evolve defenses against parasitoid infection. These defenses include immune responses against the parasitoid offspring (4), as well as behavioral defenses that hosts use to avoid becoming infected or to cure themselves after they become infected (5), (6). The most well-documented kind of immune response against parasitoid eggs/larvae is a hemocyte (blood cell) -based response termed encapsulation (7). In an encapsulation response, hundreds or thousands of blood cells rush to the foreign body, spread over it, and adhere to each other to form a tight capsule. The capsule can physically prevent the parasitoid from hatching/moving, limit molecular transfer between the host and parasitoid, and guide toxin deposition onto the parasitoid. Other types of host immune responses against parasitoids are possible as well, including the humoral production of anti-parasitoid proteins, or the production of anti-parasitoid toxins by protective symbionts (8).

Immune responses to parasitoids are not perfect and can have detrimental effects on the host. Even though an immune response like encapsulation can be induced, the cellular and genetic apparatuses that allow production of thousands of hemocytes are costly to maintain, and most immune responses lead to some autoimmune damage. For example, host organisms that are artificially selected to be strong encapsulators suffer trade-offs with other components of fitness, and quickly lose their encapsulation ability once the selection pressure is lost (9).

Behavioral defenses can be an adaptive alternative to immune responses (6). Like anti-predator behaviors, anti-parasitoid behaviors can evolve to help the host avoid parasitoid interactions, or

escape once an interaction is initiated. However, given that many parasitoids do not kill their hosts immediately, a separate class of post-infection behaviors can evolve to free the hosts of the parasitoids living on or in them.

## 16.3 Anti-parasitoid host behaviors

Many examples of host behaviors that limit successful parasitoid infection have been documented. While the specific behaviors may be unique to a certain species, there are certain behavioral themes that emerge, which can be divided into three categories. Here, we highlight a few representative examples of a broader diversity of anti-parasitoid host behaviors (Figure 16.1, Table 16.1).

### 16.3.1 Parental behaviors

Given that parasitoids often infect the juvenile stages of their hosts, host parents can play a major role in modifying the risk of offspring infection. Their decisions result in various forms of transgenerational protection. Parents can protect their offspring by altering where and when they lay their eggs, for example by withholding egg laying at places and times where parasitoids are most common. Parents can also protect their offspring by altering how they lay their eggs, for example by laying their eggs in places that are more inaccessible to parasitoids (Figure 16.1A). Finally, parents can protect their offspring by remaining at the oviposition site after they have laid their eggs, in order to ward off parasitoids physically during the early developmental stages of their offspring.

### 16.3.2 Pre-infection behaviors

Once hosts have left the egg stage and are developing autonomously, they can display their own anti-parasitoid behaviors, which may include pre-infection behaviors and post-infection behaviors. There are generally two types of pre-infection behaviors: avoidance and escape. Note that a pre-infection behavior, like a parental behavior, shows that hosts have sensory systems for detecting parasitoids in their external environment. Avoidance behaviors are those that decrease the ability of parasitoids to locate their hosts in time or space,

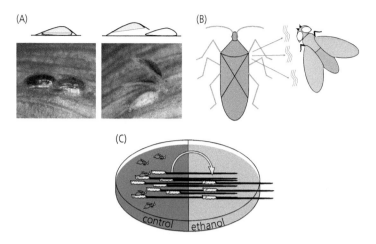

**Figure 16.1** Representative examples of anti-parasitoid host behaviors. (A) Drawings and photographs of stacked seed beetle eggs demonstrating the parental defensive behavior of laying decoy eggs. The bottom viable egg is protected from parasitoid wasps by the top inviable egg(s)(63). In the images to the right, the protective eggs have been lifted off for better viewing of the viable egg. Images courtesy of Joseph Deas. (B) A cartoon demonstrating the squash bug pre-infection defensive behavior of emitting a foul odor. The odor, released from glands on the underside of the squash bug thorax, repels and sickens infecting parasitoid flies (76). (C) A cartoon demonstrating the fruit fly post-infection defensive behavior of ingesting ethanol. Fly larvae preferentially consume toxic levels of ethanol after being infected by parasitoid wasps as a means of self-medication (red dots represent wasp eggs) (54).

whereas escape behaviors are those that take place after a parasitoid has located its host but prior to infection (Figure 16.1B).

### 16.3.3 Post-infection behaviors

Once hosts are infected, there will generally be a battle between the immune mechanisms of the host and the virulence mechanisms of the parasitoid. However, hosts can also induce behaviors at this stage to aid their immune defenses. There are generally two types of post-infection behaviors: altered feeding and altered thermoregulation. Hosts may alter their feeding behavior in order to consume specific types of food that boost their immune system, or hosts may also switch to a more toxic diet to harm the parasitoid directly (Figure 16.1C). Hosts may also behaviorally alter their temperatures when infected in order to harm the development of the parasitoids living inside them. For post-infection behaviors, it is not immediately clear whether neurological sensory systems are required to initiate the behavioral changes as the presence of the parasitoid in or on the host body may be recognized by the immune system instead. Because it is common for parasitoids to manipulate the behavior of the host they are

infecting to their own benefit, care must be taken to determine whether an altered host behavior benefits the host or the parasitoid (1), (2).

All the anti-parasitoid behaviors described above are plastic, meaning the host organisms undertake the adaptive behavior only when they identify that parasitoids are present. Of course, it is quite common for hosts to display some of these same types of behaviors constitutively. For example, the parental generation of hosts tend to deposit their offspring in places unattractive or inaccessible to parasitoids, and hosts often prefer consuming lower-quality, yet more toxic, foods. Unfortunately it is difficult to discern whether the past selective pressures that led to these evolved behaviors were parasitoid-mediated, or whether protection from parasitoids was a beneficial side effect of some other selection pressure. A common assumption, which often goes untested, is that induced behaviors are associated with fitness costs that make their constitutive deployment impractical.

Because these anti-parasitoid host behaviors fall into a handful of consistent categories, it is possible that the sensory systems and/or neuronal mechanisms that trigger the behaviors may be shared across lineages and impose some limits

**Table 16.1** A representative sample of induced anti-parasitoid host behaviors

| Behaviors | Host | Parasitoid | References |
|---|---|---|---|
| **Parental behaviors** | | | (Reviewed in (5),(58),(59)) |
| *Where to oviposit* | | | |
| Deeper in water column | Water strider (*Aquarius paludum*) | Wasp (*Tiphodytes gerriphagus*) | (60) |
| Deeper in water column | Damselfly (*Lestes sponsa*) | Wasps (*Aprostocetus* and *Tetrastichus* spp.) | (61) |
| Sites without parasitoid odors | Fruit fly (*Drosophila* spp.) | Wasps (*Leptopilina* spp.) | (35) |
| Sites with higher ethanol concentrations | Fruit fly (*Drosophila* spp.) | Wasp (*Leptopilina heterotoma*) | (14) |
| Sites with higher atropine concentrations | Fruit fly (*Drosophila suzukii*) | Wasps (*Trichopria cf. drosophilae* and *Asobara japonica*) | (62) |
| *When to oviposit* | | | |
| Reduce oviposition rates | Fruit fly (*Drosophila* spp.) | Wasps (*Leptopilina spp.*) | (15,16) |
| Reduce oviposition rates | Golden egg bug (*Phyllomorpha laciniata*) | Wasp (*Gryon bolivar*) | (55) |
| Reduce oviposition rates | Seed beetle (*Mimosestes amicus*) | Wasp (*Uscana semifumipennis*) | (56) |
| *How to oviposit* | | | |
| Lay decoy (inviable) eggs on top of viable offspring | Seed beetle (*Mimosestes amicus*) | Wasp (*Uscana semifumipennis*) | (63) |
| *Offspring protection* | | | |
| Shudder, scrape, and kick parasitoids | Tropical bug (*Antiteuchus tripterus*) | Wasps (*Trissolcus bodkini* and *Phanuropsis semiflaviventris*) | (64) |
| Lunge against parasitoids | Webspinner (*Antipaluria urichi*) | Wasp (*Embidobia urichi*) | (65) |
| Attack parasitoids | Assassin bug (*Zelus* sp.) | Wasp (*Telenomus sp.*) | (66) |
| **Pre-infection** | | | (Reviewed in (5),(58),(67)) |
| *Avoidance behaviors* | | | |
| Avoid areas with wasp odors | Fruit fly (*Drosophila* spp.) | Wasps (*Leptopilina* spp.) | (35) |
| Reduce foraging when parasitoids are present | Leaf-cutter ant (*Atta cephalotes*) | Fly (*Neodohrniphora curvinervis*) | (68) |
| *Escape behaviors* | | | |
| Larvae roll to dislodge wasp ovipositor | Fruit fly (*Drosophila melanogaster*) | Wasp (*Leptopilina boulardi*) | (18) |
| Larvae thrash and bite wasp appendages | Gum-leaf skeletonizer moth (*Uraba lugens*) | Wasps (*Cotesia urabae* and *Dolichogenidea eucalypti*) | (69) |
| Larvae head-jerk to knock parasitoids away | Butterfly (*Euphydryas phaeton*) | Wasp (*Apanteles euphydryidis*) | (70) |
| Larvae hold still, squirm, or regurgitate gut contents | Cabbage moth (*Mamestra brassicae*) | Wasp (*Dolichovespula media*) | (71) |
| Pupae move (jump) in pupal case | Wasp (*Bathyplectes anurus*) | Wasp (*Dibrachys cavus*) | (72) |
| Pupae move (twist) in pupal case | Wasp (*Dendrocerus carpenter*) | Wasp (*Dendrocerus carpenter*) | (73) |
| Pupae move (spin) in pupal case | Gypsy moth (*Lymantria dispar*) | Wasp (*Brachymeria intermedia*) | (74) |
| Secrete sticky, viscous fluid | Mole cricket (*Neocurtilla hexadactyla*) | Wasp (*Larra bicolor*) | (75) |
| Emit foul odor | Squash bug (*Anasa tristis*) | Fly (*Trichopoda pennipes*) | (76) |
| Jerk their abdomens and walk away | Aphid (*Myzus persica*) | Wasp (*Ephedrus cerasicola*) | (77) |
| Jump off host plant | Aphid (*Schizaphis graminum*) | Wasp (*Lysiphlebus testaceipes*) | (78) |
| Use wings and legs to fight off parasitoids | Aphid (*Hyperomyzus lactucae*) | Wasp (*Aphidius sonchi*) | (79) |

*continued*

**Table 16.1** *Continued*

| Behaviors | Host | Parasitoid | References |
|---|---|---|---|
| Bite, kick, and run away | Lady beetles (*Harmonia axyridis* and *Coleomegilla maculata*) | Wasp (*Dinocampus coccinellae*) | (80) |
| **Post-infection** | | | (Reviewed in (5)) |
| *Altered feeding* | | | |
| Ingest different yeast species | Fruit fly (*Drosophila melanogaster*) | Wasp (*Asobara tabida*) | (81) |
| Ingest more toxic plant compounds | Woolly bear caterpillar (*Apantesis incorrupta*) | Fly (*Exorista mella*) | (82) |
| Ingest more toxic host plant | Tiger moth caterpillar (*Platyprepia virginalis*) | Fly (*Thelaira americana*) | (83) |
| Ingest increased ethanol concentrations | Fruit fly (*Drosophila melanogaster*) | Wasps (*Leptopilina heterotoma* and *L. boulardi*) | (54) |
| *Altered thermoregulation* | | | |
| Seek cold temperature to slow and/or kill parasitoid | Bumblebee (*Bombus terrestris*) | Fly (Conopidae) | (84) |

on behavioral diversity. Identification and cross-species comparisons of these neuronal components will be required to gain a holistic understanding of host–parasitoid interactions.

## 16.4 *Drosophila* as a model for anti-parasitoid behavior mechanisms

The fruit fly *Drosophila melanogaster* is a model organism for uncovering the genetic basis of diverse phenotypes. It was first chosen due to the ease and quickness with which it can be propagated in the lab, and its ability to withstand inbreeding, which allowed the creation of isogenic stocks. Its use continues to expand as the *Drosophila* community incorporates each new technique into the *Drosophila* genetic toolbox. *D. melanogaster* was chosen for whole genome sequencing in 1990, and the first complete genome draft was released 10 years later (10), making it one of the first large multicellular organisms to be sequenced. Flies were found to have orthologs to roughly 60% of the genes in humans, including disease-causing genes (11), (12), showing that discoveries in *Drosophila* can be widely relevant to diverse organisms.

Despite its long use as a genetic model organism, there is still much to learn about the natural history of *D. melanogaster*. Nevertheless, several parasitoid wasp species that infect *D. melanogaster* in nature have now been identified, and dozens more

are known to use other *Drosophila* species as hosts (13). Furthermore, the widespread availability of *Drosophila* parasitoid cultures has hastened the use of the *Drosophila*-parasitoid system in evolutionary ecology, immunology, neurobiology, and genetics studies, and several anti-parasitoid behaviors have recently been described from *D. melanogaster* (Table 16.1). Numerous genetic tools are available in *Drosophila* for manipulating and interrogating gene and cell function as a way of elucidating the mechanisms underlying these anti-parasitoid behaviors (Box 16.1). As the fly genetic toolbox has become more powerful, our understanding of the morphology, physiology, and neuronal circuitry in the fly brain has also advanced (Box 16.2).

In the following sections, we will investigate the neurogenetic underpinnings of *Drosophila* defense behaviors against parasitoid wasps. Much of this work has focused on two of the behaviors that adult flies use to protect their offspring: oviposition at more alcoholic (toxic) oviposition sites when wasps are present, and reduced oviposition in the forced presence of wasps (Figure 16.2) (14)–(16). Footholds have been made at different steps in the adult fly's parasitoid-response circuits, but many gaps remain to be filled. The neurogenetics of fly larvae behavioral defenses against parasitoids has also been investigated (17), (18), but as yet these studies are limited in comparison to studies of adult fly behaviors.

---

### Box 16.1 The fly genetic toolbox

The simplest way to understand a gene's function is to disrupt the gene sequence and assess the phenotypic consequences. In early work, X-rays were used to induce DNA breaks and chemical mutagens were used to cause base pair changes, but the follow up analyses required to map these random mutations were time consuming. Later, transposable elements were induced to hop in and out of loci leaving indels in their wake, which provided a faster method to identify where the genomic perturbation occurred. *Drosophila* mutant strains targeting nearly every one of the ~15,000 genes in the fly genome now exist in stock centers, and for most genes multiple types of mutant strains exist (19).

One of the most widely used *Drosophila* genetic tools is the binary *GAL4/UAS* expression system (20). In its typical use, a transgenic *Drosophila* strain carries the yeast *GAL4* transcription factor with an upstream fly promoter sequence to drive its expression. A second transgenic *Drosophila* strain carries the GAL4-specific binding site, *Upstream Activation Sequence* (*UAS*), linked to a gene of interest. When these two strains are crossed, the resulting offspring will possess both elements, allowing GAL4 to bind to the *UAS* element to drive expression of the downstream gene of interest in whatever expression pattern the promoter upstream of *GAL4* determines (e.g., in a particular brain region). This allows *Drosophila* researchers to express any gene in a variety of spatial and temporal patterns. Other benefits of the system are that the *GAL4* and *UAS* components can be maintained as separate, phenotypically wild-type fly stocks until the experimental cross, and that different *GAL4* and *UAS* strains can be mixed and matched. Thousands of fly strains containing unique *UAS*-effectors have been generated and are available in stock centers. One of the most commonly used stock-type expresses hairpin RNA targeting individual *Drosophila* genes—these hairpins activate the fly RNA interference (RNAi) system to suppress expression of native fly genes (21). Other *UAS*-effectors can label cells with fluorescent reporters, alter cellular functions including neuronal activity, and ablate cells entirely (22).

---

### Box 16.2 The fly brain

Despite its small size, the *Drosophila* brain contains about 100,000 neurons, processes diverse external and internal stimuli, and initiates a range of complex behavioral outputs (23). In the central brain, the neuron cell bodies are located on the periphery, called the rind, while their projections extend inward forming synaptic connections and making up the various neuropil structures. Some structures are quite distinct, such as the optic lobes (which process visual stimuli), antennal lobes (which process olfactory stimuli), mushroom body (associated with learning and memory), and central complex (for orientation and locomotor behaviors), and each of these structures has multiple defined substructures as well. Numerous techniques have been developed for uncovering the neural circuitry that relays sensory information through various fly brain neuropil structures and into the central brain, then back out to the periphery to control fly behaviors.

One of the main goals in the field of behavioral neurobiology is circuit mapping, that is to identify the specific groups of neurons that interact with each other to convey signals through the brain, all the way from sensory input to behavioral output. Importantly, numerous *GAL4* strains have been developed to express constructs in unique expression patterns across the nervous system (24). This allows researchers to interrogate the functions of specifically defined neurons in various behaviors. For example, these tools have been used in neuronal activation screens to link particular neuronal groups to various locomotor and social behaviors (25). A new advancement in our ability to track *Drosophila* neural circuits is the release of the adult female brain connectome (26). The production of extremely thin sections of a single fly's brain for electron microscopy imaging and computer 3D tracing of neurons allowed for the reconstruction of individual cell morphologies and the location of synapses made by each individual neuron. This information can be used to identify candidate neuronal groups connected to a known neuronal group, which can be tested for importance in a particular output behavior using genetic approaches.

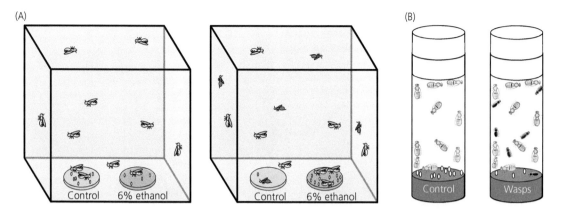

**Figure 16.2** Behavioral assays for altered *Drosophila* oviposition behavior following parasitoid exposure. (A) Flies are given a choice of oviposition substrates, either control food or food supplemented with ethanol, and egg placement is recorded when parasitoid wasps are absent (left) or present (right). Image reproduced from (14). (B) Flies are kept in vials without (left) or with (right) parasitoid wasps present, and the total number of eggs laid in each vial is recorded.

## 16.5 Sensory systems

Fruit flies have all the basic sensory capabilities that other animals do. So how might they sense the presence of parasitoid wasps? Unlike microbial parasites, parasitoids are large enough to be visible to their hosts. They also have distinct chemical signals on their body surface that can be transmitted through the air or on surfaces they have touched, and potentially recognized by olfactory, gustatory, and/or ionotropic receptors. Parasitoids also often have distinct wing-beat frequencies, so their flight sounds might be recognized by a host's auditory system (27). Finally, hosts can use their sense of touch to determine when parasitoids are laying their eggs in or on them (18). So far, adult flies have been found to sense parasitoid wasp presence and alter their oviposition strategies using vision and olfaction, and thus we will focus on these sensory mechanisms next.

### 16.5.1 Vision

*Drosophila* vision, like that of other insects, is controlled by two bilateral compound eyes with the outer retina layer consisting of repeating arrays of ommatidia (independent optical units), with about 750 ommatidia per eye. Each ommatidium has eight photosensitive cells, called R1–8, that detect light or color, defining the first step in motion and color

vision, and the axons from these photoreceptor neurons project into the optic lobe of the fly brain. The optic lobe is composed of four neuropil structures, called the lamina, medulla, lobula, and lobula plate (Figure 16.3A). The spatial orientation of the information coming from each ommatidium is maintained throughout these four optic neuropils. Because each photoreceptor is directed towards a specific point in three-dimensional space, considerable processing occurs deeper in the central brain to reconstruct an accurate image of the environment (28).

Numerous studies have shown that sight-defective flies fail to alter their oviposition behaviors in response to the presence of parasitoid wasps. Flies that had either disrupted eye development (16) or mutant photoreceptor function (29) were unable to reduce oviposition rates when co-housed with wasps. Furthermore, these same mutant flies, when given the choice of where to oviposit, failed to switch to more ethanol-rich foods when wasps were present (14). To demonstrate further that these results are explained by loss of vision rather than peculiarities in the mutant genetic backgrounds, wild-type flies were co-housed with parasitoid wasps in complete darkness and they too failed to reduce oviposition (29). These results demonstrate that fly hosts visually detect wasp parasitoids: flies know what parasitoids look like. They recognize different parasitoid wasp species,

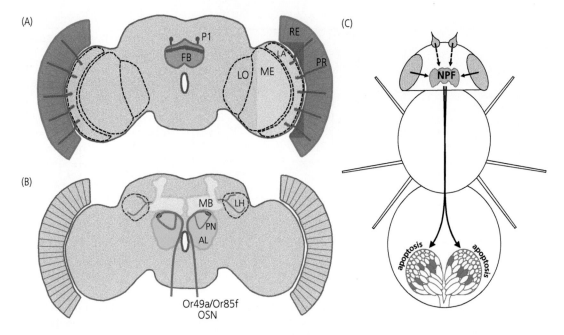

**Figure 16.3** Schematics of the fly brain and the regions involved in anti-parasitoid behaviors. (A) The visual system. The motion-sensitive photoreceptors (PR, dark red) in the retina (RE, red) project into the first neuropil of the optic lobe, the lamina (LA). The other regions of the optic lobe include the medulla (ME), lobula (LO), and lobula plate (positioned behind the LO, not shown). The two large P1 cells in the fan-shaped body (FB, purple) use visual input to regulate NPF signaling. (B) The olfactory system. The OSNs expressing Or49a and Or85f project into the antennal lobe (AL, blue) where they synapse with projection neurons (PN, dark grey). These PNs then project to the mushroom body (MB, yellow) and lateral horn (LH). The importance of the colored regions have been confirmed in wasp-exposure assays, while those regions marked by dashed lines are strong candidates but have not been tested directly. (C) Multimodal sensory integration causes reduced oviposition rates by inducing apoptosis in the ovaries. Vision and olfaction are both required for reduced egg laying, and visual wasp sensing is known to influence this apoptosis through NPF signaling.

and in some cases even seem to be able to distinguish infecting female wasps from non-infecting male wasps (14). These data open up numerous questions: for example, what is the wasp-mediated search image that flies recognize, and how does this image activate a parasitoid-specific visual circuit deeper in the fly's optic lobe? Do flies recognize the size and shape and color of wasps, or is it an aspect of wasp locomotion (i.e., motion vision) that triggers recognition (or a combination)? The answers to these questions could guide ecological and evolutionary studies of host–parasitoid interactions.

### 16.5.2 Olfaction

The olfactory system begins with olfactory sensory neurons (OSNs) positioned in the third antennal segments and maxillary palps on the fly head. The dendrites of these neurons project into different types of sensilla with small pores that allow the odor molecules to enter the sensilla lymph. Each OSN expresses one or few olfactory receptors (ORs) with specificity to different odorants. After binding their ligand, these specific receptors interact with a common odorant coreceptor (Orco) to open ion channels and stimulate the neuron (30). Flies have 60 *Or* genes; these receptors can detect a range of odorants including attractive odors, such as food or mates, as well as repulsive odors, such as geosmin, a by-product of bacteria (31)–(33).

All OSNs project their axons to a region in the central brain called the antennal lobe (Figure 16.3B). This structure consists of smaller glomeruli in which synapses connect the OSNs to local interneurons as well as to projection neurons that extend their axons deeper into the fly brain. Interestingly, while the OSNs are distributed across the antenna and maxillary palps, those neurons that express the same OR all connect to the same antennal lobe glomerulus, so that specific information (presence of an odor molecule) can be summarized before being trafficked deeper into the brain.

One way to determine whether olfaction is required for fly detection of parasitoid wasps is to eliminate olfaction entirely. A loss of the OR coreceptor, *Orco*, eliminates nearly all OSN function (34). Some work showed that *Orco* mutant flies failed to respond to wasp presence (16), (35), but other work showed that these anosmic flies maintained their normal change in oviposition behavior in the presence of wasps (14), (29). The reason for this discrepancy may have to do with the particular *Orco* mutant strains used or the way in which behaviors were assayed, but it does appear that flies know what parasitoids smell like. This opens up numerous downstream questions: for example, what is the odorant that flies smell and the odorant receptor(s) they use to smell it, and how might this parasitoid-responsive olfaction circuit coordinate with a parasitoid responsive vision circuit?

To begin to answer these questions, fly larvae and adults were exposed to odors from parasitoid wasp body washes (35). It was found that fly larvae avoid areas that smell like wasps and that adult flies also avoid ovipositing on these areas. It was further shown that the OSNs expressing the odorant receptors Or49a and Or85f can detect the wasp odorants (–)-iridomyrmecin, nepetalactol, and (R)-actinidine, and that information is transferred from these OSNs via projection neurons to other brain regions (Figure 16.3B). These odorants appear to be specific to parasitoid wasps in the family Figitidae as odors from other parasitoid wasp families did not activate these OSNs (35). However, flies still recognize both Figitid and Braconid wasp families and alter their oviposition behavior (14), suggesting that flies may have alternative sensory mechanisms for identifying Braconid parasitoids. Overall, these results suggest that flies may eavesdrop on wasp chemical communications, and provide us with a foothold into a fly's parasitoid-detection olfactory circuit from which we can move further downstream.

*D. melanogaster* uses both vision and olfaction to detect parasitoid wasps. However, these two sensory systems are entirely separated at the periphery of the brain (Figure 16.3). Are the two signals independently relayed to the intended target tissues (e.g., ovary and/or uterus) to alter oviposition behaviors? Or do the vision and olfaction circuits integrate in the central brain before output signals are sent to the reproductive tract? In the following sections, we will discuss what is known about higher order brain regions involved in the oviposition changes that adult *Drosophila* make after sensing parasitoid wasps.

## 16.6 Higher order neurons

When flies sense key parasitoid cues (e.g., visual and/or olfactory), the sensory neurons send signals to higher order neurons for information processing. General neuronal pathways that connect fly sensory systems to different regions of the central brain are known, but virtually nothing is known about the specific circuitry involved in parasitoid-response behaviors. Here, we will discuss how different central brain regions were implicated in anti-parasitoid behaviors and assess how they can be linked to sensory systems.

### 16.6.1 Learning and memory in the mushroom body

One aspect of the anti-parasitoid behaviors that we have described from adult flies is that there is a memory component to the behaviors. The search for more alcoholic oviposition sites and the reduced oviposition following wasp exposure are behaviors that play out over time, even after the wasp signal is removed (14), (29), (36), (37). This suggested that brain regions involved in learning and memory might play a role in anti-parasitoid behaviors. Downstream of the *Drosophila* antennal lobe is a structure called the mushroom body (MB) (Figure 16.3B). This structure has been associated with learning and memory through classical olfactory conditioning, where flies learn to associate a specific odor with another stimulus (usually an electric shock), and then can recall that odor at a later time point and avoid it. Long-term memory is usually assayed at least 24 hours after exposure to the stimulus, and can persist for days (38).

The MB consists of connections between the input neurons, called Kenyon cells, the output neurons, called mushroom body output neurons (MBONs), and a few other types of cells. The ~2,000 Kenyon cells receive olfactory input from the projection neurons of the antennal lobes, making them third order neurons in the olfactory system. The axons of the Kenyon cells project anteriorly to form the different vertical and horizontal MB lobes.

Flies remember being exposed to wasps and maintain their altered oviposition behavior—either oviposition preference for ethanol-rich foods or reduced oviposition rate—in the following days (14), (29), (36), (37). Interestingly, the behavioral changes that flies undergo post-wasp exposure may provide a new assay for memory that doesn't require associative learning. The MB was a clear candidate region to test for a role in memory of parasitoid exposure, so the *GAL4/UAS* system was used to block the release of neurotransmitters specifically from the MB. Although these flies showed normal altered oviposition during the wasp-exposure phase (suggesting that they recognized wasp presence), they could not maintain the altered behavior after wasps were removed (29), (37). These results confirm the role of MB function in long-term memory retrieval in general (39), but also specifically in response to the ecologically relevant stimulus of parasitoid wasps.

Specific genes associated with different temporal components of memory were also tested in wasp exposure assays (38). Fly mutants for the genes *rutabaga*, *Adf1*, *amnesiac*, *FMR1*, and *Orb2* all showed wild-type oviposition behavior changes when directly exposed to wasps, but immediately reverted back to an un-induced state when wasps were removed (14), (29), (37). These results mimicked the results of the experiments that blocked MB function, suggesting these canonical memory genes function within the MB. Indeed, knockdown of *Orb2* specifically in the MB mirrored these findings (29), (37). Computational methods were used to identify novel genes in the MB that are also responsible for memory of wasp exposure, but how they function within the MB remains unknown (36), (40). In all these experiments, a broad MB *GAL4* driver was used to knock down candidate gene expression. It will be interesting in future work to test more

restricted *GAL4* expression patterns to localize the neuronal pathways further through the MBs that are responsive to wasp presence. It may also be useful to pursue the timescales over which fly hosts remember parasitoid encounters in nature, and how this influences community-level oviposition decisions.

Olfactory signals are not solely processed through the MB. The antennal lobe projection neurons also connect to a region of the brain called the lateral horn (LH), sometimes bypassing the Kenyon cells of the MB entirely (Figure 16.3B) (41). The lateral horn has been associated with innate olfactory behaviors in *Drosophila* (42), and may control the immediate (non-memory) display of anti-parasitoid behaviors. The LH and MB do show some connectivity; there are instances of lateral horn output neurons receiving stimulation from MBONs, suggesting that there is a learning and memory modification to innate olfactory behavior (43). Furthermore, compared to the MB, the LH shows an increased range of indirect sensory input including visual, mechanosensory, and gustatory signals (43), suggesting that this could be a region where vision and olfaction signals are integrated. However, the lateral horn neurons have not yet been deeply investigated for roles in anti-parasitoid behaviors (but see (35)). At the organismal level, it will be interesting to further contrast immediate versus more long-term fly anti-parasitoid behavioral defenses.

### 16.6.2 Peptidergic signaling from the fan-shaped body

Thus far we have described different brain regions and specific neurons in these brain regions that respond to parasitoid exposure, but the connections between the different brain regions are still elusive. Neurons typically signal to one another in a localized fashion across synapses using fast-acting small-molecule neurotransmitters. On the other hand, modulatory neuropeptides can have long-lasting effects and can be produced at a great distance from their target cells.

Analysis of the fly genome identified about 45 G-protein coupled neuropeptide receptors, as well as roughly 40 genes that encode neuropeptide precursor transcripts, many of which can be processed into numerous different individual neuropeptides (44). The only neuropeptide thus far examined for its role in anti-parasitoid defenses is neuropeptide F (NPF). Knockdown of *NPF* or its receptor, *NPFR*, caused flies to preferentially oviposit on ethanol-rich foods even when wasps were absent, while overexpression of *NPF* or *NPFR* resulted in flies that failed to switch to more alcoholic food even in the presence of wasps (14, 45). Within the central nervous system, there are roughly 30 cells that express *NPF* (46). A pair of large cells, called P1, send projections into a structure in the center of the brain called the fan-shaped body (FB) (Figure 16.3A). Following wasp exposure, there was an observable decrease in fly NPF protein levels as revealed by immunostaining of the P1 cell bodies and FB (14), (45). Combined with the genetic manipulation results, these data suggest that wasp exposure alters fly NPF expression and/or localization in the FB, leading to the alcohol-seeking oviposition behavior.

The FB is part of a larger group of singular brain structures that span the midline called the central complex. The other components include the protocerebral bridge, ellipsoid body, and noduli. The neurons comprising these neuropil send projections to numerous other brain regions, suggesting that they may be involved in processing a range of stimuli, including from the visual system (47). While wild-type flies showed a decrease in NPF levels in the paired P1 neurons of the FB after wasp exposure, blind mutant flies showed no such reduction (14), demonstrating that vision affects NPF signaling. Given its known role in regulating other behaviors such as locomotion, visual place learning, flight, and courtship (47), the central complex may play an important role in *Drosophila* anti-parasitoid behaviors. However, there is no direct visual connectivity to the central complex in *Drosophila* (47), suggesting visual signals are first sent to other brain regions. Future work should seek to connect the sensory systems and the central complex. Given that flies never previously exposed to wasp parasitoids display immediate behavioral responses to them raises the question as to what brain regions are responsible for an organism's innate sense of danger—the lateral horn and central complex are obvious candidates for encoding such innate anti-parasitoid behaviors (Figure 16.3).

## 16.7 Behavioral output mechanisms

The brain is a complex organization of neuronal connections with numerous signal filtering and processing steps that occur before a final output signal is generated and transmitted back out to the periphery. We know relatively little about how output signals are transmitted through the peripheral nervous system or via neuropeptides to the muscles and other tissues in order to control resulting behaviors. However, some progress has been made in determining how the fly physiologically reduces oviposition following wasp exposure.

### 16.7.1 Oviposition reduction through embryo degradation

After parasitoid wasp exposure, changes occur in the fly reproductive tract to alter the oviposition rate. The paired *Drosophila* ovaries are each composed of 15 to 20 egg-producing stalks, called ovarioles. The tips of the ovarioles house the germline stem cells in a region called the germarium. The oocytes progress through 14 developmental stages as they pass down the length of the ovariole, and upon completion a single egg is released into the uterus where it is fertilized before being expelled during oviposition. During development of the oocyte, there are three checkpoints to ensure it is developing properly and that environmental or parental conditions are favorable for offspring survival (48).

Wasp exposure appears to be one such environmental stimulus that triggers activation of a checkpoint mechanism. It was shown that fly effector caspases, such as Drice and Dcp-1, are activated in the middle checkpoint during stages 7 to 8 of oocyte development to induce apoptosis and resorption of the developing eggs (29), (45), a mechanism that appears to be conserved across many *Drosophila* species (49). Flies still lay some eggs during wasp exposure, suggesting that induction of the oogenesis checkpoint is not absolute.

The induction of the mid-stage checkpoint in the ovaries was shown to be regulated by NPF signaling in the nervous system (Figure 16.3C). Knockdown of *NPF* or *NPFR* increased the proportion of apoptotic stage 7 to 8 oocytes in the ovaries independently of wasp exposure, while overexpression of *NPF* suppressed oocyte apoptosis in wasp-exposed flies (45). These data suggest that NPF may serve as a general, long-range defense signal, and that the two described adult fly anti-parasitoid behaviors—preference for alcoholic oviposition sites and reduced oviposition—are not necessarily separable phenotypes. It was also shown that wasp exposure induces maternal epigenetic imprinting of the *NPF* locus, resulting in heritable changes in fly offspring behavior such that they also prefer alcoholic oviposition sites despite never being exposed to wasps themselves (45), (50), although the mechanism as to how this occurs is unknown. It will be interesting in future work to determine whether NPF completely controls fly egg resorption following parasitoid exposure, or whether other neuropeptides or the peripheral nervous system are also involved. Furthermore, future work should seek to understand the ecological consequences of this egg resorption, to determine the benefits and drawbacks of this behavior for fly reproductive fitness.

## 16.8 Conclusions and future directions

*D. melanogaster* is a widely used genetic model organism but there is still a lot about the natural history of this organism that we do not know. Parasitoid wasps are an important selection pressure on flies in nature; infections are often lethal and upwards of 50% of fly larvae in local populations can be infected at any one time (51). To combat this threat, flies have evolved various defense mechanisms, from traditional immune responses like encapsulation to behavioral defenses like wasp avoidance and self-medication. Adult flies, which are not themselves infected by parasitoids, have even evolved plastic oviposition behaviors to preemptively protect their offspring from parasitoid infection.

It appears that continuous interaction between fly hosts and wasp parasitoids has led to the evolution of dedicated fly visual and olfactory circuits to recognize parasitoid wasp presence in order to trigger appropriate behavioral responses. Likewise, we assume that parasitoid wasps are under strong selection pressure to avoid being recognized by

their fly hosts. The fly-wasp system offers a unique opportunity to understand the ecological genetics of a host–parasite system. Although our community is only beginning to gain footholds into the neurogenetic underpinnings of fly behavioral defenses against parasitoids, our hope is that use of the model *D. melanogaster* will uncover conserved mechanisms of broad relevance to other host–parasitoid systems.

### 16.8.1 More work to do

At the genetic level, only a handful of genes have been shown to play a role in regulating fly behavioral defenses against parasitoids. It is likely that many more are involved due to the multilevel systems involved in behavioral defenses, from sensory signals to the neuronal and non-neuronal tissue changes that must occur. Unbiased mutagenesis screens would be helpful at uncovering the genetic regulators, but behavioral phenotypic assays like oviposition reduction are time consuming and not amenable to such high-throughput screens. A more targeted approach focused on candidate subsets of genes may be a more robust strategy, similar to how many *Drosophila* learning and memory genes were identified (40).

Given the array of new tools developed to study *Drosophila* neurobiology, uncovering the neuronal basis of anti-parasitoid defense behaviors will likely be easier than uncovering the genetic basis. For example, the circuit architecture for male fly courtship song behavior has been described, from gustatory sensory input to specific muscle groups activated by motoneurons. This circuit can now be investigated for other neuronal inputs (such as olfactory cues), as well as its evolution across different fly species (52). Many questions remain as to the parasitoid cues that flies detect (35), including whether gustatory and auditory signals are important, how the sensory circuits are integrated in the brain, and how oviposition and locomotor behaviors are executed. We also do not yet know whether *Drosophila* larvae, which do not have the same complexity as adult visual or olfactory systems, use the same sensory modalities to respond to wasp exposure.

Finally, outside of mechanisms, there is still much to learn about the evolutionary ecology of fly–wasp interactions. How well do fly defense behaviors work against parasitoids in natural settings, and how do parasitoids overcome them? With respect to adult fly behaviors, the preference for ethanol-rich foods as oviposition sites suggests that the most alcoholic parts of rotting fruits will be an attractive location for flies when wasps are present, but we assume that the toxicity trade-off affecting the host will likely cause flies to avoid the same oviposition sites when wasps are absent (53), (54). Furthermore, we assume that reducing oviposition in the presence of wasps is beneficial, if oviposition rates rebound when flies find wasp-free oviposition sites later (15). From the parasitoid side of the equation, if host sensory mechanisms are attuned to parasitoid visual and odor patterns, we assume there is strong selection pressure on parasitoids to alter how they look and smell. If hosts use environmental toxins as infection deterrents (as *D. melanogaster* does with alcohol), we assume that there is strong selection pressure on the more specialist parasitoids to evolve toxin resistance. All of these assumptions could be tested in field and lab-based studies. We feel that field-based studies of fly and wasp behaviors and population dynamics would provide especially important new insights into how fly–parasitoid interactions work, and the potential for behavioral feedbacks between hosts and parasitoids.

### 16.8.2 Mechanisms regulating anti-parasitism behaviors in other organisms

As described earlier in this chapter, there are numerous examples of plastic defense behaviors that diverse hosts use to protect themselves or their offspring from parasitoids (Table 16.1). Those are likely just a drop in the bucket, and we do not doubt that whole new categories of anti-parasitoid behaviors will be uncovered. In *D. melanogaster* alone several distinct anti-parasitoid behaviors have been described, and more are in the process of being described (Davis and Schlenke, unpublished data). Some of the known *D. melanogaster* defense behaviors have been shown to be shared across multiple other *Drosophila* species (16), (35), (49) and non-*Drosophila* species, too (55), (56).

Ever-expanding new technologies will make it easier to study the mechanistic bases of anti-parasitoid defense behaviors in non-model species, using the *D. melanogaster* work as a guide. The CRISPR-Cas9 system has already made genetic engineering feasible in many organisms, both to generate gene knockout mutations but also to insert transgenes that could recapitulate many of the neurobiological tools already at use in *D. melanogaster* (57). Given the broad conservation of brain morphological structures, genetic elements responsible for neuronal identities, and the shared logic of neuronal networks across insects, we feel it is possible that the scientific community can gain a general mechanistic understanding of anti-parasitoid defense behaviors. Combined with organismal and community-level studies of host–parasitoid interactions, a holistic understanding of host–parasitoid interactions is within reach.

### Acknowledgments

We thank the four reviewers for their constructive comments on earlier versions of this chapter. This work was supported by the University of Arizona BIO5 Institute Postdoctoral Fellowship to SMD and NSF grant IOS-1257469 to TAS.

### References

1. Vinson SB and Iwantsch GF. Host regulation by insect parasitoids. *Q Rev Biol.* 1980;55:143–65.
2. Weinersmith KL. What's gotten into you?: A review of recent research on parasitoid manipulation of host behavior. *Curr Opin Insect Sci.* 2019;33:37–42.
3. LaSalle J. Parasitic hymenoptera, biological control and biodiversity in LaSalle J and Gaul ID (eds), *Hymenoptera and Biodiversity* Oxford: Oxford University Press, 1993, 197–215.
4. Carton Y, Poirie M, and Nappi AJ. Insect immune resistance to parasitoids. *Insect Sci.* 2008;15:67–87.
5. de Roode JC and Lefèvre T. Behavioral immunity in insects. *Insects* 2012;3:789–820.
6. Parker BJ, Barribeau SM, Laughton AM, de Roode JC, and Gerardo NM. Non-immunological defense in an evolutionary framework. *Trends Ecol Evol.* 2011;26: 242–8.
7. Dubovskiy IM, Kryukova NA, Glupov VV, and Ratcliffe NA. Encapsulation and nodulation in insects. *Invertebr Surviv J.* 2016;13:229–46.

8. Oliver KM and Perlman SJ. Toxin-mediated protection against natural enemies by insect defensive symbionts. *Adv In Insect Phys*. 2020;58:277–316.

9. Kraaijeveld AR and Godfray HCJ. Trade-off between parasitoid resistance and larval competitive ability in *Drosophila melanogaster*. *Nature* 1997;389:278–80.

10. Adams MD, Celniker SE, Holt RA, Evans CA, Gocayne JD, Amanatides PG, et al. The genome sequence of *Drosophila melanogaster*. *Science* 200;287:2185–95.

11. Reiter LT, Potocki L, Chien S, Gribskov M, and Bier E. A systematic analysis of human disease-associated gene sequences in *Drosophila melanogaster*. *Genome Res*. 2001;11:1114–25.

12. Ugur B, Chen K, and Bellen HJ. *Drosophila* tools and assays for the study of human diseases. *Disease Models and Mechanisms* 2016;9:235–44.

13. Carton Y, Bouletreau M, van Alphen, JJM, and van Lenteren JC. The *Drosophila* parasitic wasps in Ashburner M, Novitski E, Carson HL, and Thompson JN (eds), *The Genetics and Biology of Drosophila*. Cambridge, MA: Academic Press, 1986, 347–94.

14. Kacsoh BZ, Lynch ZR, Mortimer NT, and Schlenke TA. Fruit flies medicate offspring after seeing parasites. *Science* 2013;339:947–50.

15. Lefèvre T, de Roode JC, Kacsoh BZ, and Schlenke TA. Defence strategies against a parasitoid wasp in *Drosophila*: Fight or flight? *Biol Lett*. 2012;8:230–3.

16. Lynch ZR, Schlenke TA, and de Roode JC. Evolution of behavioural and cellular defences against parasitoid wasps in the *Drosophila melanogaster* subgroup. *J Evol Biol* 2016;29:1016–29.

17. Ohyama T, Schneider-Mizell CM, Fetter RD, Aleman JV, Franconville R, Rivera-Alba, M, et al. A multilevel multimodal circuit enhances action selection in *Drosophila*. *Nature* 2015;520:633–9.

18. Hwang RY, Zhong L, Xu Y, Johnson T, Zhang F, Deisseroth K, et al. Nociceptive neurons protect *Drosophila* larvae from parasitoid wasps. *Curr Biol*. 2007;17:2105–16.

19. Bellen HJ, Levis RW, He Y, Carlson JW, Evans-Holm M, Bae E, et al. The *Drosophila* Gene Disruption Project: Progress using transposons with distinctive site specificities. *Genetics* 2011;188:731–43.

20. Brand AH and Perrimon N. Targeted gene expression as a means of altering cell fates and generating dominant phenotypes. *Development* 1993;118:401–15.

21. Perkins LA, Holderbaum L, Tao R, Hu Y, Sopko R, McCall K, et al. The transgenic RNAi project at Harvard medical school: Resources and validation. *Genetics* 2015;201:843–52.

22. Venken KJT, Simpson JH, and Bellen HJ. Genetic manipulation of genes and cells in the nervous system of the fruit fly. *Neuron* 2011;72:202–30.

23. Simpson JH. Mapping and manipulating neural circuits in the fly brain. *Adv Genet*. 2009;65:79–143.

24. Jenett A, Rubin GM, Ngo TTB, Shepherd D, Murphy C, Dionne H, et al. A GAL4-driver line resource for *Drosophila* neurobiology. *Cell Rep*. 2012;2:991–1001.

25. Robie AA, Hirokawa J, Edwards AW, Umayam LA, Lee A, Phillips ML, et al. Mapping the neural substrates of behavior. *Cell* 2017;170:393–406.

26. Zheng Z, Lauritzen JS, Perlman E, Robinson CG, Nichols M, Milkie D, et al. A complete electron microscopy volume of the brain of adult *Drosophila melanogaster*. *Cell* 2018;174:730–43.

27. Djemai I, Casas J, and Magal C. Matching host reactions to parasitoid wasp vibrations. *Proc R Soc B: Biol Sci*. 2001;268:2403–8.

28. Paulk A, Millard SS, and van Swinderen B. Vision in *Drosophila*: Seeing the world through a model's eyes. *Annu Rev Entomol*. 2013;58:313–32.

29. Kacsoh BZ, Bozler J, Ramaswami M, and Bosco G. Social communication of predator-induced changes in *Drosophila* behavior and germline physiology. *Elife* 2015;4:1–36.

30. Gomez-Diaz C, Martin F, Garcia-Fernandez JM, and Alcorta E. The two main olfactory receptor families in *Drosophila*, ORs and IRs: A comparative approach. *Front Cell Neurosci*. 2018;12:1–15.

31. Stensmyr MC, Dweck HKM, Farhan A, Ibba I, Strutz A, Mukunda L, et al. A conserved dedicated olfactory circuit for detecting harmful microbes in *Drosophila*. *Cell* 2012;151:1345–57.

32. Laissue PP and Vosshall LB. The olfactory sensory map in *Drosophila*. *Adv Exp Med Biol*. 2008;628:102–14.

33. Hallem EA and Carlson JR. Coding of odors by a receptor repertoire. *Cell* 2006;125:143–60.

34. Larsson MC, Domingos AI, Jones WD, Chiappe ME, Amrein H, and Vosshall, LB. Or83b encodes a broadly expressed odorant receptor essential for *Drosophila* olfaction. *Neuron* 2004;43:703–14.

35. Ebrahim SA, Dweck HK, Stokl J, Hofferberth JE, Trona F, Weniger K, et al. *Drosophila* avoids parasitoids by sensing their semiochemicals via a dedicated olfactory circuit. *PLoS Biol*. 2015;13:e1002318.

36. Kacsoh BZ, Greene CS, and Bosco G. Machine learning analysis identifies *Drosophila* Grunge/Atrophin as an important learning and memory gene required for memory retention and social learning. *G3 Genes, Genomes, Genet*. 2017;7:3705–18.

37. Kacsoh BZ, Bozler J, Hodge S, Ramaswami M, and Bosco G. A novel paradigm for nonassociative long-term memory in *Drosophila*: Predator-induced changes in oviposition behavior. *Genetics* 2015;199:1143–57.

38. Margulies C, Tully T, and Dubnau J. Deconstructing memory in *Drosophila*. *Curr Biol*. 2005;15:R700–R713.

39. Dubnau J, Grady L, Kitamoto T, and Tully T. Disruption of neurotransmission in *Drosophila* mushroom body blocks retrieval but not acquisition of memory. *Nature* 2001;411:476–80.

40. Kacsoh BZ, Barton S, Jiang Y, Zhou N, Mooney SD, Friedberg I, *et al.* New *Drosophila* long-term memory genes revealed by assessing computational function prediction methods. *G3 Genes, Genomes, Genet.* 2019;9:251–67.

41. Ito K, Shinomiya K, Ito M, Armstrong JD, Boyan G, Hartenstein V, *et al.* A systematic nomenclature for the insect brain. *Neuron* 2014;81:755–65.

42. de Belle JS and Heisenberg M. Associative odor learning in *Drosophila* abolished by chemical ablation of mushroom bodies. *Science* 1994;263: 692–5.

43. Dolan MJ, Frechter S, Bates AS, Dan C, Huoviala P, Roberts RJV, *et al.* Neurogenetic dissection of the *Drosophila* lateral horn reveals major outputs, diverse behavioural functions, and interactions with the mushroom body. *Elife* 2019;8:1–45.

44. Nässel DR and Winther ÅME. *Drosophila* neuropeptides in regulation of physiology and behavior. *Prog Neurobiol.* 2010;92:42–104.

45. Bozler J, Kacsoh BZ, and Bosco G. Transgenerational inheritance of ethanol preference is caused by maternal NPF repression. *Elife* 2019;8:1–18.

46. Shao L, Saver M, Chung P, Ren Q, Lee T, Kent CF, *et al.* Dissection of the *Drosophila* neuropeptide F circuit using a high-throughput two-choice assay. *PNAS.* 2017;114:E8091–E8099.

47. Pfeiffer K and Homberg U. Organization and functional roles of the central complex in the insect brain. *Annu Rev Entomol.* 2014;59:165–84.

48. McCall K. Eggs over easy: Cell death in the *Drosophila* ovary. *Dev Biol.* 2004;274:3–14.

49. Kacsoh BZ, Bozler J, and Bosco G. *Drosophila* species learn dialects through communal living. *PLoS Genet.* 2018;14:1–32.

50. Bozler J, Kacsoh BZ, and Bosco G. Maternal priming of offspring immune system in *Drosophila*. *G3 Genes, Genomes, Genet.* 2020;10:165–15.

51. Fleury F, Ris N, Allemand R, Fouillet P, Carton Y, and Bouletreau M. Ecological and genetic interactions in *Drosophila*-parasitoids communities: A case study with *D. melanogaster, D. simulans* and their common *Leptopilina* parasitoids in south-eastern France. *Genetica* 2004;120:181–94.

52. Sato K, Tanaka R, Ishikawa Y, and Yamamoto D. Behavioral evolution of *Drosophila*: Unraveling the circuit basis. *Genes* 2020;11:1–13.

53. Lynch ZR, Schlenke TA, Morran LT, and de Roode JC. Ethanol confers differential protection against generalist and specialist parasitoids of *Drosophila melanogaster*. *PLoS One* 2017;12:1–19

54. Milan NF, Kacsoh BZ, and Schlenke TA. Alcohol consumption as self-medication against blood-borne parasites in the fruit fly. *Curr Biol.* 2012;22:488–93.

55. Carrasco D and Kaitala A. Egg-laying tactic in *Phyllomorpha laciniata* in the presence of parasitoids. *Entomol Exp Appl.* 2009;131:300–7.

56. Deas JB. and Hunter MS. Delay, avoidance and protection in oviposition behaviour in response to fine-scale variation in egg parasitism risk. *Anim Behav.* 2013;86:933–40.

57. Matthews BJ and Vosshall LB. How to turn an organism into a model organism in 10 "easy" steps. *J Exp Biol.* 2020;223:1–10.

58. Fatouros NE, Cusumano A, Bin F, Polaszek A, and van Lenteren JC. How to escape from insect egg parasitoids: A review of potential factors explaining parasitoid absence across the Insecta. *Proc R Soc B: Biol Sci.* 2020;287:20200344.

59. Gross P. Insect behavioral and morphological defenses against parasitoids. *Annu Rev Entomol.* 1993;38:251–73.

60. Hirayama H and Kasuya E. Oviposition depth in response to egg parasitism in the water strider: High-risk experience promotes deeper oviposition. *Anim Behav.* 2009;78:935–41.

61. Harabis F, Dolny A, Helebrandova J, and Ruskova T. Do egg parasitoids increase the tendency of *Lestes sponsa* (Odonata: Lestidae) to oviposit underwater? *Eur J Entomol.* 2015;112:63–8.

62. Poyet M, Eslin P, Chabrerie O, Prud'homme SM, Desouhant E, and Gibert P. The invasive pest *Drosophila suzukii* uses trans-generational medication to resist parasitoid attack. *Sci Rep.* 2017;7:43696.

63. Deas JB and Hunter MS. Mothers modify eggs into shields to protect offspring from parasitism. *Proc R Soc B: Biol Sci.* 2012;279:847–53.

64. Eberhard WG. The ecology and behavior of a subsocial pentatomid bug and two scelionid wasps: Strategy and counterstrategy in a host and its parasites. *Smithsonian Institution Press* 1975;205:1–39.

65. Edgerly JS. Maternal behaviour of a webspinner (order Embiidina). *Ecol Entomol.* 1987;12:1–11.

66. Ralston JS. Egg guarding by male assassin bugs of the genus *Zelus* (Hemiptera: Reduviidae). *Psyche A J Entomol.* 1977;84:103–7.

67. Abram PK, Brodeur J, Urbaneja A, and Tena A. Nonreproductive effects of insect parasitoids on their hosts. *Annu Rev Entomol.* 2019;64:259–76.

68. Orr MR. Parasitic flies (Diptera: Phoridae) influence foraging rhythms and caste division of labor in the leaf-cutter ant, *Atta cephalotes* (Hymenoptera: Formicidae). *Behav Ecol Sociobiol.* 1992;30:395–402.

69. Allen GR. Influence of host behavior and host size on the success of oviposition of *Cotesia urabae* and *Dolichogenidea eucalypti* (Hymenoptera: Braconidae). *J Insect Behav*. 1990;3:733–49.

70. Stamp NE. Behavioral interactions of parasitoids and Baltimore checkerspot caterpillars (*Euphydryas phaeton*). *Environ Entomol*. 1982;11:100–4.

71. Tautz J and Markl H. Caterpillars detect flying wasps by hairs sensitive to airborne vibration. *Behav Ecol Sociobiol*. 1978;4,101–10.

72. Day WH. The survival value of its jumping cocoons to *Bathyplectes anurus*, a parasite of the alfalfa weevil. *J. Econ. Entomol*. 1970;63:586–9.

73. Bennett AW and Sullivan DJ. Defensive behaviour against tertiary parasitism by the larva of *Dendrocerus carpenteri* an aphid hyperparasitoid. *J New York Entomol Soc*. 1978;86:153–60.

74. Rotheray GE and Barbosa P. Host related factors affecting oviposition behavior in *Brachymeria intermedia*. *Entomol Exp Appl*. 1984;35:141–5.

75. Castner JL. Suitability of *Scapteriscus* spp. mole crickets (Ort, Gryllotalpidae) as hosts of *Larra bicolor* (Hym, Sphecidae). *Entomophaga* 1984;29:323–9.

76. Dietrick EJ and van den Bosch R. Insectary propagation of the squash bug and its parasite *Trichopoda pennipes* Fabr. *J Econ Entomol*. 1957;50:627–9.

77. Hofsvang T and Hagvar EB. Oviposition behaviour of *Ephedrus cerasicola* (Hym.: Aphidiidae) parasitizing different instars of its aphid host. *Entomophaga* 1986;1: 261–7.

78. Ruth WE, McNew RW, Caves DW, and Elkenbary RD. Greenbugs (Hom.: Aphididae) forced from host plants by *Lysiphlehus testaceipes* (Hym.: Braconidae). *Entomophaga* 1975;20:65–71.

79. Shu-sheng L, Morton R, and Hughes RD. Oviposition preferences of a hymenopterous parasite for certain instars of its aphid host. *Entomol Exp Appl*. 1984;35: 249–54.

80. Firlej A, Lucas É, Coderre D, and Boivin G. Impact of host behavioral defenses on parasitization efficacy of a larval and adult parasitoid. *BioControl* 2010;55: 339–48.

81. Anagnostou C, LeGrand EA, and Rohlfs M. Friendly food for fitter flies? Influence of dietary microbial species on food choice and parasitoid resistance in *Drosophila* . *Oikos* 2010;119:533–41.

82. Singer MS, Mace KC, and Bernays EA. Self-medication as adaptive plasticity: Increased ingestion of plant toxins by parasitized caterpillars. *PLoS One* 2009;4: 1–8.

83. Karban R and English-Loeb G. Tachinid parasitoids affect host plant choice by caterpillars to increase caterpillar survival. *Ecology* 1997;78:603–12.

84. Muller CB and Schmid-Hempel P. Exploitation of cold temperature as defense against parasitoids in bumblebees. *Nature* 1993;363:65–7.

# CHAPTER 17

# The behavior of infected hosts: Behavioral tolerance, behavioral resilience, and their implications for behavioral competence

Jessica F. Stephenson and James S. Adelman

## 17.1 Introduction

How animals behave during infection will directly affect their fitness and that of their pathogens and parasites (hereafter "parasites"). In the short term, individuals that maintain or increase the expression of fitness-enhancing behaviors despite infection are likely to benefit from doing so (1)–(5), and may simultaneously enhance direct transmission of parasites by maintaining higher contact rates with uninfected hosts (6). However, maintaining such behaviors during infection is likely to carry costs that could trade off with these short-term benefits, potentially manifesting later in life. Throughout this chapter, we explore such behavioral strategies in the context of tolerance of infection, or the ability to minimize the fitness losses caused by a given parasite load (7), (8). Tolerance contrasts with the more familiar concept of resistance, which is simply the killing or clearing of parasites.

Because the definitions of tolerance and related terms are central to our arguments, we address them here and in the glossary (Box 17.1). Broadly, tolerance is most easily defined as a reaction norm or slope between fitness and pathogen load (Figure 17.1A). Using this definition, tolerance is not a binary trait. Rather, some individuals, populations, or species are more or less tolerant

than others (see also Box 17.2 for measuring tolerance at individual versus group levels). Recently, Adelman and Hawley (5) proposed that among animals, the causes and consequences of tolerance are best understood when we break tolerance into two main components: tissue-specific tolerance, which places tissue pathology (or damage) along the y-axis in lieu of fitness (Figure 17.1B); and behavioral tolerance, which places the expression of fitness-enhancing behaviors, like foraging, mate attraction, or parental care, on the y-axis (Figure 17.1C). Of course, specific host behaviors could promote overall infection tolerance (quantified using direct fitness metrics such as fecundity on the y-axis) by facilitating tissue-specific tolerance rather than behavioral tolerance: for example, increased parental food provisioning can increase juvenile mockingbird tolerance of parasitic flies (9). While our focus is specifically on the infected host's ability to express behaviors that directly enhance fitness (Figure 17.1C), many of our predictions should also hold true for behaviors that promote tissue-specific tolerance.

Several additional terms will prove critical in our analysis of the causes and consequences of behavioral tolerance (Box 17.1). Before infection, individuals and groups can differ in their baseline levels of fitness, or the y-intercept in graphs

Jessica F. Stephenson and James S. Adelman, *The behavior of infected hosts*. In: *Animal Behavior and Parasitism*. Edited by Vanessa O. Ezenwa, Sonia Altizer, and Richard J. Hall, Oxford University Press. © Oxford University Press (2022). DOI: 10.1093/oso/9780192895561.003.0017

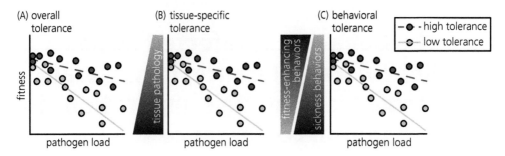

**Figure 17.1** Most generally, tolerance is defined as the slope of the relationship between pathogen load and fitness (A). However, in animals, tolerance can be broken down further using tissue-specific and behavioral traits that influence fitness, termed tissue-specific (B) and behavioral tolerance (C), respectively. See glossary definitions below. Modified from Adelman and Hawley (5), Copyright © Elsevier, 2017; used under STM Permissions Guidelines.

of tolerance (Box 17.1, Figure 1A). This is often referred to as vigor, with behavioral vigor being pre-infection levels of fitness-enhancing behaviors. During infection (Box 17.1, Figure 1B, C), an animal's competence, or its ability to transmit

pathogens successfully, is key to understanding how tolerance impacts disease dynamics. Behavioral competence then encompasses all behaviors that facilitate an animal's ability to transmit, typically by enhancing its rates of contact with new

### Box 17.1  Glossary

**Behavioral competence** (Figure 1C): The behavioral contribution to host competence, the likelihood a host will transmit its infection.

**Behavioral resilience** (Figure 1D): Torres *et al.* (18) and Louie *et al.* (19) introduced physiological resilience as the ability of the host to return to its original health following an infection. We here make the case that behavioral resilience is an important component of animal defense against parasites. We define behavioral resilience as the host's ability to regain its pre-infection expression of fitness-enhancing behavior after having cleared an infection.

**Behavioral resistance**: Behavioral changes during infection that facilitate the killing of parasites, either through reallocation of physiological resources to immune defense (i.e., sickness behaviors: see below) or behaviors that directly remove parasites (e.g., grooming). Please note that as we are discussing exclusively the behavior of infected hosts, behavioral resistance that encompasses avoidance behavior is not our focus.

**Behavioral tolerance** (Figure 1B): Minimizing fitness losses during infection by maintaining, or even increasing, fitness-enhancing behaviors; calculated at the individual

level as the slope between the expression of these behaviors and parasite load, using pre- and during-infection infection data.

**Behavioral vigor** (Figure 1A): An animal's pre-infection expression of the fitness-enhancing behaviors used to calculate behavioral tolerance and resilience.

**Parasite/pathogen**: An organism that completes all or part of its lifecycle in/on another host organism and reduces the fitness of that host.

**Resistance** (Figure 1B): The ability to kill or limit the growth of parasites during infection.

**Sickness behaviors**: A subset of behavioral resistance (above), including lethargy, anorexia, anhedonia, and asociality.

**Tissue-specific tolerance**: Minimizing fitness losses during infection by limiting tissue damage; calculated at the individual level as the slope between pathology and parasite load, using pre- and during-infection infection data.

**Tolerance**: The ability to minimize per-parasite fitness losses during infection; calculated at the individual level as the slope between fitness and parasite load, using pre- and during-infection data.

**Box 17.1** *Continued*

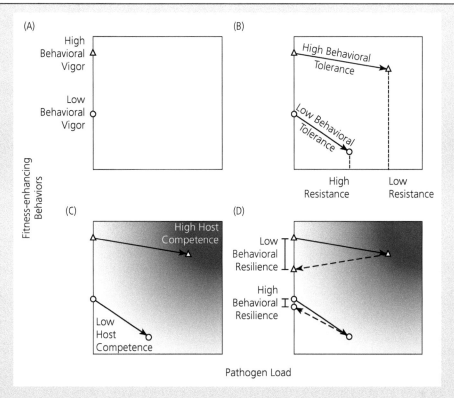

**Box 17.1, Figure 1** Commonly used terms, defined in this box, are depicted in this series of graphs, progressing across pre- (A), during (B–C), and post-infection (D) stages. Triangles represent a host with high behavioral vigor and high behavioral tolerance, but low resistance and behavioral resilience. Circles represent a host with the opposite traits: low behavioral vigor and behavioral tolerance, but high resistance and behavioral resilience. Solid lines denote pre- to during-infection transitions, whereas dashed lines denote during to post-infection transitions. Shading for competence assumes a directly transmitted parasite.

hosts or vectors. For example, if behavioral tolerance maintains behaviors like foraging or mating that bring infected animals into direct contact with others, it will also enhance their behavioral competence for directly transmitted parasites. Finally, an animal's resilience describes how well it returns to pre-infection levels of fitness, with behavioral resilience specifically referring to the expression of fitness-enhancing behaviors (Box 17.1, Figure 1D). Although resilience is only apparent after an infection, it can nonetheless prove important for future disease dynamics, potentially predicting hosts' trajectories into competence during subsequent infections.

A conceptual framework encompassing behavioral vigor, tolerance, resilience, and their interactions will help us understand the dramatic impact that individual variation in infected host behavior can have on disease dynamics (10). Here, we provide an initial draft of such a framework. We then suggest how internal and external pressures could shape investment in behavioral tolerance and resilience and offer evidence from previous studies that these may trade off. Finally, we discuss putative ways in which behavioral tolerance and resilience may differ in their contribution to behavioral competence, and thus parasite fitness. We also provide practical guidance on how observations of

## Box 17.2 Measuring and analyzing behavioral vigor, tolerance, and resilience

The principal difference among behavioral vigor, tolerance, and resilience is one of time: they concern host behavior before, during, and after infection, respectively.

**Measurements required**: The first crucial metric in assessing behavioral tolerance or resilience is parasite load, quantified before (vigor), before and during (tolerance), or after (resilience) infection. Such methods will be highly system-specific, but could rely on myriad techniques from visual inspection for ectoparasites to nucleic acid quantification for microparasites. Second is the choice of behavior(s) to measure at each of these timepoints. Best choices would include behaviors with direct links to fitness enhancing processes, including, but not limited to, mate acquisition, offspring provisioning, vigilance, and predator avoidance, or social bond formation/maintenance. Assessing both parasite load and behavior at as many time points as possible will increase both the analytical options available and the precision of estimates generated. Within-individual behavioral repeatability, at least among uninfected animals, is a key prerequisite of focal behaviors for the study of behavioral tolerance; it is otherwise impossible to attribute behavioral changes to infection rather than noise (Figure 1). These requirements necessitate a thorough understanding of the focal system, and ideally preliminary data on the host's response to a given behavioral assay (i.e., do animals need to be pre-exposed to the assay to ensure repeatability? Figure 1A, B), the repeatability of the behavior among uninfected animals (Figure 1A-C), and the optimal frequency of behavioral (Figure 1C) and pathogen load observations.

**Analytical options**: If one can collect data at more than two time points, random-slopes mixed effects models can estimate variation (and uncertainty) in the relationship between pathogen load and behavior at both the group level (such as treatments or sexes) and the individual level (25). Moreover, extending such techniques to multivariate mixed models can ask how variation in tolerance slopes relate to other quantities, including vigor, resilience (such as would be necessary to identify a trade-off), and competence. Several caveats do exist, however. First, these models tend to shrink estimates of individual slopes toward group values (i.e., reduced magnitude compared to individual-level regressions (59)). Second, relatively large sample sizes can be required to ask whether slopes vary substantially across individuals. For example, Martin *et al*. (60) suggested using over 200 individuals to evaluate individual-level random effects. The simple alternatives to such models would be calculating slopes from individual regressions or using best linear unbiased predictors (BLUPs) from mixed effects models to generate metrics for each animal. While such methods may be the only ones available for small data sets, their inability to estimate

uncertainty can lead to exaggerated conclusions when used to predict other metrics (61). As such, they are not preferred, and we would urge conservative interpretation if they are the only methods available.

A promising alternative to likelihood-based mixed effects models is Bayesian modeling that incorporates Markov chain Monte Carlo methods (62). These models allow estimating a mean and distribution of slopes for individuals, allowing uncertainty to be built into subsequent analyses that use estimates of vigor, tolerance, or resilience to predict other relevant processes, like transmission rates or infectiousness.

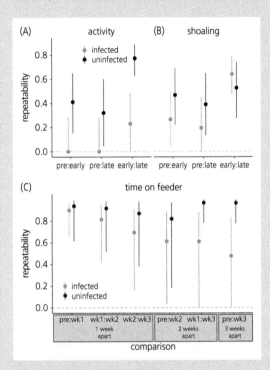

**Box 17.2, Figure 1** Behavioral repeatability varies with infection, between specific host behaviors such as activity level (A) and social behavior (shoaling; B), and between host species (A, B: data from guppies, *Poecilia reticulata* infected with *Gyrodactylus turnbulli*; C: data from house finches, *Haemorhous mexicanus* infected with *Mycoplasma gallisepticum*). Repeatability also depends on the time between behavioral measurements (C), and obtaining accurate estimates of an individual's behavior may require exposing the animals to the assay before collecting data (A: comparisons involving the first, "pre" timestep are less repeatable than that comparing between subsequent timesteps). Points give the mean ± 95% confidence interval; where these overlap 0 (dashed line), the behavior is not significantly repeatable. Repeatability here is calculated as the intraclass correlation coefficient using our published empirical data (11, 63).

hosts before, during, and after infection are critical in the quantification of behavioral resilience, behavioral tolerance, and behavioral competence. Rigorously analyzing these traits and the relationships between them will be key in revealing how behavior-parasite feedback drives both transmission and host–parasite coevolution.

## 17.2  Hypothesized costs and trade-offs associated with behavioral tolerance

Empirical studies demonstrate that individuals vary in their ability to maintain fitness-enhancing behaviors during infection (4), (11) (i.e., behavioral tolerance) and that this impacts both the immediate survival and reproduction components of their fitness (12), (13). However, there is also empirical evidence that this behavioral tolerance comes at a cost in terms of reduced resistance or the ability to kill and clear parasites: in many disease systems, hosts expressing typical sickness behaviors like lethargy, anorexia, and asociality, and therefore not expressing behavioral tolerance (Figure 17.1C), recover faster and more completely from infection, perhaps due to more effective immune defenses (3), (8), (14). Sickness behaviors may therefore promote host fitness in the longer term (4). Hosts can thus employ two opposite behavioral responses during infection, each of which can promote fitness, albeit on different timescales: the maintenance of fitness-enhancing behaviors (behavioral tolerance) or the expression of sickness behaviors that promote resistance. We predict that hosts employing the resistance strategy will also display faster and more complete recovery of fitness-enhancing behaviors following infection, in other words higher behavioral resilience (Box 17.1; Figure 17.2).

Although explicit investigation of behavioral tolerance is still in its infancy (5), and behavioral resilience is, to our knowledge, totally unexplored, previous findings suggest that they should trade off. One assumption underlying our hypothesized trade-off is that expressing sickness behaviors is necessary for hosts to return to pre-infection expression of fitness-enhancing behavior. Under conditions of typical or low resource availability, most hosts cannot energetically afford to be both behaviorally tolerant during infection (by

suppressing sickness behaviors) and behaviorally resilient. This assumption is supported by the fact that *Listeria monocytogenes*-infected mice prevented from expressing anorexia have lower survival and recovery rates than control animals (14). Similarly, ectotherms prevented from developing behavioral fever during bacterial or viral infection suffer from higher mortality (15), (16). Further, female burying beetles that continue to care for their offspring despite infection suffer a high probability of mortality (13). These examples indicate that experimental and ecological conditions that force the expression of behavioral tolerance could promote fitness in the short term, while reducing survival in the longer term, thereby reducing behavioral resilience. Additionally, there is some evidence that behavioral tolerance is energetically costly: birds maintaining feeding behavior and activity levels under both dietary stress and immune challenge lost more mass than those exposed to just one of these stressors (17). Because of these costs, individuals expressing behavioral tolerance are likely to be unable to maintain their behavior long term, and thus will be less behaviorally resilient than their behaviorally intolerant counterparts. Direct tests of this trade-off are sorely needed to help form a comprehensive framework of infected host behavior.

In addition to the indirect evidence that behavioral tolerance and resilience trade off, different conditions intuitively favor the expression of each. At the extremes, behavioral tolerance must be beneficial to hosts experiencing lifelong, chronic infection, whereas behavioral resilience may be more beneficial to hosts suffering short-term, acute infections. If we consider infections as falling along a continuum between acute and chronic, the duration of infection at which the relative fitness benefit of behavioral tolerance versus resilience switches is likely to vary both within and among individuals, populations, and species (Figure 17.2A). We suggest that residual reproductive value (RRV) is another important determinant of when behavioral tolerance is favored over resilience: high RRV individuals, populations, or species invest relatively more in future reproduction, and thus should prioritize a faster and fuller recovery of pre-infection behavioral expression levels (Figure 17.2B). Those with low RRV should instead invest in current reproduction by using a tolerance strategy to maintain normal expression of fitness-enhancing behaviors. Our assumed trade-off between behavioral tolerance and resilience is therefore broadly akin to the trade-off between reproductive effort and somatic defense (20), with investment in behavioral tolerance early in infection being similar to a terminal investment strategy (21), (22).

As well as their implications for host fitness, behavioral tolerance and behavioral resilience, through their impacts on host movement, contact rates, and habitat use, are likely to affect the behavioral components of parasite transmission, or a host's "behavioral competence" (Box 17.1, Figure 1). For example, highly social and behaviorally tolerant hosts should be more competent at transmitting directly transmitted parasites that rely on close contacts (6). Additionally, hosts that are highly social prior to infection (high behavioral vigor) are particularly likely to be exposed to parasites in the first place (23). Similarly, behaviorally resilient hosts that regain their pre-infection expression of social or sexual behavior may be more likely to become infected again, with the same or a different parasite, and could be important for onward transmission (24). Thus, both the intercept and slope of the host's behavioral reaction norm during infection (25) are important determinants of parasite fitness (Box 17.1,

Figure 1). The role of behavioral tolerance in host competence is intuitive but surprisingly understudied (5), and we suggest that future work addressing this knowledge gap should additionally consider the role of behavioral vigor and resilience.

## 17.3 Factors shaping behavioral resilience and behavioral tolerance

A key assumption of our proposed framework of infected host behavior is that investment in behavioral tolerance will trade off with resistance of infection and behavioral resilience (Figure 17.2). This relationship is likely to be dynamic and dependent on factors, both intrinsic and extrinsic to the infected host, that influence its expectation for future progeny (i.e., RRV). As the maintenance of behaviors such as courtship and parental care is crucial for reproduction, behavioral tolerance can be considered a component of current reproductive effort. Similarly, as sickness behaviors promote survival and recovery of pre-infection levels of behavior, behavioral resilience may be considered part of somatic defense, and therefore investment in future reproductive effort. The conditions affecting an infected host's relative investment in behavioral tolerance and resilience might therefore be comparable to the conditions affecting relative investment in reproductive effort versus somatic defense

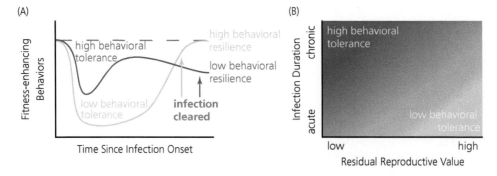

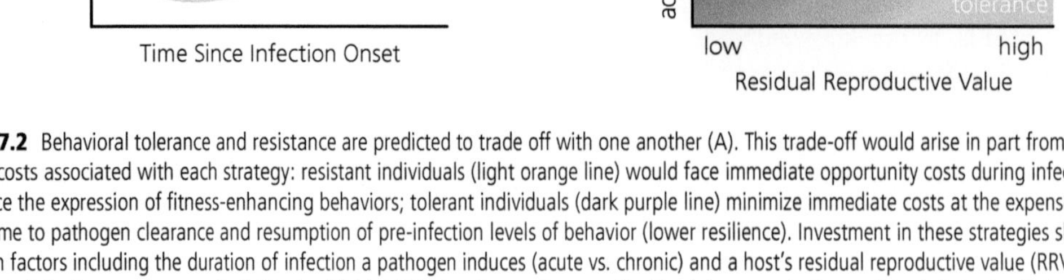

**Figure 17.2** Behavioral tolerance and resistance are predicted to trade off with one another (A). This trade-off would arise in part from the timing of costs associated with each strategy: resistant individuals (light orange line) would face immediate opportunity costs during infection as they reduce the expression of fitness-enhancing behaviors; tolerant individuals (dark purple line) minimize immediate costs at the expense of delayed time to pathogen clearance and resumption of pre-infection levels of behavior (lower resilience). Investment in these strategies should depend on factors including the duration of infection a pathogen induces (acute vs. chronic) and a host's residual reproductive value (RRV) (B). During acute infections, and among individuals with high RRV, i.e., those who can afford to delay reproduction, resistant strategies (light shading) should be favored; under the opposite conditions, tolerant (dark shading) strategies should be favored (B).

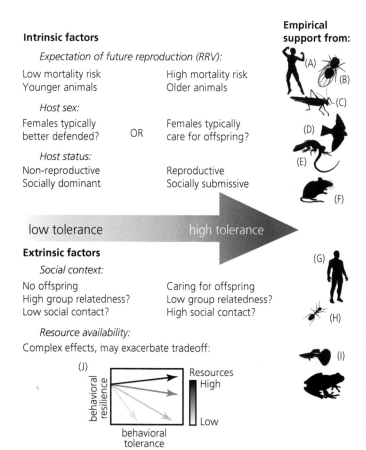

**Intrinsic factors**

*Expectation of future reproduction (RRV):*

Low mortality risk          High mortality risk
Younger animals             Older animals

*Host sex:*

Females typically           Females typically
better defended?      OR    care for offspring?

*Host status:*

Non-reproductive            Reproductive
Socially dominant           Socially submissive

low tolerance        high tolerance

**Extrinsic factors**

*Social context:*

No offspring                Caring for offspring
High group relatedness?     Low group relatedness?
Low social contact?         High social contact?

*Resource availability:*

Complex effects, may exacerbate tradeoff:

**Empirical support from:**

(A) (B) (C) (D) (E) (F) (G) (H) (I)

**Figure 17.3** The expression of behavioral tolerance likely depends on factors both intrinsic and extrinsic to the infected host. We present key examples here, along with the systems in which illustrative patterns have been observed (A–I; see main text for references). Question marks denote where the directionality of these effects remains unclear. Some factors may not directly affect the expression of these behavioral defenses, but instead act to exacerbate or alleviate the trade-off between them, as illustrated here for resource availability (J).

more broadly. In this section, we discuss existing research supporting these ideas, and highlight future research foci, summarized in Figure 17.3.

Host RRV is likely to be a strong positive predictor of investment in behavioral resilience: the higher RRV, the more the host should invest in future reproductive events and thus behavioral resilience. The simplest test of the importance of host RRV to investment in behavioral tolerance or behavioral resilience would be to compare the investment made by host populations and species with semelparous versus iteroparous reproductive strategies. Infected semelparous animals should heavily bias investment towards behavioral tolerance, whereas iteroparous should be more likely to invest in behavioral resilience, and therefore future reproduction (though some conditions may promote terminal investment among iteroparous individuals (21), (22)).

Following this reasoning, we can make specific predictions about host investment in these two strategies under predation pressure. On the one hand, because sickness behaviors are associated with decreased anti-predator responses (26), (27), predation pressure should favor investment in behavioral tolerance. Additionally, high adult mortality limits RRV and thus favors investment in current reproduction. However, in systems in which hosts can effectively hide from predators, higher levels of predation pressure may favor the expression of sickness behaviors. Indeed, higher predation pressure experienced by men than women in human hunter-gatherer societies is one explanation for the evolution of lower behavioral tolerance among men ("man flu," (28); Figure 17.3A).

In many organisms, the probability of survival and RRV both decrease with age (29); intuitively, we should therefore expect older individuals to be more behaviorally tolerant of infection. Indeed, animals across taxa have been found to invest more in reproduction as they age (22), and some studies have tested how this depends on parasitic infection. For example, *Drosophila nigrospiracula* males infected with mites invested more in courtship behaviors, and this effect was strongest among older males (30); that is older males were more behaviorally tolerant of their infections (Figure 17.3B). Copeland and Fedorka (31) found a similar pattern in crickets injected with lipopolysaccharide (LPS: a component of bacteria cell membrane commonly used to elicit immune responses in the absence of a pathogen; Figure 17.3C). Contrary to this pattern among male animals, Palacios *et al.* (32) report that older female tree swallows injected with LPS demonstrate increased sickness behaviors relative to their younger counterparts, though this effect was not consistent across study years (Figure 17.3D). This observation highlights the possibility that the factors affecting host investment in behavioral resilience and tolerance, some of which we discuss in this section, may interact in complex ways, and studies addressing multiple factors simultaneously would be particularly valuable.

Host sex is likely to impact relative investment in behavioral tolerance and resilience. Males and females interact differently with their parasites (33), ultimately because of their divergent life-history strategies (34). Typically, males maximize fitness by mating with as many females as possible; they therefore invest more in male–male competition and mating, and generally have shorter life spans than females (35), (36). By contrast, females maximize fitness by lengthening life span and producing more broods; to do so, they invest more in self-maintenance and somatic defense, such as that against parasites (34). Consistent with this model, females typically have higher levels of tissue-specific tolerance (37), (38), and tend to be less infected than males across taxa (33), (39) (Figure 17.3E). Based on our proposal that investment in behavioral resilience is akin to investment in somatic defense, this model may suggest that females should invest more in behavioral resilience, while males should prioritize behavioral tolerance. However, behavior-specific considerations include that for males, maintaining advertisement or mating behaviors may be futile if, as is common, females can detect and avoid infection in their mates (40). In such cases, males should favor behavior resilience. Similarly, when females are the sex that invests more in parental care, which prevents the expression of sickness behaviors (13), (41), females may favor behavioral tolerance. Indeed, data from animals across taxa, including humans, suggest that females may be the more behaviorally tolerant sex (11), (27), (28), (35).

Host dominance or reproductive status is likely to interact with the factors discussed so far in determining investment in each defense strategy. In general, socially dominant individuals are likely to be less behaviorally tolerant than their submissive counterparts, perhaps because they are better able to prioritize recuperative behavior (42), (43), and this effect may depend on host sex (44). Host sex is also likely to interact with reproductive status to determine investment in defense strategies. During the breeding season, particularly among populations and species in which these are short, there should be strong selection for tolerance strategies that enable animals to maintain reproductive behaviors. Indeed, when ambient temperatures were cold enough to be life-threatening to pups, lactating mice behaviorally tolerated infection (simulated by LPS injection) and continued nest building activities, but at higher ambient temperature they expressed sickness behaviors (41) (Figure 17.3F).

Reproductive status is also likely to impact the relationship between behavioral defense strategies: reproductively active hosts may be more likely to experience a trade-off between behavioral tolerance and behavioral resilience because of the additional demands on their limited resources (45). For example, Bonneaud *et al.* (46) found that house sparrows injected with LPS were more likely to abandon nestlings than control birds while they were mounting an immune response, consistent with low tolerance of infection. However, the nestlings in replacement broods laid by these LPS-treated birds had higher growth rates than those laid by control birds, indicating that the LPS-treated birds partly compensated for their lack of tolerance with resilience. While the authors suggest that these birds are likely to suffer reduced survival or lower reproductive output in subsequent breeding seasons due to this high breeding effort (46), the relative costs of tolerance and resilience are unclear: these birds may have conserved enough resources during the abandonment of their first broods to compensate for the high level of provisioning they provided to the replacement brood.

The presence of offspring is just one example of a social context that we predict should strongly influence host investment in behavioral tolerance, depending on the behavior considered (1). The presence and identity of conspecifics strongly influences infected host behavior. For example, injection with LPS induced a strong reduction in activity among zebra finches held in isolation, but not those held in a colony setting (47), or exposed to a potential mate (48). The mechanism underlying the effect of social context on behavioral tolerance is unclear and is likely to vary between systems. In some systems, individual hosts may pay direct costs, such as increased mortality, to express behavioral tolerance (13), but in others, including humans, social interactions themselves can promote recovery (49), so their maintenance may instead be a component of behavioral resistance (Box 17.1; Figure 17.3G). Some of the most dramatically changed behaviors among infected hosts occur in social insects: the high inclusive fitness associated with the continued health of the rest of the colony dictates that infected individuals should remove themselves, and signal their infection status to colony mates, and there is good evidence for both the existence and adaptive value of this lack of behavioral tolerance (50) (Figure 17.3H). More broadly, sickness behaviors may be part of an "inclusive behavioral immune system" that reduces transmission between an infected host and its healthy kin (51). Few studies have directly tested this idea, and in fact, those that do account for relatedness suggest that hosts are more likely to tolerate infection behaviorally in the presence of kin (52), (53).

One factor that is highly likely to interact with many others we consider here is host body condition, as influenced by resource availability and intake. Across systems, host body condition affects physiological defenses against parasites (54), and

these are correlated with behavior (11), (55). For example, infection with the amphibian chytrid fungus (*Batrachochytrium dendrobatidis*) is associated with a 40% reduction in calling probability among poor condition male frogs, and a 30% increase in calling probability among good condition males, relative to uninfected frogs (56) (Figure 17.3I). Similarly, empirical studies indicate that food supplementation and better body condition mitigate some of the behavioral consequences of infection (2) (but see (57)). However, theory and the better-developed empirical literature examining how resource availability may impact tissue-specific tolerance indicates that the relationship is highly nuanced and strongly interacts with other factors (54).

Limited resource availability is likely to exacerbate the trade-off between behavioral tolerance and behavioral resilience, as it does for other life-history traits (21) (Figure 17.3J). Host body condition mediates reproductive investment (22), and therefore the optimal balance of resilience and tolerance. It is also likely to be an important determinant of how other intrinsic and extrinsic factors impact that balance. For example, if condition-dependent mechanisms such as tissue repair are key components of both behavioral tolerance (1) and behavioral resilience, infected hosts in good condition may have sufficient resources to clear infection and maintain tissue repair throughout, effectively employing both behavioral tolerance and resilience successfully. Because many systems experience seasonal variation in resource availability, hosts may display seasonal fluctuations in the relative expression of behavioral tolerance and resilience, as has been observed in other defense components (58). Indeed, whether it is mediated through resource availability or other seasonal changes such as investment in reproduction, sickness behaviors do vary between seasons in some systems (2).

## 17.4 Linking behavioral tolerance/resilience to host competence

Because an infected host's behavior can alter its contact rate with other hosts or vectors, behavior represents an integral component of host competence

(23), which we have termed behavioral competence. However, precisely how behavioral tolerance and resilience impact overall competence will depend on several additional considerations.

First, overall competence will reflect not only behavioral tolerance during infection but also its relationship with both pre-infection behavior (behavioral vigor) and post-infection behavior (behavioral resilience; Box 17.1, Figure 1). For instance, in a directly transmitted disease system, hosts whose initial behaviors yield high rates of contact (high y-intercepts in Box 17.1, Figure 1A), are more likely to become infected (i.e., super-receivers or super-attractors (23)). Sickness behaviors may often reduce contact (6), so even a moderate degree of behavioral tolerance during infection could maintain high levels of contact, transitioning such individuals from super-receivers to superspreaders (63). However, if a super-receiver expressed very low behavioral tolerance, its contact rate would fall dramatically during infection, essentially transforming it into a dead-end host (or super-diluter (23)). In contrast, hosts whose pre-infection behaviors already limit contact rates (low y-intercepts in Figure 17.1 and Box 17.1, Figure 1A) would be less likely to become infected. Only with very high tolerance (e.g., a positive slope) would such hosts achieve competence during infection.

Interactions between behavioral vigor and resilience become important when considering a host's competence in future epidemics, either involving a different parasite or, if host immunity wanes, the same parasite. After infection, a highly resilient individual would return to its original behavioral baseline, setting it up to play the same role in later epidemics that it played in the current one. In contrast, after infection, a minimally resilient host will express lower levels of behaviors likely to enhance contact rate, making it less likely to act as a super-receiver in the future, regardless of the role it played in the current epidemic. Because such differences in resilience will alter continuing/future disease dynamics, they provide a pathway through which parasite–behavior feedback can impact host–parasite coevolution (Box 17.3).

Second, because tissue-specific tolerance can allow hosts to sustain heavy infection loads for longer periods, the ways in which behavioral tolerance correlates with tissue-specific tolerance will have important consequences for overall competence (5). Moreover, the effect of tissue-specific tolerance on parasite shedding, which is likely to differ across host–parasite systems, will be critical to such relationships (64). For example, if tissue-specific tolerance not only extends the infectious period but enhances shedding, a positive correlation with behavioral tolerance will greatly enhance competence. However, if tissue-specific tolerance impedes shedding, a positive correlation with behavioral tolerance may mean that highly tolerant hosts show low levels of competence (5). As above, a host's behavioral vigor will affect the magnitude of any such correlations, while its behavioral resilience will influence the role this host plays in future epidemics.

## 17.5 Conclusions and future research

Although behavioral vigor, tolerance, and resilience should all have important consequences for parasite–behavior feedback and infectious disease dynamics in general, research into these concepts is still in its earliest stages. Here, we have explored the important possibility that behavioral tolerance trades off with resistance and resilience (Figure 17.2), drawing from the better-developed literature on non-behavioral components of defense strategies. While there is evidence that the non-behavioral components of tolerance and resistance can trade off (70), this pattern has been shown to vary across host–parasite pairings and is likely to depend on the relative costs of each type of defense (70), (71). Further, the precise shape that such trade-offs take can have important consequences for epidemiological dynamics (72). Whether the same factors and costs influence trade-offs between behavioral tolerance and resilience, or the shape of the relationship between the two, remains to be seen. To investigate these details fully, future work must use live parasites—rather than the non-replicating antigens more commonly used to induce sickness behaviors—to test how sickness behaviors truly interact with parasite load to shape behavioral tolerance, resistance, and resilience.

**Box 17.3  Behavior during infection affects host–parasite interactions through both ecological and evolutionary processes**

Whether infected hosts express behavioral tolerance or behavioral resilience, how they behave directly impacts their fitness and simultaneously dictates behavioral competence, and thus parasite fitness. Eco-evolutionary feedback between infected host behavior and parasite population size, structure, and parasite evolution are therefore inevitable (65). Here, we focus on the evolution of one key parasite trait, virulence, which we define as the fitness cost inflicted on the host, including both the infection-induced behavioral changes with negative impacts for host fitness and infection-induced host tissue damage. For simplicity, we here ignore any potential influence of the behavior of naive hosts, though this is likely to be substantial, as highlighted elsewhere in this volume (66)–(68).

We propose several routes by which infected host sociality (including any interaction between conspecific animals) and parasite virulence may affect one another through ecological and evolutionary processes (Figure 1). For example, increased host sociality will increase transmission of directly transmitted parasites, while also increasing the diversity and size of the parasite communities within and among hosts. These factors both select for higher parasite virulence, because co-infecting parasites compete (69), and provide genetic variation on which that selection can act. Resultant

evolutionary increases in parasite virulence will increase host mortality, most likely disproportionately among highly susceptible hosts. Across systems, highly susceptible hosts tend to be less social (11), (55), so this differential mortality may cause an evolutionary increase in host sociality.

However, increasing parasite virulence will cause increased tissue damage to infected hosts, likely to increase the severity of symptoms and thus perhaps reducing the host's ability to behaviorally tolerate infection (i.e., favoring resilience), decreasing infected host sociality through ecological processes. Thus, increased parasite virulence may result in either increased or decreased sociality among infected hosts, which could either reverse or reinforce the processes described above. Tests of such hypotheses are likely to be fruitful: Jacot *et al.* (57) found that LPS-injected crickets ("high virulence" proxy) showed an enduring decline in calling rate, whereas saline-injected ("low virulence" proxy) control males regained calling rates rapidly after a short decline. Improving our understanding of how infected hosts behave is therefore key to understanding how host sociality and parasite virulence affect one another, and thus disease dynamics in natural systems.

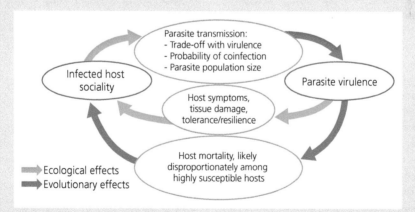

**Box 17.3, Figure 1**  Infected host sociality drives parasite virulence evolution, which feeds back on infected host social behavior through both ecological (green arrows) and evolutionary processes (blue arrows). All suggested links could positively or negatively affect one another, depending on factors such as parasite transmission route.

Research clarifying the physiological mechanisms underpinning behavioral tolerance and resilience are likely to help reveal how and why they may be correlated with one another, and with other aspects of host defense against parasites. For instance, it is likely that inflammation is an important determinant of tolerance (73), (74). Systemic inflammation, which can be measured through pro-inflammatory cytokines in vertebrates or other proxies like fever, is often highly correlated with the expression of sickness behaviors across scales: among individuals (75), populations (12), or species (76). Inflammation is also often associated with damage to the host tissue during infection, which points toward reduced inflammation as a promising mechanism underlying tissue-specific tolerance of infection (5), (73), (74). Given this shared mechanism, behavioral tolerance of infection may well be correlated with tissue-specific tolerance. However, there is substantial variation in host behavioral and physiological responses to infection across scales, and through contexts, which could lead to behavioral and tissue-specific tolerance acting independently (5).

Given the likely energetic costs of both resistance and behavioral tolerance, the availability and distribution of resources should help shape the trade-off between them and their consequences for host competence. First, if high resource availability could ameliorate the trade-off (77) (Figure 17.3J), allowing hosts to be both resistant and behaviorally tolerant, host–parasite coevolutionary dynamics would certainly shift (Box 17.3). Further, the spatial distribution of resources are likely to affect links among behavioral tolerance, parasite transmission, and the evolution of virulence. For example, using individual-based models, Franz and colleagues (78) found that when resources are aggregated and limited, reduced movements during infection (lower behavioral tolerance) actually enhanced both parasite transmission and the evolution of virulence. Because the availability and distribution of resources changes through time, it is likely that the shape and strengths of trade-offs between defense components also fluctuate spatially and temporally (54). Empirical studies testing the role of behavioral vigor, tolerance, and resilience

under different resource conditions are therefore critical (78).

Through such empirical tests, and complementary modeling efforts, we can begin building a comprehensive framework of how behavioral changes during infection alter both host and parasite fitness. We hope the tools and ideas highlighted here provide an overview of what such a framework could look like and that they stimulate future research in this area.

## References

1. Lopes PC. When is it socially acceptable to feel sick? *Proc R Soc Lond B: Biol Sci* 2014;281:20140218.
2. Owen-Ashley NT and Wingfield JC. Seasonal modulation of sickness behavior in free-living northwestern song sparrows (*Melospiza melodia morphna*). *J Exp Biol.* 2006;209: 3062–70.
3. Hart BL. Biological basis of the behavior of sick animals. *Neurosci Biobehav Rev.* 1988;12: 123–37.
4. Adelman JS and Martin LB. Vertebrate sickness behaviors: Adaptive and integrated neuroendocrine immune responses. *Integr Comp Biol.* 2009;49: 202–14.
5. Adelman JS and Hawley DM. Tolerance of infection: A role for animal behavior, potential immune mechanisms, and consequences for parasite transmission. *Horm Behav.* 2017;88:79–86.
6. Lopes PC, Block P, and König B. Infection-induced behavioural changes reduce connectivity and the potential for disease spread in wild mice contact networks. *Sci Rep.* 2016;6:31790.
7. Råberg L. How to live with the enemy: Understanding tolerance to parasites. *PLoS Biol.* 2014;12:e1001989.
8. Medzhitov R, Schneider DS, and Soares MP. Disease tolerance as a defense strategy. *Science* 2012;335:936–41.
9. McNew SM, Knutie SA, Goodman GB, Theodosopoulos A, Saulsberry A, Yepez J, *et al.* Annual environmental variation influences host tolerance to parasites. *Proc R Soc London B: Biol Sci.* 2019;286:20190049.
10. Lloyd-Smith JO, Schreiber SJ, Kopp PE, and Getz WM. Superspreading and the effect of individual variation on disease emergence. *Nature* 2005;438:355–9.
11. Stephenson JF. Parasite-induced plasticity in host social behaviour depends on sex and susceptibility. *Biol Lett.* 2019;15:20190557.
12. Adelman JS, Bentley GE, Wingfield JC, Martin LB, and Hau M. Population differences in fever and sickness

behaviors in a wild passerine: A role for cytokines. *J Exp Biol.* 2010;213:4099–109.

13. Ratz T, Monteith KM, Vale PF, and Smiseth PT. Carry on caring: Infected females maintain their level of parental care despite suffering high mortality. *Behav Ecol* 2021 32 738–46.

14. Murray MJ and Murray AB. Anorexia of infection as a mechanism of host defense. *Am J Clin Nutr.* 1979;32:593–6.

15. Kluger, M. J, Ringler, D. H. and Anver, M. R. Fever and survival. *Science* 1975;188:166–8.

16. Boltaña, S, Rey S, Roher N, Vargas R, Huerta M, Huntingford FA, *et al*.Behavioural fever is a synergic signal amplifying the innate immune response. *Proc Biol Sci.* 2013;280:20131381.

17. Wilsterman K, Alonge MM, Ernst DK, Limber C, Treidel LA, and Bentley GE. Flexibility in an emergency life-history stage: Acute food deprivation prevents sickness behaviour but not the immune response. *Proc R Soc London B: Biol Sci.* 2020;287:20200842.

18. Torres BY, Oliveira JHM, Thomas Tate A, Rath P, Cumnock K, and Schneider DS. Tracking resilience to infections by mapping disease space. *PLoS Biol.* 2016;14:e1002436.

19. Louie A, Song KH, Hotson A, Thomas Tate A, and Schneider DS. How many parameters does it take to describe disease tolerance? *PLoS Biol.* 2016;14:e1002435.

20. Lochmiller RL and Deerenberg C. Trade-offs in evolutionary immunology: Just what is the cost of immunity? *Oikos* 2000;88:87–98.

21. Clutton-Brock TH. Reproductive effort and terminal investment in iteroparous animals. *Am Nat.* 1984;123:212–29.

22. Duffield KR, Bowers EK, Sakaluk SK, and Sadd BM. A dynamic threshold model for terminal investment. *Behav Ecol Sociobiol.* 2017;71:1–17.

23. Martin LB, Addison B, Bean AGD, Buchanan KL, Crino OL, Eastwood JR, *et al*. Extreme competence: Keystone hosts of infections. *Trends Ecol Evol.* 2019;34: 303–14.

24. López S. Acquired resistance affects male sexual display and female choice in guppies. *Proc R Soc London B: Biol Sci.* 1998;265:717–23.

25. Dingemanse NJ and Dochtermann NA. Quantifying individual variation in behaviour: Mixed-effect modelling approaches. *J Anim Ecol.* 2013;82:39–54.

26. Adelman JS, Mayer C, and Hawley DM. Infection reduces anti-predator behaviors in house finches. *J Avian Biol.* 2017;48:519–28.

27. Stephenson JF, Kinsella C, Cable J, and Van Oosterhout C. A further cost for the sicker sex? Evidence for male-biased parasite-induced vulnerability to predation. *Ecol Evol.* 2016;6:250–15.

28. Sue K. The science behind "man flu." *BMJ* 2017;359:j5560.

29. Pianka ER and Parker WS. Age-specific reproductive tactics. *Am Nat.* 1975;109:453–64.

30. Polak M and Starmer WT. Parasite-induced risk of mortality elevates reproductive effort in male *Drosophila*. *Proc R Soc London B: Biol Sci.* 1998;265:2197–201.

31. Copeland EK and Fedorka KM. The influence of male age and simulated pathogenic infection on producing a dishonest sexual signal. *Proc R Soc London B: Biol Sci.* 2012;279:4740–6.

32. Palacios MG, Winkler DW, Klasing KC, and Hasselquist D. Consequences of immune system aging in nature: A study of immunosenescence costs in free-living Tree Swallows. *Ecol.* 2011;92:952–66.

33. Zuk M and McKean KA. Sex differences in parasite infections: Patterns and processes. *Int J Parasitol.* 1996;26:1009–23.

34. Rolff J. Bateman's principle and immunity. *Proc R Soc London B: Biol Sci.* 2002;269:867–72.

35. Zuk M. The sicker sex. *PLoS Pathog.* 2009;5:e1000267.

36. López-Sepulcre A, Gordon SP, Paterson IG, Bentzen P, and Reznick DN. Beyond lifetime reproductive success: The posthumous reproductive dynamics of male Trinidadian guppies. *Proc R Soc London B: Biol Sci.* 2013;280:20131116.

37. Ruiz-Aravena M, Jones ME, Carver S, Estay S, Espejo C, Storfer A, *et al*. Sex bias in ability to cope with cancer: Tasmanian devils and facial tumour disease. *Proc R Socf Lond B: Biol Sci.* 2018;285:20182239.

38. Stephenson JF, Van Oosterhout C, and Cable J. Pace of life, predators and parasites: Predator-induced life-history evolution in Trinidadian guppies predicts decrease in parasite tolerance. *Biol Lett.* 2015;11:20150806.

39. Moore SL and Wilson K. Parasites as a viability cost of sexual selection in natural populations of mammals. *Science* 2002;297:2015–18.

40. Able DJ. The contagion indicator hypothesis for parasite-mediated sexual selection. *PNAS.* 1996;93:2229–33.

41. Aubert A, Goodall G, Dantzer R, and Gheusi G. Differential effects of lipopolysaccharide on pup retrieving and nest building in lactating mice. *Brain Behav Immun.* 1997;11:107–18.

42. Moyers SC, Kosarski KB, Adelman JS, and Hawley DM. Interactions between social behaviour and the acute phase immune response in house finches. *Behaviour* 2015:152:2039–58.

43. Cohn DWH and De Sá-Rocha LC. Differential effects of lipopolysaccharide in the social behavior of dominant and submissive mice. *Physiol Behav.* 2006;87: 932–7.

44. Hawley DM, Jennelle CS, Sydenstricker KV, and Dhondt AA. Pathogen resistance and immunocompetence covary with social status in house finches (*Carpodacus mexicanus*). *Funct Ecol*. 2007;21:520–7.

45. Reznick D. Costs of reproduction: An evaluation of the empirical evidence. *Oikos* 1985;44:257–67.

46. Bonneaud C, Mazuc J, Gonzalez G, Haussy C, Chastel O, Faivre B, et al. Assessing the cost of mounting an immune response. *Am Nat*. 2003;161:367–79.

47. Lopes PC, Adelman J, Wingfield, JC, and Bentley GE. Social context modulates sickness behavior. *Behav Ecol Sociobiol*. 2012;66:1421–8.

48. Lopes PC, Chan H, Demathieu S, González-Gómez PL, Wingfield JC, et al. The impact of exposure to a novel female on symptoms of infection and on the reproductive axis. *Neuroimmunomodulation* 2013;20:348–60.

49. Snyder-Mackler N, Burger JR, Gaydosh L, Belsky DW, Noppert GA, Campos FA, et al. Social determinants of health and survival in humans and other animals. *Science* 2020;368:368:eaax9553.

50. Stroeymeyt N, Grasse AV, Crespi A, Mersch DP, Cremer S, and Keller L. Social network plasticity decreases disease transmission in a eusocial insect. *Science* 2018;362:941–5.

51. Shakhar K and Shakhar G. Why do we feel sick when infected—can altruism play a role? *PLoS Biol*. 2-15;13:e1002276.

52. Stockmaier S, Bolnick DI, Page RA, and Carter GG. Sickness effects on social interactions depend on the type of behaviour and relationship. *J Anim Ecol*. 2020;1365–2656:13193–218.

53. Poirotte C and Charpentier MJE. Unconditional care from close maternal kin in the face of parasites. *Biol Lett*. 2020;16:20190869.

54. Budischak SA and Cressler CE. Fueling defense: Effects of resources on the ecology and evolution of tolerance to parasite infection. *Front Immunol*. 2018;9:2453.

55. Zylberberg M, Klasing KC, and Hahn TP. House finches (Carpodacus mexicanus) balance investment in behavioural and immunological defences against pathogens. *Biol Lett*. 2013;9:20120856.

56. Roznik EA, Sapsford SJ, Pike DA, Schwarzkopf L, and Alford RA. Condition-dependent reproductive effort in frogs infected by a widespread pathogen. *Proc R Soc London B: Biol Sci*. 2015;282:20150694.

57. Jacot A, Scheuber H, and Brinkhof MWG. Costs of an induced immune response on sexual display and longevity in field crickets. *Evol*. 2004;58:2280–6.

58. Onishi KG, Maneval AC, Cable EC, Tuohy MC, Scasny AJ, Sterina E, et al. Circadian and circannual timescales interact to generate seasonal changes in immune function. *Brain Behav Immun*. 2020;83:33–43.

59. Westneat DF, Araya-Ajoy YG, Allegue H, Class B, Dingemanse N, Dochtermann NA, et al. Collision between biological process and statistical analysis revealed by mean centering. *J Anim Ecol*. 2020; 89(12):2813–2824.

60. Martin JGA, Nussey DH, Wilson AJ, and Réale D. Measuring individual differences in reaction norms in field and experimental studies: A power analysis of random regression models: Power analysis of random regression models. *Methods Ecol Evol*. 2011;2: 362–74.

61. Hadfield JD, Wilson AJ, Garan, D, Sheldon BC, and Kruuk LEB. The misuse of BLUP in ecology and evolution. *Am Nat*. 2010;175:116–25.

62. Nakayama S, Laskowski KL, Klefoth T, and Arlinghaus R. Between- and within-individual variation in activity increases with water temperature in wild perch. *Behav Ecol*. 2-16;27:676–83.

63. Adelman JS, Moyers SC, Farine DR, and Hawley DM. Feeder use predicts both acquisition and transmission of a contagious pathogen in a North American songbird. *Proc R Soc London B: Biol Sci*. 2015;282:20151429.

64. Henschen AE and Adelman JS. What does tolerance mean for animal disease dynamics when pathology enhances transmission? *Integr Comp Biol*. 2019;59: 1220–30.

65. Ezenwa VO, Archie EA, Craft ME, Hawley DM, Martin LB, Moore J, et al. Host behaviour-parasite feedback: An essential link between animal behaviour and disease ecology. *Proc R Soc London B: Biol Sci*. 2016;283:20153078.

66. Poirotte C and Charpentier MJE. Inter-individual variation in parasite avoidance behaviors and its epidemiological, ecological, and evolutionary consequences. In Ezenwa VO, Altizer S, and Hall RJ (eds), *Animal Behavior and Parasitism*. Oxford: Oxford University Press, 2022.

67. Lopes PC, French SS, Woodhams DC, and Binning SA. Infection avoidance behaviors across vertebrates. In Ezenwa VO, Altizer S, and Hall RJ (eds), *Animal Behavior and Parasitism*. Oxford: Oxford University Press, 2022.

68. Pirrie A, Chapman H, and Ashby B. Parasite-mediated sexual selection: To mate or not to mate?. In Ezenwa VO, Altizer S, and Hall RJ (eds), *Animal Behavior and Parasitism*. Oxford: Oxford University Press, 2022.

69. De Roode JC, Pansini R, Cheesman SJ, Helinski MEH, Huijben S, Wargo AR, et al. Virulence and competitive ability in genetically diverse malaria infections. *PNAS*. 2005;102:7624–8.

70. Balard A, Jarquín-Díaz VH, Jost J, Mittné V, Böhning F, Ďureje L', et al. Coupling between tolerance and

resistance for two related *Eimeria* parasite species. *Ecol Evol.* 2020: 13938–48.

71. Restif O and Koella JC. Concurrent evolution of resistance and tolerance to pathogens. *Am. Nat.* 2004;164:E90–102.

72. Best A, White A, and Boots M. Maintenance of host variation in tolerance to pathogens and parasites. *PNAS.* 2008;105:20786–91.

73. Sears BF, Rohr JR, Allen JE, and Martin LB. The economy of inflammation: When is less more? *Trends Parasitol.* 2011;27:382–7.

74. Råberg L, Graham AL, and Read AF. Decomposing health: Tolerance and resistance to parasites in animals. *Philos Trans R Soc Lond B: Biol Sci.* 2009;364:37–49.

75. Dantzer R, O'Connor JC, Freund GG, Johnson RW, and Kelley KW. From inflammation to sickness and depression: When the immune system subjugates the brain. *Nat Rev Neurosci.* 2008;9:46–56.

76. Martin LB, Weil ZM, and Nelson RJ. Fever and sickness behaviour vary among congeneric rodents. *Funct Ecol.* 2008;22:68–770.

77. Van Noordwijk AJ and De Jong G. Acquisition and allocation of resources: Their influence on variation in life history tactics. *Am Nat.* 1986;128:137–42.

78. Franz M, Kramer-Schadt S, Greenwood AD, and Courtiol A. Sickness-induced lethargy can increase host contact rates and pathogen spread in water-limited landscapes. *Funct Ecol.* 2018;32:2194–204.

# Emerging Frontiers

# CHAPTER 18

# Emerging frontiers in animal behavior and parasitism: Integration across scales

Sarah Guindre-Parker, Jenny Tung, and Alexander T. Strauss

## 18.1 Introduction

Animal behavior and parasitology are discrete fields with their own professional societies, publishing venues, and intellectual histories. And yet, as the chapters in this volume demonstrate, there is inherent synergy in linking these two disciplines (1). The behavior of animals in their natural environment influences infection, disease progression, and transmission rates. Meanwhile, parasites shape the external environment of their animal hosts and influence animal physiology, energetics, and behavior (2), (3). Because of the need to consider biological processes across scales—from subcellular changes in immune response mechanisms to trophic cascades across species—research at the interface between animal behavior and parasitology is therefore inherently interdisciplinary.

Here, we build on this tradition to highlight further opportunities in which methods or conceptual frameworks from allied fields can push forward the frontiers of research on animal behavior and parasitism (Figure 18.1). We focus specifically on the three focal areas of our own research: organismal biology, molecular ecology and genomics, and community and ecosystem ecology. These areas are closely linked to animal behavior and parasitism because the behavior of animals profoundly impacts genetic structure and demography as well as their interactions with other species in communities and ecosystems. In turn, host genetics and

energetics influence the physiological response of a host to parasite infection, and demographic and community structure define the ecological context in which hosts and parasites interact. Thus, not only can ideas from these fields catalyze research on behavior–parasite interactions, but studying the relationship between animal behavior and parasitism should yield reciprocal advances in each of these fields.

In this chapter we identify key points of intersection between research in animal behavior and parasitism and each of the three highlighted disciplines, with a particular focus on novel stressors, species assemblages, and environments associated with the Anthropocene. While these intersections bridge from methodological to conceptual, many relate to themes that emerge repeatedly in this volume. For instance, Chapter 7 and Chapter 17 point out that hosts are often infected by multiple species (4), (5), highlighting the importance of studying coinfection (i.e., when hosts are infected by a pathogen/parasite community, which may generate non-linear interactions that affect host outcomes ((6); Box 18.1). Advances in genomic technology allow us to quantify coinfection more precisely, while theory from community ecology gives us the terms and models necessary to predict its consequences. Meanwhile, prior history of infection may play a crucial role in understanding interindividual variation in susceptibility to subsequent parasite invasion.

Sarah Guindre-Parker, Jenny Tung, and Alexander T. Strauss, *Emerging frontiers in animal behavior and parasitism*. In: *Animal Behavior and Parasitism*. Edited by Vanessa O. Ezenwa, Sonia Altizer, and Richard J. Hall. Oxford University Press. © Oxford University Press (2022). DOI: 10.1093/oso/9780192895561.003.0018

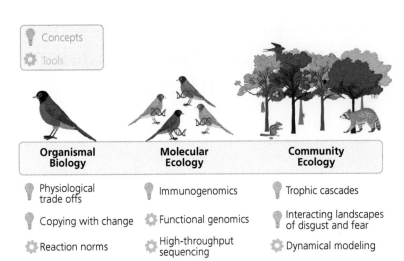

**Figure 18.1** A summary of the key concepts and tools from organismal biology, molecular ecology, and community ecology that can help transform our understanding of animal behavior and parasitism. Figure made in BioRender.

---

**Box 18.1 Coinfection: What it is, and why does it matter?**

Coinfection refers to the simultaneous infection of a single host individual by two or more parasite species or strains. Parasites are often studied in isolation, either through controlled inoculation in laboratory environments or by focusing on a particular species from field samples. In reality, hosts are frequently coinfected (7)–(10). Importantly, singly infected versus coinfected hosts can behave differently (11), (12), express different phenotypes (13), (14), and respond differently to environmental gradients (15). At the organismal level, coinfected hosts often exhibit more severe disease symptoms than singly infected hosts (8), (13). For example, mortality from tuberculosis is well-known to be much higher for humans infected with Human Immunodeficiency Virus (HIV) (14). Genomic profiling (e.g., a microbiome obtained from high-throughput sequencing) now shows that individual animals often host hundreds if not thousands of species of eukaryotic parasites and single-celled microbes in any given niche, necessitating a new set of tools and concepts to grapple with the complexity of their interactions (16). From a parasite community ecology perspective, infection by one species can mediate the risk of infection by others (7), (13), (17), potentially via altered host behavior. For example, risk of infection by strongyle nematodes in Barbary macaques (*Macaca sylvanus*) increases with both concurrent infection by helminths and interactions with social partners (17). Meanwhile, feline immunodeficiency virus infection in African lions (*Panthera leo*) exacerbates host aggressive behavior, potentially increasing their risk of secondary infections (13). Thus, grappling with the drivers and consequences of coinfection represents a frontier of research in parasitism and animal behavior at multiple scales of biological organization.

---

## 18.2 Organismal biology

Animal behavior and parasites quite naturally integrate into organismal biology, which is the study of structure and function at the level of the individual. Both behavior and parasitism play a role in shaping the way host individuals experience their environment and in shaping fitness outcomes (as discussed in greater detail in Chapter 2 of this book: (3)). Here, we highlight emerging research opportunities that fall within three broad concepts: complex trade-offs, phenotypic plasticity, and anthropogenic challenges. These concepts are closely intertwined with existing research in organismal biology, animal behavior, and disease ecology and evolution, but nonetheless remain priority areas for future work.

### 18.2.1 Trade-offs

Organismal form and function are shaped by complex selective pressures that often result in trade-offs among traits (18). Trade-offs are foundational to host–parasite interactions: hosts can trade off parasite exposure risk with foraging efficiency (19), hosts can trade off immune defenses for sexually selected ornaments (20), and parasites can trade off virulence—defined as the harm the parasite inflicts on the host—and transmission rate with host survival and period of infectiousness (21). Less is understood from the host perspective about trade-offs at play during parasite coinfections (22), including how host behavior is affected by mixes of beneficial and harmful symbionts. In brown rats (*Rattus norvegicus*), for example, coinfection with multiple helminth parasites alters anxiety and motor behavior differently than animals infected by a single parasite (12). Recent work indicates that favorable microbiome composition can mitigate infection risk by more pathogenic species, suggesting substantial scope for the entire within-host community of symbionts to shape host behavioral responses (23).

Hosts must cope with a diversity of stressors, including not only parasites but also social conflict, predators, harsh weather, and fluctuating food availability. Exploring how diverse physiological systems trade off with one another, as well as how parasitism shifts these trade-offs (15), would improve our understanding of the mechanisms that underlie the interplay between parasitism and behavior. For example, animals that mount stronger glucocorticoid stress responses may be better equipped to cope with other environmental stressors. Because of the immunomodulatory effects of glucocorticoids, however, these hosts may sacrifice some ability to resist pathogens and parasites (24). Future experimental work should explore how glucocorticoids and distinct components of host immunity interact with one another to alter host behavior and the consequences of parasitism (25), especially under changing environments.

## 18.2.2 Phenotypic plasticity

Phenotypic plasticity—the ability of one individual (or genotype) to adjust its phenotype under different environmental conditions—is key to predicting rapid organismal responses to novel conditions (26). Plasticity is best studied via reaction norms, which are functions used to describe or visualize changes in phenotypes when an individual is exposed to a range of environmental conditions (27). Reaction norms help us to improve quantification of key strategies for coping with pathogens such as tolerance to infection, where hosts prioritize minimizing the fitness costs of infection, rather than avoiding or preventing the infection in the first place. Rather than thinking about host behavioral tolerance as a binary trait (e.g., a tolerant host versus an intolerant one), behavioral tolerance could be more flexibly defined as the slope of a reaction norm between pathogen load and fitness-enhancing behaviors (5).

While reaction norm perspectives are increasingly used in work at the intersection of animal behavior and parasitism for both hosts and parasites (5), (28), (29), this framework has further scope to improve our understanding of parasite–behavior feedbacks: when host behavior drives parasitism and parasites reciprocally shape host behavior. For example, reaction norm approaches have demonstrated that environmental variation in parasite density shapes susceptibility, transmission, and infection rates across different genotypes of zooplankton hosts (30), and that not all individual hosts show the same behavioral adjustments in response to increasing risk of parasitism (31). Quantifying variation in the plastic behavioral responses of different host individuals, populations, or species could in turn facilitate a better understanding of the causes of this variation (e.g., past infection or coinfection status, genotype, early life environment, immune defenses) and its fitness consequences (Box 18.2). Indeed, considering individual variation among hosts improves epidemiological models for many diseases (32). Reaction norms would provide one approach to quantifying this host variability and would improve our understanding of behavior–parasite feedbacks (28), (33). For example, house finches (*Haemorhous mexicanus*) foraging at more urban sites show elevated poxvirus and coccidial infection prevalence relative to rural birds (34), and extending a reaction norm approach could show whether birds with elevated parasite loads also show plasticity in their foraging behavior. We note that organismal biologists could also benefit from incorporating parasite infection status, loads, or gradients of exposure into their work on phenotypic plasticity: parasitism is one possible cause of variation in behavioral plasticity among host individuals that remains poorly understood (31), (35).

**Box 18.2 Reaction norm approaches to better understand behavior–parasite plasticity**

Reaction norms can be used to explore host behavioral responses to environmental conditions, including parasite densities. Figure 1 depicts an empirical example of plasticity in behavior—in this example, foraging behavior—for individual zooplankton hosts exposed to a density gradient of a trophically transmitted parasite (*Metschnikowia bicuspidata*); adapted from Strauss *et al* (31). Each black line shows the response of one of 19 *Daphnia dentifera* genotypes, with three genotypes displayed in color to highlight different patterns of behavioral plasticity. Genotype 1 (green) shows constant foraging behavior despite variation in parasite densities, genotype 2 (blue) shows a slight linear decrease in foraging behavior at elevated parasite densities, and genotype 3 (orange) shows a steep exponential decline in foraging behavior with increasing parasite density. Genotype 3 shows the greatest behavioral plasticity, as indicated by a steeper slope of the reaction norm function.

In addition to their role in characterizing behavior-parasite relationships in a quantitative manner (Figure 2A), reaction norm approaches can be used to understand the causes and consequences of variation across individuals or genotypes. As an example, the behavioral plasticity of each animal may be explained by variation in host immune defenses, where behavior and immune defense act as two alternative tactics for coping with high risk of parasitism (Figure 2B). In this theoretical example, individual 1 (orange) shows the greatest behavioral plasticity under elevated parasite density but might have the weakest immune defenses, whereas individuals 3 (green) and individual 2 (blue) might show weaker behavioral plasticity because they can rely on their immune systems under elevated risk of parasitism. In addition, reaction norm approaches can help to understand the consequences of variation in plasticity for host fitness (Figure 2C). Here, greater behavioral plasticity in individual 3 (orange) in response to increasing parasite densities could result in elevated fitness via avoiding costs of parasitism. Such a scenario would be expected to result in selection for increased behavioral plasticity.

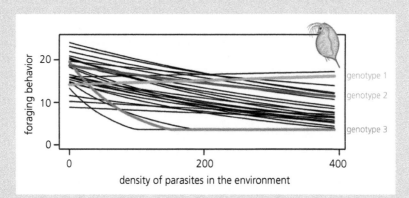

**Box 18.2, Figure 1** Plasticity in *Daphnia dentifera* host foraging rate in response to parasite density. Foraging rate in these non-selective filter feeders is measured as the change in concentration of algal food before and after foraging (expressed in units of ml per day). Empirical figure adapted from Strauss *et al*. (31) with *Daphnia* image from BioRender.

**Box 18.2** *Continued*

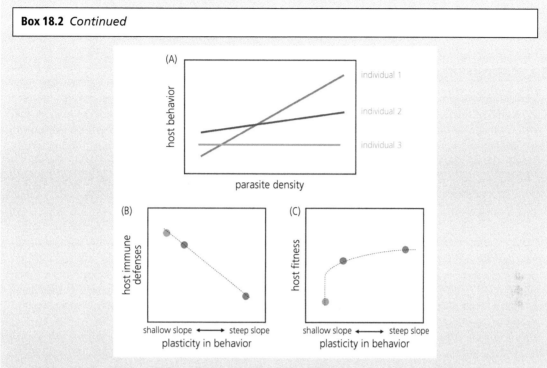

**Box 18.2, Figure 2** Conceptual reaction norms (A) that quantify behavioral plasticity in response to parasite densities, and possible extensions of this approach to uncover the causes (B) and consequences (C) of individual variation in plasticity.

### 18.2.3 Anthropogenic challenges

The Anthropocene has introduced substantial changes to the environmental conditions faced by parasites, vectors, and hosts alike. Organisms from urban populations often show distinct, repeatable differences in their behaviors compared to their rural counterparts, including bolder, more exploratory, and less neophobic behavior. Exploring the consequences of these behavioral differences for parasitism across an urban to rural gradient has already provided new insight into the role of urbanization in altering behaviorally mediated host–parasite dynamics. For example, bat communities in Brazil take advantage of human-built structures in urban spaces for roosting at high densities, which could increase parasite transmission (36). Similarly, a meta-analysis found that some urban mammalian hosts were less likely to be infected by parasites with complex lifecycles relative to rural hosts, suggesting that urbanization can disrupt parasite transmission in taxa with lifecycles that require multiple hosts (37).

Abiotic conditions are also changing in many parts of the world and their effects on hosts, vectors, parasites, and their interactions are of increasing interest (38). Predicting how climate change will impact arthropod disease vectors such as mosquitoes and in turn shape disease transmission requires an understanding of their behavior (e.g., behavioral thermoregulation) (39). Warming temperatures alter the physiology and development of individual parasites, vectors, or hosts (40). Researchers have also found an increase in chytridiomycosis in amphibians under fluctuating temperatures compared to animals experiencing constant conditions (41). An important future direction in forecasting how climate change will alter parasitism will require understanding how both changes in temperature paired with increasing variability in temperatures shape behaviorally mediated host–parasite dynamics (42).

## 18.3 Molecular ecology and genomics

Molecular approaches have a long history in the study of animal behavior and parasitism, as they have the potential to shed light on the precise targets of selection that shape parasite invasion strategies, host defenses, and their consequences for behavior (43). In doing so they link individual-level variation in behavior and parasite/pathogen susceptibility (a focus of the organismal perspective outlined earlier) to population-level variation, including the reasons that populations and species differ in their reactions to parasites. Recent advances in genomics and molecular ecology—especially those built on high-throughput sequencing—now have the potential to push this frontier further by contributing to key outstanding questions about coinfection, parasite-mediated sexual selection, and genetic contributions to the immune response highlighted in this volume (3), (43)–(45).

### 18.3.1 Technological advances in profiling parasite/pathogen infection

At the simplest level, genomic research has the potential to accelerate research on parasite or pathogen presence, abundance, and diversity because it can circumvent requirements for species-specific assay or culture conditions. By combining species-agnostic polymerase chain reaction (PCR) amplification of molecular "barcodes" (e.g., hypervariable regions of 16S rRNA genes in bacteria, 18S rRNA genes in eukaryotes, or the ribosomal internal transcribed spacer (ITS) region in fungi) with high-throughput sequencing, or by sequencing the entire contents of a sample (i.e., metagenomic shotgun sequencing) it is now possible to rapidly profile the presence and abundance of parasites and microbes—including bacteria, archaea, and viruses—in host samples (46)–(48). Such approaches have already established sequencing methods as the default technology for studying microbiomes, enabling hundreds to thousands of

samples to be processed in a single study (e.g., (49), (50)). The results of this work have begun to reveal both the importance of host behavior in shaping microbial communities and the types of microbes that are most strongly tied to behavior (51), (52). For instance, sequencing-based methods have demonstrated that microbial social transmission is common in many animals, from bees to great apes (53). In a few cases, they have also revealed its functional consequences. For example, in bumble bees, socially transmitted gut microbiota are essential to host defense against the trypanosomatid parasite *Crithidia bombi*, which can otherwise reduce queen fitness by up to 50% (54).

Increased application of high-throughput sequencing methods to surveying parasite abundance may therefore help us understand and quantify "what's there"—including what conventional methods for profiling parasite abundance have missed. For example, in recent surveys of nonhuman primate species, 18s rRNA barcoding identified a median of 8.5 parasite families more than previously described (55). And in an overlapping set of primate species, metagenomic shotgun sequencing revealed viral bacteriophage communities that exhibit phylogenetic structure between species and evidence for social transmission within species (55). While descriptive, this information is foundational for addressing outstanding questions about coinfection dynamics, parasite sharing among host species, and how these processes influence, and are influenced by, host behavior. Indeed, by exhaustively introducing all combinations of the five core gut microbes found in the *Drosophila* gut to germ-free flies, one study demonstrated extensive pairwise, three-way, four-way, and five-way interactions between microbial taxa that influence both microbial abundance and host life history (56). Similarly, another study showed that combinations of yeast and bacteria in the fruit fly gut produce non-linear effects on host egg-laying behavior (57). While such studies take advantage of the relatively simple, experimentally tractable fruit fly gut microbiome, they provide important motivation for describing the complete microbial community in other species—a task that will depend heavily on genomic analysis.

## 18.3.2 Genetic contributions to immune defense

Genomic approaches can also provide improved resolution on host genetic variation. As highlighted in this volume, if mate choice involves optimization of genetic benefits, the pool of potential mates must harbor fitness-associated genetic variation, most likely in immune genes (43), (45). To date, studies of reproductive success and mate choice have overwhelmingly focused on the role of diversity and/or dissimilarity in a small subset of genes, especially those in the vertebrate major histocompatibility complex (MHC) (43). Results from this work provide some support for pathogen/parasite-mediated sexual selection (i.e., the hypothesis that animals, usually females, choose mates based on heritable traits that signal improved pathogen or parasite resistance (58)). However, it remains unclear whether these candidate genes *causally* affect mate choice, or whether they are simply markers for correlated genetic diversity across the genome as a whole (59). Notably, work in humans suggests that the genetic architecture of immune traits rarely involves one or a few loci of large effect (with the notable exception, in some cases, of genes in the MHC) (60). Expanding genetic analyses of pathogen-mediated sexual selection to the full genome therefore constitutes a crucial next step. Importantly, such work will not only yield more robust tests of the role of genes that have been well-studied thus far, but also enable new discovery—including whether traits that signal "good genes" are as polygenic in nature as other complex traits.

Genomic approaches can also help identify the genetic variants that contribute to differences in immune function. Studies of pathogen/parasite-mediated sexual selection generally do not distinguish between functionally silent genetic variants, which have no ability to influence fitness, and those that can affect susceptibility, resistance, or tolerance; that is, the subset of variants capable of conferring genetic benefits. Functional genomic approaches, which probe gene and protein function beyond the DNA sequence level, have potential to open this black box. For example, recent studies in humans have demonstrated how genotype information can be combined with *ex vivo* experimental pathogen challenges to reveal the genetic determinants of the gene expression response to infection. Variants that predict this response ("response expression quantitative trait loci," or "reQTL") are, in turn, enriched for signatures of recent natural selection (61), (62) (Box 18.3). Because such studies have not included a behavioral component, it remains unclear whether mate choice and sexual selection contribute to these signatures. However, this methodology is highly generalizable to other species, including in settings where behavioral data can also be brought to bear (e.g., (63)). Consequently, it may soon become possible to refine analyses of mate choice, gene flow, and immunogenetic diversity to focus specifically on the variants that matter to immunocompetence.

Finally, a recurring theme in this volume involves the mutual entanglement between hosts and parasites, which can give rise to parasite-behavior feedbacks (3). In as much as these changes involve shifts in mating behavior, migration patterns, or the relative survival of different genotypes, they will also determine how genetic variation is structured within and between host and parasite populations. Indeed, recent work in *Drosophila* has demonstrated that short-term changes in selection pressure can be detected by monitoring cyclical changes in allele frequencies across seasons (64), (65). This example focuses on cyclical abiotic selection pressures. However, studies of behaviorally mediated host–parasite feedback loops present an exciting opportunity to quantify, in parallel, effects on host and parasite population genetic variation, a key aspect of understanding the consequences of pathogen-mediated fluctuating selection in nature.

## 18.4 Community and ecosystem ecology

Much of this volume has focused on individual parasite taxa and their impacts on the behaviors of individual host taxa, and vice versa. In reality, as outlined earlier, most hosts are coinfected by many symbionts (Box 18.1). Simultaneously, hosts are embedded in ecological communities where they interact with their resources, their competitors, and their predators. All of these species' interactions can alter host behavior in ways to cascade to shape disease (Table 18.1). Furthermore, these communities are embedded in ecosystems, where parasites and hosts can both influence and be influenced by fluxes of elements and energy. Several research

## Box 18.3 Identifying genetic contributions to gene regulatory variation in the immune response

When animals encounter parasites, pathogens, or molecular signals that indicate infection, their cells respond by rapidly mobilizing an immune response. This process involves coordinated changes in cell type-specific gene regulation—gene expression, chromatin accessibility, and a variety of epigenetic marks—that are often detectable within hours (66), (67). Understanding sources of variation in the gene regulatory response to pathogens can therefore provide insight into both the genetic and environmental determinants of variation in host immune function, including the role of social behavior (68).

Importantly, gene regulatory phenotypes are also highly amenable to trait mapping (69), (70). Thousands of "expression quantitative trait loci" (eQTL) have been mapped in human tissues (71), each of which point to an individual genetic variant associated with quantitative variation in gene expression levels (Box Figure 1A). Because of relatively modest sample size demands compared to mapping organism-level traits, similar approaches have also been successfully applied in animal populations (63), (72), (73). The result is a catalogue of gene-genetic variant pairs that can be used to study the evolutionary history of gene regulation, prioritize variants for mechanistic studies, or predict the genetic contribution to organism-level traits.

Because the presence and magnitude of eQTL (as well as other types of regulatory QTL) often depend on cellular state (70), eQTL mapping can also be used to understand how host genotypes influence the response to immune stimuli. "Response" eQTL studies (abbreviated reQTL, or sometimes called "dynamic eQTL") typically test cells from each study subject in baseline and stimulated conditions to identify cases where the response to the stimulant is stratified by genotype (Box Figure 1B). reQTL in humans have already been shown to influence the response to tuberculosis, influenza, *Listeria*, *Salmonella*, and a variety of pathogen-associated molecular patterns (PAMPs) such as bacterial lipopolysaccharide (62), (74)–(76).

Notably, reQTL carry signatures of selection between populations and between humans and other species, pointing to their importance in host-pathogen evolution (61), (62). Hosts that carry different genotypes might therefore engage in different parasite/pathogen avoidance behavior (as highlighted in section 18.2 of this chapter). Or, sexual selection on traits that signal immunocompetence might be linked to reQTL that influence the immune response. Integrating reQTL studies with rigorous analyses of behavior and parasite/pathogen loads represents one potential avenue for testing these ideas.

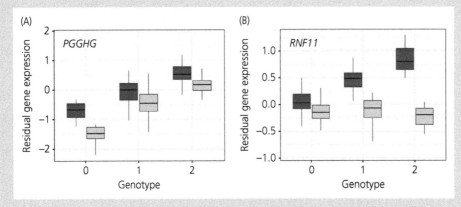

**Box 18.3, Figure 1** (A) Example eQTL for the gene *PGGHG*, where immune stimulation does not alter the effect of genotype (x-axis) on gene expression. (B) Example reQTL for *RNF11*, a gene involved in terminating NFkB-mediated inflammatory signaling (NFkB is a transcription factor complex often referred to as a "master regulator" of inflammation and other innate immune processes). There is no detectable effect of genotype in the baseline condition, but a clear difference in the responsiveness of different genotypes after stimulation with bacterial lipopolysaccharide (LPS), which strongly induces NFkB activity. In both panels, genotype is represented as 0/1/2 based on the count of alternative alleles in that genotype, and gene expression is shown after controlling for batch and other technical effects. Data in both examples are from a natural population of baboons in the Amboseli ecosystem of Kenya (Anderson and Tung, unpublished data).

frontiers and opportunities emerge when parasite–behavior feedbacks are viewed through the lenses of community and ecosystem ecology.

## 18.4.1 Within-host communities

As discussed earlier in this chapter, coinfections are important from an organismal perspective, and genomic advances now enable us to more easily describe the diversity of parasites and other symbionts inside hosts. A community ecologist might then ask, how does host behavior shape the assembly of symbiont communities and maintenance of their diversity? Hosts live in "landscapes of disgust," where the decisions they make are influenced by their perception of parasites in the environment and their risk of infection (19). If parasites co-occur in microhabitats or share transmission modes, avoidance of one parasite could protect a host from others (19), (77), (78). For example, social distancing in response to the COVID-19 pandemic has indirectly slowed transmission of influenza (79), (80). On the other hand, if different parasites are transmitted via different pathways (9), or occupy different microhabitats in a "landscape of disgust," avoiding one parasite could come at the cost of increased exposure to another. For example, vervet monkeys with more grooming partners may benefit from lower risk of ectoparasite infection, but also suffer from higher rates of infection by hookworms, likely due to more close contact with conspecifics (81). Thus, parasites can inhibit or facilitate one another via changes in host behavior.

Parasite manipulation of host behavior can also introduce antagonism or facilitation among parasites. For most horizontally or vertically transmitted parasites, predation of their host is a dead end for transmission; yet for trophically transmitted parasites, it is required. Coinfection by both types of parasites could therefore create conflict, with each parasite benefiting from opposing behavioral changes that make the host more versus less susceptible to predation (82). Finally, parasites can also alter host behavior in ways that benefit one another and promote coinfection. For example, infection by one virus can induce changes in squash plants that attract aphid vectors capable of transmitting

additional viruses (83). Thus, host and vector behaviors introduce a suite of indirect pathways for parasites to antagonize or facilitate one another—making coinfection more or less likely. Such behaviors could emerge as critical forces shaping the assembly of parasite communities.

## 18.4.2 Species interactions

Interactions of hosts or vectors with their predators, competitors, or resources can be critical drivers of behavior-parasite feedbacks ((3); Table 18.1). Such feedbacks would require (i) that parasite-mediated behaviors of hosts or vectors alter their cost of competition or risk of predation, and (ii) that predators or competitors impact the likelihood of infection. For example, trophic transmission and behavioral manipulation of intermediate hosts would satisfy these requirements. How might predators alter transmission of other parasites via changes in host behavior? Tadpoles behaviorally avoid trematode parasites when possible, but in the presence of predator cues, they prefer to risk infection rather than predation (84). Aphids transmit pea enation mosaic virus (PEMV) among pea plants, but exposure to predator cues alters aphid preference and reduces transmission efficiency (85). These examples illustrate that predators can induce behavioral changes that increase or decrease transmission. Analogous results may emerge from competitive interactions. Just as a landscape of disgust can be integrated within a landscape of fear (i.e., perception of the risk of predation over space) (86), so too could it be integrated within a "landscape of strife." If hosts compete for resources or habitat use, their perception of risk and reward across a landscape could depend on resource availability, the severity of competition with heterospecifics, the degree of overlap in their parasite communities, and the potential for cross-species transmission (87), (88). Despite the ubiquity of competition in ecological communities, its potential role in mediating behavioral responses to parasitism has received little attention.

Direct interactions with resources can also mediate hosts' decisions to avoid parasites or risk infection. In consumer-resource dynamics, the

**Table 18.1** Species interactions induce changes in host or vector behavior that unleash or inhibit infectious disease

| Species interaction | Focal species | Interacting species | Altered behavior of focal species | Disease outcome | Citation |
|---|---|---|---|---|---|
| Host–parasite | Humans | Parasite: SARS-CoV-2 | Social distancing in response to SARS-CoV-2 | Lower incidence of influenza strains A and B due to social distancing | (80) |
| | Vervet Monkey (*Chlorocebus aethiops*) | Parasite: Ectoparasites | Grooming behavior to reduce ectoparasite burden | Greater likelihood of infection by hookworm with more grooming partners | (81) |
| Predator-Prey | Pea Aphid (*Acyrthosiphon pisum*) | Predator: Lady beetle | Altered foraging behavior, including probing, insertion, salivation, or ingestion with predator cues | Decreased transmission of PEMV from aphids to pea plants with altered foraging behavior | (85) |
| | Northern Leopard Frog (*Lithobates pipiens*) tadpoles | Predator: Dragonfly larvae | Decreased foraging activity in presence of predator cues | Greater exposure to trematode parasites with altered foraging decisions | (84) |
| Competition | North American deer mouse (*Peromyscus maniculatus*) | Competitor: Voles and Shrews | Greater contact rates among conspecifics with higher densities of dominant competitors | Greater transmission of Sin Nombre hantavirus with increased conspecific contacts | (87) |
| | Eurasian Red Squirrel (*Sciurius vulgaris*) | Competitor: Eastern Gray Squirrels (*Sciurus carolinensis*) | Decreased levels of activity in presence of invasive gray squirrels | More infection by helminths with less activity | (88) |
| Consumer-Resource | Elk (*Cervus elaphus*) | Resource: natural forage or hay | Greater aggregation in response to anthropogenic resource subsidies | Higher seroprevalence of brucellosis associated with aggregations | (90) |
| | Zooplankton (*Daphnia dentifera*) | Resource: Algae | Higher concentrations of algae cause slower foraging rates (functional response) | Reduced transmission of fungal parasites with slower foraging rates | (89) |

"functional response" of a consumer describes its foraging behavior over a gradient of the abundance of its resource. For example, with a type-II functional response, higher abundances of resources cause consumers to decrease their foraging rates, because consumers can more easily obtain the resources they need. When hosts encounter parasites while foraging, higher abundance of resources can therefore slow host foraging and inhibit transmission (89). On the other hand, higher concentration of resources in a specific location (e.g., anthropogenic resource subsidies such as birdfeeders, dumpsters, or feedlots) can accelerate transmission of directly transmitted parasites by inducing aggregating behavior and contacts for groups of hosts (90). Thus, resources, competitors, and predators can all induce behavioral changes in hosts that promote or inhibit disease (Table 18.1).

### 18.4.3 Ecosystem dynamics

Finally, parasite–behavior feedbacks can manifest at the ecosystem scale. Indeed, host manipulation is increasingly recognized as an important regulator of the flow of trophic energy through ecosystems (91). Effects of parasite–mediated behavior on ecosystem function are most likely when common parasites strongly affect hosts with important roles in ecosystems (92). These roles may often include consumption of primary producers or detritus. Consumption rates can either increase with parasitism due to compensatory feeding (93) or increased boldness in the presence of predators (94), or decrease with parasitism due to costs of infection (95) or sickness behaviors (96). For example, trematode infections slow the per-capita grazing of keystone snail hosts, which promotes the resistance of salt marshes to drought and slows habitat loss of this ecosystem (95). Fungal parasites also slow per-capita foraging of infected zooplankton (31), (97). This parasite-mediated behavior releases algae from top-down grazer control, and—with a surprising twist—increases the density of hosts, despite ongoing epidemics of the infectious fungus, due to elevated primary productivity (97). In both these cases, infection in dominant herbivores slows their consumption of primary

producers and triggers bottom-up trophic cascades. In turn, parasites are likely to benefit from the cascading ecosystem changes (e.g., realized as increased ecosystem resilience (95) or higher host density (97)). While these ecosystem-scale feedbacks are triggered by particular parasites, their consequences have the potential to impact many other species that reside in the same ecosystem. Such ecosystem-scale feedbacks remain an exciting emerging research frontier.

## 18.5 Conclusions, with an eye to the future

This chapter highlights exciting research frontiers that integrate methodological and conceptual ideas from adjacent fields to derive new insight into host behavior–parasite interactions. We emphasize that nearly all topics covered in this volume and this chapter are sensitive to another area in which interdisciplinary research is critical: anthropogenic environmental change (98). From the level of individual hosts or parasites to allelic changes in populations, to communities and ecosystems, anthropogenic changes are certain to alter behavior–parasite interactions. For example, drought could warp risk-reward relationships in a landscape of disgust (19) and alter patterns of animal migration (4). Urbanization introduces an entirely new set of challenges and opportunities for populations of hosts and their parasites, including elevated thermal stress, increased anthropogenic resource subsidies, altered social behavior, and restricted gene flow. Among-population comparisons along gradients of anthropogenic disturbance provide rich opportunities to explore the interplay between parasites and host behavior in a changing world. Warmer temperatures could accelerate parasite development time and intensify their impact on host foraging, while eutrophication and biodiversity loss are restructuring food webs and the roles of parasites in them. Finally, increasingly extreme or fluctuating weather could differentially alter the physiology of parasites, vectors, or hosts. In some cases, global changes will drive directional evolutionary responses, and in others they will select for greater phenotypic plasticity. As the world experiences accelerating and multifaceted environmental change, the intersection of animal behavior and parasitism will overlap with an increasingly broad set of disciplines, expanding its own research frontier (Box 18.4).

---

**Box 18.4 New opportunities for research on animal behavior and parasitism across scales**

Here, we summarize directions in research on behavior and parasitism that draw on the intersection of disciplines discussed in this chapter.

- *Genomic approaches* will increasingly allow us to characterize the full parasite/symbiont community of animal hosts, facilitating studies on how coinfections or symbiont communities shape host behavior and vice versa.
- *Research at the organismal level* will reveal how infection and coinfection shape host behavioral reaction norms, lead to trade-offs with other physiological systems, and influence host capacity to cope with multiple stressors.
- *Studies that disentangle the environmental and genetic predictors* of host immunity will allow us to test hypotheses about infection and behavior that assume immune trait heritability, and investigate immunological interactions or trade-offs against physiology and behavior.
- *Emergent community or ecosystem-level dynamics* will be connected to behavior–parasite feedbacks, and highlight processes at scales beyond the level of hosts, vectors, or parasites.
- *Together*, these approaches will contribute to predictive modeling of behavior–parasite dynamics under anthropogenic environmental change.

---

## Acknowledgments

We thank the participants of the Research Frontiers in Animal Behavior and Parasitism symposium (May 2021) for a fruitful discussion and inspiration for this chapter. We also thank the Center for the Ecology of Infectious Diseases at the University of Georgia for hosting the virtual symposium and facilitating the discussion. Finally, we thank J Anderson for contributing to the data and layout of Box 18.3, Figure 1, as well as the editors and reviewers for constructive feedback on an earlier draft of this chapter.

## References

1. Ezenwa VO, Altizer S, and Hall RJ. Animal behavior and parasitism: Where have we been, where are we going?. In Ezenwa VO, Altizer S, and Hall RJ (eds), *Animal Behavior and Parasitism*. Oxford: Oxford University Press, 2022. DOI: 10.1093/oso/9780192895561.003.0001.
2. Godfrey SS and Poulin R. Host manipulation by parasites: From individual to collective behavior. In Ezenwa VO, Altizer S, and Hall RJ (eds), *Animal Behavior and Parasitism*. Oxford: Oxford University Press, 2022. DOI: 10.1093/oso/9780192895561.003.0012.
3. Hawley DM and Ezenwa VO. Parasites, host behavior, and their feedbacks. In Ezenwa VO, Altizer S, and Hall RJ (eds), *Animal Behavior and Parasitism*. Oxford: Oxford University Press, 2022. DOI: 10.1093/oso/9780192895561.003.0002.
4. Hall RJ, Altizer S, Peacock SJ, and Shaw AK. Animal migration and infection dynamics: Recent advances and future frontiers. In Ezenwa VO, Altizer S, and Hall RJ (eds), *Animal Behavior and Parasitism*. Oxford: Oxford University Press, 2022. DOI: 10.1093/oso/9780192895561.003.0007.
5. Stephenson JF and Adelman JS. The behavior of infected hosts: Behavioral tolerance, behavioral resilience, and their implications for behavioral competence. In Ezenwa VO, Altizer S, and Hall RJ (eds), *Animal Behavior and Parasitism*. Oxford: Oxford University Press, 2022. DOI: 10.1093/oso/9780192895561.003.0017.
6. Viney ME and Graham AL. Patterns and processes in parasite co-infection. *Adv Parasitol*. 2013;82:321–69.
7. Fountain-Jones NM, Packer C, Jacquot M, Blanchet FG, Terio K, and Craft ME. Endemic infection can shape exposure to novel pathogens: Pathogen co-occurrence networks in the Serengeti lions. *Ecol Lett*. 2019;22(6):904–13.
8. Sweeny AR, Albery GF, Becker DJ, Eskew EA, and Carlson CJ. Synzootics. *J Anim Ecol*. 2021;90:2744–54.
9. Leu ST, Sah P, Krzyszczyk E, Jacoby A-M, Mann J, and Bansal S. Sex, synchrony, and skin contact: Integrating multiple behaviors to assess pathogen transmission risk. *Behav Ecol*. 2020;31(3):651–60.
10. Pedersen AB and Fenton A. Emphasizing the ecology in parasite community ecology. *Trends Ecol Evol*. 2007;22(3):133–9.
11. Eidelman A, Cohen C, Navarro-Castilla Á, Filler S, Gutiérrez R, Bar-Shira E, *et al*. The dynamics between

limited-term and lifelong coinfecting bacterial parasites in wild rodent hosts. *J Exp Biol.* 2019;222(15): jeb 203562.

12. Queiroz ML, Viel TA, Papa CH, Lescano SA, and Chieffi PP. Behavioral changes in *Rattus norvegicus* coinfected by *Toxocara canis* and *Toxoplasma gondii*. *Rev Inst Med Trop Sao Paulo*. 2013;55(1):51–3.

13. Broughton H, Govender D, Serrano E, Shikwambana P, and Jolles A. Equal contributions of feline immunodeficiency virus and coinfections to morbidity in African lions. *Int J Parasitol Parasites Wildl.* 2021;16:83–94.

14. Corbett EL, Watt CJ, Walker N, Maher D, Williams BG, Raviglione MC, *et al.* The growing burden of tuberculosis: Global trends and interactions with the HIV epidemic. *Arch Intern Med.* 2003;163(9):1009–21.

15. Budischak SA, Sakamoto K, Megow LC, Cummings KR, Urban JF, Jr, and Ezenwa VO. Resource limitation alters the consequences of co-infection for both hosts and parasites. *Int J Parasitol.* 2015;45(7):455–63.

16. Rynkiewicz EC, Pedersen AB, and Fenton A. An ecosystem approach to understanding and managing within-host parasite community dynamics. *Trends Parasitol.* 2015;31(5):212–21.

17. Müller-Klein N, Heistermann M, Strube C, Franz M, Schülke O, and Ostner J. Exposure and susceptibility drive reinfection with gastrointestinal parasites in a social primate. *Functional Ecology.* 2019;33(6):1088–98.

18. Garland T, Jr. Trade-offs. *Curr Biol.* 2014;24(2):R60–R1.

19. Weinstein SB, Moura CW, Mendez JF, and Lafferty KD. Fear of feces? Tradeoffs between disease risk and foraging drive animal activity around raccoon latrines. *Oikos.* 2018;127(7):927–34.

20. Mougeot F, Irvine JR, Seivwright L, Redpath SM, and Piertney S. Testosterone, immunocompetence, and honest sexual signaling in male red grouse. *Behav Ecol.* 2004;15(6):930–7.

21. Day T, Graham AL, and Read AF. Evolution of parasite virulence when host responses cause disease. *Proc Biol Sci.* 2007;274(1626):2685–92.

22. Spoel SH, Johnson JS, and Dong X. Regulation of trade-offs between plant defenses against pathogens with different lifestyles. *PNAS.* 2007;104(47):18842–7.

23. Mockler BK, Kwong WK, Moran NA, and Koch H. Microbiome structure influences infection by the parasite *Crithidia bombi* in bumble bees. *Appl Environ Microbiol.* 2018;84(7).

24. Schoenle LA, Moore IT, Dudek AM, Garcia EB, Mays M, Haussmann MF, *et al.* Exogenous glucocorticoids amplify the costs of infection by reducing resistance and tolerance, but effects are mitigated by co-infection. *Proc Biol Sci.* 2019;286(1900):20182913.

25. Graham SP, Kelehear C, Brown GP, and Shine R. Corticosterone-immune interactions during captive stress in invading Australian cane toads (*Rhinella marina*). *Horm Behav.* 2012;62(2):146–53.

26. Piersma T and Drent J. Phenotypic flexibility and the evolution of organismal design. *Trends Ecol Evol.* 2003;18(5):228–33.

27. Arnold PA, Kruuk LE, and Nicotra AB. How to analyse plant phenotypic plasticity in response to a changing climate. *New Phytologist.* 2019;222(3): 1235–41.

28. Mideo N and Reece SE. Plasticity in parasite phenotypes: Evolutionary and ecological implications for disease. *Future Microbiol.* 2012;7(1): 17–24.

29. Poulin R. Parasite manipulation of host personality and behavioural syndromes. *J Exp Biol.* 2013;216(Pt 1):18–26.

30. Gervasi SS, Civitello DJ, Kilvitis HJ, and Martin LB. The context of host competence: A role for plasticity in host–parasite dynamics. *Trends Parasitol.* 2015;31(9):419–25.

31. Strauss AT, Hite JL, Civitello DJ, Shocket MS, Caceres CE, and Hall SR. Genotypic variation in parasite avoidance behaviour and other mechanistic, nonlinear components of transmission. *Proc Biol Sci.* 2019;286(1915):20192164.

32. Paull SH, Song S, McClure KM, Sackett LC, Kilpatrick AM, and Johnson PT. From superspreaders to disease hotspots: Linking transmission across hosts and space. *Front Ecol Environ.* 2012;10(2):75–82.

33. Lefevre T, Ohm J, Dabire KR, Cohuet A, Choisy M, Thomas MB, *et al.* Transmission traits of malaria parasites within the mosquito: Genetic variation, phenotypic plasticity, and consequences for control. *Evol Appl.* 2018;11(4):456–69.

34. Giraudeau M, Mousel M, Earl S, and McGraw K. Parasites in the city: Degree of urbanization predicts poxvirus and coccidian infections in house finches (*Haemorhous mexicanus*). *PLoS One.* 2014;9(2): e86747.

35. Stephenson JF. Parasite-induced plasticity in host social behaviour depends on sex and susceptibility. *Biol Lett.* 2019;15(11):20190557.

36. Nunes H, Rocha FL, and Cordeiro-Estrela P. Bats in urban areas of Brazil: Roosts, food resources and parasites in disturbed environments. *Urban Ecosyst.* 2017;20(4):953–69.

37. Werner CS and Nunn CL. Effect of urban habitat use on parasitism in mammals: A meta-analysis. *Proc R Soc B: Biol Sci.* 2020;287(1927):20200397.

38. Altizer S, Ostfeld RS, Johnson PT, Kutz S, and Harvell CD. Climate change and infectious diseases: From evidence to a predictive framework. *Science.* 2013;341(6145):514–9.

39. Couper LI, Farner JE, Caldwell JM, Childs ML, Harris MJ, Kirk DG, et al. How will mosquitoes adapt to climate warming? *Elife.* 2021;10:e69630.

40. Shapiro LLM, Whitehead SA, and Thomas MB. Quantifying the effects of temperature on mosquito and parasite traits that determine the transmission potential of human malaria. *PLoS Biol.* 2017;15(10):e2003489.

41. Raffel TR, Halstead NT, McMahon TA, Davis AK, and Rohr JR. Temperature variability and moisture synergistically interact to exacerbate an epizootic disease. *Proc Biol Sci.* 2015;282(1801):20142039.

42. Rohr JR and Cohen JM. Understanding how temperature shifts could impact infectious disease. *PLoS Biology.* 2020;18(11):e3000938.

43. Winternitz JC and Abbate JL. The genes of attraction: Mating behavior, immunogenetic variation and parasite resistance in Ezenwa VO, Altizer S, and Hall RJ (eds), *Animal Behavior and Parasitism.* Oxford: Oxford University Press, 2022. DOI: 10.1093/oso/9780192895561.003.0011.

44. Davis S, Schlenke T. Behavioral defenses against parasitoids: Genetic and neuronal mechanisms. In Ezenwa VO, Altizer S, and Hall RJ (eds), *Animal Behavior and Parasitism.* Oxford: Oxford University Press, 2022. DOI: 10.1093/oso/9780192895561.003.0016.

45. Pirrie A, Chapman H, Ashby B. Parasite-mediated sexual selection: To mate or not to mate?. In Ezenwa VO, Altizer S, and Hall RJ (eds), *Animal Behavior and Parasitism.* Oxford: Oxford University Press, 2022. DOI: 10.1093/oso/9780192895561.003.0009.

46. Hadziavdic K, Lekang K, Lanzen A, Jonassen I, Thompson EM, and Troedsson C. Characterization of the 18S rRNA gene for designing universal eukaryote specific primers. *PLoS one.* 2014;9(2):e87624.

47. Knight R, Vrbanac A, Taylor BC, Aksenov A, Callewaert C, Debelius J, et al. Best practices for analysing microbiomes. *Nature Rev Microbiol.* 2018;16(7):410–22.

48. Nilsson RH, Anslan S, Bahram M, Wurzbacher C, Baldrian P, and Tedersoo L. Mycobiome diversity: High-throughput sequencing and identification of fungi. *Nature Rev Microbiol.* 2019;17(2):95–109.

49. Grieneisen L, Dasari M, Gould TJ, Björk JR, Grenier J-C, Yotova V, et al. Gut microbiome heritability is nearly universal but environmentally contingent. *Science.* 2021;373(6551):181–6.

50. Youngblut ND, Reischer GH, Walters W, Schuster N, Walzer C, Stalder G, et al. Host diet and evolutionary history explain different aspects of gut microbiome diversity among vertebrate clades. *Nature Commun.* 2019;10(1):1–15.

51. Archie EA, Tung J. Social behavior and the microbiome. *Curr Op Behav Sci.* 2015;6:28–34.

52. Vuong HE, Yano JM, Fung TC, and Hsiao EY. The microbiome and host behavior. *Annu Rev Neurosci.* 2017;40:21–49.

53. Sarkar A, Harty S, Johnson KV-A, Moeller AH, Archie EA, Schell LD, et al. Microbial transmission in animal social networks and the social microbiome. *Nature Ecol Evol.* 2020;4(8):1020–35.

54. Koch H and Schmid-Hempel P. Socially transmitted gut microbiota protect bumble bees against an intestinal parasite. *PNAS* 2011;108(48):19288–92.

55. Gogarten JF, Calvignac-Spencer S, Nunn CL, Ulrich M, Saiepour N, Nielsen HV, et al. Metabarcoding of eukaryotic parasite communities describes diverse parasite assemblages spanning the primate phylogeny. *Mol Ecol Resour* 2020;20(1):204–15.

56. Gould AL, Zhang V, Lamberti L, Jones EW, Obadia B, Korasidis N, et al. Microbiome interactions shape host fitness. *PNAS.* 2018;115(51):E11951–E60.

57. Fischer CN, Trautman EP, Crawford JM, Stabb EV, Handelsman J, and Broderick NA. Metabolite exchange between microbiome members produces compounds that influence *Drosophila* behavior. *Elife.* 2017;6:e18855.

58. Hamilton WD and Zuk M. Heritable true fitness and bright birds: A role for parasites? *Science.* 1982;218(4570):384–7.

59. Radwan J, Biedrzycka A, and Babik W. Does reduced MHC diversity decrease viability of vertebrate populations? *Biol Con.* 2010;143(3):537–44.

60. Boyle EA, Li YI, and Pritchard JK. An expanded view of complex traits: From polygenic to omnigenic. *Cell.* 2017;169(7):1177–86.

61. Nedelec Y, Sanz J, Baharian G, Szpiech ZA, Pacis A, Dumaine A, et al. Genetic ancestry and natural selection drive population differences in immune responses to pathogens. *Cell.* 2016;167(3):657–69 e21.

62. Quach H, Rotival M, Pothlichet J, Loh YE, Dannemann M, Zidane N, et al. Genetic adaptation and Neandertal admixture shaped the immune system of human populations. *Cell.* 2016;167(3):643–56 e17.

63. Lea AJ, Akinyi MY, Nyakundi R, Mareri P, Nyundo F, Kariuki T, et al. Dominance rank-associated gene expression is widespread, sex-specific, and a precursor to high social status in wild male baboons. *PNAS.* 2018;115(52):E12163–E71.

64. Bergland AO, Behrman EL, O'Brien KR, Schmidt PS, and Petrov DA. Genomic evidence of rapid and stable adaptive oscillations over seasonal time scales in *Drosophila. PLoS Genet.* 2014;10(11):e1004775.

65. Machado HE, Bergland AO, Taylor R, Tilk S, Behrman E, Dyer K, et al. Broad geographic sampling reveals the shared basis and environmental correlates of seasonal adaptation in *Drosophila. Elife.* 2021;10.

66. McDowell IC, Barrera A, D'Ippolito AM, Vockley CM, Hong LK, Leichter SM, et al. Glucocorticoid receptor recruits to enhancers and drives activation by motif-directed binding. *Genome Resear.* 2018;28(9):1272–84.

67. Pacis A, Mailhot-Léonard F, Tailleux L, Randolph HE, Yotova V, Dumaine A, et al. Gene activation precedes DNA demethylation in response to infection in human dendritic cells. *PNAS.* 2019;116(14):6938–43.

68. Snyder-Mackler N, Sanz J, Kohn JN, Brinkworth JF, Morrow S, Shaver AO, et al. Social status alters immune regulation and response to infection in macaques. *Science.* 2016;354(6315):1041–5.

69. Nica AC and Dermitzakis ET. Expression quantitative trait loci: Present and future. *Philos Trans R Soc Lond B Biol Sci.* 2013;368(1620):20120362.

70. Umans BD, Battle A, and Gilad Y. Where are the disease-associated eQTLs? *Trends Genet.* 2021;37(2):109–24.

71. Consortium G. The GTEx Consortium atlas of genetic regulatory effects across human tissues. *Science.* 2020;369(6509):1318–30.

72. Jasinska AJ, Zelaya I, Service SK, Peterson CB, Cantor RM, Choi O-W, et al. Genetic variation and gene expression across multiple tissues and developmental stages in a nonhuman primate. *Nature Genetics.* 2017;49(12):1714–21.

73. Tung J, Zhou X, Alberts SC, Stephens M, and Gilad Y. The genetic architecture of gene expression levels in wild baboons. *Elife.* 2015;4:e04729.

74. Barreiro LB, Tailleux L, Pai AA, Gicquel B, Marioni JC, and Gilad Y. Deciphering the genetic architecture of variation in the immune response to *Mycobacterium tuberculosis* infection. *PNAS.* 2012;109(4):1204–9.

75. Kim-Hellmuth S, Bechheim M, Pütz B, Mohammadi P, Nédélec Y, Giangreco N, et al. Genetic regulatory effects modified by immune activation contribute to autoimmune disease associations. *Nature Commun.* 2017;8(1):1–10.

76. Lee MN, Ye C, Villani A-C, Raj T, Li W, Eisenhaure TM, et al. Common genetic variants modulate pathogen-sensing responses in human dendritic cells. *Science.* 2014;343(6175).

77. Lopes PC, French SS, Woodhams DC, and Binning SA. Infection avoidance behaviors across vertebrate taxa: Patterns, processes and future directions. In Ezenwa VO, Altizer S, and Hall RJ (eds), *Animal Behavior and Parasitism.* Oxford: Oxford University Press, 2022. DOI: 10.1093/oso/9780192895561.003.0014.

78. Poirotte C, Charpentier MJE. Inter-individual variation in parasite avoidance behaviors and its epidemiological, ecological, and evolutionary consequences. In Ezenwa VO, Altizer S, and Hall RJ (eds), *Animal Behavior and Parasitism.* Oxford: Oxford University Press, 2022. DOI: 10.1093/oso/9780192895561.003.0015.

79. Cheng W, Yu Z, Liu S, Sun W, Ling F, Pan J, et al. Successful interruption of seasonal influenza transmission under the COVID-19 rapid response in Zhejiang Province, China. *Public Health.* 2020;189:123–5.

80. Pierce A, Haworth-Brockman M, Marin D, Rueda ZV, and Keynan Y. Changes in the incidence of seasonal influenza in response to COVID-19 social distancing measures: An observational study based on Canada's national influenza surveillance system. *Can J Public Health.* 2021;112(4):620–8.

81. Wren BT, Remis MJ, Camp JW, and Gillespie TR. Number of grooming partners is associated with hookworm infection in wild vervet monkeys (*Chlorocebus aethiops*). *Folia Primatol (Basel).* 2016;87(3):168–79.

82. Hafer N. Conflicts over host manipulation between different parasites and pathogens: Investigating the ecological and medical consequences. *BioEssays.* 2016;38(10):1027–37.

83. Salvaudon L, De Moraes CM, and Mescher MC. Outcomes of co-infection by two potyviruses: Implications for the evolution of manipulative strategies. *Proc Biol Sci.* 2013;280(1756):20122959.

84. Koprivnikar J and Penalva L. Lesser of two evils? Foraging choices in response to threats of predation and parasitism. *PLoS One.* 2015;10(1):e0116569.

85. Lee BW, Basu S, Bera S, Casteel CL, and Crowder DW. Responses to predation risk cues and alarm pheromones affect plant virus transmission by an aphid vector. *Oecologia.* 2021;196(4):1005–15.

86. Doherty J-F and Ruehle B. An integrated landscape of fear and disgust: The evolution of avoidance behaviors amidst a myriad of natural enemies. *Front Ecol Evol.* 2020;8:317.

87. Eleftheriou A, Kuenzi AJ, and Luis AD. Heterospecific competitors and seasonality can affect host physiology and behavior: Key factors in disease transmission. *Ecosphere.* 2021;12(6):e03494.

88. Santicchia F, Wauters LA, Piscitelli AP, Van Dongen S, Martinoli A, Preatoni D, et al. Spillover of an alien parasite reduces expression of costly behaviour in native host species. *J Anim Ecol.* 2020;89(7):1559–69.

89. Hall SR, Sivars-Becker L, Becker C, Duffy MA, Tessier AJ, and Caceres CE. Eating yourself sick: Transmission of disease as a function of foraging ecology. *Ecol Lett.* 2007;10(3):207–18.

90. Cross PC, Edwards WH, Scurlock BM, Maichak EJ, and Rogerson JD. Effects of management and climate on elk brucellosis in the Greater Yellowstone ecosystem. *Ecol Appl.* 2007;17(4):957–64.

91. Sato T, Iritani R, and Sakura M. Host manipulation by parasites as a cryptic driver of energy flow through food webs. *Curr Opin Insect Sci.* 2019;33:69–76.

92. Lafferty KD and Kuris AM. Ecological consequences of manipulative parasites in Hughes DP, Brodeur J, and Thomas F (eds), *Host Manipulations by Parasites*. Oxford: Oxford University Press, 2012, 158–68.

93. Davis AK and Prouty C. The sicker the better: Nematode-infected passalus beetles provide enhanced ecosystem services. *Biol Lett.* 2019;15(5):20180842.

94. Reisinger LS and Lodge DM. Parasites alter freshwater communities in mesocosms by modifying invasive crayfish behavior. *Ecology.* 2016;97(6):1497–506.

95. Morton JP and Silliman BR. Parasites enhance resistance to drought in a coastal ecosystem. *Ecology.* 2020;101(1):e02897.

96. Hite JL, Pfenning AC, and Cressler CE. Starving the enemy? Feeding behavior shapes host–parasite interactions. *Trends Ecol Evol.* 2020;35(1):68–80.

97. Penczykowski RM, Hall SR, Shocket MS, Ochs JH, Lemanski BC, Sundar H, *et al.* Virulent disease epidemics can increase host density by depressing foraging of hosts. *bioRxiv.* 2020.

98. Labaude S, Rigaud T, and Cezilly F. Host manipulation in the face of environmental changes: Ecological consequences. *Int J Parasitol Parasites Wildl.* 2015;4(3):442–51.

# CHAPTER 19

# Parallels in parasite behavior: The other side of the host–parasite relationship

Emlyn J. Resetarits, Lewis J. Bartlett, Cali A. Wilson, and Anna R. Willoughby

## 19.1 Introduction

Parasitism is the most common lifestyle on the planet, with parasites comprising approximately half of all species. With this diversity comes a wide array of parasite behaviors, many that parallel free-living organisms (Table 19.1). In fact, a single parasite may exhibit a multitude of behaviors underpinning transmission, virulence, and interactions with the host and other parasites during its life cycle. For example, *Aspergillus* spp. include fungi which parasitize plants, arthropods, and vertebrates, including humans. *Aspergillus* spp. exhibit a diverse set of behaviors, including extra-host movement via arthropod vectoring or weather-dependent dispersal of hardy spores, chemotaxic-like infection behaviors, intra-host movement, defense against competitors, and both sexual reproduction and asexual reproduction (Figure 19.1).

Although studies of the relationship between parasites and behavior have typically focused on host or vector behaviors, parasites behave in ways equally consequential for disease outcomes and transmissibility. Entomopathogenic nematodes (EPNs; Figure 19.2A) provide an excellent and economically relevant example of why studying parasite behavior is important. EPNs are parasites of arthropods and have been commercially used as biopesticides against diverse crop pests, including Coleoptera (e.g., citrus root weevil, corn rootworm), Lepidoptera (e.g., tomato leafminer, navel orangeworm), and Diptera (e.g., fruit flies, fungus gnats), since the 1980s (25). EPNs use a variety of behaviors to move around and forage for suitable hosts. Some species like *Steinernema carpocapsae* and *S. scapterisci* act as sit-and-wait/ambush predators, staying near the surface and infecting mobile hosts by either jumping onto the host or raising over 95% of their body vertically to latch on (a behavior called nictation). Other species such as *S. glaseri* and *Heterorhabditis bacteriophora* are cruisers, actively dispersing and distributing themselves throughout the soil to find and infect less mobile hosts. Chemotaxis in response to stimuli emitted from arthropod hosts (7) and pest-damaged plants (26) is a key mechanism EPNs use to affect these diverse foraging strategies. To effectively use EPNs as biocontrol agents, researchers select for strains with dispersal and host-finding behaviors that optimize the ability to locate specific hosts (27). Correspondingly, this process has improved the biocontrol potential of various EPN strains already in culture, demonstrating the applied relevance of a basic understanding of parasite behavior.

In this chapter, we highlight the diversity of parasite behaviors and use key themes emphasized in this book (social behavior, behavioral defenses, mating behavior and sexual selection, and movement behavior) to explore the importance of parasite behavior for understanding disease ecology and evolution. We show that parasite behavior

Emlyn J. Resetarits et al., *Parallels in parasite behavior*. In: *Animal Behavior and Parasitism*. Edited by Vanessa O. Ezenwa, Sonia Altizer, and Richard J. Hall, Oxford University Press. © Oxford University Press (2022). DOI: 10.1093/oso/9780192895561.003.0019

**Table 19.1** Representative examples of parasite behaviors studied thus far.

| Behavior Class | Specific Behavior | Parasite Taxa | Parasite species | Host | Citations |
|---|---|---|---|---|---|
| **Social Behavior** | Eusociality: caste division of labor | Trematode | e.g., *Himasthla* sp. B, *Euhaplorchis californiensis* | Snail | (1), (2) |
| | Eusociality: caste division of labor | Arthropod | Parasitoid wasp (*Copidosoma floridanum*) | Moth | (3) |
| | Eusociality: caste division of labor | Arthropod | Gall-inhabiting aphids (*Pemphigus spyrothecae*) | Poplar | (4), (5) |
| | Prudence (low rates of resource exploitation) | Bacteria | e.g., *Plasmodium chabaudi*, baculovirus | e.g., Rodents, Caterpillars | reviewed in (6) |
| | Cooperation | Bacteria | e.g., *Pseudomonas aeruginosa* | e.g., Humans, Caterpillars | reviewed in (6) |
| | Spite | Bacteria | e.g., *Photorhadbus* spp., *Xenorhadbus* spp. | e.g., Caterpillars | reviewed in (6) |
| **Behavioral defenses** | Host discrimination, preference for naïve hosts | Nematode | *Steinernema* spp. | Insects | (7) |
| | Host discrimination, preference for naïve hosts | Arthropod | Parasitoid wasp (*Leptopilina heterotoma*) | *Drosophila melanogaster* | (8) |
| | Avoid attaching to hosts at sites where ticks have previously fed | Arthropod | Ticks | Rodents | (9), (10) |
| **Mating Behavior** | Plasmid exchange | Bacteria | e.g., *Escherichia coli*, *Klebsiella pneumoniae* | e.g., Humans | (11) |
| | Haplodiploidy | Arthropod | Varroa mite | Honey bee | (12), (13) |
| | Polygyny | Trematode | *Austrobilharzia variglandis* | Songbirds; Human | (14) |
| | Polygyny | Trematode | *Schistosoma mansoni* | Mice | (15) |
| | Female mate choice | Trematode | *Schistosoma mansoni* | Human | (16), (17) |
| | Male-male competition | Trematode | *Schistosoma mansoni* | Human | (17) |
| **Between-host movement** | Cercariae host-seeking behavior | Trematode | *Maritrema misenensis* | Amphipod | (18) |
| | Chemotaxis (Miracidae host-seeking behavior) | Trematode | *Schistosoma japonicum* | Gastropod | reviewed in (19) |
| | Chemotaxis (Cercariae host-seeking behavior) | Trematode | *Hypoderaeum conoideum* | Gastropod | reviewed in (19) |
| | Photo-orientation (Cercariae host-seeking behavior) | Trematode | *Echinostoma* spp. | Gastropod, Amphibian | (20) |
| | Thermotaxis | Nematode | *Necator americanus* | Human | reviewed in (19) |
| | Thermotaxis | Nematode | *Ancylostoma duodenale* | Human, Dog, Cat | reviewed in (19) |
| | Infected juveniles jump in response to external stimuli | Nematode | *Steinernema carpocapsae* | Insects | (21) |
| **Within-host movement** | Tissue-seeking behavior within host | Nematode | *Trichinella spiralis* | Mice | (22) |
| | Intra-host site selection and quorum-sensing | Virus | propagative plant viruses | Plants | (23) |
| | Use host circadian rhythms to coordinate movement within host tissue | Nematode | *Microfilaria* spp. | Vertebrates | (24) |

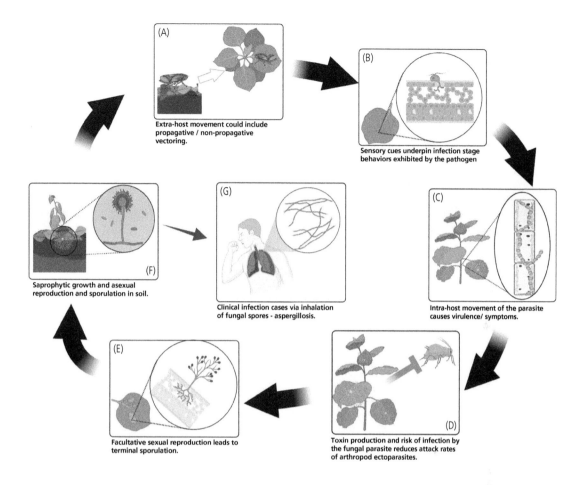

**Figure 19.1** Throughout their life cycles parasites exhibit a multitude of behaviors underpinning transmission, virulence, and interactions with the host and other parasites. For example, *Aspergillus* spp. fungi, exhibit: (A) extra-host movement of environmental spores, plausibly via vectoring and secondary hosts; (B) chemotaxic-like infection behaviors, which may be part of arms-race dynamics with hosts evolving new resistance mechanisms; (C) intra-host movement, which may underpin the pathology and virulence (outcome) of infection; (D) defense against competitors, for example by producing secondary toxins; (E) sexual reproduction, which may lead to rapid evolution of resistance against control agents; and (F) asexual reproduction, allowing rapid expansion into new patches, driving epidemics, and contributing to other ecological processes such as nutrient cycling.

underpins multiple aspects of the host–parasite interaction, ranging from transmission dynamics to the evolution of resistance and virulence. We demonstrate that like hosts and vectors, parasites display a full range of behaviors traditionally considered in behavior-parasite studies, and we argue that these parasite behaviors are equally important for understanding disease outcomes. We conclude by discussing how parasite behaviors can create important feedbacks with host behaviors, with implications for parasite transmission and coevolution between hosts and parasites.

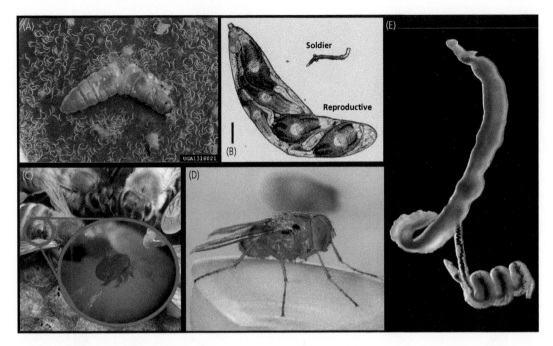

**Figure 19.2** Examples of the diversity of parasite types engaging in complex behaviors. (A) Entomopathogenic nematodes emerging from a wax moth cadaver to seek out new hosts (image from Peggy Greb, CC 3.0); (B) Caste division of labor in digenetic trematode species, *Himasthla* sp. B. Soldiers (small morph) defend the parasite colony, while reproductives (large morph) produce free-swimming parasite stages (i.e., cercariae; image from (1)); (C) Varroa mite feeding on a honey bee (image from Jennifer Berry, UGA Bee Lab); (D) Parasitoid wasp, *Ormia ochracea,* eavesdropping on its cricket host mating call (image from Jpaur, CC 3.0); (E) male–male competition in adult trematodes, *Schistosoma mansoni.* Here, one male is removing a female from the other male's ventral groove (image from (17)).

## 19.2 Social behavior and behavioral defenses

Host social behavior impacts parasite transmission in diverse ways (28), (29). Larger group size is related to increased levels of parasitism, whereas collective anti-parasite behaviors, such as allogrooming, can result in lower levels of ectoparasites (29), (30). Hosts, however, are not alone in their proclivity for social behavior: some parasites also exhibit forms of social behavior that have implications for disease outcomes. Perhaps the most extreme example of this is the occurrence of eusociality, or reproductive division of labor, in parasitic group-living, colonial organisms such as parasitoid wasps (3), gall-inhabiting aphids (4), and digenean trematodes (1), (2). In many trematode species, colonies of reproductive larvae form in the first intermediate host, often a snail, which produce dispersive larvae known as cercariae that leave

the snail and seek out the next host. However, in the case of eusocial trematode species, colonies also produce soldiers, which are smaller, more motile, and use their mouthparts to ingest larvae from other colonies that might be parasitizing the same snail host (Figure 19.2B). This behavior of the soldiers helps established colonies defend their hosts by outcompeting invading pathogens (31).

Free-living eusocial species have been shown to dynamically alter their colony composition (i.e., the proportion of soldiers and reproductive morphs) in response to various biotic threats, such as competition (32). Similarly, eusocial trematodes appear to shift their colony composition as a result of parasite pressure (33). Previous research has shown that at sites with high infection prevalence in snail hosts, which indicates an increased risk of coinfection and competition in snails, trematode colonies have increased investment in soldiers (33). Increased soldier investment during multiple infections is also related to decreased output of reproductive larvae (i.e., cercariae; 31). Consequently, in areas of high infection prevalence, trematode colonies with a caste division of labor may have lower per-capita transmission rates due to declines in cercarial emergence. While we know that host diversity can decrease disease transmission (34), (35), parasite diversity may also dampen the transmission of competing species by causing those parasites to invest more in defense than reproduction.

In addition to fighting off conspecifics from other colonies, recent work suggests that trematode soldiers also fight off other parasite taxa like bacteria and fungi. For example, Macleod *et al.* (36) found that trematode-infected snails with experimentally inflicted shell damage produced more soldiers and these soldiers congregated at the host wound site consuming fungal and bacterial pathogen propagules. Interestingly, this behavior of the soldiers might protect snail hosts from infection, acting as a pseudo-immune system (36). This raises the question of whether this behavior can be considered an extended phenotype of the host, and by extension, whether trematode colonies can be considered as a form of "extended" social immunity (37). In this case, the social behavior of the trematodes protects both conspecific group members and heterospecific snail hosts from infection.

To date, 15 of 16 trematode species (out of roughly 10,000) studied have displayed a caste division of labor, suggesting that sociality may be the rule rather than the exception in this group (38). However, work on eusociality in trematodes is still in its infancy and is not without controversy (39). Nevertheless, the study of eusociality in trematodes and other parasites offers a fruitful avenue for exploring the implications of parasite sociality for disease outcomes.

## 19.3 Mating behavior and sexual selection

Control of parasites at every scale, including treating individual hosts, reducing transmission opportunities between hosts, or slowing invasive parasite movement across landscapes, remains a critical applied scientific frontier. One emblematic example is the ongoing battle against the evolution of antibiotic resistance in bacterial pathogens—a process underpinned by mating behavior in the form of plasmid exchange or "bacterial sex."(11) Earlier in this book, Pirrie and colleagues explored how host mating systems interact with parasite pressure, including how host mating systems determine the evolution of resistance to parasites in hosts (40). In parallel to this, parasite mating systems can also crucially determine their ability to evolve resistance to control agents.

Facultative sexual reproduction can greatly increase the rate at which parasite populations can evolve resistance to control agents. For example, the cyclical parthenogenesis exhibited by aphids leads to more rapid pesticide resistance evolution, corresponding to aphids remaining a pervasive parasite problem in agriculture (41). Even in obligately sexually reproducing species, the presence of sexual selection can increase the rate at which pesticide resistance fixes in the population, as shown in the red flour beetle *Tribolium castaneum* (42). A current example of the importance of mating systems for parasite control is illustrated by the ongoing efforts to combat invasive, industry-devastating *Varroa* mites (Figure 19.2C), which are parasites of managed honey bees. On the one hand, *Varroa*'s highly inbred haplodiploid mating system limits gene flow and reduces the rates at which resistance alleles can spread, resulting in slower rates of chemical pesticide resistance evolution (12). However, these same inbred haplodiploid mating behaviors, where foundress females asexually produce a single male who then mates with all of his sexually produced sisters, render our ability to use gene drive control technologies mostly impotent (13). This is because haploid males prevent the spread of lethal alleles in the population, while full-sibling inbreeding greatly reduces the rate at which non-lethal alleles (e.g., alleles associated with loss of acquired acaricide resistance) can effectively spread. Understanding the mating strategies of parasites is thus an important part of assessing the strength of long-term disease control strategies.

Parasite-mediated sexual selection in hosts has been a central focus of research at the intersection of animal behavior, parasitism, and evolutionary biology (40). Although not much is known about sexual selection in parasites themselves, there is evidence to suggest that this can occur (e.g., 17). For example, *Schistosoma mansoni* is a trematode parasite that sexually reproduces in humans and causes the disease schistosomiasis. During sexual reproduction, a female fits into the ventral groove of a large, muscular male, stimulating her sexual maturation and egg production (43). Female schistosomes are capable of mate switching and have been shown to do so when mate switching increases genetic dissimilarity between the mating pair (16), thereby increasing heterozygosity in their offspring. This preference for disassortative mating should maintain genetic diversity in the parasite population and may improve the ability of parasite offspring to counter host resistance (16). On the other hand, male schistosomes can compete by actively removing females from the ventral groove of other males (17, Figure 19.2E) or by holding multiple females in their ventral groove at the same time (14). While studies have found high variance in mating success of schistosome males, with larger males siring more offspring, this variance was not due to polygamy (17). Because schistosome males have relatively high investment in offspring (44) and are generally monogamous, these male behaviors instead may suggest a role of male mate choice in this system (17). Although work in *S. mansoni* provides compelling evidence that sexual selection can occur in parasites, a number of questions remain on who is doing the choosing, what traits are being chosen, and the repercussions for *S. mansoni* evolution and disease outcomes.

## 19.4 Movement

For hosts, local dispersal and long-distance migration influence host–pathogen interactions by affecting both host exposure and susceptibility to parasites. For example, energetically costly migrations can dampen host immune responses, increasing susceptibility to infection, or can allow hosts to leave behind infected habitats and individuals, lowering infection risk (45), (46). Furthermore, non-migratory movements, like dispersal, allow the genetic diversification of host populations, impacting exposure, susceptibility, and host–parasite coevolution (47). For parasites, movement is also a fundamental part of life. Parasites move between

hosts and within hosts, and these movements influence key processes dictating disease outcomes, including transmission rates (48), virulence (49), and disease pathology (50).

Many parasites have free-living life stages that are responsible for transmission between hosts. Parasites find their hosts using a variety of behaviors that, if understood, could facilitate efforts to mitigate disease transmission (51). For example, trematode parasites have multiple free-living life stages (miracidium, cercaria) that exhibit host-seeking behaviors crucial for transmission. Cercariae are short-lived, free-swimming stages that actively search for intermediate hosts (e.g., tadpoles, fish, mollusks), which can vary dramatically in their spatial (e.g., benthic or pelagic) and temporal (e.g., nocturnal or diurnal) niches. Cercariae display species-specific movement behaviors that facilitate transmission by exposing them to suitable hosts at the right time and in the right place. For example, *Maritrema misenensis* cercariae disperse from a marine snail to a terrestrial amphipod by actively swimming upward and clinging to the air–water surface interface, allowing waves to carry them up to the shore, where they can come into contact with their amphipod hosts (18). In general, cercariae can also respond to host chemical cues, water turbulence, touch, and dark shadows (19), (52). Historically, cercarial movement behaviors have been studied to minimize human exposure risk to trematodes that use humans as definitive hosts (e.g., schistosomiasis) or that infect humans incidentally (e.g., cercarial dermatitis or "Swimmer's Itch"). For example, in Northern Michigan, Lindblade found that risk factors for acquiring cercarial dermatitis included swimming in the mornings, in the month of July, and in shallow areas (53). These risk factors for humans are likely adaptations by the parasite to increase the probability of finding a shorebird definitive host (54). Understanding cercarial movement and the cues used to seek out and infect hosts can help minimize human exposure and potentially interrupt transmission (51).

Vectoring of pathogens by arthropods is of critical importance to understanding the transmission and control of many diseases of public health, wildlife, and agricultural concern. A crucial part of the vectoring process is the intra-host movement of pathogens inside the arthropod vector that enables onward transmission (55). For example, infection may occur in the vector gut but then requires transmission via salivary glands. Most pathogens achieve this by infecting and then replicating inside their vectors. Well-established theory predicts that when a pathogen has to maintain infectivity in two hosts—namely a definitive host and a vector—evolution should favor minimizing virulence in the vector, at the expense of higher virulence in the definitive host (56). Simply, this is because vectors typically have higher between-patch mobility than definitive hosts, making equivalent virulence in the vector relatively more costly than in the less-mobile definitive host, biasing the pathogen towards higher virulence in the latter. Infection of vectors by pathogens therefore determines how virulent pathogens evolve to be. However, pathogens can achieve the intra-host movement required to be effectively vectored without actively infecting the vector host. This is illustrated by comparing propagative and non-propagative vectoring of plant viruses by arthropods (57).

Propagative plant viruses such as *Phytoreovirus* spp. move within their insect vectors by infecting the vector, and then concentrating in the salivary glands through a multi-generation process of site-specific replication preferentially targeting certain cells in the insect (23). We can assume, as discussed earlier, that this leads to higher virulence in the definitive host (the plant) as the virus balances virulence in multiple hosts, favoring minimal virulence in the mobile vector. We can contrast this to non-propagative vectored plant viruses, which congregate in the salivary glands of their vectors despite an inability to replicate in the insect host. For example, some plant viruses formerly categorized as the "*Luteoviridae*" (taxonomy since revised) rely on aphid vectors but do not replicate in the aphid "host," instead using a complex series of cellular binding and transport mechanisms to move from the insect gut to the salivary glands (58). These viruses may be able to achieve lower virulence in the definitive plant host by avoiding the constraints of also maintaining the ability to infect an insect vector. Comparative studies examining the differential virulence of insect-vectored plant viruses based

on whether they achieve intra-host movement in their insect vectors via propagative versus non-propagative means would significantly expand our understanding of virulence evolution. The behavioral basis of this intra-host movement, which affects host and pathogen evolution, is thus an exciting frontier. Further, the comparative evolution of the insect vectors (parasites in their own right) in these cases would be a new frontier in understanding the ecological role of vector evolution in disease transmission in response to pathogen behavioral strategy.

## 19.5 Behavior begets behavior: Host–parasite feedbacks

Behavior–parasite feedbacks are bidirectional interactions between host behavior and a parasite trait, such as prevalence, load, or virulence, where a parasite trait influences a host behavior, which in turn affects that same parasite trait (59), (60). These feedbacks can also operate the other way, from host behavior to parasite trait, back to host behavior. So far in this chapter, we have demonstrated that parasites have a diverse assemblage of behaviors that are important for host–parasite interactions and disease outcomes. In this section, we argue that parasite traits capable of eliciting feedbacks with host behavior include not just load, prevalence, and virulence (60) but also parasite behaviors. Specifically, we posit that parasite behaviors are intertwined with host behaviors, resulting in feedbacks where behavior begets behavior (BBB) across ecological and evolutionary time scales. Here, we showcase a number of examples, including two (mating calls, grooming) described earlier in this book (60), that demonstrate BBB feedbacks and discuss additional scenarios where these feedbacks are likely to occur.

Hawley and Ezenwa (60) describe one such feedback between the parasitoid wasp, *Ormia ochracea*, and the field cricket, *Teleogryllus oceanicus*. In Hawaii, male field crickets use mating calls to attract female crickets. A lethal parasitoid wasp has evolved the ability to eavesdrop on these mating calls to locate its cricket hosts (61; Figure 19.2D). This results in rapid declines in the cricket population, acting as a strong selective pressure for changes in male calling behavior, resulting in the loss or change of this sexual signal (61)–(63). In this case, the host behavior (calling) evolved for mating but influenced parasite behavior by giving parasites a fitness advantage when eavesdropping, which in turn caused changes in host mating call behavior (Figure 19.3A). Whether parasite eavesdropping behavior has subsequently evolved in response to these altered host calls is unknown.

BBB feedbacks may also arise from host defensive behaviors, like grooming. For example, self-grooming behavior by impala may cause ticks (which become immobile once attached) to settle (i.e., habitat selection) in areas that are harder for the host to reach, such as on the head, neck and perianal region (64). This, in turn, can generate a positive feedback, favoring the evolution of reciprocal allogrooming of these areas that are difficult for individuals to reach themselves (65; Figure 19.3B). Host allogrooming and parasite movement may feedback on each other on shorter timescales as well, such as during a single grooming session. For example, fleas can migrate in response to active grooming, resulting in a back-and-forth of host grooming strategies and parasite movement counterstrategies (e.g., 66).

Finally, a number of parasite behaviors described in this chapter, such as host-seeking behaviors, lend themselves to similar BBB feedbacks. Many parasites, such as trematodes, ticks, and fleas, use a variety of cues and movement behaviors to seek out suitable hosts (18), (19), (52), (67). On the other hand, host–parasite avoidance behaviors, such as moving away from parasite cues and increasing activity levels, help hosts actively avoid those parasites (e.g., 68, 69). For parasites that seek out hosts through olfactory cues (67), hosts may even manipulate their own olfactory cues through changes in diet choice or habitat use. These strategies employed by hosts should decrease the risk of infection and result in strong selection on parasites for novel or exaggerated host-seeking behaviors that facilitate finding and infecting the correct hosts (Figure 19.3C). While both parasite and host behaviors are likely acting in tandem in many systems, studying these behaviors within a BBB feedback framework has yet to be done.

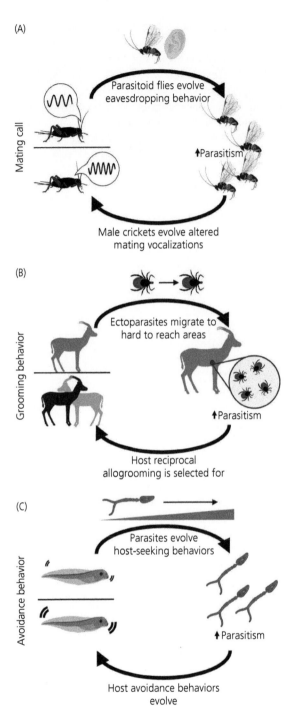

**Figure 19.3** Three examples highlighting how host and parasite "behavior begets behavior" in ways that can potentially initiate a feedback loop of behavioral complexity. (A) Feedback between host mating calls and parasitoid eavesdropping. (B) Feedback between host grooming behaviors and parasite migration. (C) Feedback between host avoidance behavior and parasite seeking behavior.

## 19.6 Future directions and concluding remarks

Behavior is fundamental and is the means by which organisms turn their genes into fitness. Why should behavior be any less important for parasites, and subsequently, for disease outcomes? In this chapter, we have used a broad suite of examples to demonstrate how parasite behaviors are as varied and integral to understanding disease ecology and evolution as host behaviors. Parasites, across a range of taxa, are social, exhibit complex mating behavior and sexual selection, undergo nuanced movement behaviors both outside and inside hosts, and employ strategies to protect themselves from other parasites. These behaviors merit study for the contributions they can make to our fundamental understanding of animal behavior, such as using social parasites to understand the evolution of eusociality (70). However, these parasite behaviors also influence transmission rates, parasite genetic diversity, susceptibility to control efforts, and the evolution of virulence. Parasite behaviors are thus not only equally important for disease dynamics as host behaviors but also coevolve with host behaviors to create feedbacks where behavior begets behavior. While this chapter has demonstrated that parasite behaviors can be important for understanding disease dynamics and outcomes, there are substantial fundamental and applied questions remaining (Box 19.1).

Despite the importance of parasite behavior for disease ecology and evolution, parasites remain understudied, with estimates as high as 95% of undocumented parasite species in certain taxa (71), (72). The first step in documenting parasite behavior is species identification. With advances in rapid and scalable genomic techniques, parasite species discovery is relatively low-hanging fruit for biodiversity research (73), (74). However, beyond genetic identification, studies of parasite behavior require collection, culturing, and rearing of parasites for phenotypic study. For example, identifying and culturing new strains and species of EPNs and characterizing their foraging behaviors and host specificity is vital for developing biocontrol agents that work for more agricultural systems. Likewise, understanding the behavior of emerging infectious pathogens may help determine what types of

---

**Box 19.1 Twelve outstanding questions in parasite behavior**

1. How does parasite sociality alter reproductive output, transmission, and virulence?
2. Can parasite competition and behavioral defenses act as a pseudo-immune system for the host? What does this mean for the evolution of parasite immune evasion strategies?
3. How does host breadth constrain or expand parasite behaviors?
4. Is mating behavior a significant evolutionary force in maintaining parasite virulence?
5. How common is sexual selection in parasites? Can this result in runaway or directional sexual selection?
6. What are the consequences of parasite sexual selection for transmission and virulence?
7. Are females the choosy sex? What types of traits are selected upon?
8. How much individual trait variation is there within natural parasite populations? How does this variation influence infection outcomes?
9. How does variation in parasite morphology (e.g., body size) promote variation in behavior (e.g., habitat selection, movement, and foraging rates)?
10. What are differences in evolutionary outcomes (e.g., virulence levels) based on modes of intra- and extra-host movement for parasites?
11. Can we produce "evolution-proof" control measures for parasites by better understanding their sexual reproduction and mating behaviors?
12. Can parasite habitat selection be framed through optimal foraging theory to explain host breadth and infection patterns?

---

control efforts (gene drive, biocides, etc.) are feasible. We hope this chapter has demonstrated the importance of studying host and parasite behaviors in tandem for understanding disease ecology, evolution, and control. We urge behavioral ecologists and disease ecologists alike to start investigating parasite behaviors in their own study systems.

## References

1. Hechinger RF, Wood AC, and Kuris AM. Social organization in a flatworm: Trematode parasites form soldier and reproductive castes. *Proc R Soc Biol Sci.* 2011 Mar 7;278(1706):656–65.

2. Leung TLF and Poulin R. Small worms, big appetites: Ratios of different functional morphs in relation to interspecific competition in trematode parasites. *Int J Parasitol*. 2011 Aug 15;41(10):1063–8.

3. Harvey JA, Corley LS, and Strand MR. Competition induces adaptive shifts in caste ratios of a polyembryonic wasp. *Nature*. 2000 Jul 13;406(6792): 183–6.

4. Benton T and Foster W. Altruistic housekeeping in a social aphid. *Proc R Soc B: Biol Sci*. 1992;247: 199–202.

5. Pike N, Whitfield JA, and Foster WA. Ecological correlates of sociality in Pemphigus aphids, with a partial phylogeny of the genus. *BMC Evol Biol*. 2007;7(1): 185–185.

6. Buckling A and Brockhurst MA. Kin selection and the evolution of virulence. *Heredity*. 2008;100(5):484–8.

7. Baiocchi T, Lee G, Choe D-H, and Dillman AR. Host seeking parasitic nematodes use specific odors to assess host resources. *Sci Rep*. 2017;7(1):6270.

8. Ruschioni S, Loon JJA van, Smid HM, and van Lenteren JC. Insects can count: Sensory basis of host discrimination in parasitoid wasps revealed. *PLoS One*. 2015;10(10):e0138045.

9. Trager W. Acquired Immunity to ticks. *J Parasitol*. 1939;25(1):57.

10. Eberle JU and Voehringer D. Role of basophils in protective immunity to parasitic infections. *Semin Immunopathol*. 2016;38(5):605–13.

11. Millan AS. Evolution of plasmid-mediated antibiotic resistance in the clinical context. *Trends Microbiol*. 2018;26(12):978–85.

12. Beaurepaire AL, Krieger KJ, and Moritz RFA. Seasonal cycle of inbreeding and recombination of the parasitic mite *Varroa destructor* in honeybee colonies and its implications for the selection of acaricide resistance. *Infect Genetics Evol*. 2017;50:49–54.

13. Faber NR, Meiborg AB, McFarlane GR, Gorjanc G, and Harpur BA. A gene drive does not spread easily in populations of the honey bee parasite *Varroa destructor*. *Apidologie* 2021;52:1112–27.

14. Chu GWTC and Cutress CE. Austrobilharzia variglandis (Miller and Northup, 1926) Penner, 1953, (Trematoda: Schistosomatidae) in Hawaii with notes on its biology. *J Parasitol*. 1954;40(5):515.

15. Armstrong JC. Mating behavior and development of Schistosomes in the mouse. *J Parasitol*. 1965;51(4): 605–16.

16. Beltran S, Cézilly F, and Boissier J. Genetic dissimilarity between mates, but not male heterozygosity, influences divorce in schistosomes. *PLoS One*. 2008;3(10):e3328.

17. Steinauer ML. The sex lives of parasites: Investigating the mating system and mechanisms of sexual selection of the human pathogen Schistosoma mansoni. *Int J Parasitol*. 2009;39(10):1157–63.

18. Bartoli P. Rôle favorisant de l'hydrodynamisme dans la contamination des deuxièmes hôtes intermédiaires par les cercaires de Maritrema misenensis (A. Palombi, 1940) (Digenea, Microphallidae). *Ann De Parasitol Humaine Et Comparée*. 1986;61(1):35–41.

19. Haas W. Parasitic worms: Strategies of host finding, recognition and invasion1. *Zoology*. 2003;106(4): 349–64.

20. Haas W, Körner M, Hutterer E, Wegner M, and Haberl B. Finding and recognition of the snail intermediate hosts by 3 species of echinostome cercariae. *Parasitol*. 1995;110(2):133–42.

21. Campbell JF and Kaya HK. How and why a parasitic nematode jumps. *Nature*. 1999;397(6719):485–6.

22. Sukhdeo MVK and Croll NA. The location of parasites within their hosts: Factors affecting longitudinal distribution of Trichinella spiralis in the small intestine of mice. *Int J Parasitol*. 1981;11(2):163–8.

23. Mao Q, Liao Z, Li J, Liu Y, Wu W, Chen H, *et al*. Filamentous structures induced by a phytoreovirus mediate viral release from salivary glands in its insect vector. *J Virol*. 2017;91(12):e00265–17.

24. Hawking F. The 24-hour periodicity of microfilariae: Biological mechanisms responsible for its production and control. *Proc R Soc B: Biol Sci*. 1967;169(1014): 59–76.

25. Koppenhöfer AM, Shapiro-Ilan DI, and Hiltpold I. Entomopathogenic nematodes in sustainable food production. *Frontiers Sustain Food Syst*. 2020;4:125.

26. Rasmann S, Buri A, Lavallée MG, Joaquim J, Purcell J, and Pellissier L. Differential allocation and deployment of direct and indirect defences by Vicia sepium along elevation gradients. *J Ecol*. 2014 Jul 1;102(4): 930–8.

27. Shapiro D, Gouge D, and Koppenhöfer A. Factors affecting field efficacy: Analysis of case studies in cotton, turf, and citrus in Gaugler R (ed), *Entomopathogenic Nematology*. Wallingford: CABI Publishing, 2002, 333–56.

28. Hawley DM, Etienne RS, Ezenwa VO, and Jolles AE. Does animal behavior underlie covariation between hosts' exposure to infectious agents and susceptibility to infection? implications for disease dynamics. *Integr Comp Biol*. 2011;51(4):528–39.

29. Sadoughi B, Anzà S, Defolie C, Manin V, Müller-Klein N, Murillo T, *et al*. Parasites in a social world: Lessons from primates. In Ezenwa VO, Altizer S, and Hall RJ (eds), *Animal Behavior and Parasitism*. Oxford:

Oxford University Press, 2022. DOI: 10.1093/oso/9780192895561.003.0003.

30. Keiser CN. Collective behavior and parasite transmission. In Ezenwa VO, Altizer S, and Hall RJ (eds), *Animal Behavior and Parasitism*. Oxford: Oxford University Press, 2022. DOI: 10.1093/oso/9780192895561.003.0005.

31. Lagrue C, MacLeod CD, Keller L, and Poulin R. Caste ratio adjustments in response to perceived and realised competition in parasites with division of labour. *J Anim Ecol*. 2018 Jun 29;87(5):1429–39.

32. Passera L, Roncin E, Kaufmann B, and Keller L. Increased soldier production in ant colonies exposed to intraspecific competition. *Nature*. 1996;379(6566):630–1.

33. Resetarits EJ, Torchin ME, and Hechinger RF. Social trematode parasites increase standing army size in areas of greater invasion threat. *Biol Letters*. 2020;16(2):20190765.

34. Schmidt KA and Ostfeld RS. Biodiversity and the dilution effect in disease ecology. *Ecol*. 2001;82(3):609–19.

35. Civitello DJ, Cohen J, Fatima H, Halstead NT, Liriano J, McMahon TA, *et al.* Biodiversity inhibits parasites: Broad evidence for the dilution effect. *PNAS*. 2015 Jul 14;112(28):8667–71.

36. MacLeod C, Poulin R, and Lagrue C. Save your host, save yourself? Caste-ratio adjustment in a parasite with division of labor and snail host survival following shell damage. *Ecol Evol*. 2018 Jan 3;8(3):1615–25.

37. Cremer S, Armitage SAO, and Schmid-Hempel P. Social immunity. *Curr Biol*. 2007 Aug;17(16):R693–702.

38. Poulin R, Kamiya T, and Lagrue C. Evolution, phylogenetic distribution and functional ecology of division of labour in trematodes. 2019; Jan 4:1–10.

39. Galaktionov KV, Podvyaznaya IM, Nikolaev KE, and Levakin IA. Self-sustaining infrapopulation or colony? Redial clonal groups of Himasthla elongata (Mehlis, 1831) (Trematoda: Echinostomatidae) in Littorina littorea (Linnaeus) (Gastropoda: Littorinidae) do not support the concept of eusocial colonies in trematodes. *Folia Parasit*. 2015;62:1–14.

40. Pirrie A, Chapman H, and Ashby B. Parasite-mediated sexual selection: To mate or not to mate?. In Ezenwa VO, Altizer S, and Hall RJ (eds), *Animal Behavior and Parasitism*. Oxford: Oxford University Press, 2022. DOI: 10.1093/oso/9780192895561.003.0009.

41. Simon J-C and Peccoud J. Rapid evolution of aphid pests in agricultural environments. *Curr Opin Insect Sci*. 2018;26:17–24.

42. Jacomb F, Marsh J, and Holman L. Sexual selection expedites the evolution of pesticide resistance. *Evolution*. 2016;70(12):2746–51.

43. Standen O. The relationship of sex in Schistosoma Mansoni to migration within the hepatic portal system of experimentally infected mice. *Ann Trop Med Parasitol*. 1953;47(2):139–45.

44. Cornford E and Fitzpatrick A. The mechanism and rate of glucose transfer from male to female schistosomes. *Mol Biochem Parasitol*. 1985;17:131–41.

45. Altizer S, Bartel R, and Han BA. Animal migration and infectious disease risk. *Science*. 2011;331(6015):296–302.

46. Hall RJ, Altizer S, Peacock SJ, and Shaw AK. Animal migration and infection dynamics: Recent advances and future frontiers. In Ezenwa VO, Altizer S, and Hall RJ (eds), *Animal Behavior and Parasitism*. Oxford: Oxford University Press, 2022. DOI: 10.1093/oso/9780192895561.003.0007.

47. Boulinier T, Kada S, Ponchon A, Dupraz M, Dietrich M, Gamble A, *et al.* Migration, prospecting, dispersal? What host movement matters for infectious agent circulation? *Integr Comp Biol*. 2016;56(2):330–42.

48. Pérez-Tris J and Bensch S. Dispersal increases local transmission of avian malarial parasites. *Ecol Lett*. 2005;8(8):838–45.

49. Shimogawa MM, Ray SS, Kisalu N, Zhang Y, Geng Q, Ozcan A, *et al.* Parasite motility is critical for virulence of African trypanosomes. *Sci Rep*. 2018;8(1):9122.

50. King IL and Li Y. Host–parasite interactions promote disease tolerance to intestinal helminth infection. *Front Immunol*. 2018;9:2128.

51. Haas W, Haberl B, Kalbe M, and Stoll K. Traps for schistosome miracidia/cercariae. *Parasitol Int*. 1998;47:47.

52. Haas W. Physiological analysis of cercarial behavior. *J Parasitol*. 1992;78(2):243–55.

53. Lindblade KA. The epidemiology of cercarial dermatitis and its association with limnological characteristics of a northern Michigan lake. *J Parasitol*. 1998;84(1):19–23.

54. Cort W. Studies on schistome dermatitis: XI. Status of knowledge after more than twenty years. *Am J Epidem*. 1950;52(3):251–307.

55. Reitmayer CM, Evans MV, Miazgowicz KL, Newberry PM, Solano N, Tesla B, *et al.* Mosquito-virus interactions in Drake JM, Bonsall M, Strand M (eds), *Population Biology of Vector-Borne Diseases*. Oxford: Oxford University Press, 2020, 191–214.

56. Elliot SL, Adler FR, and Sabelis MW. How virulent should a parasite be to its vector? *Ecology*. 2003;84(10):2568–74.

57. Blanc S, Uzest M, and Drucker M. New research horizons in vector-transmission of plant viruses. *Curr Opin Microbiol*. 2011;14(4):483–91.

58. Brault V, Herrbach É, and Reinbold C. Electron microscopy studies on luteovirid transmission by aphids. *Micron*. 2007;38(3):302–12.

59. Ezenwa VO and Snider MH. Reciprocal relationships between behaviour and parasites suggest that negative feedback may drive flexibility in male reproductive behaviour. *Proc R Soc B: Biol Sci*. 2016;283(1831):20160423.

60. Hawley DM and Ezenwa VO. Parasites, host behavior and their feedbacks. In Ezenwa VO, Altizer S, and Hall RJ (eds), *Animal Behavior and Parasitism*. Oxford: Oxford University Press, 2022. DOI: 10.1093/oso/9780192895561.003.0002.

61. Zuk M, Rotenberry JT, and Tinghitella RM. Silent night: Adaptive disappearance of a sexual signal in a parasitized population of field crickets. *Biol Lett*. 2006;2(4):521–4.

62. Tinghitella RM. Rapid evolutionary change in a sexual signal: Genetic control of the mutation "flatwing" that renders male field crickets (Teleogryllus oceanicus) mute. *Heredity*. 2008;100(3):261–7.

63. Tinghitella RM, Broder ED, Gurule-Small GA, Hallagan CJ, and Wilson JD. Purring crickets: The evolution of a novel sexual signal. *Am Nat*. 2018;192(6):773–82.

64. Hart BL. Behavioural defense against parasites: Interaction with parasite invasiveness. *Parasitology*. 1994;109(S1):S139–51.

65. Hart BL and Hart LA. Reciprocal allogrooming in impala, Aepyceros melampus. *Anim Behav*. 1992;44(6):1073–83.

66. Stewart PD and Macdonald DW. Badgers and badger fleas: Strategies and counter-strategies. *Ethology*. 2003;109(9):751–64.

67. Krasnov BR, Khokhlova IS, and Shenbrot GI. Density-dependent host selection in ectoparasites: An application of isodar theory to fleas parasitizing rodents. *Oecologia*. 2003;134(3):365–72.

68. Taylor CN, Oseen KL, and Wassersug RJ. On the behavioural response of Rana and Bufo tadpoles to echinostomatoid cercariae: Implications to synergistic factors influencing trematode infections in anurans. *Can J Zool*. 2004;82(5):701–6.

69. Rohr JR, Swan A, Raffel TR, and Hudson PJ. Parasites, info-disruption, and the ecology of fear. *Oecologia*. 2009;159(2):447–54.

70. Whyte BA. The weird eusociality of polyembryonic parasites. *Biol Letters*. 2021;17(4):20210026.

71. Chapman A. Number of Living Species in Australia and the World. Canberra, Australia: Department of the Environment, Water, Heritage and the Arts; 2009.

72. Carlson CJ, Dallas TA, Alexander LW, Phelan AL, and Phillips AJ. What would it take to describe the global diversity of parasites? *Proc R Soc B: Biol Sci*. 2020;287(1939):20201841.

73. Anthony SJ, Epstein JH, Murray KA, Navarrete-Macias I, Zambrana-Torrelio CM, Solovyov A, et al. A strategy to estimate unknown viral diversity in mammals. *Mbio*. 2013;4(5):e00598–13.

74. Pinacho-Pinacho CD, García-Varela M, Sereno-Uribe AL, and de León GP-P. A hyper-diverse genus of acanthocephalans revealed by tree-based and non-tree-based species delimitation methods: Ten cryptic species of Neoechinorhynchus in Middle American freshwater fishes. *Mol Phylogenet Evol*. 2018;127:30–45.

# Index

Note: Tables, figures, and boxes are indicated by an italic *t*, *f*, and *b* following the page number.